Plant Nutrition and Soil Science

Plant Nutrition and Soil Science

Edited by Brian Bechdal

SYRAWOOD
PUBLISHING HOUSE
New York

Published by Syrawood Publishing House,
750 Third Avenue, 9th Floor,
New York, NY 10017, USA
www.syrawoodpublishinghouse.com

Plant Nutrition and Soil Science
Edited by Brian Bechdal

International Standard Book Number: 978-1-68286-386-2 (Hardback)

Cataloging-in-publication Data

Plant nutrition and soil science / edited by Brian Bechdal.
p. cm.
Includes bibliographical references and index.
ISBN 978-1-68286-386-2
1. Plants--Nutrition. 2. Soil management. 3. Soil science. 4. Soil biodiversity. I. Bechdal, Brian.
QK867 .P53 2017
581.13--dc23

Printed in China.

TABLE OF CONTENTS

Preface ... **IX**

Chapter 1 **Differential Nutrient Limitation of Soil Microbial Biomass and Metabolic Quotients (qCO_2): Is there a Biological Stoichiometry of Soil Microbes?** ... **1**
Wyatt H. Hartman, Curtis J. Richardson

Chapter 2 **Fungal Community Structure in Disease Suppressive Soils Assessed by 28S LSU Gene Sequencing** ... **15**
C. Ryan Penton, V. V. S. R. Gupta, James M. Tiedje, Stephen M. Neate, Kathy Ophel-Keller, Michael Gillings, Paul Harvey, Amanda Pham, David K. Roget

Chapter 3 **Foliar Spray with Vermiwash Modifies the Arbuscular Mycorrhizal Dependency and Nutrient Stoichiometry of Bhut Jolokia (*Capsicum assamicum*)** ... **27**
Mohammad Haneef Khan, Mukesh K. Meghvansi, Rajeev Gupta, Vijay Veer, Lokendra Singh, Mohan C. Kalita

Chapter 4 **Dynamics of Soil Bacterial Communities in Response to Repeated Application of Manure Containing Sulfadiazine** ... **35**
Guo-Chun Ding, Viviane Radl, Brigitte Schloter-Hai, Sven Jechalke, Holger Heuer, Kornelia Smalla, Michael Schloter

Chapter 5 **Spatial Variability and Stocks of Soil Organic Carbon in the Gobi Desert of Northwestern China** ... **45**
Pingping Zhang, Ming'an Shao

Chapter 6 **Tree Species Composition Influences Enzyme Activities and Microbial Biomass in the Rhizosphere: A Rhizobox Approach** ... **57**
Shengzuo Fang, Dong Liu, Ye Tian, Shiping Deng, Xulan Shang

Chapter 7 **Rice-Straw Mulch Reduces the Green Peach Aphid, *Myzus persicae* (Hemiptera: Aphididae) Populations on Kale, *Brassica oleracea* var. *acephala* (Brassicaceae) Plants** ... **68**
Reinildes Silva-Filho, Ricardo Henrique Silva Santos, Wagner de Souza Tavares, Germano Leão Demolin Leite, Carlos Frederico Wilcken, JoséEduardo Serrão, JoséCola Zanuncio

Chapter 8 **Environmental Fate of Double-Stranded RNA in Agricultural Soils** ... **74**
Samuel Dubelman, Joshua Fischer, Fatima Zapata, Kristin Huizinga, Changjian Jiang, Joshua Uffman, Steven Levine, David Carson

Chapter 9 **Multiple Exposure and Effects Assessment of Heavy Metals in the Population near Mining Area in South China** ... **81**
Ping Zhuang, Huanping Lu, Zhian Li, Bi Zou, Murray B. McBride

Chapter 10 **Influence of 20-Year Organic and Inorganic Fertilization on Organic Carbon Accumulation and Microbial Community Structure of Aggregates in an Intensively Cultivated Sandy Loam Soil** **92**
Huanjun Zhang, Weixin Ding, Xinhua He, Hongyan Yu, Jianling Fan, Deyan Liu

Chapter 11 **Impact of Shortened Crop Rotation of Oilseed Rape on Soil and Rhizosphere Microbial Diversity in Relation to Yield Decline** **103**
Sally Hilton, Amanda J. Bennett, Gary Keane, Gary D. Bending, David Chandler, Ron Stobart, Peter Mills

Chapter 12 **Assessment of the Soil Organic Carbon Sink in a Project for the Conversion of Farmland to Forestland** **115**
Lan Mu, Yinli Liang, Ruilian Han

Chapter 13 **Effects of Straw Incorporation on Soil Organic Matter and Soil Water-Stable Aggregates Content in Semiarid Regions of Northwest China** **124**
Peng Zhang, Ting Wei, Zhikuan Jia, Qingfang Han, Xiaolong Ren, Yongping Li

Chapter 14 **Evaluation of Chloropicrin as a Soil Fumigant against *Ralstonia solanacarum* in Ginger (*Zingiber officinale* Rosc.) Production in China** **135**
Liangang Mao, Qiuxia Wang, Dongdong Yan, Taotao Ma, Pengfei Liu, Jin Shen, Yuan Li, Canbin Ouyang, Meixia Guo, Aocheng Cao

Chapter 15 **Long-Term Effect of Agricultural Reclamation on Soil Chemical Properties of a Coastal Saline Marsh in Bohai Rim, Northern China** **142**
Yidong Wang, Zhong-Liang Wang, Xiaoping Feng, Changcheng Guo, Qing Chen

Chapter 16 **Tungsten Distribution in Soil and Rice in the Vicinity of the World's Largest and Longest-Operating Tungsten Mine in China** **149**
Chunye Lin, Ruiping Li, Hongguang Cheng, Jing Wang, Xiao Shao

Chapter 17 **Impacts of Agricultural Management and Climate Change on Future Soil Organic Carbon Dynamics in North China Plain** **157**
Guocheng Wang, Tingting Li, Wen Zhang, Yongqiang Yu

Chapter 18 **The Impact of Selective-Logging and Forest Clearance for Oil Palm on Fungal Communities in Borneo** **167**
Dorsaf Kerfahi, Binu M. Tripathi, Junghoon Lee, David P. Edwards, Jonathan M. Adams

Chapter 19 **Evaluation of the CENTURY Model using Long-Term Fertilization Trials under Corn-Wheat Cropping Systems in the Typical Croplands of China** **175**
Rihuan Cong, Xiujun Wang, Minggang Xu, Stephen M. Ogle, William J. Parton

Chapter 20 **Can Biochar and Phytoextractors be Jointly used for Cadmium Remediation?** **188**
Huanping Lu, Zhian Li, Shenglei Fu, Ana Méndez, Gabriel Gascó, Jorge Paz-Ferreiro

Chapter 21 **Phytoavailability of Cadmium (Cd) to Pak Choi (*Brassica Chinensis* L.) Grown in Chinese Soils: A Model to Evaluate the Impact of Soil Cd Pollution on Potential Dietary Toxicity** **195**
Muhammad Tariq Rafiq, Rukhsanda Aziz, Xiaoe Yang, Wendan Xiao, Peter J. Stoffella, Aamir Saghir, Muhammad Azam, Tingqiang Li

Chapter 22 **Root Interactions in a Maize/Soybean Intercropping System Control Soybean Soil-Borne Disease, Red Crown Rot** 204
Xiang Gao, Man Wu, Ruineng Xu, Xiurong Wang, Ruqian Pan, Hye-Ji Kim, Hong Liao

Chapter 23 **Illumina Amplicon Sequencing of 16S rRNA Tag Reveals Bacterial Community Development in the Rhizosphere of Apple Nurseries at a Replant Disease Site and a New Planting Site** 213
Jian Sun, Qiang Zhang, Jia Zhou, Qinping Wei

Chapter 24 **SoilGrids 1km — Global Soil Information based on Automated Mapping** 221
Tomislav Hengl, Jorge Mendes de Jesus, Robert A. MacMillan, Niels H. Batjes, Gerard B. M. Heuvelink,, Eloi Ribeiro, Alessandro Samuel-Rosa, Bas Kempen, Johan G. B. Leenaars, Markus G. Walsh, Maria Ruiperez Gonzalez

Permissions

List of Contributors

Index

PREFACE

This book traces the progress of plant nutrition and soil science, highlighting some of the key concepts and applications. Plant nutrition deals with the study of various chemical compounds and elements which are required for proper growth and sustainability of plants. Soil Science plays a significant role in plant nutrition. The aim of this book is to delve into the relationship of these two fields and understand their interdisciplinary aspects. Such selected concepts that redefine these disciplines have been presented in this book. It will serve as a valuable source of relevance for agronomists, botanists, students & researchers associated with these fields.

Various studies have approached the subject by analyzing it with a single perspective, but the present book provides diverse methodologies and techniques to address this field. This book contains theories and applications needed for understanding the subject from different perspectives. The aim is to keep the readers informed about the progress in the field; therefore, the contributions were carefully examined to compile novel researches by specialists from across the globe.

Indeed, the job of the editor is the most crucial and challenging in compiling all chapters into a single book. In the end, I would extend my sincere thanks to the chapter authors for their profound work. I am also thankful for the support provided by my family and colleagues during the compilation of this book.

Editor

1

Differential Nutrient Limitation of Soil Microbial Biomass and Metabolic Quotients (qCO_2): Is There a Biological Stoichiometry of Soil Microbes?

Wyatt H. Hartman*, Curtis J. Richardson

Duke University Wetland Center, Nicholas School of the Environment, Duke University, Durham, North Carolina, United States of America

Abstract

Background: Variation in microbial metabolism poses one of the greatest current uncertainties in models of global carbon cycling, and is particularly poorly understood in soils. Biological Stoichiometry theory describes biochemical mechanisms linking metabolic rates with variation in the elemental composition of cells and organisms, and has been widely observed in animals, plants, and plankton. However, this theory has not been widely tested in microbes, which are considered to have fixed ratios of major elements in soils.

Methodology/Principal Findings: To determine whether Biological Stoichiometry underlies patterns of soil microbial metabolism, we compiled published data on microbial biomass carbon (C), nitrogen (N), and phosphorus (P) pools in soils spanning the global range of climate, vegetation, and land use types. We compared element ratios in microbial biomass pools to the metabolic quotient qCO_2 (respiration per unit biomass), where soil C mineralization was simultaneously measured in controlled incubations. Although microbial C, N, and P stoichiometry appeared to follow somewhat constrained allometric relationships at the global scale, we found significant variation in the C:N:P ratios of soil microbes across land use and habitat types, and size-dependent scaling of microbial C:N and C:P (but not N:P) ratios. Microbial stoichiometry and metabolic quotients were also weakly correlated as suggested by Biological Stoichiometry theory. Importantly, we found that while soil microbial biomass appeared constrained by soil N availability, microbial metabolic rates (qCO_2) were most strongly associated with inorganic P availability.

Conclusions/Significance: Our findings appear consistent with the model of cellular metabolism described by Biological Stoichiometry theory, where biomass is limited by N needed to build proteins, but rates of protein synthesis are limited by the high P demands of ribosomes. Incorporation of these physiological processes may improve models of carbon cycling and understanding of the effects of nutrient availability on soil C turnover across terrestrial and wetland habitats.

Editor: Melanie R. Mormile, Missouri University of Science and Technology, United States of America

Funding: This work was funded by a Graduate Research Fellowship from the National Science Foundation awarded to Wyatt Hartman, with additional support from the Duke Wetland Center Graduate Student Endowment. The funders had no role in study design, data collection and analysis, decision to publish, or preparation of the manuscript.

Competing Interests: The authors have declared that no competing interests exist.

* E-mail: wyatt.hartman@duke.edu

Introduction

Variation in heterotrophic microbial metabolism poses a critical uncertainty in our current understanding of soil carbon (C) cycling in terrestrial and wetland soils, and improved understanding of microbial mediation of soil C and nutrient cycling is needed to predict ecosystem responses to human alteration of land use, climate, and nutrient availability [1,2,3,4]. Soil carbon turnover in terrestrial and wetland ecosystems may closely depend on the availability of major nutrients like nitrogen (N) and phosphorus (P), and ratios of these elements relative to microbial demand strongly influences C and nutrient mineralization during decomposition [2,5,6]. However, the processes governing the relative demand of microbes for C, N, and P in soils are poorly understood, and little data is available to characterize variation in the stoichiometry and metabolism of soil microbes [6,7,8].

At global scales, the relative demand of soil microbes for C, N, and P are thought to occur in broadly fixed ratios, reflected by patterns in the elemental stoichiometry of both microbial biomass pools and soil enzyme activities [7,9,10]. However, in ecosystem studies similar large-scale stoichiometric patterns may obscure considerable variability among habitats [11] and stoichiometric variation is routinely observed in terrestrial and aquatic ecosystems among habitats, and among species of autotrophs and heterotrophs [8]. We might expect similar stoichiometric variation in soil microbes, given widely reported differences in the C:N:P biomass ratios of aquatic and cultured microbes among ecosystems, habitats, and taxonomic groups [12,13,14,15]. Variation in the stoichiometry of microbes may be coupled with differences in growth rates like other heterotrophs [14], and might influence C use efficiency during decomposition in terrestrial ecosystems [6].

Metabolic relationships between carbon and nutrient cycling in heterotrophs can be understood as a function of the biochemical composition of the cellular machinery. Particularly, the Growth Rate Hypothesis (GRH) describes a relationship between cellular growth rate and P concentrations that results from the requirement of growing cells for P-rich ribosomes to produce new proteins [16,17]. Relationships between organismal stoichiometry and growth rate described by the GRH appear consistently across heterotrophs and autotrophs spanning several orders of magnitude in size, and have broad implications at ecosystem scales (called Biological Stoichiometry) linked to trophic status and functional differentiation of communities and whole ecosystems [11,16,18,19,20,21,22,23,24]. While relationships between stoichiometry and metabolism are essentially untested in soils [8,25], biochemical mechanisms described by the GRH may suggest microbial metabolism in soils could be particularly sensitive to ecosystem P availability.

The availability of phosphorus may be particularly important for the growth and metabolism of microbes in soils, with significant implications for decomposition and global carbon cycling. On average, the relative P demand of soil microbial biomass (C:N:P = 60:7:1) is considerably greater than the relative availability of P in soils (C:N:P = 186:13:1) and plant litter (C:N:P = 3144:45:1) inputs [10]. Moreover, the global stoichiometry of enzyme activities (C:N:P = 1:1:1) in terrestrial and aquatic ecosystems [9] may suggest microbial allocation to P uptake is considerably greater than that of C and N when compared to requirements for the growth of biomass. Although the influence of P availability on soil C cycling has been understudied compared to variation in N availability and C:N ratios, recent findings indicate that P availability may influence soil C cycling even in ecosystems traditionally considered to be N limited [2,6,26,27,28].

To assess whether the influence of P availability on soil C cycling might reflect underlying metabolic mechanisms like the GRH, we collected all available data from published studies to evaluate the effects of N and P availability on the stoichiometry, biomass, and metabolic rates of soil microbes. Our cross-ecosystem dataset included measurements of soil and microbial C, N, and P pools, and C mineralization rates in soils spanning global variation soil conditions and nutrient availability in terrestrial and wetland ecosystems, and we explicitly considered study location factors of climate, land use, and vegetation as predictor variables. Soil C mineralization data were obtained only from controlled laboratory incubations in studies that simultaneously measured microbial element pools. We indexed the metabolism of microbes using the metabolic quotient qCO_2 [29], calculated as the rate of C mineralization per unit of microbial biomass C, and also referred to as mass specific respiration [30]. Where data were available, we also assessed the influence of inorganic P availability [31] and soil pH on microbial stoichiometry, growth, and metabolism.

Results and Discussion

Global variation in the stoichiometry of soil resources

Although a previous study of forest and grasslands suggested that pools of soil C, N, and P are closely linked with allometric relationships describing nearly fixed stoichiometry [10], our broader data set indicated considerable differences in the C:N:P stoichiometry of terrestrial and wetland soils, especially with respect to the relative availability of P. Carbon and nitrogen pools appeared closely coupled in soils (Fig. 1A), although we found some evidence that C and N pools did not increase isometrically as in a previous study [10]. Across all global soils, the allometric slope of the relationship between soil C and N was 0.88 (Table 1), which falls below the 1:1 line representing constant C:N ratios (Fig. 1A). This relationship represents a subtle decline in the relative availability of N with increasing accumulation of soil C pools. However, we note that soil C and N scaling was significantly different in litter layers and organic soils, where soil C:N ratios increased dramatically with soil C (Fig. 1A, Fig. S1). Although scaling of C and N appeared to be isometric (slope = 1) when considering only forest and pasture soils (Table S1), closer analysis revealed a slight, but significant increase in C:N ratios with increasing C in forest and pastures (Fig. S1, Table S1) like the global pattern.

The stoichiometry of soil phosphorus appeared considerably more variable than N across global soils, with much wider divergence in soil C:P and N:P ratios from the Redfield ratios (Fig. 1, Fig. S1). In contrast to previous findings in forests and pastures [10], our analysis across a broader range of global ecosystems showed marked departure in soil C:P and N:P from the previously observed isometric relationships (Fig. 1B–C, Table 1, Table S1). The additional ecosystem types included in our analysis could be considered "disturbed" outliers, with lower C:P in crops due to tillage increasing soil P, and higher C:P in boreal forests, wetland organic soils, and litter resulting from relatively undecomposed substrates. However, these "outlier" ecosystems appear as endpoints in nearly continuous relationships in soil C:P and N:P ratios, which increased directly as a function of soil C (Fig. S1, Table S2). This dependence of soil C:P and N:P ratios on soil C may arise from dilution of soil P concentrations with soil C accumulation, as total soil P does not increase concomittantly with soil C to nearly the extent that N does (Fig. 1).

Variation in N:P ratios across global soils is shaped by fundamental differences in the ecosystem sources of soil N (from atmospheric fixation by soil heterotrophs) and P (from mineral weathering) [32]. While soil N pools appeared closely linked with soil C accumulation, soil P pools were only weakly related to soil C, and were highly variable within and among ecosystem types (Fig. 1). Dilution of soil P concentrations by soil C (and N) accumulation appeared to be a primary driver of variation in soil N:P ratios, which were tighly linked with C:P ratios across all global soils ($r^2 = 0.88$), but only weakly varied with C:N ratios in leaf litter and soil humic horizons (Fig, S2, Table S2).

Although comparison of soil stoichiometry to the Redfield ratios might suggest that soil N appears more consistently limiting across global ecosystems than soil P (Fig. 1, Fig. S1), we note that the autotroph-based Redfield ratios may not appropriately describe microbial stoichiometry [14]. Soil microbial biomass N:P ratios (9.0:1) were considerably lower (more P rich) than the Redfield ratio (16:1) or N:P ratios of soils (21:1, Table 1). Although this excess of P compared to the Redfield ratio could be interpeted as luxury uptake (reflecting N limitation), accumulation of excess P by soil microbes appears to be of minor importance, and coincides primarily with extreme P limitation [33]. We posit that microbial P demand is intrinsically greater than the Redfield ratio based on their small size and higher rates of metabolism compared to multicellular organisms [16,18,34,35,36]. If microbial biomass P quotas are intrinsically greater than the Redfield ratio, P availability in soils would appear to be limiting to microbes relative to N across a considerably broader range of ecosystem types than suggested by the Redfield ratio, including ecosystems typically considered to be N limited with respect to plant growth (Fig. S1C).

Variation in soil microbial stoichiometry

In comparison to the wide variation we observed in the C:N:P ratios of soils, the stoichiometry of soil microbes appeared to be largely constrained (Fig. 1), although our results suggest potentially important biological and ecological sources of variation. Scaling

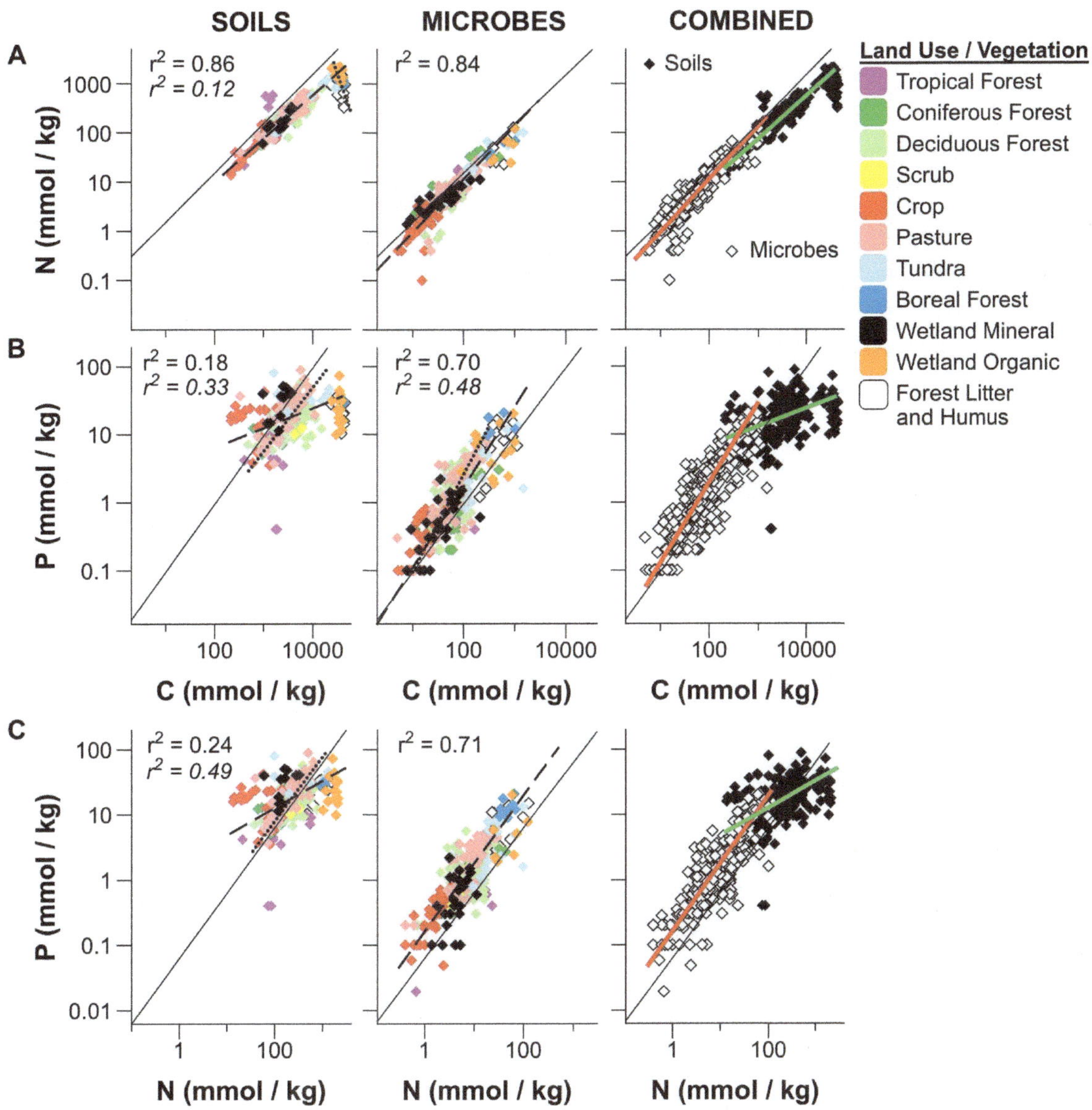

Figure 1. Global stoichiometric scaling of C, N, and P contents of soil and microbial biomass pools. Relationships in plots show variation in A) C:N ratios B) C:P ratios C) N:P ratios of soils, microbial biomass, and combined data. Data were $\log_{10}$ transformed transformed to improve normality and plotted to express size dependent relationships in comparison to the Redfield (1958) ratios (solid black lines). Dashed lines are regression fits for all global soils, with correlation coeficients in plain text and parameters estimated by SMA given in Table 1. Global relationships were compared with fits obtained using different subsets of habitat types, and where slopes were significantly different we plotted fits as dotted lines, with correlation coeficients given in italics. Soil C:N scaling (A) was different in litter and organic soils (wetland organic, boreal forest, and humic horizons), while forest and pasture soils were different from global relationships in soil and microbial C:P (B) and soil N:P (C). SMA regression parameters for these relationships using subsets of our data are given in Table S1.

relationships between C and N pools were generally similar in microbes and soils, although the slopes of these relationships were somewhat different (Table 1). In contrast, scaling of P pools with C and N differed dramatically among soils and microbes, with C:P and N:P ratios appearing more isometric (slopes closer to the 1:1 line) in microbes than in soils (Fig. 1, Table 1). Similar to the results of a previous study [10], we found that microbial C:N, C:P, and N:P ratios were not correlated with the corresponding element ratios in soils (Fig. S3, Table S2). The relative inflexibility of microbial biomass compared to highly variable resource ratios in soils has been suggested to reflect a homeostatic maintenance of microbial C:N:P ratios with nearly fixed stoichiometry [10]. While our results appear to similarly illustrate Resource Homeostasis in soil microbes, we emphasize that this need not imply soil microbial stoichiometry is inherently biologically or ecologically inflexible (i.e. Strict Homeostasis [37]).

Our results directly showed that the C:N:P stoichiometry of soil microbes does not occur in strictly fixed isometric ratios. At global scales, we observed slight increases in the proportions of N and P with increasing soil microbial biomass C pools, as indicated by allometric slopes significantly greater than one (Fig. 1, Table 1). This size-dependent effect was particularly pronounced for microbial P contents, and even more prominent in forest and pasture soils (Fig. 1, Table S1). These increases in average cellular N and P content with increasing microbial biomass in soils may indicate that microbial nutrient use efficiency (NUE) declines as

Table 1. Summary of SMA regressions of log_{10}-transformed C, N, and P contents in soil and microbial pools, along with predictors of soil C mineralization (CO_2) and microbial metabolism (qCO_2).

Analysis	*x*	*y*	*n*	r^2	Int.	Slope	x:y ratio	x:y mean			CV
Soil allometry	*C	N	280	0.86	−0.77	**0.88**	C:N	18.8	±	0.8	0.7
	*C	P	261	0.18	0.23	**0.29**	C:P	445	±	43	1.6
	*N	P	261	0.24	0.27	**0.42**	N:P	21.4	±	1.7	1.3
Microbial allometry	MBC	MBN	237	0.84	−1.11	**1.08**	mC:N	10.7	±	0.6	0.8
	*MBC	MBP	298	0.70	−2.07	**1.18**	mC:P	87.2	±	13.2	2.6
	MBN	MBP	267	0.71	−0.79	1.04	mN:P	9	±	1.2	2.2
Microbial biomass	C	MBC	289	0.75	−1.34	**0.88**	MBC:C	2.01%	±	0.08	0.6
	N	MBC	266	0.73	−0.52	0.99					
	*P	MBC	247	0.132	0.08	**1.37**					
Respiration and metabolism	MBC	CO_2	92	0.7	−4.17	**1.57**					
	**mC:P	qCO_2	92	0.21	1.97	**−1.15**					
	**Pi	qCO_2	38	0.44	0.51	**0.65**					

Bivariate relationships were significant ($P<0.001$) for all relationships shown. Slopes significantly different from one ($P>0.05$) are shown in boldface font. Slopes not different from one (not bold) indicate an isometric (linear) relationship among parameters. The geometric mean and standard errors (SE) of stoichiometric ratios (x:y ratio, x:y mean) are given for reference, but are not representative where allometric slopes are different than 1. The coefficient of variation (CV) of these stoichiometric ratios is provided as a dimensionless index of dispersion about the mean. Single asterisks (*) indicate where different slopes are observed by considering only forest and pasture soils (Table S1), and ** indicates relationships fit for all soils excluding litter and humus.

microbial population size increases, a finding broadly analogous to size-dependent decreases in NUE (especially with respect to P) in aquatic bacteria [38]. However, we did not find a significant size-dependent scaling relationship between microbial N and P pools (allometric slope = 1.04) analogous to those observed in aquatic microbes [38] and across species of plants and animals [11,16,23,39,40].

Stoichiometric analysis of soil microbial pools differs fundamentally from that of higher organisms, as the unit of observation is the elemental average of mixed communities rather than individual organisms [41,42]. In stoichiometric analysis of ecosystem element pools, global scaling relationships reflecting broad stoichiometric constraints may obscure important variation among habitats [11], which may reflect both shifts in species composition, as well as the coexistence of species mixtures with different stoichiometry [41,42]. We found that although microbial stoichiometry appeared on average to converge on broadly constrained ratios at a global scale, the N:P ratios of soil microbes varied significantly among vegetation and land use types across global soils and litter layers (Fig. 2). Differences in soil microbial stoichiometry have also been observed previously with land use change in temperate forest and grassland ecosystems [10,43,44,45,46,47]. While we note that the stoichiometry of microbes is highly variable within habitat types, our results show vegetation and land use may broadly influence the N:P ratios of soil microbes across the global range of terrestrial ecosystems (Fig. 2).

Our analyses did not clearly identify factors associated with vegetation and land use types that might underlie variability in microbial stoichiometry across global soils. While vegetation and land use categories were associated with soil C pools and soil stoichiometry (Table S3), these factors were not correlated with microbial stoichiometry, nor were any other soil chemical or biological parameters (Table S4). We suggest instead that the differences we observed in microbial N:P ratios among land use and vegetation types might be linked with variation in size-dependent scaling relationships.

The stoichiometric scaling relationships of microbes differed significantly among land use and vegetation types (Tables S5, S6, S7, Fig. S4), and we directly compared habitat specific differences in these scaling relationships (Fig. 3). Size-dependent scaling of soil microbial C:N and C:P (but not N:P) like that observed in global relationships (Fig. 1, Table 1) was present in some but not all habitats, with variation in the degree of deviation of allometric slopes from the isometric Redfield ratios among land use and vegetation types (Fig. 3). When comparing the slopes of stoichiometric relationships among land use and vegetation categories, size-dependent scaling relationships of soil microbial biomass C:N and C:P appeared to be somewhat asymptotic to the Redfield ratio as the size of the biomass pools increased (Fig. 3A–B).

Size-dependent slopes of C:N and C:P scaling relationships appeared closely coupled in forests, crops, and pasture soils (Fig. 3A–B, Tables S5, S6, S7), and these parallel decreases in NUE of N and P appeared to "cancel out" size-dependent relationships in microbial N:P scaling (Fig. 3, Fig. 1). However, we note that while the slopes of microbial C:N scaling relationships among habitats approached the Redfield ratio from below (higher C:N) as biomass C increased, microbial C:P scaling approached the Redfield ratio from above (lower C:P). These generally higher C:N and lower C:P ratios (relative to Redfield) in habitats where C:N and C:P scaling was size-dependent could help explain variation in microbial N:P ratios by habitat, despite their lack of size-dependent relationships (Fig. 3).

Global scale variation in the stoichiometry of autotrophs may reflect broad patterns based on climate [40,48,49,50], and we found some indication of a similar climatic influence on the stoichiometry of microbes. Although microbial stoichiometry did not vary closely with latitude or climate categories (Table S4), we found a significant effect of climate on stoichiometric scaling relationships of microbes (Tables S8, S9, S10) like that of vegetation and land use (Fig. 3). While our climate categories did not represent physiochemical factors like soil temperature or moisture, these factors may appear linked with seasonal variation

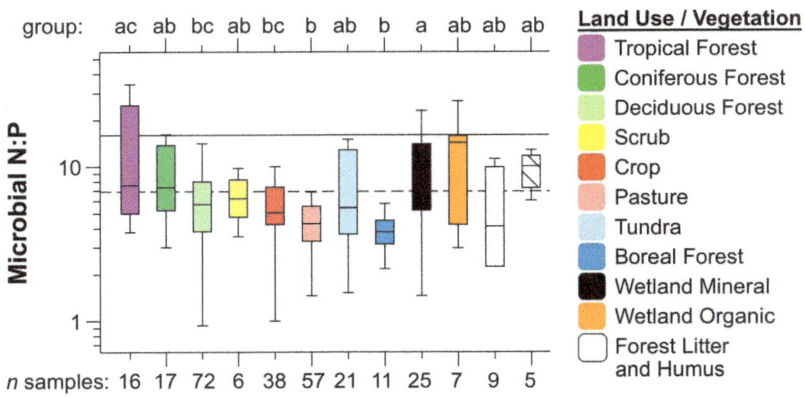

Figure 2. Differences in N:P stoichiometry of soil microbial biomass among global vegetation and land use categories. Letters on x-axis above the plot show group differences among vegetation types (using Tukey's tests), and number of samples for each vegetation type are given on the lower x-axis. Overall variance described by vegetation ($R^2 = 0.193$, $p < 0.001$) was determined using a general linear model. Solid horizontal line is the Redfield (1958) ratio N:P = 16:1, dashed line is average microbial N:P (6.9) reported in [10].

in microbial stoichiometry within habitats [51,52,53,54,55]. Variation in microbial stoichiometry could also be linked with other factors not captured by our dataset, such as soil C quality [7,56] or microbial communities [57,58,59].

Our results showed important variation in the stoichiometry of soil microbes, including size-dependent scaling of C:N and C:P, and differences in the N:P ratios of soil microbial biomass among land use and vegetation types that might arise from size-dependent differences in NUE among habitats. We also found some indication of non-homeostatic responses to soil stoichiometry in a few vegetation and climate categories, albeit with small sample sizes (Tables S11, S12). These results generally indicate that the stoichiometry soil microbes is not inherently fixed, but instead may exhibit some stoichiometric flexibility like cultured and aquatic microbes, which similarly show non-homeostasis among habitats and size-dependent scaling patterns [12,14,15,38,42,60].

Limitation of microbial biomass pool size

Although we found the relative availability of P varied more dramatically than N across global soils, pools of microbial biomass carbon (MBC) were not closely related to soil P, but rather more closely reflected C and N pools in soils (Fig. 4). Although MBC was closely related to both soil C and soil N, the relationship between MBC and N was linear, while increases in MBC with soil C were non-linear, slightly lagging accumulation of soil C (Fig. 4A–B). While MBC clearly increased with soil C, the allometric slope (0.88, Table 1) of this relationship falls below the 1:1 line (isometric ratio based on geometric mean), indicating diminishing growth return of MBC with increasing soil C, which also appeared in forest and pasture soils (Table S1). These findings might suggest the availability of nutrients or labile C may limit biomass growth at higher soil C [7]. In contrast, we found that soil MBC pools were linearly related to soil N pools (allometric slope 0.99, Table 1, Fig. 4B) consistently across soil types, which could suggest that N

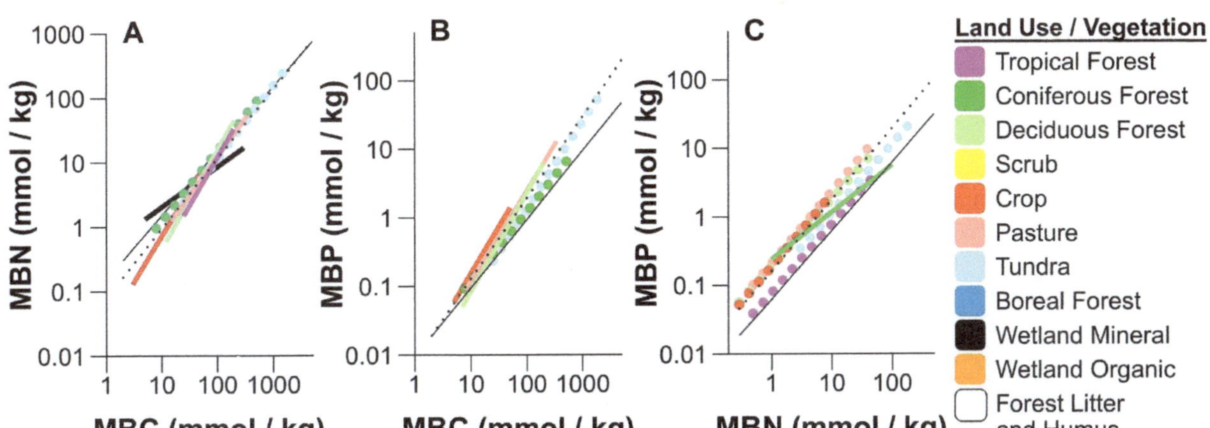

Figure 3. Estimated differences in stoichiometric scaling of microbial C, N, and P by land use/vegetation categories. Estimated SMA regression fit lines for each land use and vegetation category are shown to express habitat level differences in scaling of microbial biomass A) C:N ratios, B) C:P ratios, and C) N:P ratios, with data $\log_{10}$ transformed for normality. Bold lines are colored by land use and vegetation category, and treatments without significant fits are not shown. Colored solid lines indicate relationships where slopes were not equal to 1, while slopes not significantly different from 1 are are displayed as bold colored dotted lines. Thin black dotted lines show the regression fits for all groups combined (the same as in Fig. 1), while thin black solid lines indicate the Redfield ratios (C:N:P = 106:16:1). Individual plots for each regression fit by land use and vegetation categories are given in Figure S4, with parameters estimated by SMA provided in Tables S3, S4, S5, along with results of intercept and slope tests, and multiple comparisons of these parameters.

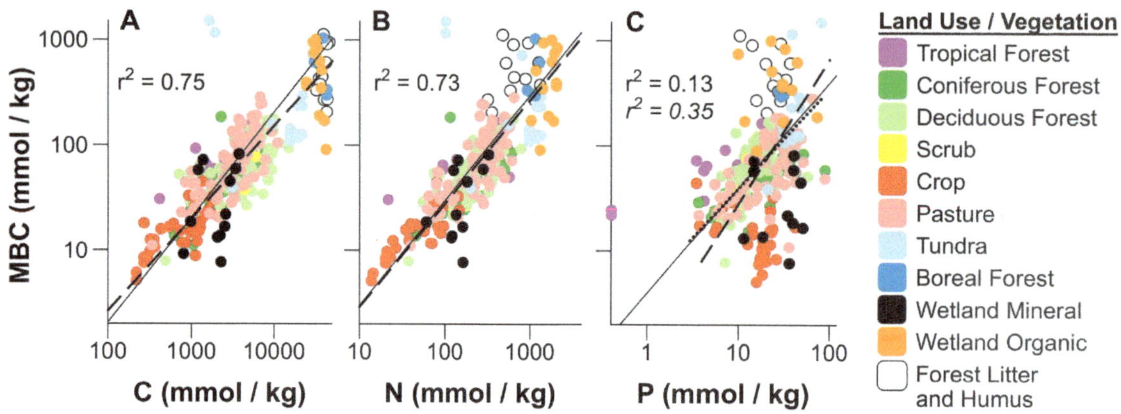

Figure 4. Scaling of soil microbial biomass C (MBC) with soil C, N, and P pools. Relationships in plots show variation in MBC with A) soil C, B) soil N, and C) soil P. Outliers from the general relationship between MBC and soil C in A), including floodplain mineral soils [95] and arctic tundra [96] were removed prior to fitting regressions, and data were $\log_{10}$ transformed to improve normality. Solid lines are the 1:1 isometric lines based on the geometric mean ratio of each scaling relationship. Dashed lines are regression fits for all global soils, with correlation coeficients in plain text and parameters estimated by SMA given in Table 1. Global relationships were compared with fits obtained using only data from forests and pastures, and where slopes were significantly different from all combined treatments we plotted fits as dotted lines, with correlation coeficients given in italics. SMA regression parameters fit using only forest and pasture soils are given in Table S1. Regressions were also tested seperately with only litter and organic soils data, but were these relationships were not significant.

availability more directly constrains the size of microbial C pools across global soils.

Comparison of soil and microbial stoichiometry may further suggest potential N limitation of microbial biomass pools, as the average N availability in soils (mean C:N = 19:1) is about half of that of microbial demand (mean C:N = 11:1, Table 1, Fig. 1). Moreover, microbial demand for N appears to generally increase in a size dependent manner with biomass C (and soil C pools), while soil N availability may slightly decline with soil C accumulation, and is particularly deficient in litter layers (Fig. 1A, Fig. S1A). Nitrogen limitation of soil microbial biomass pools been predicted by multiple element limitation models [61], and would appear consistent with the growth rate hypothesis (GRH), which links biomass growth to the synthesis of N rich proteins [16,17].

While our results may appear to suggest soil microbes might be more limited by N than by C availability, this interpretation is challenged by empirical findings showing close association between microbial C and N acquisition. Soil N may be tightly bound with recalcitrant humic C in soils, and N mineralization may be dependent on enzymatic mining of recalcitrant soil C [62,63]. As a result, addition of inorganic N to soils may decrease decomposition and soil respiration rates [63,64], thought to be associated with suppression of enzymes breaking down humic C [7]. Although enzymatic models suggest that microbial biomass may increase in response to inorganic N additions [61], empirical data indicates soil microbial biomass may consistently decrease with N addition [65] potentially as the result of shifts in microbial community composition [66,67]. These findings may ultimately support the hypothesis that microbial N mining drives decomposition of recalcitrant C in soils, and we posit that these responses to inorganic N additions may reflect preferential demand for N rather than C in soil microbial communities under ambient conditions where N is bound with humic C.

At global scales, soil microbial biomass C pools were not closely related to ecosystem P pools (Fig. 4C) or relative P availability. While microbial biomass was somewhat correlated with soil C:P and N:P ratios (Table S3), this may likely reflect covariation of C:P and N:P element ratios with soil C (Fig. S1), which was in turn more closely related to MBC (Table S3). Although a weak general relationship between soil total P and MBC could be described for pastures and forest soils (Fig. 4C, Table S1), MBC appeared decoupled from soil P pools at the extremes of soil C and MBC, which were lowest in crop soils, and highest in organic wetland soils, boreal forests, and litter and humus layers, despite similar concentrations of soil P (Fig. 4C). In contrast to tightly linked pools of C and N in soils and microbes, decoupling of microbial growth from soil P pools may present stark contrasts in microbial P demand relative to soil total P pools at the extremes of soil C accumulation (Fig. 1B–C, Fig. 4C). Soil microbial communities may have evolved several mechanisms to cope with these large variations in the relative availability of P across terrestrial habitats, including exudation of phosphatase enzymes, differential rates of P uptake and cycling [7,9,68], and stoichiometric variation (Fig. 2–3).

Factors shaping microbial metabolic quotients

Microbial turnover of soil carbon pools may vary as a factor of both the size of biomass pools of soil microbes, and their rate of metabolism per unit biomass. Soil carbon mineralization rates in controlled laboratory incubations were closely linked with the size of microbial biomass C pools (Fig. 5A), although we found increases in soil C mineralization with microbial biomass were not linear (Table 1). This non-linear relationship between CO_2 and MBC was essentially the same when comparing all global soils (including litter—Table 1) to results obtained only from forest and pasture soils (Table S1). The allometric slope of the relationship between CO_2 and MBC lies above the 1:1 line of constant proportions (slope = 1.57, Table 1), indicating an exponential growth of C mineralization with MBC pool size. This exponential relationship between CO_2 flux and microbial biomass C pools indicates the rate of metabolism *per unit biomass* increased with total biomass of soil microbes in laboratory incubations.

To assess controls on microbial metabolic rates, we determined the microbial metabolic quotient qCO_2 [29], essentially an analysis of the residual variation in CO_2 flux after accounting for differences in soil microbial biomass ($qCO_2 = CO_2/MBC$). We acknowledge that the metabolic quotient qCO_2 is an imperfect proxy for microbial growth rates, which does not describe biomass

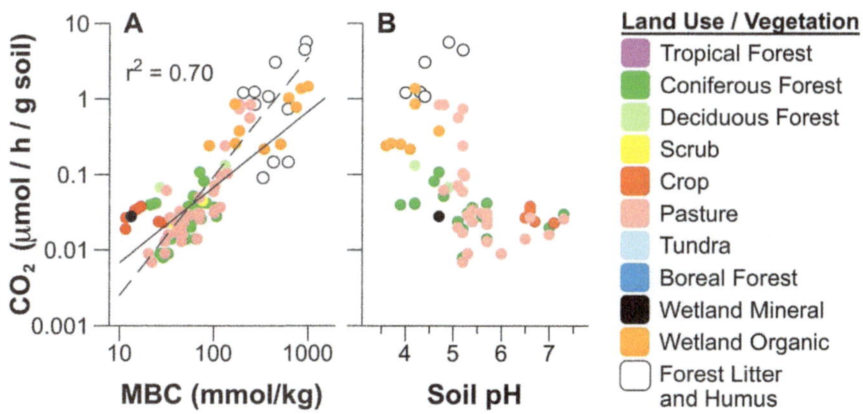

Figure 5. Carbon mineralization rates (CO_2) varied with A) microbial biomass C (MBC) and B) soil pH. C mineralization rates were measured in glass jar incubations in studies with concurrent measurements of microbial biomass C, N, and P. Dashed line in A) is the regression fit, and solid line is the 1:1 isometric line based on the geometric mean ratio of CO_2 to MBC (mean qCO_2). Parameters estimated by SMA are given in Table 1. SMA regressions fit using only forest and pasture soils (parameters in Table S1) are not shown as they were essentially the same as the global relationships.

accumulation rates (δ MBC) or potential variation in the proportion of C incorporated into biomass compared to CO_2 respired [6,26,69]. Although qCO_2 may be interpreted as reflecting microbial C use efficiency (CUE) [70], derived from the proportion of C incorporated into biomass : C respired (CUE = δ MBC/(δ MBC+δ CO_2)) [71,72], soil microbial growth rates are difficult to measure, and without δ MBC data, qCO_2 is not an appropriate metric of CUE [72]. Given the lack of data on MBC accumulation rates in soils, we view qCO_2 as a readily determined index of microbial metabolic rates (respiration *per unit biomass*), for which existing data can be used to explore empirical support for the GRH in soil microbes.

The growth rate hypothesis would suggest microbial metabolic rates should be closely coupled with biomass C:P ratios [16,17], and we found a weak but significant negative relationship between microbial C:P ratios and qCO_2 in existing data from soil incubations (Fig. 6A). The negative sign of the regression relationship between microbial biomass C:P ratios and qCO_2 indicates higher rates of metabolism (qCO_2) corresponded with greater microbial P concentrations (lower C:P), as the GRH would suggest (Fig. 6A, Table 1, Table S1). We also tested multivariate regression models of CO_2 and qCO_2, and found that although microbial C:P ratios were a persistent factor in the most robust models, microbial C:P only predicted soil qCO_2 strongly in combination with additional factors including soil stoichiometry, pH, vegetation and climate (Tables S13, S14).

We found that soil available inorganic phosphorus (P_i) was the single strongest predictor of microbial metabolic quotients (qCO_2) in soil incubations from a wide range of terrestrial ecosystems spanning deserts, crop lands, pastures, coniferous and deciduous forests, and organic wetland soils (Fig. 6B). Addition of inorganic P to soils has been shown to accelerate decomposition rates in temperate forest soils [2], and differences in inorganic P concentrations were recently found to describe variation in rates of C mineralization among land use types including crops, pastures, and pine and hardwood forests [28]. However, to our knowledge the present work is the first to show shifts in soil microbial metabolic quotients (qCO_2) with inorganic P availability. This result may suggest that the effects of soil inorganic P on C cycling recently observed by others [2,6,28] could arise from changes in mass-specific rates of microbial metabolism with P availability (Fig. 6B) rather than shifts in microbial biomass, which appeared to be limited by N and C pools in soils (Fig. 4A–B).

Our results also showed soil pH strongly affected both rates of carbon mineralization (CO_2) and microbial metabolism (qCO_2) across terrestrial soils, with non-linear changes in respiration and metabolism appearing to shift about a pH value of 5.5 (Figs. 5B, 6C). Variation in soil qCO_2 with soil pH has been described in previous studies as the result of increased of maintenance respiration required for metabolic responses to pH stress [29,73]. However, shifts in microbial metabolism with pH might also be related to changes in soil microbial communities, as both fungal:bacterial ratios [74,75] and the taxonomic composition of soil bacteria vary with pH across a wide range of soils, and may particularly shift about a soil pH value near 5.5 [57,58,76,77,78], similar to patterns we observed in qCO_2. Importantly, the relative abundance of some bacterial taxonomic groups appear to be linked with soil respiration rates, suggesting r - vs. K - selection of microbial life strategies based on differences in growth rates [79].

Is there a Biological Stoichiometry of the soil microbial biomass?

The growth rate hypothesis (GRH) has been shown to link the elemental composition of organisms to their metabolic rates in higher autotrophs and heterotrophs, with important implications for trait differentiation, community structure, and element cycling in terrestrial and aquatic ecosystems [11,16,18,19,20,21,22,23,24]. Similarly, relationships between bacterial metabolism, RNA content, and biomass C:P supporting the GRH have been observed in cultured bacterial strains and mixed aquatic assemblages [14,18,80]. Importantly, both biomass stoichiometry and the degree of homeostatic response to nutrient availability appear to vary among different species of cultured bacteria [13,14,80], and with the degree of nutrient limitation [15,42]. In culture and aquatic microcosms, mixed species assemblages may facilitate greater stoichiometric flexibility and less resource homeostasis at the community level, presumably as the result of shifts in community composition [14,41,42,60]. However, different nutrient use strategies may coexist due to niche partitioning in mixed communities, potentially driving nutrient co-limitation (and apparent homeostasis) of mixed assemblages [41,42].

We postulate that similar processes might govern the stoichiometry of microbial communities in soils, which have lower

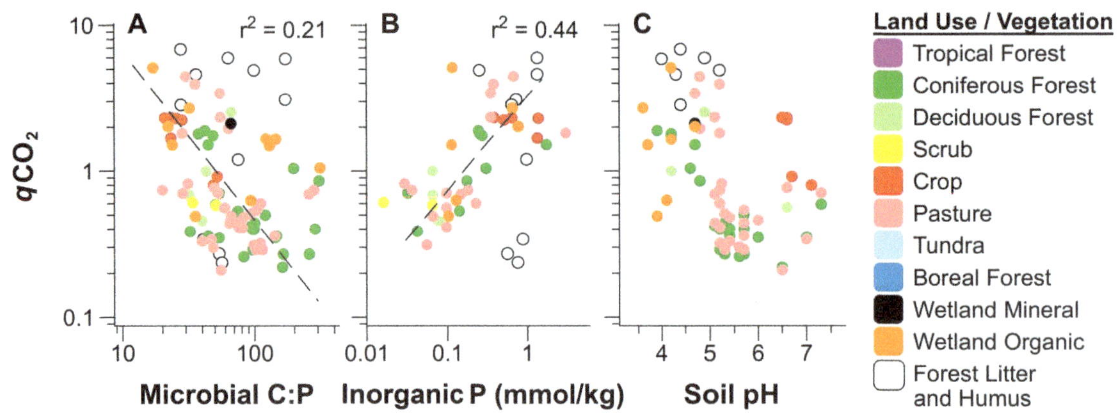

Figure 6. Factors influencing microbial metabolic quotients (qCO_2) of soil incubations. Significant factors included A) microbial C:P ratios, B) available inorganic P, and C) soil pH. qCO_2 was calculated as the mol/mol ratio of C mineralization rates measured in glass jar incubations per unit microbial biomass C obtained from the same soils, with units of mmol CO_2-C/h/mol MBC-C/g soil. Relationships of qCO_2 with microbial C:P (A), and inorganic P (B) were fit without data from litter or soil humic horizons. Data were log_{10} transformed for normality and parameters estimated by SMA are given in Table 1. SMA regressions fit using only forest and pasture soils (parameters in Table S1) are not shown as they were essentially the same as the global relationships.

relative P availability and C lability than aquatic habitats [81]. While we acknowledge the potential for measurement error in soil microbial biomass pools, we observed considerable variation in microbial C:N (up to one order of magnitude), and especially C:P and N:P ratios (up to two orders of magnitude) across global ecosystems (Fig. S3). This stoichiometric variaton was comparable to biological variation in bacterial cultures and aquatic microcosms [15,38,42], which also show greater flexibility in biomass C:P and N:P ratios than C:N ratios [15,38,82,83].

However, our results showed microbial stoichiometry largely did not appear responsive to resource ratios in soils [10], unlike non-homeostatic patterns often found in microbes in culture and aquatic microcosms [15,41,42,60]. This apparent resource homeostasis might be related to community level observations inherent in analysis of soil microbes [25], and the potential for niche partitioning to drive community homeostasis [41,42]. However, cultured and aquatic microbes may show strict homeostasis only under P limitation [41,42,82], and the homeostatic responses of soil microbes could conceivably be viewed as reflecting widespread limitation by soil P availablilty (Fig. S1).

Although we found only limited evidence that microbial stoichiometry was not homeostatic with respect to soil resources (Tables S11, S12), several of our findings suggest soil microbes are not Strictly Homeostatic (i.e. fixed, isometric stoichiometry [37]). Our results showed patterns of size-dependent scaling and habitat specific variation in soil microbial stoichiometry closely analogous to those in cultured and aquatic microbes [12,14,15,38,42,60]. These size-dependent and habitat specifc differences in microbial stoichiometry may have important implications for ecosystem element retention and fluxes [38,42], including microbial turnover of nutrients and C in soils and litter [6,25,26]. Concommitant changes in microbial stoichiometry and metabolism along land use gradients [44,45,47,84], and dynamic variation in soil microbial stoichiometry with season, soil moisture [51,52,53,54,55], and wetland innundation [85,86,87], could also suggest variation in biomass stoichiometry is linked with shifts in soil microbial activity.

We found microbial stoichiometry and metabolic rates appeared directly linked in soils, with a weak but significant relationship between microbial biomass C:P ratios and metabolic quotients (qCO_2) across a wide range of global soils. This finding parallels the association between rates of cellular growth and biomass C:P ratios described by the GRH [16,17], although the analogy is imperfect as our observations represent mixed communites rather than individual organisms, and we used a crude index of microbial metabolism (qCO_2) as a proxy for growth rates.

We also found that microbial metabolic quotients (qCO_2) were most strongly associated with inorganic P availability across global soils, consistent with increased rates of C cycling associated with P availability and P fertilization, even in soils considered to be N limited [2,28]. In contrast, our results suggested that soil microbial biomass pools appeared constrained primarily by soil N. Although inorganic N additions to soil appear inconsistent with our results, we contend these manipulations may critically alter ecosystem processes and soil communities relative to ambient conditions [63,64,66,67].

We suggest differential nutrient limitation of microbial biomass and metabolism across terrestrial and wetland soils may broadly reflect biochemical mechanisms described by the GRH, where relationships between growth rates and organismal stoichiometry arise from the control of P-rich ribosomes and rRNA pools over rates of synthesis of N-rich proteins [16,17], This mechanism essentially implies N limitation of the structural components of cells (biomass) and P limitation of their metabolic rates [16,17], and our analogous findings in soil microbes may implicate these basic processes of cellular metabolism in the cycling and retention of soil nutrients and C at global scales.

Our findings suggest further study is needed to investigate potential linkages between microbial stoichiometry, metabolism, and community composition [25,80] in soils. We note that the primary controls over microbial community composition across terrestrial and wetland soils are land use and pH [57,58,59,76,78], which were associated with respective shifts in microbial stoichiometry and metabolism in our study. We observed non-linear patterns in inflection points of soil CO_2 flux and metabolic quotients near soil pH 5.5, a value strikingly similar to widely observed change points in soil microbial community composition and diversity. Specific groups of soil bacteria have been linked to variation in soil C mineralization rates across terrestrial soils, which may indicate an energetic basis of ecological strategies in soil microbes [79] that also appears to describe microbial community responses to soil N addition [67,88]. The ecological strategies of soil bacteria may also appear directly related to differences in their ribosome copy number [56,89], suggesting GRH-like mechanisms could differentiate functional groups and shape microbial community structure and function in soils.

Materials and Methods

To determine the influence of resource availability on microbial growth and metabolism in soils, we compiled data on soil and microbial stoichiometry and respiration, and metabolism from existing published studies, focusing exclusively on studies including measurements of microbial C, N, and P pools in soils. Our approach to literature review and data collection was broadly similar to that of Cleveland and Lipzin [10], although we used a different search approach to obtain a greater number of publications encompassing a broader range of soil habitats (including wetlands and crops), and explicitly categorized the major habitats, vegetation types, and land uses from which samples were obtained. Importantly, we collected additional data parameters not included in the previous study [10], particularly measurements of soil C mineralization and qCO_2, along with measurements of available inorganic P and soil pH where data was available in studies simultaneously measuring microbial biomass C,N, and P pools.

Literature Review

To efficiently obtain publications with complete microbial biomass C, N, and P data, all literature searches included the term "microbial biomass phosphorus" (including quotes), as studies of microbial biomass P (MBP) often include data on microbial C and N pools, while more prevalent studies of microbial C:N stoichiometry often do not analyze microbial P [10]. Rather than use a citation based literature search based upon common methods for determining microbial P pools [10], we used a general search for "microbial biomass phosphorus" to obtain publications that might cite the authors' earlier work rather than the original methods papers. Literature searches were conducted with Google Scholar, and search results were compiled in bulk using the Zotero plug-in (www.Zotero.org) for the Firefox browser [90]. Redundant citations were removed from the search results database and publications were screened based on the inclusion of data on microbial biomass P and other microbial biomass pools.

To obtain a dataset with the most comprehensive coverage of global terrestrial ecosystems, we conducted a stratified literature search, with general searches followed by searches for specific ecosystem types. All searches contained the text "microbial biomass phosphorus," with additional search strings describing terrestrial biomes and land uses along with wetland habitats. A general search was conducted using the string ""“microbial biomass phosphorus," "C/P", and "soil""". In place of "soil" additional stratified searches were used to capture results in different ecosystem types (11 terms, e.g. "tropical", "desert", "boreal," etc.) and wetland habitats (9 terms, e.g. "wetland", "peatland", "bog", "fen", "salt marsh", "floodplain," etc.). However, these additional terms yielded no unique results beyond the general search using "soils". To find studies measuring both MBP and soil C cycling, the search ""“microbial biomass phosphorus," and "C/P""", was repeated with additional terms "CO2" and "qCO2" in separate iterations.

Data collection and processing

Our search results yielded 238 unique journal articles, from which we obtained 107 candidate publications with potentially relevant data on microbial biomass element pools. However, fewer studies (66) were used to obtain the final data set due to the exclusion of studies lacking required data layers, or with data presented in unusable formats, or from non-soils or unusually disturbed soils (Fig. S5). Publications used as data sources are listed in (Table S15), and citations for these data references are provided as Text S1. Data was obtained only from publications measuring microbial biomass C, N, and P concentrations using standard chloroform fumigation-extraction methods (0.5 M KCl to extract biomass C and N, and 0.5 M $NaHCO_3$ to extract biomass P) following [10]. In contrast to the prior study [10], we obtained data from as many soils as possible, including soil samples representing altered human land use regimes (e.g. crops, pastures and grazed savannas). We further cross-classified soils and soil datasets by climate, land use, and vegetation (Tables S16, S17).

Our final dataset included measurements from 355 soils published in 66 studies (Table S18), a substantial increase from the prior study [10], which obtained 186 soils from 48 studies. To standardize data and allow for comparison of elemental stoichiometry, all measurements of microbial biomass C, N, and P were converted into units of mmol/kg soil. Where data presented were not calculated using conversions for extraction efficiency, standard correction coefficients (0.45 for microbial C and N, 0.40 for microbial P) were applied [91]. Data were also collected for total pools of soil C, N, and P, and converted to mmol/kg soil. Extractable inorganic phosphorus data was obtained from studies using the Olsen extraction (0.5 M $NaHCO_3$), which is thought to reflect soil P availability [31], and data was converted to mmol P/kg soil. However, soil extractable inorganic N was not included in our dataset, as few studies included this variable. Soil pH data was collected where available, as soil pH is known to strongly affect the community composition of soil microbes [57,78].

We also collected available data on soil C mineralization rates where controlled experiments were conducted concurrently with measurements of microbial biomass C, N, and P pools in soils. Soil C mineralization data were only collected from studies that determined respiration rates in jars using soil and litter incubations with standardized moisture and temperature, with the average hourly respiration over a ten hour period [92] used to obtain an average rate of C mineralization expressed as mmol CO_2-C/h/g soil. Metabolic rates of soil microbial communities were indexed by calculating the metabolic quotient qCO_2 [29,92], by dividing C mineralization rates (per g soil) by microbial biomass (per g soil), to yield qCO_2 values expressed in mmol CO_2-C/mol MBC/h.

Statistical methods

Nutrient concentrations in the environment are often distributed log-normally [93], and measurements of element pools, fluxes, and ratios in soils were $\log_{10}$ transformed to improve normality. Stoichiometric relationships in soils and microbes were determined using a size-dependent approach, which describes allometric relationships based on the power function $y = a\ x^b$ [10,11]. Log transformation of this power relationship yields the linear function $\log y = a + b\ (\log x)$, allowing the use of linear regressions of $\log_{10}$ transformed data to determine stoichiometric relationships among nutrients in soils and microbes. Bivariate relationships between soil and microbial element pools, and ecosystem C fluxes were determined using standardized major axis (Type II) regression using the SMATR 3.0 package [94] in the open source *R* statistical software program (www.r-project.org), which we used to test whether slopes of allometric relationships were isometric (Slope = 1). Results of SMA regressions of each pairwise combination of all variables in our dataset are given in (Tables S3, S4).

Variation in the stoichiometry of soils and microbes, along with soil CO_2 flux and metabolic quotients (qCO_2) were compared among climate regions and vegetation by one-way ANOVA, and interactions with allometric slopes were tested using SMATR. Pairwise differences among habitat types were determined by Tukey's tests, and general linear models were used to determine the proportion of variance in soils, microbes, and metabolism explained by climate and vegetation groups. We also compared multivariate general linear models of factors influencing soil CO_2 flux and qCO_2, using an exhaustive search of all combinations of factors in our dataset (Tables S13, S14).

Supporting Information

Figure S1 Scaling of soil stoichiometric ratios with soil carbon (C) accumulation. Stoichiometric variation of A) soil C:N, B) C:P, and C) N:P ratios are shown as a function of soil C, and soil stoichiometry and C data were $\log_{10}$ transformed for normality. Dashed lines indicate SMA regression fits, and solid lines indicate the Redfield (1958) ratios (C:N:P = 106:16:1), while dotted horizontal lines show the average C:N:P ratios of soil microbes in this study (Table 1). Soil C:N (A) was fit separately for organic soil horizons and forest litter, and for all other soils. Litter and organic soils did not show relationships with C:P and N:P ratios as a function of soil C. All SMA regression parameters are given in Table S2.

Figure S2 Contribution of soil C:N and C:P ratios to variation in N:P ratios across global soils. Relationships of soil N:P with A) soil C:N and B) soil C:P were determined seperately for only litter and humic horizons, and for all other soils exclusive of forest litter and humus. SMA fits for litter and humic horizons are shown with dotted black lines with correlation coefficients in italics. Dashed black lines show SMA fits for all soils less litter and humic horizons, with correlation coefficients in plain text. Soil stoichiometric data were $\log_{10}$ transformed for normality, and parameter estimates for SMA regressions are given in Table S2.

Figure S3 Comparisons of microbial biomass C:N:P stoichiometric ratios with corresponding soil C:N:P stoichiometry. Relationships between microbial and soil stoichiometry are shown for A) C:N ratios, B) C:P ratios, and C) N:P ratios. Soil and microbial stoichiometry were $\log_{10}$ transformed for normality. No regression fits are shown as none of these relationships were significant (Table S2).

Figure S4 Stoichiometric scaling of microbial biomass C, N, and P by land use and vegetation categories. Data and SMA regression fit lines are shown for each land use and vegetation category to express habitat level differences in scaling of microbial biomass A) C:N ratios, B) C:P ratios, and C) N:P ratios. Microbial biomass element pool data were $\log_{10}$ transformed for normality prior to fitting SMA regressions for each treatment simultaneously, and treatments without significant fits are not shown. Open circles indicate data where regressions were not individually significant, while significant relationships are plotted with data as solid points. Regression fits are shown with colored lines, with red solid lines indicating slopes were not equal to 1, and blue lines showing slopes not significantly different from 1. Thin black dotted lines show the regression fits for all groups combined (the same as in Fig. 1), while thin black solid lines indicate the Redfield (1958) ratios (C:N:P = 106:16:1). Parameters estimated by SMA regressions are provided in Tables S5, S6, S7, along with results of intercept and slope tests and multiple comparisons of these parameters.

Figure S5 Flow of included studies for used as data sources for the meta-analysis.

Table S1 Summary of SMA regressions of $\log_{10}$-transformed C, N, and P contents in soil and microbial pools, along with predictors of soil C mineralization (CO_2) and microbial metabolism (qCO_2), considering only data from forest and pasture soils. Bivariate relationships were significant ($P<0.001$) for all relationships shown. Slopes significantly different from one ($P>0.05$) are shown in boldface font. Slopes not different from one (not bold) indicate an isometric (linear) relationship among parameters. The geometric mean and standard errors (SE) of stoichiometric ratios (x:y ratio, x:y mean) are given for reference, but are not representative where allometric slopes different from 1. The coefficient of variation (CV) of stoichiometric ratios is given as a dimensionless index of dispersion about the mean. Single asterisks (*) indicate where different slopes are observed by considering only forest and pasture soils compared to the full range of sites (presented in Table 1), and ** indicates a different relationship was tested for only litter and organic soils (wetland organic, boreal forest, and humic horizons).

Table S2 Summary of SMA regressions of $\log_{10}$-transformed relationships among soil C and soil stoichiometry, paired analysis of soil stoichiometric ratios, and comparisons of soil and microbial stoichiometry. SMA regression fits for these relationships correspond with data presented in Figs. S1, S2, S3. Bivariate relationships were significant ($P<0.001$) for all relationships shown., unless otherwise noted (n.s.). Slopes significantly different from one ($P>0.05$) are shown in boldface font. Slopes not different from one (not bold) indicate an isometric (linear) relationship among parameters. Analysis of relationships between soil C:N ratios and soil C accumulation were divided by habitat given different relationships observed among different vegetation types (Fig. S1). These groupings corresponded with soil C content greater than 24% C (20,000 mmol/kg) for habitats including boreal forests, wetland organic soils, and litter, and soil C content less than 24% (all other habitats).

Table S3 Results of all pairwise SMA regressions among $\log_{10}$-transformed study variables including soil and microbial C, N, and P pools, and soil stoichiometric ratios. Microbial biomass C, N, and P are abbreviated MBC, MBN, and MBP. P_i is the concentration of extractable inorganic (Olsen) P (available P), and P_i:P is the ratio of inorganic P to soil total P. Relationships among variables were compared using Standardized Major Axis (Type II) regression (SMA), except for relationships with habitat categories, which were assessed using generalized linear models (GLM). Only relationships with $P>0.05$ and r^2 (SMA) or R^2 (GLM)>0.25 are shown for clarity, except where data are displayed graphically in separate figures (boldface). Italics indicate relationships that are autocorrelated by their definition.

Table S4 Results of all pairwise SMA regressions among $\log_{10}$-transformed study variables including soil microbial stoichiometry, respiration, and metabolism

(qCO_2). Microbial biomass C:N, C:P, and N:P ratios are abbreviated by mC:N, mC:P, and mN:P, respectively. The ratio of microbial P to soil total P is abbreviated P_m:P, and P_m:P_i is the ratio of microbial P to inorganic P. Relationships among variables were compared using Standardized Major Axis (Type II) regression (SMA), except for relationships with habitat categories, which were assessed using generalized linear models (GLM). Only relationships with P>0.05 and r^2 (SMA) or R^2 (GLM)>0.25 are shown for clarity, except where data are displayed graphically in separate figures (boldface). Italics indicate relationships that are autocorrelated by their definition.

Table S5 SMA parameter estimates for simultaneous fitting of microbial biomass C and N scaling relationships by land use and vegetation categories. Regression fit lines are compared by category in Fig. 3A, and data and regression fits are plotted by category in Fig. S4A. The simultaneous SMA relationships for microbial biomass C and N scaling were tested for differences in intercepts (P<0.001) and slopes (P<0.001), and significantly different intercept and slope groups were determined by multiple comparisons in SMATR v.3.0, controlling the overall error rate at p<0.05. Slopes of individual relationships significantly different from one are shown in boldface. For each category, geometric mean of C:N ratios are presented (± SE) with their coefficient of variation (CV), and grouping by multiple comparisons using Tukey's test (p<0.05) on log_{10}-transformed data.

Table S6 SMA parameter estimates for simultaneous fitting of microbial biomass C and P scaling relationships by land use and vegetation categories. Regression lines are compared by category in Fig. 3B, and data and regression fits are plotted by category in Fig. S4B. The simultaneous SMA relationships for microbial biomass C and P scaling were tested for differences in intercepts (P<0.001) and slopes (P = 0.284), and significantly different intercept and slope groups were determined by multiple comparisons in SMATR v.3.0, controlling the overall error rate at p<0.05. Slopes of individual relationships significantly different from one are shown in boldface. For each category, geometric mean of C:P ratios are presented (± SE) with their coefficient of variation (CV), and grouping by multiple comparisons using Tukey's test (p<0.05) on log_{10}-transformed data.

Table S7 SMA parameter estimates for simultaneous fitting of microbial biomass N and P scaling relationships by land use and vegetation categories. Regression lines are compared by category in Fig. 3C, and data and regression fits are plotted by category in Fig. S4C. The simultaneous SMA relationships for microbial biomass N and P scaling were tested for differences in intercepts (P<0.001) and slopes (P = 0.012), and significantly different intercept and slope groups were determined by multiple comparisons in SMATR v.3.0, controlling the overall error rate at p<0.05. Slopes of individual relationships significantly different from one are shown in boldface. For each category, geometric mean of N:P ratios are presented (± SE) with their coefficient of variation (CV), and grouping by multiple comparisons using Tukey's test (p<0.05) on log_{10}-transformed data.

Table S8 SMA parameter estimates for simultaneous fitting of microbial biomass C and N scaling relationships by climate categories. The simultaneous SMA relationships were tested for differences in intercepts (P<0.001) and slopes (P<0.001), and significantly different intercept and slope groups were determined by multiple comparisons in SMATR v.3.0, by controlling the overall error rate at p<0.05. Bivariate relationships of log_{10}-transformed data were significant (P<0.001) for all relationships shown, unless otherwise noted due to insufficent data. Slopes significantly different from one (P>0.05) are shown in boldface font. For each category, geometric mean of N:P ratios are presented (± SE) with their coefficient of variation (CV), and with grouping by multiple comparisons using Tukey's test (p<0.05) on log_{10}-transformed data.

Table S9 SMA parameter estimates for simultaneous fitting of microbial biomass C and P scaling relationships by climate categories. The simultaneous SMA relationships were tested for differences in intercepts (P<0.001) and slopes (P<0.001), and significantly different intercept and slope groups were determined by multiple comparisons in SMATR v.3.0, by controlling the overall error rate at p<0.05. Bivariate relationships of log_{10}-transformed data were significant (P<0.001) for all relationships shown, unless otherwise noted due to insufficent data. Slopes significantly different from one (P>0.05) are shown in boldface font. For each category, geometric mean of N:P ratios are presented (± SE) with their coefficient of variation (CV), and with grouping by multiple comparisons using Tukey's test (p<0.05) on log_{10}-transformed data.

Table S10 SMA parameter estimates for simultaneous fitting of microbial biomass N and P scaling relationships by climate categories. The simultaneous SMA relationships were tested for differences in intercepts (P<0.001) and slopes (P<0.001), and significantly different intercept and slope groups were determined by multiple comparisons in SMATR v.3.0, by controlling the overall error rate at p<0.05. Bivariate relationships of log_{10}-transformed data were significant (P<0.001) for all relationships shown. Slopes significantly different from one (P>0.05) are shown in boldface font. For each category, geometric mean of N:P ratios are presented (± SE) with their coefficient of variation (CV), and with grouping by multiple comparisons using Tukey's test (p<0.05) on log_{10}-transformed data.

Table S11 SMA parameter estimates for simultaneous fitting of homeostatic relationships between microbial and soil stoichiometry by vegetation categories. The simultaneous SMA relationships were tested for differences in intercepts (P<0.001) and slopes (P<0.001). Slopes significantly different from one (P>0.05) are shown in boldface font. Only significant relationships with $r^2 \geq 0.3$ and n>5 are shown.

Table S12 SMA parameter estimates for simultaneous fitting of homeostatic relationships between microbial and soil stoichiometry by climate categories. Microbial biomass C:N, C:P, and N:P ratios are abbreviated by mC:N, mC:P, and mN:P, respectively. The simultaneous SMA relationships were tested for differences in intercepts (P<0.001) and slopes (P<0.001). Slopes significantly different from one (P>0.05) are shown in boldface font. Only significant relationships with $r^2 \geq 0.3$ and n>5 are shown.

Table S13 Multivariate general linear regression models of soil C mineralization rates (CO_2) as a function of ecosystem, soil and microbial factors. General linear models were compared using an exhaustive search, but only selected models with all predictors simultaneously significant are shown. To account for differences in missing data among parameters, we computed the test statistic $\%Var = R^2 * $ (n samples in model/n total samples). Model 9 (bold) had the greatest %Var, explaining more variance in more data points. Notably, addition of soil pH to models (Models 10–14) improved fit (R^2), but at the cost of fewer observations (df), resulting in a lower % Var. Models including both pH and inorganic P (P_i) had far fewer observations (Models 15–19), and consequently lower % Var explained. Linear regression results for individual predictors of CO_2 by Standardized Major Axis regression (SMA) are given in Table S4.

Table S14 Multivariate general linear regression models of the microbial metabolic quotient *qCO₂* as a function of ecosystem, soil and microbial factors. General linear models were compared using an exhaustive search, but only selected models with all predictors simultaneously significant are shown. To account for differences in missing data among parameters, we computed the test statistic $\%Var = R^2 * $ (n samples in model/n total samples). Model 9 had the greatest %Var and lowest AIC. Linear regression results for individual predictors of qCO_2 by Standardized Major Axis regression (SMA) are given in Table S4.

Table S15 Publications used as sources of soil and microbial element pool data. Ref. no. refers to reference citation number for this article, with most data source publication references given in Text S1. Data Ref. no. indicates the numbering used in the extracted data set presented in Table S18.

Table S16 Codes for climate categories used to describe soils in the full microbial stoichiometry data set (Table S18).

Table S17 Codes used to describe land use and vegetation classification of soils in the full microbial stoichiometry data set (Table S18).

Table S18 Data set obtained for soil and microbial stoichiometry, and C mineralization across ecosystems, classified by climate (Clim) and land use (LU). Codes for climate and land use categories are given in Tables S11 and S12, respectively. All soil chemical and microbial pools are expressed as mmol/kg soil. Soil respiration rate from standardized incubations (CO_2) is expressed as μmol CO_2-C/g soil/h.

Text S1 Additional references for publications used as data sources.

Acknowledgments

We thank Stefano Manzoni, Emily Bernhardt, Dan Richter, and two anonymous reviewers for valuable comments on the manuscript.

Author Contributions

Conceived and designed the experiments: WHH CJR. Analyzed the data: WHH. Wrote the paper: WHH CJR. Conducted literature review and collected data: WHH.

References

1. Allison SD, Wallenstein MD, Bradford MA (2010) Soil-carbon response to warming dependent on microbial physiology. Nature Geoscience 3: 336–340.
2. Bradford MA, Fierer N, Reynolds JF (2008) Soil carbon stocks in experimental mesocosms are dependent on the rate of labile carbon, nitrogen and phosphorus inputs to soils. Functional Ecology 22: 964–974.
3. Chapin FS, McFarland J, McGuire AD, Euskirchen ES, Ruess RW, et al. (2009) The changing global carbon cycle: Linking plant–soil carbon dynamics to global consequences. Journal of Ecology 97: 840–850.
4. Bradford MA, Watts BW, Davies CA (2010) Thermal adaptation of heterotrophic soil respiration in laboratory microcosms. Global Change Biology 16: 1576–1588.
5. Manzoni S, Porporato A (2009) Soil carbon and nitrogen mineralization: Theory and models across scales. Soil Biology and Biochemistry 41: 1355–1379.
6. Manzoni S, Trofymow JA, Jackson RB, Porporato A (2010) Stoichiometric controls on carbon, nitrogen, and phosphorus dynamics in decomposing litter. Ecological Monographs 80: 89–106.
7. Sinsabaugh RL, Follstad Shah JJ (2011) Ecoenzymatic stoichiometry of recalcitrant organic matter decomposition: the growth rate hypothesis in reverse. Biogeochemistry 102: 31–43.
8. Sardans J, Rivas-Ubach A, Peñuelas J (2011) The elemental stoichiometry of aquatic and terrestrial ecosystems and its relationships with organismic lifestyle and ecosystem structure and function: a review and perspectives. Biogeochemistry: DOI: 10.1007/s10533-10011-19640-10539.
9. Sinsabaugh RL, Hill BH, Follstad Shah JJ (2009) Ecoenzymatic stoichiometry of microbial organic nutrient acquisition in soil and sediment. Nature 462: 795–798.
10. Cleveland CC, Liptzin D (2007) C:N:P stoichiometry in soil: is there a "Redfield ratio" for the microbial biomass? Biogeochemistry 85: 235–252.
11. Sterner RW, Anderson T, Elser JJ, Hessen DO, Hood JM, et al. (2008) Scale-dependent carbon : nitrogen: phosphorus seston stoichiometry in marine and freshwaters. Limnology and Oceanography 53: 1169–1180.
12. Gundersen K, Heldal M, Norland S, Purdie DA, Knap AH (2002) Elemental C, N, and P cell content of individual bacteria collected at the Bermuda Atlantic Time-series Study (BATS) site. Limnology and Oceanography 47: 1525–1530.
13. Bratbak G (1985) Bacterial biovolume and biomass estimations. Applied and Environmental Microbiology 49: 1488–1493.
14. Makino W, Cotner JB, Sterner RW, Elser JJ (2003) Are bacteria more like plants or animals? Growth rate and resource dependence of bacterial C:N:P stoichiometry. Functional Ecology 17: 121–130.
15. Makino W, Cotner JB (2004) Elemental stoichiometry of a heterotrophic bacterial community in a freshwater lake: implications for growth-and resource-dependent variations. Aquatic Microbial Ecology 34: 33–41.
16. Elser JJ, Dobberfuhl DR, MacKay NA, Schampel JH (1996) Organism size, life history, and N:P stoichiometry. BioScience 46: 674–684.
17. Elser JJ, Sterner RW, Gorokhova E, Fagan WF, Markow TA, et al. (2000) Biological stoichiometry from genes to ecosystems. Ecology Letters 3: 540–550.
18. Elser JJ, Acharya K, Kyle M, Cotner J, Makino W, et al. (2003) Growth rate–stoichiometry couplings in diverse biota. Ecology Letters 6: 936–943.
19. Elser JJ, Fagan WF, Kerkhoff AJ, Swenson NG, Enquist BJ (2010) Biological stoichiometry of plant production: metabolism, scaling and ecological response to global change. New Phytologist 186: 593–608.
20. Reich PB, Oleksyn J, Wright IJ, Niklas KJ, Hedin L, et al. (2010) Evidence of a general 2/3-power law of scaling leaf nitrogen to phosphorus among major plant groups and biomes. Proceedings of the Royal Society B: Biological Sciences 277: 877–883.
21. Elser JJ, Fagan WF, Denno RF, Dobberfuhl DR, Folarin A, et al. (2000) Nutritional constraints in terrestrial and freshwater food webs. Nature 408: 578–580.
22. Agren GI (2008) Stoichiometry and nutrition of plant growth in natural communities. Annual Review of Ecology, Evolution, and Systematics 39: 153–170.
23. Arrigo KR (2005) Marine microorganisms and global nutrient cycles. Nature 437: 349–355.
24. Güsewell S, Koerselman W, Verhoeven JTA (2003) Biomass N:P ratios as indicators of nutrient limitation for plant populations in wetlands. Ecological Applications 13: 372–384.
25. Hall EK, Maixner F, Franklin O, Daims H, Richter A, et al. (2010) Linking microbial and ecosystem ecology using ecological stoichiometry: A synthesis of conceptual and empirical approaches. Ecosystems 14: 261–273.
26. Hessen DO, Agren GI, Anderson TR, Elser JJ, de Ruiter PC (2004) Carbon sequestration in ecosystems: the role of stoichiometry. Ecology 85: 1179–1192.
27. Elser JJ, Bracken MES, Cleland EE, Gruner DS, Harpole WS, et al. (2007) Global analysis of nitrogen and phosphorus limitation of primary producers in freshwater, marine and terrestrial ecosystems. Ecology Letters 10: 1135–1142.
28. Strickland MS, Callaham MA, Davies CA, Lauber CL, Ramirez K, et al. (2010) Rates of in situ carbon mineralization in relation to land-use, microbial community, and edaphic characteristics. Soil Biology and Biochemistry 42: 260–269.

29. Anderson TH, Domsch KH (1993) The metabolic quotient for CO2 (qCO2) as a specific activity parameter to assess the effects of environmental conditions, such as pH, on the microbial biomass of forest soils. Soil Biology and Biochemistry 25: 393–395.
30. Bradford MA, Wallenstein MD, Allison SD, Treseder KK, Frey SD, et al. (2009) Decreased mass specific respiration under experimental warming is robust to the microbial biomass method employed. Ecology Letters 12: E15–E18.
31. Olsen SR, Cole CV, Wantabe FS, Dean LA (1954) Estimation of available phosphorus in soils by extraction with sodium bicarbonate. U.S. Department of Agriculture. Circ. 939 Circ. 939.
32. Vitousek PM, Porder S, Houlton BZ, Chadwick OA (2010) Terrestrial phosphorus limitation: mechanisms, implications, and nitrogen–phosphorus interactions. Ecological Applications 20: 5–15.
33. McDowell RW, Cade-Menun B, Stewart I (2007) Organic phosphorus speciation and pedogenesis: analysis by solution 31-P nuclear magnetic resonance spectroscopy. European Journal of Soil Science 58: 1348–1357.
34. DeLong JP, Okie JG, Moses ME, Sibly RM, Brown JH (2010) Shifts in metabolic scaling, production, and efficiency across major evolutionary transitions of life. Proceedings of the National Academy of Sciences 107: 12941–12945.
35. Vrede T, Dobberfuhl DR, Kooijma SALM, Elser JJ (2004) Fundamental connections among organism C:N:P stoichiometry, macromolecular composition, and growth. Ecology 85: 1217–1229.
36. Allen AP, Gillooly JF (2009) Towards an integration of ecological stoichiometry and the metabolic theory of ecology to better understand nutrient cycling. Ecology Letters 12: 369–384.
37. Sterner RW, Elser JJ (2002) Ecological Stoichiometry: The Biology of Elements from Molecules to the Biosphere. Princeton, NJ: Princeton University Press. 468 p.
38. Cotner JB, Hall EK, Scott JT, Heldal M (2010) Freshwater bacteria are stoichiometrically flexible with a nutrient composition similar to seston. Frontiers in Microbiology 1: 132–132.
39. Heldal M, Scanlan DJ, Norland S, Thingstad F, Mann NH (2003) Elemental composition of single cells of various strains of marine *Prochlorococcus* and *Synechococcus* using X-Ray microanalysis. Limnology and Oceanography 48: 1732–1743.
40. McGroddy ME, Daufresne T, Hedin LO (2004) Scaling of C:N:P stoichiometry in forests worldwide: Implications of terrestrial Redfield-type ratios. Ecology 85: 2390–2401.
41. Danger M, Daufresne T, Lucas F, Pissard S, Lacroix G (2008) Does Liebig's law of the minimum scale up from species to communities? Oikos 117: 1741–1751.
42. Scott JT, Cotner JB, Lapara TM (2012) Variable stoichiometry and homeostatic regulation of bacterial biomass elemental composition. Frontiers in Microbiology 3: 42–42.
43. Yeates GW, Saggar S (1998) Comparison of soil microbial properties and fauna under tussock-grassland and pine plantation. Journal of the Royal Society of New Zealand 28: 523–535.
44. Ross D, Tate KR, Scott NA, Feltham CW (1999) Land-use change: Effects on soil carbon, nitrogen and phosphorus pools and fluxes in three adjacent ecosystems. Soil Biology Biochemistry 31: 803–813.
45. Chen CR, Condron LM, Davis MR, Sherlock RR (2000) Effects of afforestation on phosphorus dynamics and biological properties in a New Zealand grassland soil. Plant and Soil 220: 151–163.
46. Chen CR, Condron LM, Davis MR, Sherlock RR (2004) Effects of plant species on microbial biomass phosphorus and phosphatase activity in a range of grassland soils. Biology and Fertility of Soils 40: 313–322.
47. Chen CR, Condron LM, Xu ZH (2008) Impacts of grassland afforestation with coniferous trees on soil phosphorus dynamics and associated microbial processes: A review. Forest Ecology and Management 255: 396–409.
48. Reich PB, Oleksyn J (2004) Global patterns of plant leaf N and P in relation to temperature and latitude. Proceedings of the National Academy of Sciences of the United States of America 101: 11001–11006.
49. Weber TS, Deutsch C (2010) Ocean nutrient ratios governed by plankton biogeography. Nature 467: 550–553.
50. Hedin LO (2004) Global organization of terrestrial plant–nutrient interactions. Proceedings of the National Academy of Sciences of the United States of America 101: 10849–10850.
51. Schmidt IK, Jonasson S, Shaver GR, Michelsen A, Nordin A (2002) Mineralization and distribution of nutrients in plants and microbes in four arctic ecosystems: Responses to warming. Plant and Soil 242: 93–106.
52. Cleveland CC, Townsend AR, Constance BC, Ley RE, Schmidt SK (2004) Soil microbial dynamics in Costa Rica: seasonal and biogeochemical constraints. Biotropica 36: 184–195.
53. Devi NB, Yadava PS (2006) Seasonal dynamics in soil microbial biomass C, N and P in a mixed-oak forest ecosystem of Manipur, North-east India. Applied Soil Ecology 31: 220–227.
54. Aponte C, Marañón T, García LV (2010) Microbial C, N and P in soils of Mediterranean oak forests: influence of season, canopy cover and soil depth. Biogeochemistry 101: 77–92.
55. Chen CR, Condron LM, Davis MR, Sherlock RR (2003) Seasonal changes in soil phosphorus and associated microbial properties under adjacent grassland and forest in New Zealand. Forest Ecology and Management 177: 539–557.
56. Goldfarb KC, Karaoz U, Hanson Ca, Santee Ca, Bradford Ma, et al. (2011) Differential growth responses of soil bacterial taxa to carbon substrates of varying chemical recalcitrance. Frontiers in Microbiology 2: 94–94.
57. Hartman WH, Richardson CJ, Vilgalys R, Bruland GL (2008) Environmental and anthropogenic controls over bacterial communities in wetland soils. Proceedings of the National Academy of Sciences 105: 17842.
58. da C Jesus E, Marsh TL, Tiedje JM, de S Moreira FM (2009) Changes in land use alter the structure of bacterial communities in Western Amazon soils. The ISME Journal 3: 1004–1011.
59. Nacke H, Thürmer A, Wollherr A, Will C, Hodac L, et al. (2011) Pyrosequencing-based assessment of bacterial community structure along different management types in German forest and grassland soils. PLoS ONE 6: e17000–e17000.
60. Güsewell S, Gessner MO (2009) N:P ratios influence litter decomposition and colonization by fungi and bacteria in microcosms. Functional Ecology 23: 211–219.
61. Schimel J, Weintraub MN (2003) The implications of exoenzyme activity on microbial carbon and nitrogen limitation in soil: a theoretical model. Soil Biology and Biochemistry 35: 549–563.
62. Moorhead DL, Sinsabaugh RL (2006) A theoretical model of litter decay and microbial interaction. Ecological Monographs 76: 151–174.
63. Craine JM, Morrow C, Fierer N (2007) Microbial nitrogen limitation increases decomposition. Ecology 88: 2105–2113.
64. Ramirez KS, Craine JM, Fierer N (2010) Nitrogen fertilization inhibits soil microbial respiration regardless of the form of nitrogen applied. Soil Biology and Biochemistry 42: 2336–2338.
65. Treseder KK (2008) Nitrogen additions and microbial biomass: a meta-analysis of ecosystem studies. Ecology Letters 11: 1111–1120.
66. Ramirez KS, Lauber CL, Knight R, Bradford Ma, Fierer N (2010) Consistent effects of nitrogen fertilization on soil bacterial communities in contrasting systems. Ecology 91: 3463–3470.
67. Ramirez KS, Craine JM, Fierer N (2012) Consistent effects of nitrogen amendments on soil microbial communities and processes across biomes. Global Change Biology: DOI: 10.1111/j.1365-2486.2012.02639.x.
68. Marklein AR, Houlton BZ (2011) Nitrogen inputs accelerate phosphorus cycling rates across a wide variety of terrestrial ecosystems. The New Phytologist 193: 696–704.
69. Doi H, Cherif M, Iwabuchi T, Katano I, Stegen JC, et al. (2010) Integrating elements and energy through the metabolic dependencies of gross growth efficiency and the threshold elemental ratio. Oikos 119: 752–765.
70. Wallenstein MD, Hall EK (2011) A trait-based framework for predicting when and where microbial adaptation to climate change will affect ecosystem functioning. Biogeochemistry 109: 35–47.
71. Keiblinger KM, Hall EK, Wanek W, Szukics U, Hämmerle I, et al. (2010) The effect of resource quantity and resource stoichiometry on microbial carbon-use-efficiency. FEMS Microbiology Ecology 73: 430–440.
72. Six J, Frey SD, Thiet RK, Batten KM (2006) Bacterial and fungal contributions to carbon sequestration in agroecosystems. Soil Science Society of America Journal 70: 555–569.
73. Blagodatskaya EV, Anderson T-H (1998) Interactive effects of pH and substrate quality on the fungal-to-bacterial ratio and QCO2 of microbial communities in forest soils. Soil Biology and Biochemistry 30: 1269–1274.
74. Rousk J, Brookes PC, Bååth E (2010) Investigating the mechanisms for the opposing pH relationships of fungal and bacterial growth in soil. Soil Biology and Biochemistry 42: 926–934.
75. Rousk J, Brookes PC, Baath E (2009) Contrasting soil pH effects on fungal and bacterial growth suggest functional redundancy in carbon mineralization. Applied and Environmental Microbiology 75: 1589–1596.
76. Chu H, Fierer N, Lauber CL, Caporaso JG, Knight R, et al. (2010) Soil bacterial diversity in the Arctic is not fundamentally different from that found in other biomes. Environmental Microbiology 12: 2998–3006.
77. Jones RT, Robeson MS, Lauber CL, Hamady M, Knight R, et al. (2009) A comprehensive survey of soil acidobacterial diversity using pyrosequencing and clone library analyses. The ISME Journal 3: 442–453.
78. Lauber CL, Hamady M, Knight R, Fierer N (2009) Pyrosequencing-based assessment of soil pH as a predictor of soil bacterial community structure at the continental scale. Applied and Environmental Microbiology 75: 5111–5120.
79. Fierer N, Bradford MA, Jackson RB (2007) Toward an ecological classification of soil bacteria. Ecology 88: 1354–1364.
80. Karpinets TV, Greenwood DJ, Sams CE, Ammons JT (2006) RNA: protein ratio of the unicellular organism as a characteristic of phosphorous and nitrogen stoichiometry and of the cellular requirement of ribosomes for protein synthesis. BMC biology 4: 30.
81. Sinsabaugh RL, Follstad Shah JJ, Hill BH, Elonen CM (2011) Ecoenzymatic stoichiometry of stream sediments with comparison to terrestrial soils. Biogeochemistry: DOI: 10.1007/s10533-10011-19676-x.
82. Franklin O, Hall EK, Kaiser C, Battin TJ, Richter A (2011) Optimization of biomass composition explains microbial growth-stoichiometry relationships. The American Naturalist 177: E29–42.
83. Chan L-K, Newton RJ, Sharma S, Smith CB, Rayapati P, et al. (2012) Transcriptional changes underlying elemental stoichiometry shifts in a marine heterotrophic bacterium. Frontiers in Microbiology 3: 159–159.
84. Sparling GP, Hart PBS, August JA, Leslie DM (1994) A comparison of soil and microbial carbon, nitrogen, and phosphorus contents, and macro-aggregate

stability of a soil under native forest and after clearance for pastures and plantation forest. Biology and Fertility of Soils 17: 91–100.
85. Li Y-S, Wu L-H, Lu X-H, Zhao L-M, Fan Q-L, et al. (2006) Soil microbial biomass as affected by non-flooded plastic mulching cultivation in rice. Biology and Fertility of Soils 43: 107–111.
86. McLatchey GP, Reddy KR (1998) Regulation of organic matter decomposition and nutrient release in a wetland soil. Journal of Environmental Quality 27: 1268–1274.
87. Brake M, Höper H, Joergensen RG (1999) Land use-induced changes in activity and biomass of microorganisms in raised bog peats at different depths. Soil Biology and Biochemistry 31: 1489–1497.
88. Fierer N, Lauber CL, Ramirez KS, Zaneveld J, Bradford Ma, et al. (2011) Comparative metagenomic, phylogenetic and physiological analyses of soil microbial communities across nitrogen gradients. The ISME Journal: DOI:10.1038/ismej.2011.1159-DOI:1010.1038/ismej.2011.1159.
89. Shrestha PM, Noll M, Liesack W (2007) Phylogenetic identity, growth-response time and rRNA operon copy number of soil bacteria indicate different stages of community succession. Environmental Microbiology 9: 2464–2474.
90. Hull D, Pettifer SR, Kell DB (2008) Defrosting the digital library: Bibliographic tools for the next generation web. PLoS Computational Biology 4: e1000204.
91. Brookes PC (2001) The soil microbial biomass: Concept, measurement and applications in soil ecosystem research. Microbes and Environment 16: 131–140.
92. Anderson T-H, Domsch KH (1990) Application of eco-physiological quotients (qCO2 and qD) on microbial biomasses from soils of different cropping histories. Soil Biology and Biochemistry 22: 251–255.
93. Qian SS (2009) Environmental and Ecological Statistics with R. Boca Raton, FL: Chapman & Hall/CRC. 421 p.
94. Warton DI, Duursma RA, Falster DS, Taskinen S (2011) SMATR 3- an R package for estimation and inference about allometric lines. Methods in Ecology and Evolution: DOI: 10.1111/j.2041-1210X.2011.00153.x.
95. Schilling EB, Lockaby BG (2005) Microsite influences on productivity and nutrient circulation within two southeastern floodplain forests. Soil Science Society of America Journal 69: 1185–1195.
96. Jonasson S, Michelsen A, Schmidt IK, Nielsen EV, Callaghan TV (1996) Microbial biomass C, N and P in two arctic soils and responses to addition of NPK fertilizer and sugar: Implications for plant nutrient uptake. Oecologia 106: 507–515.

Fungal Community Structure in Disease Suppressive Soils Assessed by 28S LSU Gene Sequencing

C. Ryan Penton[1]*, V. V. S. R. Gupta[2], James M. Tiedje[1], Stephen M. Neate[3], Kathy Ophel-Keller[4], Michael Gillings[5], Paul Harvey[2], Amanda Pham[1], David K. Roget[2]

1 Center for Microbial Ecology, Michigan State University, East Lansing, Michigan, United States of America, **2** CSIRO Ecosystem Sciences, Glen Osmond, South Australia, Australia, **3** Department of Agriculture, Fisheries and Forestry, Queensland, Leslie Research Centre, Towoomba, Queensland, Australia, **4** SARDI, Glen Osmond, South Australia, Australia, **5** Department of Biological Sciences, Macquarie University, North Ryde, New South Wales, Australia

Abstract

Natural biological suppression of soil-borne diseases is a function of the activity and composition of soil microbial communities. Soil microbe and phytopathogen interactions can occur prior to crop sowing and/or in the rhizosphere, subsequently influencing both plant growth and productivity. Research on suppressive microbial communities has concentrated on bacteria although fungi can also influence soil-borne disease. Fungi were analyzed in co-located soils 'suppressive' or 'non-suppressive' for disease caused by *Rhizoctonia solani* AG 8 at two sites in South Australia using 454 pyrosequencing targeting the fungal 28S LSU rRNA gene. DNA was extracted from a minimum of 125 g of soil per replicate to reduce the micro-scale community variability, and from soil samples taken at sowing and from the rhizosphere at 7 weeks to cover the peak *Rhizoctonia* infection period. A total of ~994,000 reads were classified into 917 genera covering 54% of the RDP Fungal Classifier database, a high diversity for an alkaline, low organic matter soil. Statistical analyses and community ordinations revealed significant differences in fungal community composition between suppressive and non-suppressive soil and between soil type/location. The majority of differences associated with suppressive soils were attributed to less than 40 genera including a number of endophytic species with plant pathogen suppression potentials and mycoparasites such as *Xylaria* spp. Non-suppressive soils were dominated by *Alternaria*, *Gibberella* and *Penicillum*. Pyrosequencing generated a detailed description of fungal community structure and identified candidate taxa that may influence pathogen-plant interactions in stable disease suppression.

Editor: Boris Alexander Vinatzer, Virginia Tech, United States of America

Funding: Financial support was provided by the Commonwealth Scientific and Industrial Research Organisation and Grains Research Development Corporation. The funders had no role in study design, data collection and analysis, decision to publish, or preparation of the manuscript.

Competing Interests: The authors have declared that no competing interests exist.

* E-mail: dobber4uf@gmail.com

Introduction

Plant-microbe-soil interactions play a vital role in maintaining plant health and productivity in agricultural and horticultural crops. Plant diseases caused by soilborne pathogens result in substantial losses to agricultural production worldwide [1]–[3]. For example, roots of cereals, pasture plants and oil seed crops are prone to attack by soilborne necrotrophic pathogens such as *Rhizoctonia solani*, *Fusarium pseudograminearum*, *Gaeumannomyces graminis* var *tritici* and *Pythium* spp. These are among the most difficult groups of plant pathogens to control due to their ability to persist in crop residues [1]–[4]. Due to the limitations in the effectiveness of fungicides and a lack of successful plant-based resistance, enhancement of soil-based natural disease suppression could be an effective option to control disease, especially if it can be achieved by in-field enhancement through crop and/or soil management practices [5]–[9].

Soil suppressiveness is the ability of a soil to prevent/suppress disease even in the presence of a pathogen, suitable host plant and favorable climatic conditions [9]–[13]. In this study we use the term 'non-suppressive' for soils that are unable to suppress disease incidence by the pathogen. Biological suppression of soilborne pathogens has been reported from a variety of cropping systems worldwide [11]–[12], [14]–[15]. In the case of wheat and barley crops, this suppression has been shown against a number of soilborne diseases including *Fusarium* wilt, Take-all and *Rhizoctonia* bare patch. In Australia, biologically-based disease suppression has been reported in long-term experimental plots and farmer fields [11], [16]–[17]. This suppression has been attributed to diverse microbial communities including bacteria, fungi and protozoa and is reported to affect pathogen survival, growth in bulk soil and rhizosphere and root infection [18]–[19]. The adoption of no-till and stubble retention practices can, in some cases, increase soilborne plant diseases in the short-term [11]–[20]. However, long-term adoption of crop management practices that supply higher levels of biologically-available carbon inputs either through crop residues or addition of composts and organic manures can support higher levels of suppression. This occurs through changes to the composition and activity of the soil microbial community [7], [21]–[23].

Rhizoctonia bare patch disease generally starts in young seedlings and the disease manifests during the first 8 weeks of crop growth causing significant crop yield losses [4]. Two complementary mechanisms are suggested to be involved in disease suppression in both the bulk soil and rhizosphere;

competition for nutrients between the pathogen and general microbial community and the activity of antagonists [7]. Interactions in bulk soil involve general competition for carbon and nutrients (fungistatis) or antibiosis (soil bacteria or fungi vs. pathogenic fungi) and mycoparasitism (pathogenic fungi vs. other soil fungi) that can affect the survival and growth of the pathogen [24]–[26]. Rhizosphere interactions can directly prevent the pathogen reaching the root or interfere with infection processes [27]. Indirectly such interactions may induce host plant resistance [14], [28].

Research on microbial communities in disease suppressive soils has mainly focused on bacteria [28]–[31]. A wide range of bacterial groups have been suggested as contributing to disease suppression through antibiosis, plant growth promotion or systemic induced resistance [13], [14], [30]. The functional diversity of soil fungi and their capacity to colonize diverse microhabitats can influence pathogen levels and play a significant role in improving plant health, e.g. *Trichoderma* spp and mycorrhizal fungi [25], [32]. The genus *Trichoderma* has been studied extensively for its biocontrol potential and a number of fungi and oomycetes are registered as biocontrol agents [33]–[34]. Soils with higher disease suppressive potential have been found to exhibit higher fungal diversity [22]. In view of the large diversity of uncultured fungi in soil, culture-independent methods are required to describe their composition and to identify community differences between soils. Recently, based on high-throughput sequencing, soils from pea fields with different degrees of disease were discriminated on the basis of their fungal communities [35].

Our objective was to determine in what way fungal communities differed between paired soils, one with long-term high disease suppression and the other with no disease suppression, at two wheat-growing locations, Avon and Minnipa, in South Australia. For comparing the fungal populations among the four sites and two sampling times, we used pyrosequencing of the 28S LSU rRNA gene in soil DNA from the four fields and RDP's Naïve Bayesian Classifier, which provides both genus identification and taxonomic placement for otherwise unclassified sequences.

Materials and Methods

Site and soil descriptions

The agricultural fields studied are located in the wheat-cropping region in South Australia and have been under continuous cropping for more than 10 years. The Avon and Minnipa locations are, respectively, 'suppressive' (SP) (S34 13.981, E138 18.586 and S32 59.066, E135 9.424) and 'non-suppressive' (NSP) (S34 13 29.06, E138 19 3.66 and S32 49.955, E135 9.595) for diseases caused by soil-borne necrotrophic pathogens (e.g. *Rhizoctonia solani* AG 8, *Fusarium pseudograminearum*) in cereal crops (**Figure S1**) based on field disease measurements and glasshouse pot assays over the previous 20 years [11], [36]–[37]. The fields at Minnipa are located at the Minnipa Agricultural Centre, a collaborative research organisation in this study and the land at Avon field sites has housed CSIRO research trials for over 25 years. The Avon and Minnipa sites are 350 km apart and the SP and NSP fields are 1.1 and 0.4 km apart respectively. The soil at both sites is Luvic Calcisol and sandy or sandy loam in texture (Lithocalcic Calcarosol, [38]–[39]). Organic carbon in the SP fields ranged from 1.1–1.6% and in the NSP fields from 0.7–1.0%, clay ranged from 8–17% (**Table S1**). The biological nature of higher disease suppression in these fields has been previously established [16]–[17], [36]–[37]. The climate is Mediterranean-type, characterized by hot dry summers and cool wet winters, with an average annual rainfall of 260–300 mm. During the 3 years prior to sampling all the fields were under cereal crops though the SP fields at both sites were under continuous cropping with stubble retention and no-till practices for longer periods (>10 y) compared to the NSP fields that were under a low-input pasture-crop rotation. Other general agronomic practices such as weed control, fertilizer addition and time of sowing were similar in both the SP and NSP fields. Soils from the SP fields also showed higher microbial biomass (by 20–35%) and more particulate organic carbon (10–20% higher).

Eight samples (10 cm deep cores at sowing and rhizosphere samples at 7 weeks) were collected in 2010 and mixed to generate a composite sample for each of the eight true replicates per field, for a total of 64 samples. From each of these eight replicates, 125 g of soil was used for DNA extraction. Soils at sowing were collected in the previous year's crop row, and at 7 weeks samples were collected from the wheat rhizosphere within the 0–10 cm depth. The rhizosphere soil was defined as the soil remaining on the root after gentle shaking; both the root and rhizosphere were used for DNA extraction. Immediately after collection, samples were stored on ice and transported to the laboratory in Adelaide. Subsamples were separated for microbial and chemical analyses and bioassay experiments.

Disease suppression potential

Surface soils collected prior to sowing in 2010 were used to measure disease suppression potential, i.e. maximum level of disease suppression that can be observed with added pathogen inoculum and under controlled environmental conditions, using a carbon amendment assay and a soil transfer assay with wheat as a host plant [17]. During the pre-incubation in both these growth chamber assays, added pathogen inoculum was allowed to interact with the native soil microbial communities, under optimal soil moisture and temperature conditions, prior to introducing the wheat [17]. Briefly, in the carbon amendment assay, Rhizoctonia root damage was measured in response to *R. solani* AG 8 inoculum (2 x, 8 mm dia disks of colonized 1/4 strength potato dextrose agar per pot) with or without the addition of sucrose (2 g granular sucrose/300 g soil per pot), a simple carbon substrate that has been shown to accentuate suppression characteristics when added to a pre-incubation [17]. Soils were pre-incubated with or without inoculum and carbon substrate for 2 weeks prior to sowing wheat (*Triticum aestivum* cv. Yitpi) and all the pots were incubated at 10°C on a 10/14 h day-night cycle. Soil moisture in the pots was maintained at 80% field capacity by adding water at 2–3 day intervals. All experiments were harvested 4–5 weeks after seedling emergence. Roots were washed carefully and scored for disease rating (0–5 scale) and plant growth [17].

For the soil transfer assay, subsamples of field soils collected at sowing were sterilized by exposure to gamma-radiation (3 cycles of 25 rads; Steritech, Victoria) and then incubated 2 weeks at 10°C with or without *Rhizoctonia solani* AG8 as inoculum and with and without addition of 10% of the original non-sterile soil. A plant assay was then performed using a method similar to that in the carbon amendment assay and roots were scored for disease severity. Carbon substrate utilization profiles of soil microbial communities were determined using specific carbon substrates selected for Australian soils through a modified [22] Microresp method [40].

DNA extraction

Field moist soils were stored at −20°C until lyophilized for DNA extraction. DNA was extracted from ~125 g of soil for each sample by the Root Disease Testing Service at SARDI (Adelaide) [41]. Subsamples (aliquots) of extracted DNA were shipped to

Michigan State University, USA, for fungal community sequencing. qPCR assays using rDNA (TaqMan) probe sequences specific to *Rhizoctonia solani* AG 8 and *Trichoderma* spp. were conducted [41], and quantified using a DNA standard of *R. solani* AG8 DNA (pg g^{-1} soil sample). Amounts of total fungal and bacterial DNA were quantified using group specific primers (FR 1/FF390 [42]; F968/R1378 [43] based on the QuantiTect SYBR Green PCR kit (Qiagen) and the PCR was carried out on a Strategene Maxpro3000P qPCR system.

Fungal community profiling using T-RFLP analysis

Fungal community DNA was amplified from 14 ng of template DNA using the ITS1F.FAM forward (5′-CTTGGTCATTTAGAGGAAGTAA-3′) and ITS4R.HEX reverse (5′-TCCTCCGCTTATTGATATGC-3′) primers [44]–[45]. PCR was carried out in a 35 μL total volume using 0.4 μM of primers, 0.2 mM of dNTPs, 1x PCR buffer (Qiagen, Australia), and 4 units of HotStarTaq DNA Polymerase (Qiagen, Australia). The PCR conditions were 94°C for 1 min; 56°C for 1 min; 72°C for 1 min and 1.5 min for 35 cycles. The products were checked for size and specificity by agarose gel electrophoresis followed by purification using the MiniElute 96 UF PCR Purification Kit (Qiagen, Australia) and 100–150 ng of purified PCR product was digested for 3 h at 37°C followed by 65°C denaturation with the restriction enzymes *Alu*I and *Cfo*I. The digested DNA was purified using SigmaSpin Post-Reaction Purification Columns (Sigma, Australia) and 10 μL of the purified T-RFs were analyzed for size by the Australian Genome Research Facility (Adelaide, Australia) using capillary separation on an ABI 3730 DNA analyser with a LIZ500 size standard. TRF size and intensity data were collected using the GeneMarker analysis software (version 1.85; SoftGenetics Inc.), with a minimum cut off of 100 intensity units. Relative abundances of TRFs were calculated and normalized against the total peak height of all TRFs in the profile. TRFLP fragment data were then analyzed using the Primer6 software package (Primer-E Ltd, Plymouth, U.K.) by cluster analysis and non-metric multidimensional scaling (NMDS).

Fungal 28S Amplification, Sequencing and Processing by RDP Classifier

Anchored by conserved regions, the 28S gene contains two hypervariable regions, denoted D1 and D2 [46]–[47]. The LR3/LR0R primer combination (http://www.biology.duke.edu/fungi/mycolab/primers.htm) was used for this study. Spanning the D1/D2 regions, these primers have been identified as the most suitable in terms of amplicon length (625 bp), resolution, and accuracy by pyrosequencing [48]. Amplification was performed per methods previously described [49] and available in supplementary information (**Text S1**). Amplicons were sequenced following Lib-L adapter ligation by Utah State University CIB Genomics Core Lab and processed using the shotgun protocol. Raw sequences were quality processed (**Text S1**) and sorted by tag through the RDP pyrosequencing pipeline (http://pyro.cme.msu.edu) and subjected to classification using the RDP naïve Bayesian fungal LSU Classifier version 1 [50] that is based on a manually archived LSU gene training set [48] at 0% "best-match" bootstrap confidence ("classification confidence"-CC) [49]. Each sample was randomly re-sampled to 4,484 sequences per sample and classifications at the phylum, class, order, family, and genus levels were treated as bins for downstream statistical analysis. Supplementary text describes sequence classification at 50% bootstrap confidence to investigate the composition of the 'unclassified' fungi (**Text S2**) and comparisons between the whole and resampled datasets were also performed (**Text S3**). Sequences were deposited in the European Nucleotide Archive under study accession PRJEB4037, sample accessions ERS253863-ERS253910.

Data Analysis

We used several multi-variate statistical analyses for community comparisons, for both the TR-F data and 28S sequence data, using the PRIMER-6 software package [51], (Primer-E Ltd, 239 Plymouth, U.K.). Firstly, Hellinger transformed data (square root of relative abundance) was used to generate Bray-Curtis (+1) dissimilarity matrices and an analysis of similarity (ANOSIM) was carried out to test whether different sites harbored distinct fungal communities [52]. Non-metric multidimensional scaling (nMDS) analysis was performed using all pairwise distances between different fungal communities (Bray-Curtis distances) and the statistical significance was tested through permutational analysis of variance (PERMANOVA) [53]. The average contribution of individual sampling units (e.g. genera OTU) to the overall Bray-Curtis distances was estimated using the similarity percentage (SIMPER) analysis from the pair wise comparisons [54]. The Shannon index (H') estimate of community richness [55] and Pielou's evenness (J') [56] were used for community diversity estimates. Differences in the disease incidence (root infection) between fields were compared by ANOVA analysis using Genstat (v14.2, VSN International Ltd.).

Results

Pathogen, microbial and disease suppression properties

Soils collected at the time of sowing from the suppressive (SP) and non-suppressive (NSP) fields at Avon and Minnipa yielded 89–244 pg of DNA g^{-1} soil of *R. solani* AG 8 (**Table 1**). These inoculum concentrations were considered within a high disease risk category based on southern Australian experience [65]. However, the analysis of root samples from 7 week old plants indicated significantly lower disease incidence in the SP field samples compared to those from the NSP fields, at both the sites (**Table 1**). *Rhizoctonia solani* AG8 is generally considered a seedling pathogen, hence root measurements were taken using seedlings 7 weeks after sowing in order to quantify differences between the two fields.

Results from the two growth chamber assays also clearly differentiated the level of disease suppression potential (DSP) in the soils from the two locations and indicated higher suppression potential in the soils from SP fields at both sites compared to NSP fields (**Figure 1A**). In the soil transfer assay, the addition of fresh soil from SP fields plus pathogen inoculum significantly (ANOVA, $P<0.001$) reduced disease incidence in sterile soil by 63.8% and 44.1% in the Avon and Minnipa soils, respectively, compared to addition of pathogen inoculum alone. In contrast, the addition of fresh soil from the NSP fields caused only a minor reduction in disease incidence in the sterile NSP soil (17.6–20.1%). In the DSP bioassay the addition of *R. solani* AG 8 inoculum increased disease levels considerably in all soils but the disease suppression potential was significantly (ANOVA, $P<0.011$) lower (mean = 27%) in the non-suppressive soil compared to suppressive soil (mean = 46%) (**Figure 1B**). A higher level of disease response to added inoculum coupled with a smaller reduction following C addition in NSP fields suggests a lower suppressive ability of these soils.

Compared to the suppressive soils, Avon non-suppressive soils showed higher levels of total fungal DNA both at sowing and at 7 weeks, but no difference in fungal DNA quantity was observed in the Minnipa soils (**Table S2**). A wider fungal to bacterial ratio was observed in the non-suppressive soil at sowing (62.8) compared to the suppressive soils (14.9). A reduction in the fungi:bacteria DNA

Table 1. Pathogen abundance and disease incidence.

Site	Field	*R. solani* AG 8 (pg DNA g^{-1} soil)	Root rating (0–5 scale)	% infected crowns
Avon	**Suppressive**	89±17 a	0.269±0.019 a	6.9±1.4 a
	Non-supp	244±25 b	3.181±0.132 c	59.6±4.0 b
Minnipa	**Suppressive**	107±32 a	0.444±0.036 a	8.4±1.4 a
	Non-supp	167±52 ab	2.181±0.148 b	59.0±4.3 b

Note: values within each column followed by the same letter are not statistically significant at P<0.05.
Amount of pathogen *R. solani* AG 8 inoculum in the surface soil at sowing and the level of disease incidence measured in 7 week old seedlings from suppressive and non-suppressive fields at Avon and Minnipa with standard errors.

ratio 'in-crop' was only seen in the Avon NSP soil but not in Minnipa soils or Avon SP soil. Although a measurable amount of *Trichoderma* spp. DNA was found in all soils (average = 56 pg±41 DNA/g soil) there was no trend based on soil type, suppression or sampling time. *Trichoderma* DNA levels were generally lower in the Minnipa soils compared to the Avon soils. PERMANOVA analysis of T-RFLP data (**Table 2**) showed significant effects at both sowing and in-crop for suppression that was supported by NMDS ordination (**Figure S2**).

Fungal 28S LSU sequence processing and classification

Of 1,260,461 fungal 28S sequences, a total of 994,430 passed quality filtering (**Table S3**) that were then classified against the RDP fungal classifier reference database with a range between 4,484 and 44,431 reads per sample. Of these, 345,066 were read from the reverse (LR3) direction. Unless explicitly stated, all results are based on the re-sampled dataset at 4,484 reads per sample. Sequences classified as *Eukaryota incertae sedis* that encompassed non-fungal eukaryotes comprised 1.6±0.3% and 5.3±1.5% of reads in the Avon and Minnipa soils, respectively, and were removed prior to analyses. Rarefaction curves indicated near-coverage saturation in some samples, with no consistent differences in coverage among sampling locations (**Figure S3**).

Best-match classification yielded a total of 917 unique genera that covered 54% of the RDP Fungal Classifier reference database. Ascomycota (78.1±9.0%), followed by Basidiomycota (9.5±4.7%), and Chytridiomycota (5.9±4.8%) dominated the fungal communities. Of the 37 fungal classes identified, the top 10 accounted for 95.5±2.8% of all sequences. Likewise, the top 10 of 104 orders contained 77.8±7.7% of all sequences, the top 20 of 298 families accounted for 75.7±6.9%, and the top 20 of 917 genera had 56.2±7.6% of all classified reads (**Figure 2**). Results and discussion of classification at 50% bootstrap confidence (**Text S2**) and differences between the whole dataset and the randomly resampled datasets (**Text S3**) are discussed in supplementary information.

Suppressive versus Non-Suppressive Fungal Community Comparisons

Based on the criterion that >60% of the replicates per site must contain a certain genus-level OTU, we determined to what degree these OTUs were shared in the samples at sowing. A total of 84 OTUs, comprising 8.3% of all sequences, were shared among all sites (**Figure 3**). A very small proportion (<0.1%) of sequences were unique to any one location while suppressive and non-suppressive fields shared 102 and 97 genera, respectively. ANOSIM two-way crossed analysis showed significant differences in fungal community composition between both suppression status and soil type at the genus, family, and order units. Global R statistics decreased at higher taxonomic levels, suggesting a lower degree of separation between the different suppression groups (**Table 3**). PERMANOVA analysis indicated significant soil type effect for W (CV = 13.89, P = 0.001) and for RS (CV = 14.13, P = 0.001) (**Table 2**). Non-metric MDS ordination illustrated the dissimilarities in fungal communities at the best genus level matches among the four sites, with smaller differences at the family level (**Figure 4**). The average contribution of individual genera OTU to the overall Bray-Curtis distances (SIMPER analysis) showed that 33 genera were responsible for 26.8% and 31.2% of the discrimination between suppressive and non-suppressive soils at sowing and in-crop, respectively (**Figure 5**). No significant differences were identified in measures of Shannon diversity (H') and Pielou's evenness (*J'*) (**Table S2**).

Plant development

NMDS ordination showed a much stronger discrimination at sowing versus at 7 weeks (data not shown). This was also reflected in the PERMANOVA results, where the coefficient of variation decreased in the site, suppression, and site x suppression factors with time (**Table 2**). Only 28% of the top 33 genera significantly contributing to the suppressive community discrimination (based on SIMPER analysis at sowing and in-crop) were shared between the two time periods. The majority of those belonged to the class Sordariomycetes, where 56% of the genera were shared. In contrast, only one genus was shared among the 10 that comprised the Agaricomycetes between the sowing and in-crop periods. *Alternaria* significantly (t-test, p<0.01) decreased with time in the non-suppressive soils from 11.7% to 8.1% relative abundance, while *Schizothecium* increased (t-test, p<0.01) 3 to 7 fold to 3.0% and 4.8% of the community in the suppressive and non-suppressive soils, respectively. A large 8-fold decrease with time (t-test, P<0.01) was associated with *Christiansenia* in both soils. Both *Ascolobus* (t-test, P<0.01) and *Corynespora* (P<0.05) decreased from 1.3% to 0.5% and 1.1% to 0.3% relative abundance in the suppressive soils but significantly increased (t-test, P<0.05) three-fold in the non-suppressive soils. *Zygopleurage* decreased from 3.6% to 1.9% (t-test, P<0.05) only in the suppressive soil. Lastly, the *Hypocrea* increased with time (t-test, P<0.05) from 0.2% to 1.2% of the community in the non-suppressive soils with no increase in the suppressive sites.

Discussion

Suppressive versus non-suppressive fungal community discrimination

Significant fungal composition differences were revealed between suppressive and non-suppressive soils at each site as well as for both sites combined, which were also shown by the

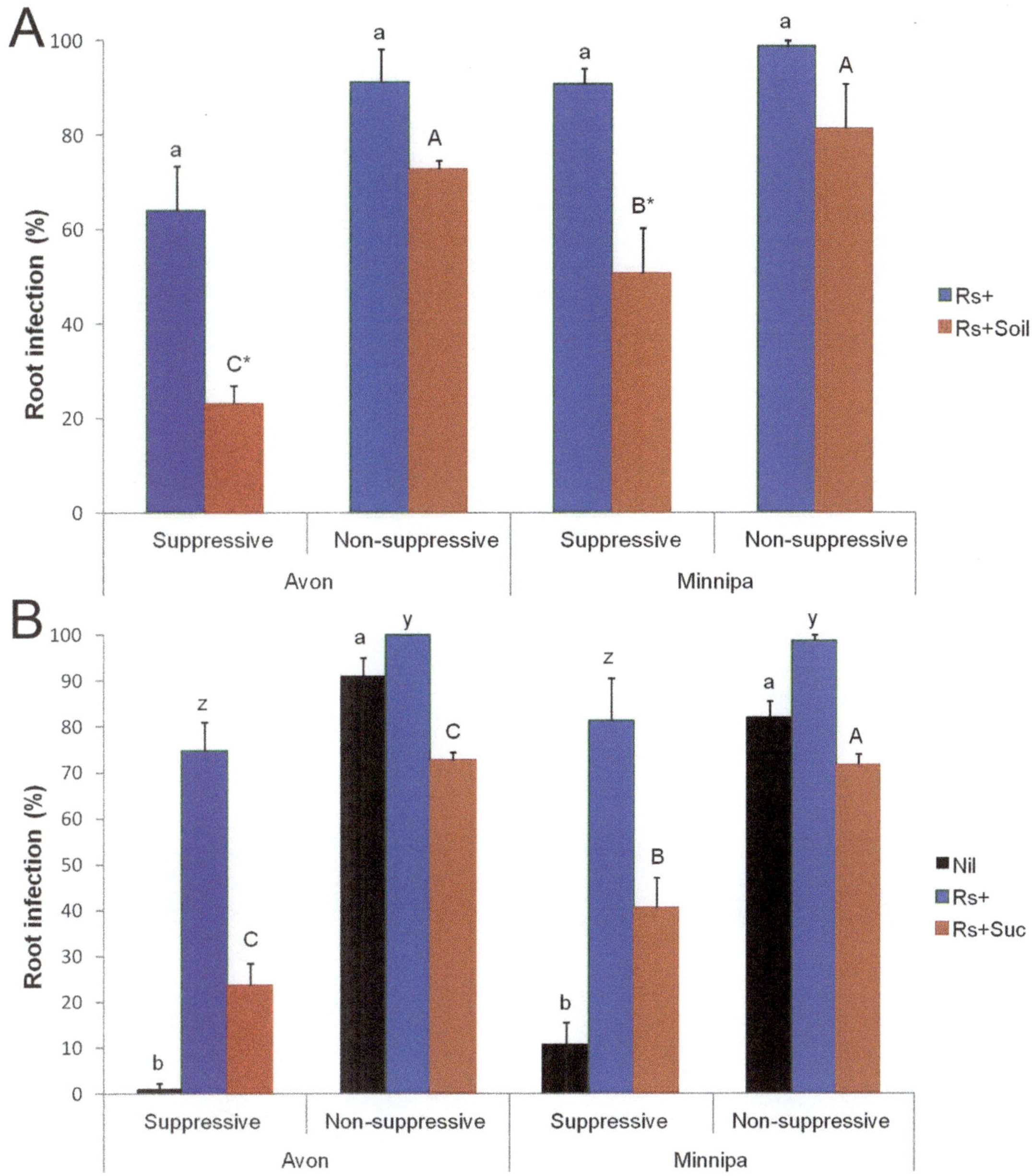

Figure 1. Rhizoctonia disease suppression in wheat seedlings. Results from a 28 day bioassay using soils collected from field sites at the start of the crop season in 2010. Soil transfer bioassay (A); sterile soil with *R. solani* AG 8 (Rs+, blue) inoculum and sterile soil + *R. solani* AG 8 (Rs+, red) inoculated with 10% of field soil. Bars of same color with different letters are significantly different at P<0.05 (ANOVA), * indicates significant reduction (P<0.008; ANOVA) in disease with field add soil addition. Disease suppression potential bioassay (B); field soil (Nil, black), field soil with *R. solani* AG 8 (Rs+, blue) and field soil +C added (Suc) + *R. solani* AG 8 (Rs+suc, red). Disease suppression potentials – 68.4 & 27.3 Avon soils; 50.1 & 27.6 Minnipa soils. Bars of similar color with different letters are significantly different at P<0.05 (ANOVA).

independent T-RFLP method. For example, according to the SIMPER results, 4.0% of the genera identified in this study accounted for 26.8% and 31.2% of the discrimination between suppressive and non-suppressive soils in the sowing and in-crop samples, respectively. A number of fungal genera with plant pathogen suppression potential were identified as well as fungi causing diseases of wheat, maize, and grapevines (**Table S4**). In this context we identify the average classification confidence (CC) and standard deviation for each genus discussed in order to provide insight into the dissimilarity between the query sequences and the reference database closest match. The inferred characteristics discussed are based on studied members of each genus, but there could be variation in phenotype within these genera that could affect suppressiveness or disease.

The genera most associated with the suppressive fields identified through the SIMPER analysis largely contained putative endophytes, saprophytes and fungi that may play a role in disease suppression (**Figure 5**). Differences were due mostly to changes in relative abundances, not presence/absence, as indicated by the similar number of shared and unique OTUs among treatments (**Figure 3**). In general, members of the Xylariaceae, Bionectriaceae and Hypocreaceae families with known antifungal capability are well represented among the dominant communities in the suppressive fields. For example, *Xylaria* (CC = 68±19%), a genus

Table 2. PERMANOVA analysis.

Dataset	Factor	Sowing		7 weeks	
		CV	P	CV	P
t-RFLP	Site	20.90	0.001	13.14	0.008
	Supp	16.20	0.002	10.70	0.033
	Site * Supp	21.50	0.003	18.50	0.006
Whole Dataset	Site	13.89	0.001	13.62	0.001
	Supp	12.63	0.001	7.36	0.001
	Site * Supp	13.93	0.001	13.56	0.002
Resample Dataset	Site	14.13	0.001	13.68	0.001
	Supp	12.90	0.001	7.61	0.004
	Site * Supp	14.23	0.001	13.93	0.001

Statistical comparisons of the t-RFLP, the 28S whole dataset, and the 28S re-sampled (4484 sequences per sample) datasets. CV = Component of variation.

reported with primarily endophytic and some saprotrophic species [57], exhibited the largest gain (t-test, $P<0.05$) in abundance in the suppressive community of any genus. Endophytes such as these are known to stimulate plant growth, improve the ability of plants to withstand environmental stresses, and increase disease resistance [58]. Specifically, *Xylaria* has also been reported to exhibit antifungal activity against plant pathogenic fungi probably through secondary metabolites such as xylarinic acids [59]–[60]. The *Bionectria* (CC = 96±13%) that were also enriched, though not significantly, contains species that are saprotrophic, necrotrophic, or biotrophic. The genus also includes mycoparasites that are used as biocontrol agents of fungal plant pathogens [61]. Also higher (t-test, $P<0.05$) in suppressive soil were the *Eutypa* (CC = 77±20%), which include species responsible for grapevine dieback disease [62] and the *Anthostomella* (CC = 50±14%), which are primarily saprotrophic [63]. Some other fungi that trended towards higher abundance in suppressive soils with reported antifungal activity include *Chaetomium*, *Corynascus* and *Microdiplodia* spp. [64]–[66].

In contrast to the suppressive fields, the majority of putative fungal pathogens had higher abundances in the non-suppressive fields. *Gibberella* (CC = 90±14), whose anamorph is *Fusarium* sp., the cause of crown rot of wheat, a common disease in Australian wheat fields was found in higher abundance, though not significantly, in the non-suppressive fields. This genus may also include non-pathogenic isolates of *Fusarium oxysporum*, identified as a biocontrol agent [67]–[68]. *Alternaria* (CC = 93±15%), the causative endophytic fungus for a leaf blight and black point of wheat, which has also been routinely isolated as an endophyte of leaves from maize and wheat [69] was associated with non-suppressive ($P<0.05$) soils. It is the most abundant endophyte in wheat cultivars [70] and in maize [71]. Likewise, *Podospora* (CC = 58±16%) and *Penicillium* (CC = 95±13%) were associated ($P<0.05$) with the non-suppressive soil, in contradiction to another

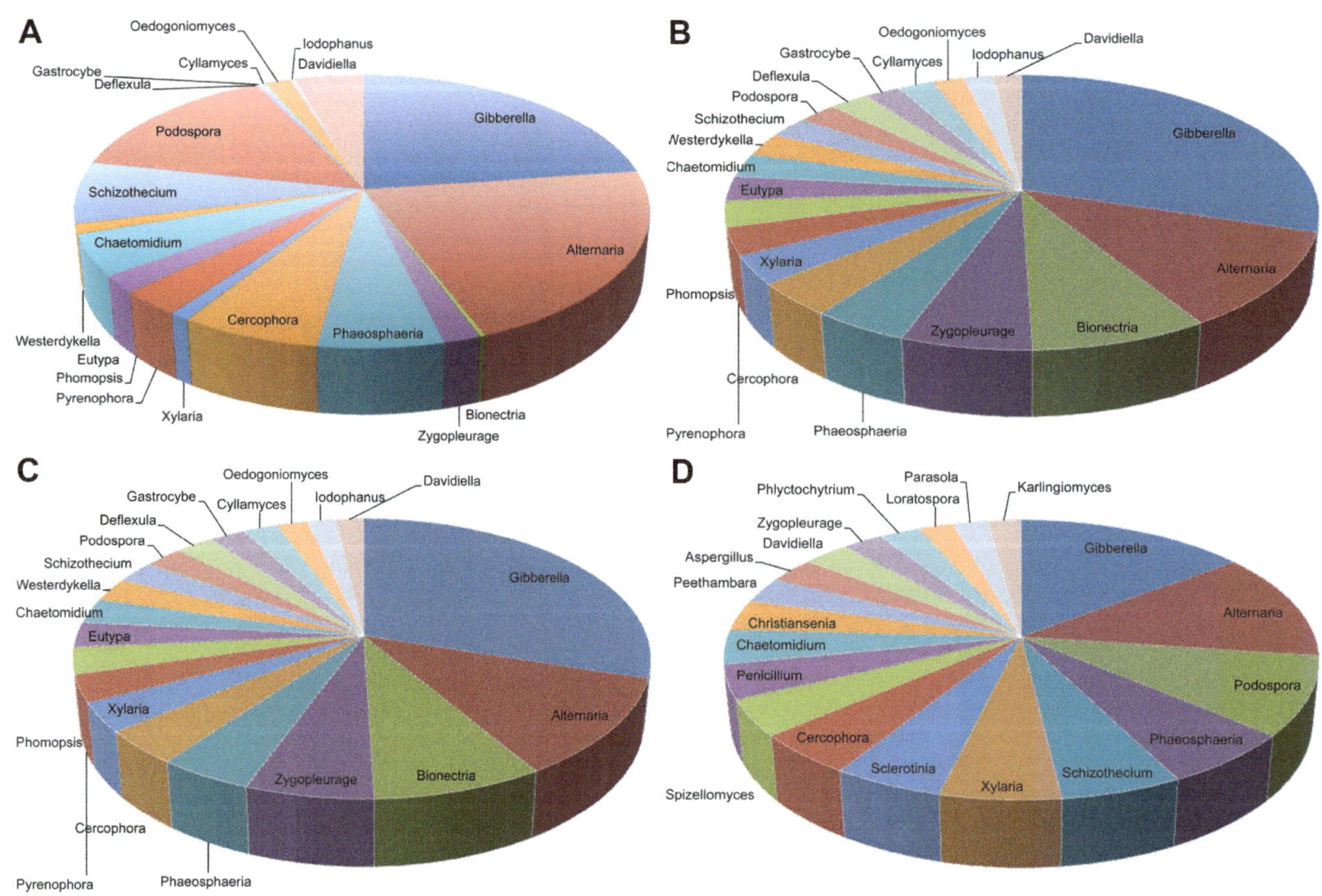

Figure 2. Composition of the 20 most abundant genera. Data from (A) suppressive soils, (B) non-suppressive soils, (C) Avon soil, and (D) Minnipa soil.

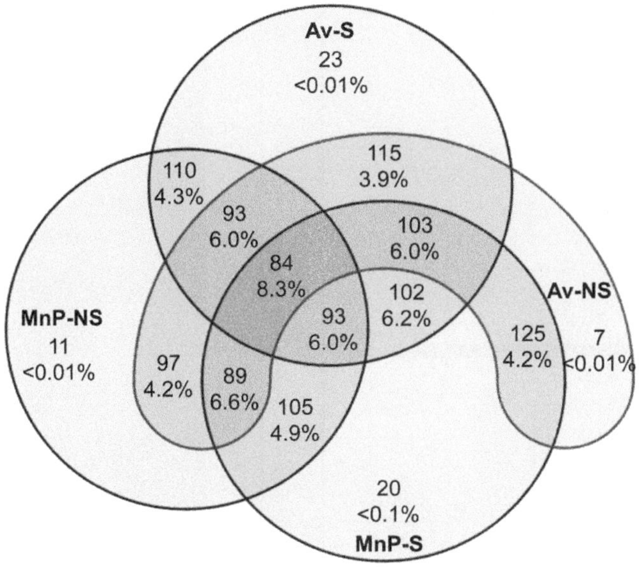

Figure 3. Venn diagram for genus-level OTUs. Number of shared and unique taxa at the genus level among four sites using the criterion that >60% of the replicates in a given sowing sample contain sequences belonging to that OTU. Percentages denote the proportion of shared reads over the total reads obtained and the sites are shown by their abbreviations. Overlapping areas indicate shared taxa among sites. Av-S (Avon suppression), Av-NS (Avon non-suppression), MnP-S (Minippa suppression), MnP-NS (Minippa non-suppression).

study involving pea crop soil in Denmark where they were associated with healthy soils [35]. *Aspergillus* sp. (CC = 62±15%), and *Penicillium* sp. are endophytes isolated from wheat leaves and roots [70], [72] and dominate in the non-suppressive soils. *Thanatephorus* (CC = 97±10%), the telomorph of the genus *Rhizoctonia*, which are saprotrophic and disease-causing endophytes, were found in slightly higher numbers in non-suppressive soils. Two other genera that contain wheat pathogens, *Pyrenophora* (cause of tan spot) and *Phaeosphaeria* (cause of glume blotch), did not significantly change between fields.

Plant development

In addition to the pathogenic fungi, the rhizosphere environment would be dominated by fungi capable of responding rapidly to carbon substrates from rhizodeposition (r-selection) whereas the fungal community in the bulk soil depends on a diverse array of carbon substrates. Therefore, the rhizosphere fungal community composition likely differs from the bulk soil [73]–[74]. Fungal communities at sowing represent the bulk soil while the 7-week (in-crop) communities also reflect the rhizosphere-induced changes [26], as seen from the change in fungal community structure shown in Figure 5. The build-up of *R. solani* AG 8 inoculum in wheat and other cereals during this early plant growth period was previously observed [75] and we also observed an increase in *R. solani* in both the suppressive and non-suppressive soils, although the increase was greater in diseased fields. *R. solani* AG 8 is an effective saprophyte hence may grow in the rhizosphere in SP fields. Despite changes in the fungal composition from sowing to 7 weeks, fungal genera with potential to contribute to antibiosis, mycoparasitism and mycorrhization remained the dominant members in the SP communities, suggesting the importance of a

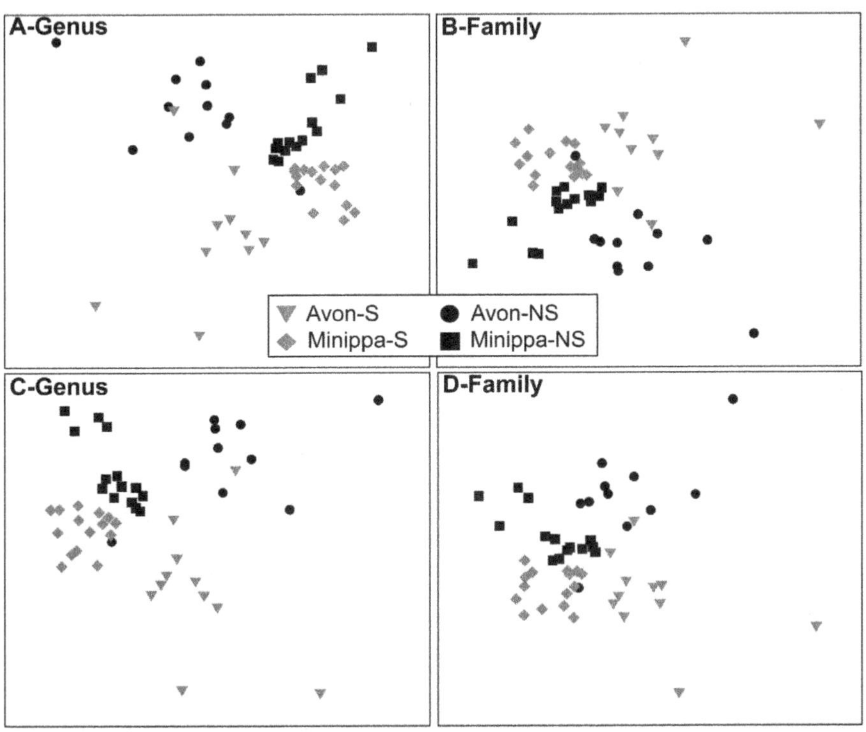

Figure 4. Non-metric dimensional scaling (NMDS) based ordinations for differences among sites and treatments. Analyses generated from Bray Curtis dissimilarity plus a dummy variable (+d) on Hellinger-transformed relative abundances for all-data at 0% bootstrap (closest match) at the genus (A) and family (B) levels and the re-sampled data at the genus (C) and family (D) levels. 2D stress values were 0.17 (A), 0.18 (B), 0.18 (C), and 0.19 (D).

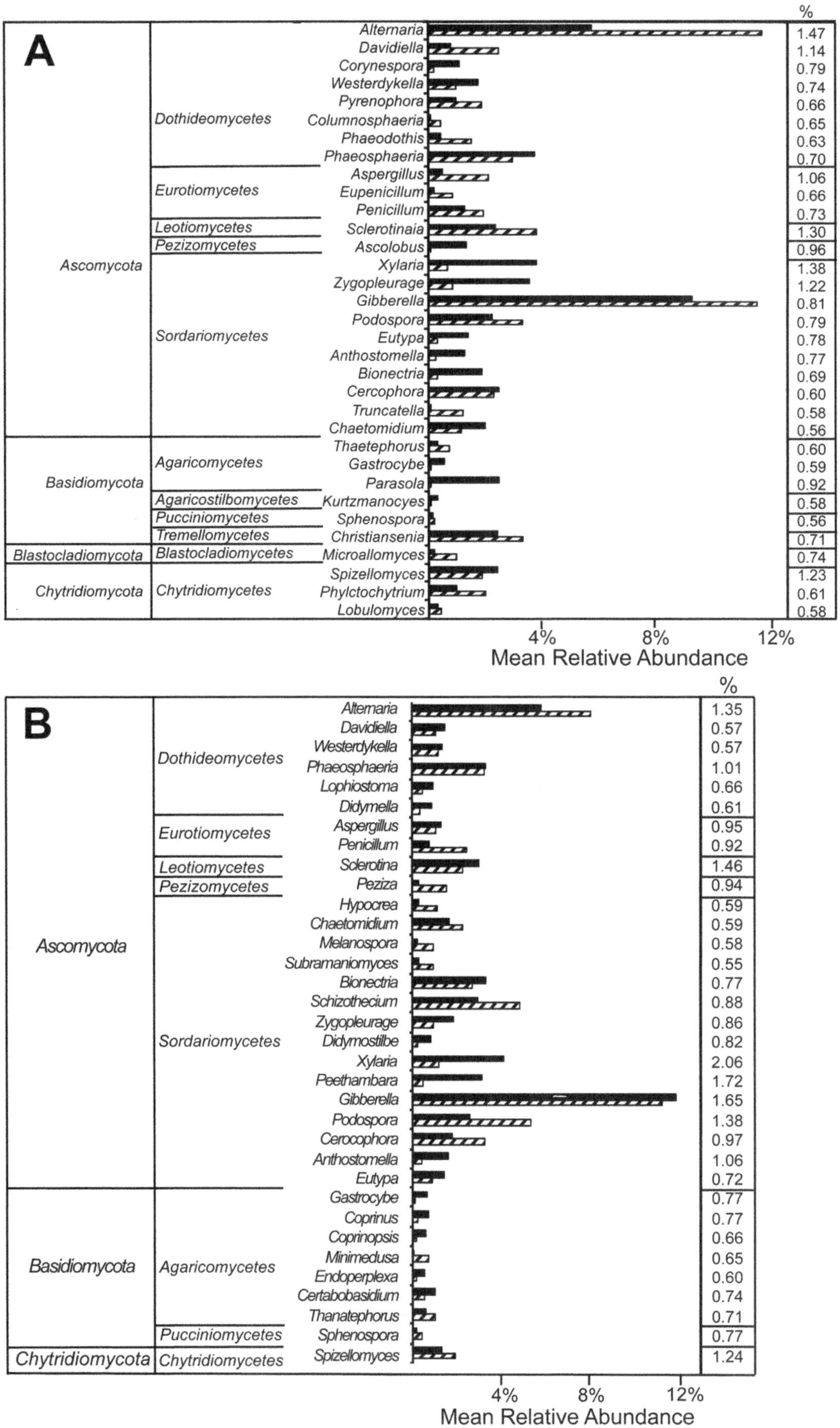
A
%
Ascomycota
Dothideomycetes
Alternaria 1.47
Davidiella 1.14
Corynespora 0.79
Westerdykella 0.74
Pyrenophora 0.66
Columnosphaeria 0.65
Phaeodothis 0.63
Phaeosphaeria 0.70
Eurotiomycetes
Aspergillus 1.06
Eupenicillum 0.66
Penicillum 0.73
Leotiomycetes
Sclerotinaia 1.30
Pezizomycetes
Ascolobus 0.96
Sordariomycetes
Xylaria 1.38
Zygopleurage 1.22
Gibberella 0.81
Podospora 0.79
Eutypa 0.78
Anthostomella 0.77
Bionectria 0.69
Cercophora 0.60
Truncatella 0.58
Chaetomidium 0.56
Basidiomycota
Agaricomycetes
Thaetephorus 0.60
Gastrocybe 0.59
Parasola 0.92
Agaricostilbomycetes
Kurtzmanocyes 0.58
Pucciniomycetes
Sphenospora 0.56
Tremellomycetes
Christiansenia 0.71
Blastocladiomycota
Blastocladiomycetes
Microallomyces 0.74
Chytridiomycota
Chytridiomycetes
Spizellomyces 1.23
Phylctochytrium 0.61
Lobulomyces 0.58
4%
8%
12%
Mean Relative Abundance
B
%
Ascomycota
Dothideomycetes
Alternaria 1.35
Davidiella 0.57
Westerdykella 0.57
Phaeosphaeria 1.01
Lophiostoma 0.66
Didymella 0.61
Eurotiomycetes
Aspergillus 0.95
Penicillum 0.92
Leotiomycetes
Sclerotina 1.46
Pezizomycetes
Peziza 0.94
Sordariomycetes
Hypocrea 0.59
Chaetomidium 0.59
Melanospora 0.58
Subramaniomyces 0.55
Bionectria 0.77
Schizothecium 0.88
Zygopleurage 0.86
Didymostilbe 0.82
Xylaria 2.06
Peethambara 1.72
Gibberella 1.65
Podospora 1.38
Cerocophora 0.97
Anthostomella 1.06
Eutypa 0.72
Basidiomycota
Agaricomycetes
Gastrocybe 0.77
Coprinus 0.77
Coprinopsis 0.66
Minimedusa 0.65
Endoperplexa 0.60
Certabobasidium 0.74
Thanatephorus 0.71
Pucciniomycetes
Sphenospora 0.77
Chytridiomycota
Chytridiomycetes
Spizellomyces 1.24
4%
8%
12%
Mean Relative Abundance

Figure 5. Similarity percentage analysis (SIMPER). Relative abundances (%) of OTUs at the genus level that contribute to the discrimination between fungal communities in suppressive and non-suppressive soils (solid bar = suppression, hatched bar = non-suppression) in the (A) sowing and (B) in-crop (7 week) samples. Numbers in the right column indicate percent contribution to discrimination by SIMPER analysis (sum = 26.8% (A) and 31.2% (B)). * indicate significant differences between suppressive and non-suppressive fields (t-test, $p<0.05$). Bold taxa indicate that they are shared between the sowing and in-crop samples.

fungal contribution to the stability of the disease suppression capacity. Plants can exploit their interaction with rhizosphere bacterial and fungal communities for protection against infection, plant growth promotion and can also benefit from an induced resistance [26], [30], [76]. The importance of the variable response by the putative plant pathogens from sowing to 7 weeks requires further investigation.

Effect of soil characteristics on the fungal community

Variation in the soil physico-chemical properties and the spatial and temporal distribution of plant residues can have a significant impact on the diversity and biomass of fungal communities. Soil type is similar at both our field sites, but there was some difference in the soil texture (% clay) especially between the SP and NSP fields. Although there was significant variation in the fungal diversity based on soil type (PERMANOVA analysis), the effect was smaller than that associated with suppression capacity. Crop and soil management (tillage) practices, which influence the quantity and quality of organic carbon sources, could also influence populations and biomass of specific fungi and thus modify diversity over time [77]–[78]. Land use and soil pH have been shown to influence the diversity of some fungal groups [79] but the effect of pH on fungi was weaker than that on bacterial diversity [80]. In one study Hypocreales made up to 5% of fungi in soils with high pH (8.0) while they were absent in soils with pH below 5.5 [80]. Members of the Hypocreales (e.g. *Hypocrea*, *Bionectrecia* and *Peethambara*) were also part of the dominant group of fungi in the alkaline pH soils at both sites we studied. Organic C and total N levels were consistently higher in the soils from SP fields at both sites mainly from the long-term (>10 y) adoption of stubble retention and reduced till cropping practices compared to <5 y for these practices in the NSP fields. Above and below ground plant residues from the annual crops are the only source of C inputs (<1 tonne C per annum) hence the stocks of decomposable particulate organic matter are lower (<30% of total C) than those generally found in tundra grassland and forest soils [81]–[82]. Overall, the observation of soil type based differences in fungal community composition illustrates the need for the assessment of multiple soil types and locations in order to decipher the signature(s) of a disease suppressive microbial community(s).

Table 3. Soil type and suppression status statistical comparisons.

Group		Original dataset			Re-sampled dataset		
		Genus	Family	Order	Genus	Family	Order
Soil Type	Global R	0.542	0.511	0.294	0.558	0.497	0.288
	p<	0.01	0.01	0.01	0.01	0.01	0.01
Suppression	Global R	0.751	0.665	0.518	0.750	0.652	0.510
	p<	0.01	0.01	0.01	0.01	0.01	0.01

Significance of differences (main effects only) at different taxonomic levels using ANOSIM two-way crossed analysis for the original dataset and re-sampled dataset.

Diversity profiling of suppressive and non-suppressive soils

Direct classification of ~1 million filtered sequences yielded 917 genera. After re-sampling and normalizing for read abundance, our genus-level richness was equivalent to or higher than other studies that used 28S pyrosequencing with clustering [83]–[84] or even with high throughput ITS sequencing [35], [73], [77], [85], [86] which resolves at species level. While such direct comparisons are difficult due to sample size effects, clustering cut-offs and relating 28S to ITS data, as a group they do indicate that diversity in these alkaline, calcareous, low organic matter soils is higher than might be expected. This may be partially attributed to the use of a 125 g soil sample for DNA extraction, since, the larger than typical (<1 g) sample size reduces small-scale spatial heterogeneity thereby decreasing replicate variability while encompassing a diverse array of microhabitats. Ophel-Keller [41] had previously noted the need for large soil samples to accurately determine changes in soilborne pathogenic fungi and relate them to disease risk categories in the field. Conversely, the high diversity may also be attributed to the soil physico-chemical properties, quality and quantity of organic plant residues and environmental factors (e.g. low rainfall) that may favor soil fungi with their ability to access large volumes of soil through hyphal networks.

The dominance of Ascomycota in these cropping soils where organic matter is concentrated primarily at the surface is similar to that found in tundra soils [87] whereas Basidiomycota was the most abundant phylum in forest soils [77], [88]. Rarefaction curves were nearly saturated in some samples, indicating that approximately 20,000 sequences was sufficient to cover the majority of genus-level diversity with diminishing returns thereafter. This saturation is highly ecosystem and depth-dependent, with near-saturation occurring at less than 4,000 sequences in permafrost soils at depth (40–50 cm) [49].

Disease suppression

The level of impact of soil-borne disease incidence on plant growth is influenced by the amount of plant pathogen present and its interactions with the plant, soil and environmental factors. A main criterion defining biological disease suppression is the ability of soil biotic communities to suppress the disease expression in the host plants even in the presence of adequate virulent inoculum and a susceptible host. *R. solani* AG 8 levels measured at sowing indicate the presence of sufficient inoculum to cause significant disease [89]. Differences in pathogen inoculum levels between suppressive and non-suppressive soils were small, indicating similar pathogen pressure. Field disease assessments showing distinct differences in the disease levels in the SP and NSP fields validate the suitability of these field soils to investigate disease suppressive communities. These field results are supported by the differences in the DS potential of suppressive and non-suppressive soils. Grain yields of crops were higher (avg. >30%) in the suppressive fields compared to non-suppressive fields confirming better plant performance partly due to the lack of disease in the SP fields.

The biological disease suppression observed at these field sites is effective against multiple soil-borne pathogens (16) and under multi-crop rotations. Hence we hypothesize that both the general and specific disease suppression may be acting simultaneously. Continued supply of crop residue carbon stimulates general microbial activity including fungal populations that can compete with *R. solani* for resources [18], [22], [90] and also may increase the build-up of antibiotic producing/antagonistic microflora including fungi [19]. TRF profiling and sequence data both show differentiation of fungal communities of SP and NSP soils from both field sites but the combined lack of consistent trends in the total fungal biomass (measured as DNA concentration) suggests that the change in community composition may play an important role in the observed suppression. Since a number of pathogenic fungi can become active saprophytes depending upon resource availability and host presence, total fungal biomass may not truly reflect the net ecological role of entire soil fungal community. The significant difference (PERMANOVA) in the fungal community structure in the soils from 7 weeks, representing the root microbiome suggests a relationship between the disease incidence (root health) and associated fungal community. While we found that the fungal component of soil communities was different in suppressive versus their paired non-suppressive soil, bacteria and their interactions with fungi and plants could also be important to biological disease suppression [13], [30], [91], and is the subject of continuing work for these soils.

Conclusions

In a comparative analysis of soils with high and low suppressive potential, from the wheat-cropping region in South Australia, we measured the fungal community composition using targeted pyrosequencing of the 28S LSU gene. Overall, the study clearly demonstrates the need to include soil fungal communities, along with bacterial based investigations, in order to obtain a full understanding of disease suppressive community. The observation of significant fungal community composition differences between SP and NSP soils prior to sowing suggest varying pathogen-fungal community interactions in the suppressive and non-suppressive soils, whereas the differences observed at 7 weeks demonstrate a significant relationship between the fungal community structure and root health. The dominance of members of the order Hypocreales with known antifungal capability in the suppressive soils highlights the importance of pathogenic fungi - general fungal community interactions in the disease suppression.

The taxonomically based assessment of fungal community structure should also be strengthened with functional characterization at species level resolution, in order to decipher the mechanisms of disease suppression. In this context, the ITS region would be preferred and thus, the development of a robust and taxonomically concise ITS-based fungal classifier is needed. Lastly, findings from this type of sequencing based study could be used to direct future isolation and manipulation studies both to understand the mechanisms of disease suppression and to develop disease control options.

Supporting Information

Figure S1 Examples of field and root conditions from non-suppressive fields. (A) Avon non-suppressive field at 16 weeks post-sowing, (B) roots from 2 week-old samples from the Avon non-suppressive field, (C) roots from the Avon suppressive field at 2 weeks.

Figure S2 Non-metric dimensional scaling (NMDS) of ITS rRNA-T-RFs. Ordination based on Bray Curtis similarity plus a dummy variable (+d) with square root transformation of ITS rRNA-T-RFs from (A) sowing and (B) in crop (7 wk) sampling. 2D stress 0.14 (A) and 0.17 (B).

Figure S3 Rarefaction curves for genus-level bins.

Figure S4 Relationship between the proportions of unclassified reads at each taxonomic level and suppression status. Data shows averages and standard errors.

Figure S5 Classification confidence (bootstrap) for all genera within classes. Data based on classes that contain >0.5% read abundance from RDP Fungal Classifier. The cumulative percent of total sequences is denoted by the graph on the right.

Figure S6 Non-metric dimensional scaling (NMDS) with all unclassified removed and only unclassified reads. Ordination based on Bray Curtis similarity plus a dummy variable (+d) with Hellinger-transformed relative abundances for all-data at 50% bootstrap at the genus level with all unclassified removed (A) and with only the unclassified reads (B). 2D stress values were 0.17 (A), 0.21(B). NS = non-suppressive soil, S = suppressive soil.

Figure S7 Correlation between original and re-sampled dataset. Relative genera OTU abundance between the original dataset and the re-sampled dataset.

Table S1 Soil characteristics of the Avon and Minippa sites. Standard errors shown. Supp = suppressive and Non-supp = non suppressive, CEC = cation exchange capacity.

Table S2 Summary of results from 28S sequencing and T-RFLP profiling data. Standard deviations are shown adjacent to means and Fisher's least significant difference test results are presented in the bottom row.

Table S3 Summary of sequence processing results using the Ribosomal Database Project (RDP) pyrosequencing pipeline. Sequencing read information with the number of raw pyrosequencing reads, processed sequences and average processed sequence length. Filters include: Primer trimmer allowing 0 mismatches, N count filter, length >400 bp filter and exponential quality filter (Q>20). Site abbreviations: Av – Avon, Min-Minippa, S-suppression, NS-non-suppression, Sow-sowing, IC-in-crop samples at 7 wks post sowing.

Table S4 Primary habitat, function and disease suppression related property for dominant fungal genera.

Text S1 Fungal 28S amplification and sequence processing.

Text S2 Unclassified fungal diversity.

Text S3 Comparisons of resampled to whole 28S sequence dataset.

Acknowledgments

Authors acknowledge the assistance from Marcus Hicks, Nady Harris and Stasia Kroker for laboratory analysis in Adelaide, Amanda Cook with field sampling at Minnipa, Robin Manley for access to the Avon field site and Derek St. Louis for laboratory work and data management at Michigan State University.

Author Contributions

Conceived and designed the experiments: VVSRG JMT SMN KOK DKR. Performed the experiments: CRP VVSRG AP DKR. Analyzed the data: CRP VVSRG SMN. Contributed reagents/materials/analysis tools: CRP VVSRG KOK PH. Wrote the paper: CRP VVSRG JMT KOK MG.

References

1. Rovira AD (1990) Ecology, epidemiology and control of take-all, Rhizoctonia bare patch and cereal cyst nematode in wheat. Aust Plant Path 19: 101–111.
2. Jeger MJ, Hide GA, van Den Boogert PHJF, Termorshuizen AJ, van Baarlen P (1996) Pathology and control of soil-borne fungal pathogens of potato. Potato Res 39: 437–469.
3. Duveiller E, Singh RP, Nicol JM (2007) The challenges of maintaining wheat productivity: pests, diseases, and potential epidemics. Euphytica 157: 417–430.
4. Neate SM (1994) Soil and crop management practices that affect root diseases of crop plants. (CSIRO: East Melbourne), 96–106.
5. Hornby D (1983) Suppressive soils. Annu Rev Phytopathol 21: 65–85.
6. Fukui R (2003) Suppression of soilborne plant pathogens through community evolution of soil microorganisms. Microbes Environ18: 1–9.
7. Gupta VVSR, Rovira AD, Roget DK (2011) Principles and management of soil biological factors for sustainable rainfed farming systems. In: Rainfed farming systems' by P. . Tow, I. . Cooper, I. . Partridge, C. . Birch, pp. 149–184, Springer Science and Business Media.
8. Kinkel LL, Bakker MG, Schlatter DC (2011) A co-evolutionary framework for managing disease-suppressive soils. Annu Rev Plant Pathol 49: 47–67.
9. Weller DM, Raaijmakers JM, Gardener BBM, Thomashow LS (2002) Microbial populations responsible for specific soil suppressiveness to plant pathogens. Annu Rev Phytopathol 40: 309–347.
10. Baker K, Cook R (1974) Biological control of plant pathogens. San Francisco, CA. W.H. . Freeman.
11. Roget DK (1995) Decline in root rot (*Rhizoctonia solani* AG-8) in wheat in a tillage and rotation experiment at Avon, South Australia. Aust J Exp Agr 35: 85–89.
12. Mazzola M (2004) Assessment and management of soil microbial community structure for disease suppression. Annu Rev Phytopathol 42: 35–59.
13. Weller DM, Landa BB, Mavrodi OV, Schroeder KL, De La Fuente L, et al. (2006) Role of 2,4-Diacetylphloroglucinol-Producing Fluorescent *Pseudomonas* spp. in the Defense of Plant Roots. Plant Biol 9: 4–20.
14. Garbeva P, vanVeen JA, vanElsas JD (2004) Microbial diversity in soil: selection of microbial populations by plant and soil type and implications for disease suppression. Annu Rev Phytopathol 42: 243–270.
15. Fernandez MR, Conner RL (2011) Root and crown rot of wheat. Prairie Soils Crops J 4: 151–157.
16. Wiseman BM, Neate SM, Ophel-Keller K, Smith SE (1996) Suppression of *Rhizoctonia solani* Anastamosis Group 8 in Australia and its biological nature. Soil Biol Biochem 28: 727–732.
17. Roget DK, Coppi JA, Herdina, Gupta VVSR (1999) Assessment of suppression to *Rhizoctonia solani* in a range of soils across SE Australia. In: Proceedings of the First Australasian SoilBorne Disease symposium, Magarey RC (Ed.), pp. 129–130, BSES, Brisbane, Australia.
18. Gupta VVSR, Neate SM (1999) Root disease incidence-A simple phenomenon or a product of diverse microbial/biological interactions. In: Proceedings of the First Australasian SoilBorne Disease symposium, R.C. . Magarey (Ed.), pp. 3–4, BSES, Brisbane, Australia.
19. Barnett SJ, Roget DK, Ryder MH (2006) Suppression of *Rhizoctonia solani* AG-8 induced disease on wheat by the interaction between *Pantoea, Exiguobacterium* and *Microbacteria*. Aust J Soil Res 44: 331–342.
20. Paulitz TC (2006) Low input no-till cereal production in the Pacific Northwest of the US: The challenges of root diseases. Eur J Plant Pathol 115: 271–281.
21. Postma J, Montanari M, van den Boogert PHJF (2003) Microbial enrichment to enhance the disease suppressive activity. Eur J Soil Biol 39: 157–163.
22. Gupta VVSR, Reddy NPE (2010) Response of soil microbial communities to stubble addition differs between disease suppressive and non-suppressive soils. Proceedings Sixth Australian Soilborne Diseases Symposium, 9–11 August 2010, G.R. . Stirling (Ed.), p.50.
23. Bonanomi G, Antignani V, Capodilupo M, Scala F (2010) Identifying the characteristics of organic soil amendments that suppress soilborne plant diseases. Soil Biol Biochem 42: 136–144.
24. de Boer W, Verheggen P, Gunnewiek P, Kowalchuk GA, van Veen JA (2003) Microbial community composition affects soil fungistasis. Appl Environ Microbiol 69: 835–844.
25. Vinale F, Sivasithamparam K, Ghisalberti EL, Marra R, Woo SL, et al. (2008) *Trichoderma*-plant-pathogen interactions. Soil Biol Biochem 40:1–10.
26. Bowen GD, Rovira AD (1999) The rhizosphere and its management to improve plant growth. Adv Agron 66: 1–102.
27. Berendsen RL, Pieterse CMJ, Bakker PAHM (2012) The rhizosphere microbiome and plant health. Trends Plant Sci 17: 478–486.
28. Van Elsas JD, Costa R, Jansson J, Sjoling S, Balley M, et al. (2008) The metagenomics of disease suppressive soil – experiences from the METACONTROL project. Trends Biotechnol 26: 591–601.
29. Benitez M, McSpadden Gardner BB (2009) Linking sequence to function in soil bacteria: Sequence-derived isolation of novel bacteria contributing to soilborne plant disease suppression. Appl Environ Microb. 75: 915–924.
30. Mendes R, Kruijt M, deBruijn I, Dekkers E, van derVoort M, et al. (2011) Deciphering the rhizosphere microbiome for disease-suppressive bacteria. Science 332: 1097–1100.
31. Rosenzweig N, Tiedje JM, Quensen JF, Meng Q, Hao JJ (2012) Microbial communities associated with potato common scab-suppressive soil determined by pyrosequencing analyses. Plant Dis 96: 718–725.
32. Smith SE, Read DJ (2008) Mycorrhizal symbiosis 3rd ed. Academic Press. 800 p.
33. Woo SL, Lorito M. Exploiting the interactions between fungal antagonists,pathogens and the plant for biocontrol. In: Vurro M, Gressel J, editors. Novel biotechnologies for biocontrol agent enhancement and management. Amsterdam:IOS Springer Press; 2007. p. 107–30.
34. Nelson EB (2004) Biological Control of Oomycete and Fungal Pathogens. In: Goodman R.M, ed. Encyclopedia of Plant and Crop Science. Marcel Dekker, Inc. pp 137–140.
35. Xu L, Ravnskov S, Larsen J, Nilsson RH, Nicolaisen M (2012) Soil fungal community structure along a soil health gradient in pea fields examined using deep amplicon sequencing. Soil Biol Biochem 46: 26–32.
36. Cook A, Wilhelm N, Dyson C (2008) Survey of soil-borne disease suppression to Rhizoctonia solani in low rainfall farming systems on upper Eyre Peninsula, South Australia. Proceedings of the 14th ASA Conference, 21–25 September 2008, Adelaide, South Australia.
37. Gupta VVSR, Roget DK, Coppi JC, Kroker SK (2009) Soil type and rotation effects on the suppression of Rhizoctonia bare patch disease in Wheat. In: 5th Australasian SoilBorne Disease Symposium – Extended abstracts, Thredbo, NSW, pp. 85–87.
38. Northcote KH, Hubble GD, Isbell RF, Thompson C.H, Bettenay E (1975) 'A Description of Australian Soils.'(CSIRO: East Melbourne.).
39. FAO. (2006) World reference base for soil resources 2006. World Soil Resources Reports No. 103. Rome: Food and Agriculture Organization of the United Nations.
40. Campbell CD, Chapman SJ, Cameron CM, Davidson MS, Potts JM (2003) A rapid microtiter plate method to measure carbon dioxide evolved from carbon substrate amendments so as to determine the physiological profiles of soil microbial communities by using whole soil. Appl Environ Microbiol 69: 3593–3599.
41. Ophel-Keller K, McKay A, Hartley D, Herdina, Curran J (2008) Development of a routine DNA-based testing service for soilborne diseases in Australia. Australasian Pl Pathol 37: 243–253.
42. Vainio E J, Hantula J (2000). Direct analysis of wood inhabiting fungi using denaturing gradient gel electrophoresis of amplified ribosomal DNA. Mycol Res 104: 927–936.
43. Smalla K, Oros-Sichler M, Milling A, Heuer H, Baumgarte S, et al. (2007) Bacterial diversity of soils assessed by DGGE, T-RFLP and SSCP fingerprints of PCR-amplified 16S rRNA gene fragments: Do the different methods provide similar results. J Microbiol Methods. 69: 470–479.
44. Gardes M, Bruns T (1993) ITS primers with enhanced specificity for basidiomycetes –application to the identification of mycorrhizae and rusts. Mol Ecol 2: 113–118.
45. White TJ, Bruns TD, Lee SB, Taylor JW (1990) Amplification and direct sequencing of fungal ribosomal RNA genes for phylogenetics. In: Innis MA, Gelfand DH, Sninsky JJ, White TJ (Eds.), PCR Protocols: a Guide to Methods and Applications. New York: Academic Press. pp. 315–322.
46. John S, Hopple J, Vilgalys R (1999) Phylogenetic relationships in the mushroom genus *Coprinus* and dark-spored allies based on sequence data from the nuclear gene coding for the large ribosomal subunit RNA: divergent domains, outgroups, and monophyly. Mol Phylogenet Evol 13: 1–19.
47. Schnare MN, Damberger SH, Gray MW, Gutell RR (1996) Comprehensive comparison of structural characteristics in eukaryotic cytoplasmic large subunit (23 S-like) ribosomal RNA. J Mol Biol 256: 701–719.
48. Liu K-L, Porras-Alfaro A, Kuske CR, Elchorst SA, Xie G (2012) Accurate, rapid taxonomic classification of fungal large-subunit rRNA genes. Appl Environ Microb 78: 1523–1533.

49. Penton CR, St. Louis D, Cole JR, Luo Y, Wu L, et al. (2013) Fungal diversity in permafrost and tallgrass prairie soils under experimental warming conditions. Appl Environ Microb 79: 7063–7072.
50. Porter TM, Golding GB (2012) Factors that affect large subunit ribosomal amplicon sequencing studies of fungal communities: Classification method, primer choice, and error. PLOS One 7(4): 1–12.
51. Clarke KR, Warwick RM (2001) Change in marine communities: an approach to statistical analysis and interpretation, 2nd ed. PRIMER-E, Plymouth.
52. Clarke KR (1993) Non-parametric multivariate analyses of changes in community structure. Aust J Ecol 18: 117–143.
53. Anderson MJ (2001) A new method for non-parametric multivariate analysis of variance. Austral Ecol 26: 32–46.
54. Warwick RM, Platt HM, Clarke KR, Agard J, Gobin J (1990) Analysis of macrobenthic and meiobenthic community structure in relationship to pollution and disturbance in Hamilton Harbour, Bermuda. J Exp Mar Biol Ecol 138: 119–142.
55. Shannon CE, Weaver W (1949) The mathematical theory of communication. Urbana: The University of Illinois Press. 117 p.
56. Pielou EC (1975) Ecological Diversity. New York: Wiley Press. 165 p.
57. Suryanarayanan TS, Murali T, Venkatesan G (2002) Occurrence and distribution of fungal endophytes in tropical forests across a rainfall gradient. Can J Bot 80: 818–826.
58. Macia-Vincente JG, Jansson H, Mendgen K, Lopez-Llorca LV (2008) Colonization of barley roots by endophytic fungi and their reduction of take-all caused by *Gaeumannomyces graminis* var. *tritici*. Can J Microbiol 54: 600–609.
59. Park J-H, Chio G-J, Lee S-W, Lee H-B, Kim K-M, et al. (2005) Griseofulvin from *Xylaria* sp. strain F0010, an endophytic fungus of *Abie holophylla* and its antifungal activity against plant pathogenic fungi. J Microbiol Biot 15: 112–117.
60. Jang Y-W, Lee I-K, Kim Y-S, Lee S, Lee H-J, et al. (2007) Xylarinic acids A and B, new antifungal polypropionates from the fruiting body of *Xylaria polymorpha*. J Antibiot 60: 696–699.
61. Schroers H-J (2001) A monograph of *Bionectria* (Ascomycota, Hypocreales, Bionectriacea) and its *Clonostachys* anamorphs. Stud Mycol 46: 1–214.
62. Moller WJ, Kasimatis AN (1978) Dieback of grapevines caused by *Eutypa armeniacae*. Plant Dis Rep 62:254–258.
63. Hyde KD, Bussaban B, Paulas B, Crous PW, Lee S, et al. (2007) Diversity of saprobic microfungi. Biodivers Conserv 16: 7–35.
64. Klich MA (1998) Soil fungi of some low-altitude desert cotton fields and ability of their extracts to inhibit Aspergillus flavus. Mycopathologia 142: 97–100.
65. Vilich V, Dolfen M, Sikora RA (1998) *Chaetomium* spp. Colonization of barley following seed treatment and its effect on plant growth and *Erysiphe graminis* f. sp. *Hordei* disease severity. J Plant Dis Protect 105: 130–139.
66. Shiono Y, Hatakeyama T, Murayama T, Koseki T (2012) Polyketide metabolites from the endophytic fungus Microdiplodia sp. KS 75–1. Nat Prod Commun 7: 1065–1068.
67. Kim H-Y, Choi GJ, Lee HB, Lee S-W, Lim HK, et al. (2007) Some fungal endophytes from vegetable crops and their anti-oomycete activities against tomato late blight. Lett Appl Microbiol 44: 332–337.
68. Paulitz TC, Park CS, Baker R (1987) Biological control of Fusarium wilt of cucumber with non pathogenic isolates of *Fusarium oxysporum*. Can J Microbiol 33: 349–353.
69. Wellacher M (1991) Die Pilz kolonisation von Maispflanzen. Diplomarbeit. Baarn, The Netherlands. Centraalbureau voor Schimmelcultures.
70. Larran S, Perelló A, Simón MR, Moreno V (2007) The endophytic fungi from wheat (*Triticum aestivum* L.). World J Microb Biot 23: 565–572.
71. Pan JJ, Baumgarten AM, May G (2008) Effects of host plant environment and *Ustilago maydis* infection on the fungal endophyte community of maize (*Zea mays*). New Phytol 178: 147–156.
72. Larran S, Perelló A, Simón MR, Moreno V (2002) Isolation and analysis of endophytic microorganisms in wheat (*Triticum aestivum* L.) leaves. World J Microb Biot 18: 683–686.
73. Xu L, Ravnskov S, Larson J, Nicolaisen M (2012) Linking fungal communities in roots, rhizosphere, and soil to the health status of *Pisum sativum*. FEMS Microbiol Ecol 82: 736–745.
74. Berg G, Smalla K (2009) Plant species and soil type cooperatively shape the structure and function of microbial communities in the rhizosphere. FEMS Microbiol Ecol 68: 1–13.
75. Gupta VVSR, McKay A, Diallo S, Smith D, Cook A, et al. (2012) *Rhizoctonia solani* AG8 inoculm levels in Australian soils are influenced by crop rotation and summer rainfall. 7th Australasian Soilborne Diseases Symposium, Sept 17–20, Freemantle, WA.
76. Conn VM, Walker AR, Franco CMM (2008) Endophytic actinobacteria induce defence pathways in *Arabidopsis thaliana*. Mol Plant Microbe In 21: 208–218.
77. Buée M, Reich M, Murat C, Morin E, Nilsson RH, et al. (2009) 454 pyrosequencing analyses of forest soils reveals unexpectedly high fungal diversity. New Phytol 184: 449–456.
78. Roper MM, Gupta VVSR (1995) Management practices and soil biota. Aust J Soil Res 33: 321–339.
79. Garnica S, Riess K, Bauer R, Oberwinkler F, Weiss M (2013) Phylogenetic diversity and structure of sebacinoid fungi associated with plant communities along an altitudinal gradient. FEMS Microbiol Ecol 83: 265–278.
80. Rousk J, Bååth E, Brookes PC, Lauber CL, Lozupone C, et al. (2010) Soil bacterial and fungal communities across a pH gradient in an arable soil. ISME J 4: 1340–1351.
81. Loss A, Pereira MG, Perin A, Coutinho FS, Anios LHC (2012) Particulate organic matter in soil under different systems in the Brazilian Cerrado. Soil Res 50: 685–693.
82. Cambarella CA, Elliott ET (1992) Particulate soil organic matter changes across a grassland cultivation sequence. Soil Sci Soc Am J 56: 777–783.
83. Gottel NR, Castro HF, Kerley M, Yang Z, Pelletier DA, et al. (2011) Distinct microbial communities within the endosphere and rhizosphere of *Populus deltoids* roots across contrasting soil types. Appl Environ Microb 77: 5934–5944.
84. Weber CF, Vilgalys R, Kuske CR (2013) Changes in fungal community composition in response to elevated atmospheric CO_2 and nitrogen fertilization varies with soil horizon. Front Microbiol 4: 78.
85. Orgiazzi A, Lumini E, Nilsson RH, Girlanda M, Vizzini A, et al. (2012) Unravelling soil fungal communities from different Mediterranean land-use backgrounds. PLOS One 7(4) e34847. doi:10.1371/journal.pone.0034847.
86. McGuire KL, Fierer N, Bateman C, Treseder KK, Turner BL (2012) Fungal community composition in neotropical rain forests: the influence of tree diversity and precipitation. Microb Ecol 63: 804–812.
87. Schadt CW, Martin AP, Lipson DA, Schmidt SK (2003) Seasonal dynamics of previously unknown fungal lineages in tundra soils. Science 301: 1359–1361.
88. O'Brien HE, Parrent JL, Jackson JA, Moncalvo JM, Vilgalys R (2005) Fungal community analysis by large-scale sequencing of environmental samples. Appl Environ Microbiol 71: 5544–5550.
89. McKay A, Roget D, Hannam R, Ophel-Keller K (2008) Root disease risk management resource manual. A. . McKay, D. . Roget, R. . Hannam, K. . Ophel-Keller eds. PIRSA publishing. 114 pp.
90. Janvier C, Villeneuve F, Alabouvette C, Edel-Hermann V, Mateille T, et al. (2007) Soil health through soil disease suppression: Which strategy from descriptors to indicators? Soil Biol Biochem 39: 1–23.
91. Gupta VVSR, Reddy NPE (2010) Response of soil microbial communities to stubble addition differs between disease suppressive and non-suppressive soils, Proceedings Sixth Australian Soilborne Diseases Symposium, 9–11 August 2010, G.R. . Stirling (Ed.), p.50.

3

Foliar Spray with Vermiwash Modifies the Arbuscular Mycorrhizal Dependency and Nutrient Stoichiometry of Bhut Jolokia (*Capsicum assamicum*)

Mohammad Haneef Khan[1]☯, Mukesh K. Meghvansi[1]☯*, Rajeev Gupta[1], Vijay Veer[1], Lokendra Singh[2], Mohan C. Kalita[3]

1 Defence Research Laboratory, Defence Research & Development Organisation, Post Bag 2, Tezpur, Assam, India, **2** Directorate General of Life Sciences, Defence Research & Development Organisation, New Delhi, India, **3** Department of Biotechnology and Bioengineering, Gauhati University, Guwahati, Assam, India

Abstract

Vermiwash (VW), a liquid extract obtained from vermicomposting beds, is used as an organic fertilizer for crop plants. The current study investigated the effect of a vermiwash foliar spray on the response of bhut jolokia (*Capsicum assamicum*) exposed to two different arbuscular mycorrhizal fungi (AMF: *Rhizophagus irregularis,* RI and *G. mosseae,* GM) in acidic soil under naturally ventilated greenhouse conditions. The VW spray significantly influenced the growth of plants receiving the dual treatment of AMF+VW. Plant growth was more prominent in the GM+VW treatment group than that in the RI+VW treatment group. The plant-AMF interactions in relation to growth and nutrient requirements were also significantly influenced by the application of VW. Interestingly, the VW treatment appeared to contribute more N to plants when compared to that under the AMF treatment, which led to changes in the C:N:P stoichiometry in plant shoots. Furthermore, the increased potassium dependency, as observed in the case of the dual treatments, suggests the significance of such treatments for improving crop conditions under salt stress. Overall, our study shows that the VW foliar spray modifies the response of a crop to inoculations of different AMF with regard to growth and nutrient utilization, which has implications for the selection of an efficient combination of nutrient source for improving crop growth.

Editor: Martin Heil, Centro de Investigación y de Estudios Avanzados, Mexico

Funding: This study was funded by Defence Research & Development Organisation (DRDO), New Delhi, India from operating funds rather than grants. The funders had no role in study design, data collection and analysis, decision to publish, or preparation of the manuscript.

Competing Interests: The authors have declared that no competing interests exist.

* E-mail: mk_meghvansi@yahoo.co.in

☯ These authors contributed equally to this work.

Introduction

Vermiwash (VW) is a brownish-red liquid extract collected during vermicomposting of organic waste. VW can also serve as a valuable foliar spray, because it is composed of excretory products and mucus secretions from earthworms and micronutrients from the organic molecules in the soil. These nutrients are absorbed and then transported to the leaves, shoots, and other parts of a plant [1]. Various experimental results have shown that the application of VW or compost tea improves plant health, yield, and nutritional quality [2,3]. Studies have shown that vermicompost tea contains plant growth-promoting substances (e.g., humic, fulvic, and other organic acids [4]; auxin-like substances [5]; and cytokinin-like substances [6]). A VW foliar spray is more advantageous from economical and environmental perspectives owing to the absence of nutrient leaching, which is often encountered when performing soil amendments.

Arbuscular mycorrhizal fungi (AMF), the most widespread root fungal symbionts associated with the vast majority of higher plants, have been shown to increase plant growth and yield by improving nutrient uptake in various agroecosystems [7,8]. In addition, AMF have been reported to enhance stress tolerance in plants under severe stress conditions, including heavy metal, drought, and salinity [9]. Although AMF species, typically, are considered nonspecific, variations in the growth response of different plant genotypes to different AMF species have been described [7,8]. Such variation in the growth response, termed mycorrhizal species sensitivity, implies that the mycorrhizal dependency of plants can also vary [10]. The determination of plant growth responses under controlled conditions, therefore, is a valuable first step in evaluating the importance of symbiotic relationships in nature and understanding the dynamic interactions within an ecosystem [11].

In intensive agriculture systems, major emphasis has been placed on increasing the input of nutrients through various approaches to achieve higher crop productivity, thus, leading to low nutrient-use efficiency and increased environmental problems. The chemical and biological processes that occur in the rhizosphere affect the mobilization and acquisition of soil nutrients and microbial dynamics; they also control the nutrient-use efficiency of crops, and therefore, profoundly influence crop productivity and sustainability [12]. Thus, researchers indicate that it is necessary to maintain greater nutrient-use efficiency through the integration of various approaches, including the combined use of organic amendments and chemical fertilizers [13] or organic amendments and AMF [14,15]. The majority of these studies have provided empirical evidences on the usefulness of such approaches with regard to soil amendments [13,15]. However, to the best of our knowledge, no reports are available on the effects of the combined use of VW, as a foliar spray, and AMF, as soil inoculants. In addition, the influence of such amendment methods on each other could be exacerbated when applied to acidic soils that have high phosphorus (P)-fixing power because of the excessive presence of iron and aluminum ions, thus, resulting in low P availability for crop production [16].

Carbon (C), Nitrogen (N), and P are the major plant nutrients that are both structurally and functionally essential to all organisms [17]. A mechanistic link between tissue elemental stoichiometry (C:N:P *per se*) and the growth rate of an organism has been suggested [18,19]. P is a critical element in the production of ribosomes; ribosomes manufacture N-rich proteins, which in turn, constitute the C and energy-harvesting organs [20]. The growth rate hypothesis proposes that higher growth rates are associated with lower C:N, C:P, and N:P ratios. Therefore, variations in the P levels in organisms are driven by differences in the allocation of P-rich ribosomal RNA (rRNA), which is required to meet the protein synthesis demands of increased growth rates [17,18,21]. Thus, the factors or physiological changes that affect growth rates could result in differences in the C:N:P ratio of biomass and cellular allocation [21]. The combined use of biological and organic amendments may differentially affect a crop with respect to nutrient-use efficiency and C:N:P stoichiometry, thus, affecting the potential benefits and limitations. For instance, the excessive use of organic sources of P can result in the suppression of the AMF community [22], thereby highlighting the significance of careful nutrient budgeting for optimizing, rather than maximizing, soil fertility.

Crop management involves a range of practices that can affect AMF associations directly (e.g., by damaging or killing AMF) and indirectly (e.g., by creating favorable or unfavorable conditions for AMF) [23]. The organic sources of nutrients (e.g., farmyard manure, compost, and crop residues) and slow release of mineral fertilizers (e.g., rock phosphate) do not typically suppress AMF and may even stimulate their growth [24]. However, the overuse of organic amendments, especially those high in P, such as chicken manure, may have a negative influence on AMF, and the precise effect of the organic amendments has been shown to be variable (i.e., dependent on the type of soil or amendment) [22]. Therefore, the role and extent of the effects of a combination of organic and biological amendments on crops should be investigated to obtain more precise data before such practices are implemented in crop fields. Furthermore, it is imperative to determine the extent to which organic amendments affect the response of a crop to different AMF species. The present study was undertaken with the objective of determining the influence of a VW foliar spray on mycorrhizal growth dependency and nutrient stoichiometry of bhut jolokia (*Capsicum assamicum*) inoculated using two different AMF, in acidic soil under pot conditions. Bhut jolokia is a highly pungent chili species native to the northeastern region of India and has tremendous ethnopharmacological potential [25].

Materials and Methods

Plants, AMF and vermiwash

Three-week-old seedlings of bhut jolokia procured from a local nursery (M/s Lakshmi Agric. Pabohi, District Sonitpur, Assam, India) were used in the pot experiment. The pure inocula of two AMF (i.e., *R. irregularis* and *G. mosseae*) were supplied by The Energy and Resources Institute (TERI), New Delhi. The AMF bulk inoculum preparation was conducted in a naturally ventilated greenhouse by sowing surface-sterilized maize (*Zea mays* L.) seeds in earthen pots with an autoclaved soil:sand mixture (1:1, w/w). The substrate containing the spores and root pieces served as the stock culture for AMF inoculum. The VW was prepared from air-dried mixed waste (i.e. vegetable waste:paddy straw:water hyacinth; 1:1:1;) according to the procedure suggested by Ismail [26] using a cylindrical metal container (30-cm height ×30-cm diameter). In brief, a tap was fixed on the lower side of the container. The bottom of the container had a 5-cm layer of broken pebbles, followed by a 5-cm layer of coarse sand. Air-dried mixed waste as mentioned above was put into the container as substrate and one thousand earthworms (*Eisenia fetida* Savigny) were introduced into the substrate. An appropriate level of moisture (60%) was maintained in the container by adding water at regular intervals. During the VW preparation, the temperature and pH inside the container ranged from 29–32°C and 7.0–7.3 respectively. The brown-colored watery extract of earthworm-worked substrate was allowed to drain out of the container and used for foliar spray on test crops. The physico-chemical composition of the VW is presented in Table 1.

Pot experiment

The pot experiment was performed from 6 May to 5 August 2011, in a naturally ventilated greenhouse at the Defence Research Laboratory (DRL), Tezpur (26°38′N, 92°48′E), District Sonitpur, Assam. Air-dried and sieved soil collected from the experimental field of the DRL was used to fill earthen pots (3.5 kg pot^{-1}). The physico-chemical composition of the experimental soil is presented in Table 1. Ten grams of mycorrhizal inoculum (containing 20 spores per g) of each AMF (i.e., *R. irregularis* and *G. mosseae*) was placed in each pot (2 cm below the seed sowing level) and covered using soil. Initially, five seedlings were grown in each pot, which were thinned to three at the start of the experiment. There were nine replicates of each treatment with a randomized block design. The plants in the pots were grown for 90 d in the greenhouse (temperature, 27–35°C; relative humidity, 70–80%) under natural illumination and watered (i.e., tap water) as needed.

Physico-chemical analysis of the soil and VW used in the experiment

The pH of diluted-water (1:5) sample was determined using a digital pH meter (Model PT10; Sartorius AG, Germany). Electrical conductivity (EC) was determined using a conductivity bridge (Model 306; Systronics, India). Organic carbon (OC) was estimated using the method of Walkley and Black [27], by using 1N potassium dichromate and back-titrated using 0.5N ferrous ammonium sulfate solution. Available nitrogen (N) was deter-

Table 1. Physico-chemical properties of the soil and vermiwash used in the study.

	Soil	Vermiwash
pH (H_2O)	6.330 ±0.072*	8.77
Electrical conductivity (dS m^{-1})	0.027 ±0.009	0.04
Available Nitrogen (mg kg^{-1})	350.816±21.831	338.88**
Available Phosphorus (mg kg^{-1})	55.041 ±0.120	57.2**
Available Potassium (mg kg^{-1})	314.27 ±19.548	154.94**
Total Organic Carbon (%)	1.160 ±0.027	1.06

*Mean ± standard error (n = 3). **Values in ppm.

mined by Kjeldahl method [28], by using an automatic nitrogen estimation apparatus (Model KelPlus Classic DX VA; Pelican Equipments, Chennai, India). Available P was determined spectrophotometrically in a 1M acidic ammonium fluoride solution [29]. Available potassium (K) was measured by the ammonium acetate method [30] using a flame photometer (Model FP114; Thermo Scientific, USA).

Plant growth parameters

Plants were uprooted after 90 d, and the data on vegetative growth, chlorophyll content (SPAD), shoot nutrients, AMF colonization in the roots, and AMF spore count in rhizospheric soil were recorded. Shoot and root dry matters were recorded after drying the sample at 70°C for 48 h. The relative amount of chlorophyll present in the leaves was determined by measuring the transmittance of the leaf in two wave bands (i.e., 600–700 nm and 400–500 nm) by using a SPAD Meter (SPAD 502 Plus; Konica Minolta, Japan). SPAD values from the midpoint region near the midrib of five fully expanded leaf samples from a single plant were taken at 1100 IST, and the mean value was calculated [31].

Shoot nutrients

For nutrient analysis of the shoots, the oven-dried samples were finely ground. Total N and C in the shoots were determined using an elemental analyzer (EA3000, Eurovector, Italy). To estimate the P, K, Na, and Ca levels in the shoots, 1 g of the finely ground sample was subjected to a wet oxidation treatment using tri-acid (HNO_3:H_2SO_4:$HClO_4$; 10:1:4) digestion in a digestion block (KELPLUS, KES12L; Pelican Equipments, Chennai, India) at 200°C. Following acid digestion, the samples were diluted and filtered for further nutrient analysis. Total P in the shoots was determined by the vanadomolybdophosphoric acid colorimetric method [32] using a spectrophotometer (Specord 200; Analytik Jena, Germany). Total shoot Ca, Na, and K were measured by the ammonium acetate method of Hanway and Heidel [30], by using a flame photometer (Model FP114; Thermo Scientific, USA). For estimating mycorrhizal colonization, freshly collected roots were washed in water, cleaned using10% KOH, acidified using 1N HCl, and stained in 0.05% trypan blue [33].

Mycorrhizal colonization and AMF spore count in rhizospheric soil

Quantification of root colonization for AMF was conducted using the gridline intersection method [34], and 100 segments from each sample were observed under a compound microscope (Leica DM750). Spores of AMF were extracted from the rhizospheric soil (50g for each treatment) before and after experimentation using the wet-sieving and decanting methods [35]. The total number of spores was estimated and the values were expressed as *x* number of AMF spores per 50 g of soil.

Calculation of mycorrhizal dependency

Mycorrhizal dependency (MD) based on shoot dry matter was calculated according to the formula suggested by van der Heijden [36] as given below:

$$MD=\left(1-\left(\frac{b_n}{\sum_1^n a_n}\right)\right)\times 100\%$$

Where *a* is the mean total shoot dry matter in a given treatment (AMF and/or VW], *n* is the number of treatments, and *b* is the mean shoot dry matter of untreated plants. Similarly, the nutrient dependencies were calculated, where the values of a given shoot nutrient were used in place of the shoot dry matter in the above-mentioned formula.

Statistical analysis of the data

Means were compared by ANOVA, followed by the least significant difference (LSD) test. Statistically significant differences were then determined at $p = 0.05$, using the software IBM SPSS Statistics version 19.

Results

Mycorrhizal colonization and AMF spore count

Root colonization (RC) at the beginning of the experiment was 26.66%, which increased up to 40% in the control plants following completion of the experiment. Furthermore, comparisons of the control to the AMF- and VW-treated plants suggests a statistically significant increase ($p \leq 0.05$) in RC; 100% RC was recorded in all the plants receiving the AMF treatments (Figure 1). Except for the single treatment with VW, there was no statistically significant difference with respect to RC among the single and dual treatments. Mycorrhizal spore count in the rhizospheric soil at the beginning of the experiment was 24.33 spores per 50 g^{-1}, which increased to 34.33 in the control following completion of the experiment. Furthermore, a comparison of the control to the AMF- and VW-treated plants revealed a significant increase ($p \leq 0.05$) in spore count, except for the single treatment with VW. Comparisons among the AMF treatments did not show significant differences with respect to the spore count in the rhizospheric soil (Figure 1).

Plant growth parameters and nutrient uptake

The shoot dry matter (SDM), root dry matter (RDM), and chlorophyll content increased significantly in the plants receiving single and combined treatments of AMF and VW in comparison to the control. However, the RDM:SDM ratio in all treated plants decreased significantly when compared to that in the control. Inoculations of RI increased the growth values (SDM and RDM) more than those of GM, and this increase was statistically significant. However, there were no significant differences between the RI and GM treatments with respect to the chlorophyll content and the RDM:SDM ratio. The pattern of growth observed for the three individual treatments followed the order RI>GM>VW. Overall, the growth values increased significantly in the dual treatments when compared to those in the single treatments. However, comparison of the RI and RI+VW treatments revealed a significant increase only in the RDM. In addition, the VW treatment considerably influenced the increase in growth of the plants exposed to dual treatments. The influence of VW on growth was more prominent in the GM+VW treatment when compared to that in the RI+VW treatment. With regard to the dual treatments, SDM, RDM, and chlorophyll content in the plants was higher under the GM+VW treatment when compared to those in the RI+VW treatment. However, a statistically significant difference in the growth parameters was noted only for the RDM and chlorophyll content (Table 2).

In comparison to the control plants, plants under single and combined treatments of AMF and VW showed significantly higher nutrient (N, C, P, K, Na, and Ca in the shoots) uptake. However, the concentration of N in the shoots of plants exposed to the single treatments of RI and GM did not significantly differ from that in the control. A comparison between the RI- and GM-inoculated plants indicated that the concentration of K in the shoots was significantly higher in the RI-inoculated plants but, there were no

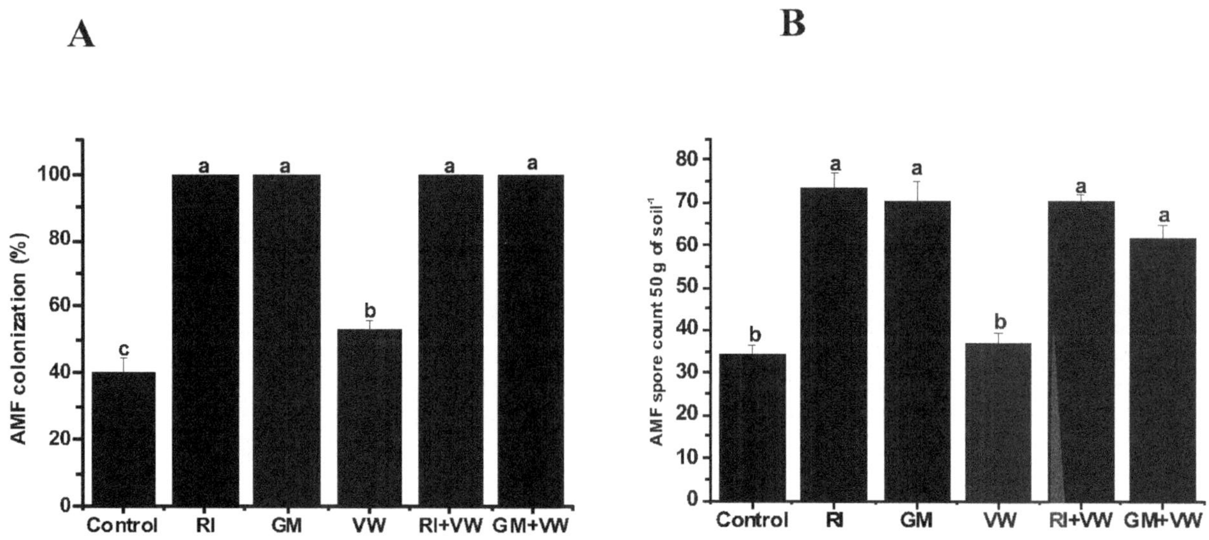

Figure 1. The AMF colonization in the roots (A) and spore count in rhizospheric soil (B). RI- *Rhizophagus irregularis*, GM- *G. mosseae*, VW- vermiwash. Values are means of nine biological replicates. Bar shows standard error of mean (SEM). Different letters on the column represent significant difference ($p < 0.05$) based on LSD test.

significant differences in the concentrations of the other shoot nutrients. The maximum concentration of K in the shoots was observed in the plants receiving the VW treatment; this finding was statistically significant when compared to the plants receiving the single AMF treatments. However, the plants receiving the VW treatment had significantly lower P when compared to that in the GM- and GI-treated plants. The maximum total C in the shoots was observed in the RI-treated plants, followed by the GM- and VW -treated plants. However, among the single treatments, a significant difference with respect to total C was found only between the RI- and VW-treated plants. A comparison of the RI and RI+VW treatments revealed a significant increase only in the K, Ca, and N contents in the shoots. However, all of the shoot nutrients were significantly higher in the GM+VW-treated plants when compared to those in the GM-treated plants. In addition, while comparing the VW treatment to the GM+VW and RI+VW treatments, the uptake of all shoot nutrients was shown to increase significantly. However, the concentration of N in the shoots of the VW-treated plants was not significantly different from that of the RI+VW-treated plants. The uptake of C, K, and Ca in the shoots significantly increased in the GM+VW-treated plants when compared to the RI+VW-treated plants (Table 2).

Mycorrhizal dependency

It was observed that the mycorrhizal growth dependency (MGD) of the RI-treated plants was significantly higher than that of the GM-treated plants, followed by the VW-treated plants. VW treatment significantly influenced the MGD in plants. The MGD of the RI treatment was not significantly different from that of the RI+VW treatment. Nevertheless, MGD of the GM+VW treatment was significantly higher than that of the RI treatment,

Table 2. Effect of foliar spray of vermiwash on response of inoculation of two different AM fungi in *Capsicum assamicum* under pot conditions.

Parameters	Control	*Rhizophagus irregularis*	*Glomus mosseae*	Vermiwash	*R. irregularis* +Vermiwash	*G. mosseae* +Vermiwash
Shoot dry matter (g)	1.390±0.074 d	3.529±0.062 ab	2.780±0.034 c	2.400±0.068 c	3.613±0.252 a	3.970±0.026 a
Root dry matter (g)	0.563±0.023 f	0.973±0.009 c	0.860±0.032 d	0.640±0.012 e	1.060±0.012 b	1.220±0.021 a
Root:shoot ratio	0.406±0.004 a	0.275±0.002 bc	0.308±0.013 b	0.268±0.006 c	0.296±0.017 b	0.307±0.003 b
Chlorophyll content*	30.400±0.740 e	35.90±0.309 bc	39.300±0.478 bc	35.233±0.165 bcd	38.233±0.320 b	44.933±2.045 a
Total shoot N (%)	2.856±0.023 abc	3.562±0.012 ab	3.432±0.023 abc	4.042±0.094 ab	4.673±0.198 a	4.959±0.549 a
Total shoot P (%)	0.139±0.002 d	0.226±0.008 ab	0.221±0.001 ab	0.177±0.007 abc	0.240±0.002 a	0.246±0.004 a
Total shoot K (%)	0.173±0.001 f	0.206±0.002 d	0.181±0.000 e	0.256±0.001 c	0.269±0.000 b	0.326±0.000 a
Total shoot C (%)	39.410±0.578 d	47.26±0.506 b	46.199±0.279 bc	44.938±0.175 bc	48.378±0.201 b	52.296±0.605 a
Total shoot Na (mg kg^{-1})	66.660±13.60 abc	150.00±23.571 a	135.00±6.236 ab	110.000±4.714 ab	161.667±3.600 a	193.333±5.443 a
Total shoot Ca (mg kg^{-1})	566.660±13.60 d	800.00±23.571 c	766.66±13.609 c	750.000±23.571 c	883.33±13.609 b	1116.667±13.609 a

Values are means ± standard errors of nine biological replicates. Values within a column not followed by the same letter are statistically. different (p<0.05). *Measured as SPAD value using chlorophyll meter (SPAD 502 Plus; Konica Minolta, Japan).
N- Nitrogen; P- Phosphorus; K- Potassium; C- Carbon; Na- Sodium; Ca- Calcium.

followed by the GM and VW treatments. Analysis of the AMF-VW growth inter-dependency for the dual treatments in comparison to the individual treatments indicated that plants exposed to RI treatment were more dependent on that AMF for growth, while the GM-treated plants were dependent on both the GM and VW for growth (Figure 2).

The inoculation of VW considerably influenced shoot N, K, and C dependencies of the test plants on two AMF, which was evident from the significantly variable response of the plants to the AMF under the VW treatment. Overall, the N and K dependencies of the plants were higher under the VW treatment when compared to those under the AMF treatment; the P, C, and Na dependencies were higher under the AMF treatment when compared to those under the VW treatment (Figure 3).

Furthermore, shoot Ca under the VW treatment appeared to influence the plant's response to AMF. However, the response of shoot Ca to AMF had lesser prominence than that observed for shoot N, K, and C. Overall, the nutrient dependency data indicate that, although the VW treatment considerably influenced plant-AMF interactions, the RI treatment affected nutrient dependencies to a lesser extent than that by the GM+VW treatment (Figure 3).

Nutrient stoichiometry

With regard to the individual treatments, both the RI- and GM-treated plants showed significantly decreased shoot C:P and N:P ratios and an increased shoot Na:K ratio; however, no significant effect on the C:N ratio was observed when compared to that in the control (Figures 4b&c). The VW-treated plants showed a significant decrease only in the shoot C:N and C:P ratios; there was no significant effect on the Na:K and N:P ratios when compared to those in the control. In the dual treatments, both the GM+VW- and the RI+VW-treated plants showed a significant decrease in the shoot C:P ratio, but there was no significant effect on the stoichiometries of C:N and N:P when compared to those in the control. However, the shoot Na:K ratio increased significantly only in the RI+VW treated plants when compared to that in the control (Figure 4).

A comparison of single RI and GM treatments indicated that there were no significant differences with respect to nutrient stoichiometry. However, a comparison of the RI and VW treatments revealed significantly greater C:P and N:P ratios under the VW treatment and a significantly greater Na:K ratio under the RI treatment. Similarly, the C:P and N:P ratios were significantly greater under the VW treatment when compared to those under the GM treatment; however, the Na:K ratio was significantly higher under the GM treatment.

A comparison between the single and dual treatments exhibited considerable and significant variation in nutrient stoichiometry. The C:N ratio was significantly higher in the individual AMF-treated plants when compared to the RI+VW-treated plants, whereas the N:P ratio was significantly higher in the GM+VW-treated plants when compared to the individual AMF-treated plants. On the other hand, the C:P ratio was significantly higher in the VW-treated plants when compared to the AMF+VW-treated plants. With regard to the two dual treatments, no significant differences were recorded across all nutrient stoichiometric parameters. Overall, the VW application significantly influenced the pattern of nutrient stoichiometry in the plants that were differentially subjected to the two AMF treatments (Figure 4).

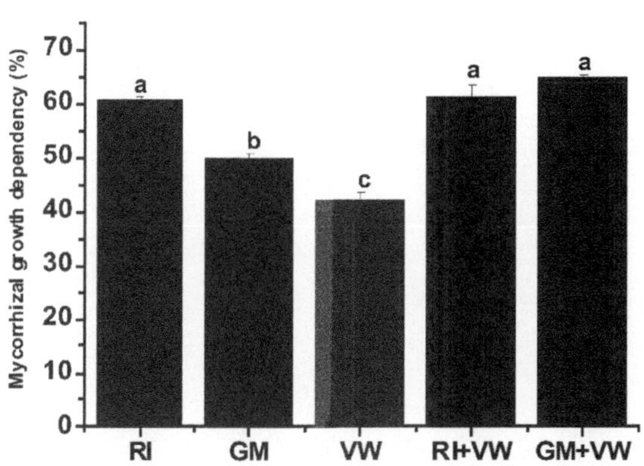

Figure 2. Mycorrhizal growth dependency of *C. assamicum*. RI- *Rhizophagus irregularis*, GM- *G. mosseae*, VW- vermiwash. Values are means of nine biological replicates. Bar shows SEM. Different letters on the column represent significant difference ($p < 0.05$) based on LSD test.

Discussion

Mycorrhizal colonization, plant growth, and nutrient uptake

The aim of the present study was to evaluate the influence of a VW foliar spray on the mycorrhizal growth dependency and nutrient stoichiometry of *Capsicum assamicum* inoculated using two different AMF in acidic soil under pot conditions. Our results suggest that 100% mycorrhizal colonization was achieved following completion of the experiment, thus, indicating the positive response of *C. assamicum* to the inoculations of the two different AMF. Although several studies have demonstrated differences in the responses of mycorrhizae to crops [7,8,37], mycorrhizal colonization and spore count in rhizospheric soil in the current study were not significantly different from the plants inoculated using the two different fungi. Furthermore, the influence of the VW treatment on mycorrhizal colonization could not be inferred as there was no significant difference between the single and dual treatments.

Our results suggest that plant growth and nutrient uptake significantly increased under the AMF and VW treatments, which is in agreement with earlier reports [2,3,7]. The positive influence of the VW on plant growth, as observed in the current study, might be ascribed to the presence of plant-growth-promoting substances in the VW, such as humic, fulvic, and other organic acids [4]; auxin-like substances [5]; and cytokinin-like substances [6]. In addition, the absorption of nutrient elements from water solutions applied to the leaves and their beneficial physiological effects on plants have been previously reported [38]. The leaf cuticle is a hydrophobic layer comprised of high molecular-weight biopolymers such as cutin and suberins and hydrophobic C14–C27 epiculticular waxes [39]. Further, recent physiological studies have identified polar aqueous pores, which facilitate the absorption of charged ions into the epidermal cells [40]. The foliar application of fertilizers has become more prevalent agricultural crop production because it is more environmentally friendly when compared to soil fertilization techniques [41].

In addition, in the current study, it was observed that the combined inoculation of AMF and VW had a significantly more positive effect on plant growth and nutrient uptake. Several studies have suggested that combined treatments of AMF and organic amendments may be more beneficial, thus, indicating a positive interaction between these components [42,43]. In our study, a synergistic interaction between AMF and VW treatments was observed, which might have contributed to enhanced growth and nutrient uptake. To the best of our knowledge, this is the first study

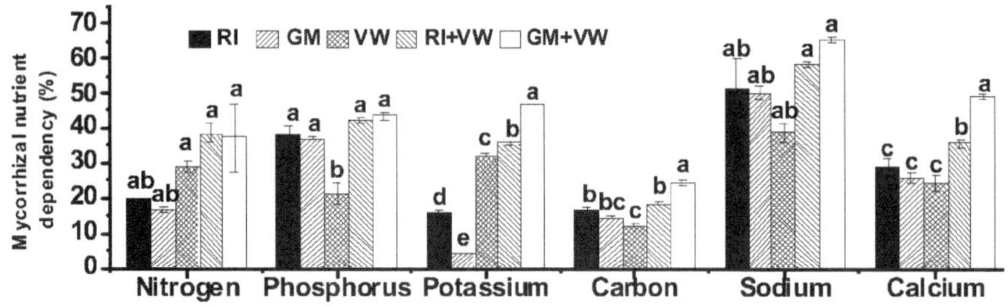

Figure 3. Mycorrhizal shoot nutrient dependency of *C. assamicum*. RI- *Rhizophagus irregularis,* GM- *G. mosseae,* VW- vermiwash. Values are means of nine biological replicates. ± Bar shows SEM. Different letters on the column represent significant difference ($p < 0.05$) based on LSD test.

on the synergistic interaction of AMF and VW treatments in chili plants.

Interestingly, the VW treatment influenced the extent of the increase in growth and nutrient uptake in plants exposed to the dual treatments. The influence of the VW on the magnitude of this increase was more prominent in the GM+VW treatment when compared to that in the RI+VW treatment. It has been shown that the changes in crop growth and yield in response to soil amendments could be the result of the differential effects of the types of amendment on the proliferation of different AMF [44]. Although we did not conduct a microbiological analysis of the VW used in the current study, it is well known that VW contains macro- and micronutrients such as N, P, K, Ca, Fe, and Mn [3] and several beneficial microbes [45]. These properties of VW may have an indirect role in improving fertility and the physical and chemical conditions of the soilless medium, thus, influencing nutrient availability and contributing to the enhanced growth, yield, and antioxidant content of plants [13,46]. However, the

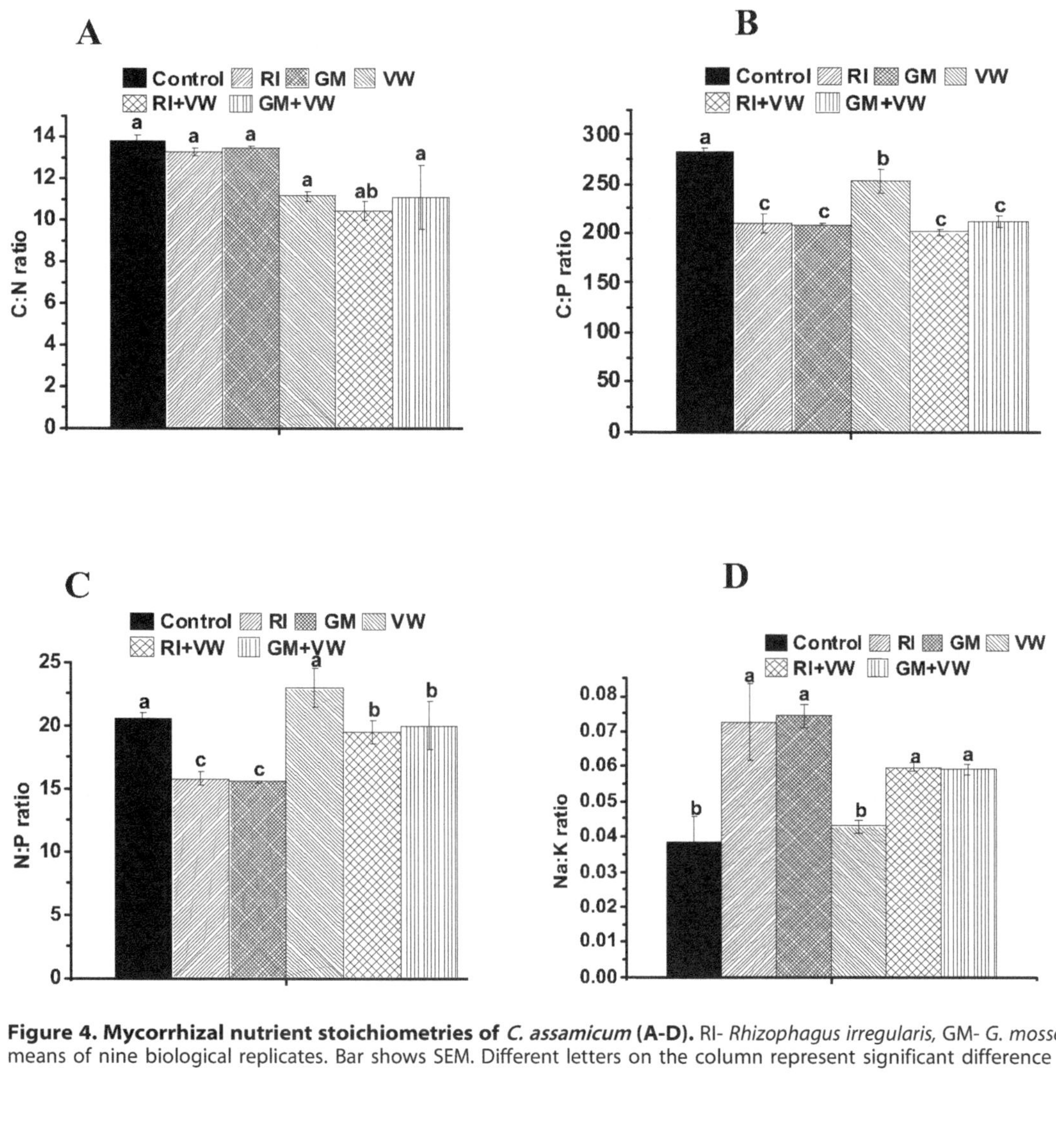

Figure 4. Mycorrhizal nutrient stoichiometries of *C. assamicum* (A-D). RI- *Rhizophagus irregularis,* GM- *G. mosseae,* VW- vermiwash. Values are means of nine biological replicates. Bar shows SEM. Different letters on the column represent significant difference (p <0.05) based on LSD test.

differential response of plants to the different treatment conditions, as observed in the current study, requires further investigation to ascertain the underlying mechanisms.

Mycorrhizal dependency

Mycorrhizal dependency, which refers to the degree that a plant benefits from AMF associations [47], has often been quantified by calculating the yield or shoot dry matter ratio between mycorrhizal plants and control plants grown in a particular soil [37,48]. In the current study, we demonstrated that the application of VW significantly influenced mycorrhizal dependency. It was also observed that the plants inoculated using RI+VW were more dependent on the AMF for growth, while the GM+VW-treated plants were dependent on both the GM and VW for growth. With respect to mycorrhizal nutrient dependency, the VW treatment appeared to influence plant responses to AMF, which was more prominent with regard to shoot N, K, and C concentrations. This may be explained by the fact that the AM symbiotic relationships and subsequent benefits to the host plant are affected by the availability/limitation of nutrients [49]. Moreover, it has been well documented that the symbiotic relationship between AMF and the host plants is not specific; however, under stress, there may be combinations of AM–host plants that are more beneficial for growth [50]. In the current study, nutrient supplementation in the form of foliar spray may have modified the degree of functional plant–AMF associations in acidic soil, which requires further exploration for ascertaining the underlying mechanisms. Interestingly, greater K dependency was observed in VW-treated plants when compared to that in AMF-treated plants, which was shown to increase significantly in the dual treatments. It has been well established that K plays a key role in osmoregulation, stomatal regulation, photosynthesis, and cell extension [37]. In addition, higher K uptake can help mitigate salt stress in plants [51]. Therefore, it can be assumed that the application of VW in combination with AMF may be more beneficial to chili crops grown under conditions of salt stress.

Nutrient stoichiometry

Ecological stoichiometry is the study of the mass balance of multiple chemical elements in living systems [18,52], which has implications for nutrient cycling, and the functioning of living organisms and communities of organisms in aquatic and terrestrial ecosystems [17,53]. The importance of the relative proportions of elements essential for the growth performance of plants has long been known to the farmers. However, there is considerable variability with regard to the optimal ratios for the observed values [20]; thus, its significance has not been widely appreciated among applied researchers. The most investigated relationshiphas been that of C:N:P because N and P commonly limit growth [54]. In addition, while C provides the structural basis for plants, it constitutes a stable percentage (i.e., 50%) of a plant's dry mass andcan act as a limiting element. These three elements are strongly coupled by their biochemical functions [20]. In the current study, the C:P ratio significantly decreased in all treated plants when compared to that in the control. Similar findings have been reported by Guo et al. [55] in maize plants inoculated using *G. mosseae* and *G. versiforme*. In our study, comparisons among the single treatments revealed significantly greater C:P and N:P ratios under the VW treatment when compared to those under the AMF treatment. This pattern indicates that there was an increase in the uptake of N and a decrease in the uptake of P in the plants treated using VW when compared to those treated using the AMF. The pattern of change in the N:P ratio has been reported as a consequence of AMF-dependent modifications in plant and soil nutrient relations [56]. However, in the current study, the VW treatment appeared to contribute more N to plants when compared to the AMF treatment, thus, leading to changes in the C:N:P stoichiometry of the plant shoots. This pattern was also supported by the fact that the N:P ratio significantly increased using the dual treatments when compared to that in the single treatments of the two AMF. N is an essential component of all enzymes and P is necessary for rRNA synthesis [57]. Therefore, it is likely that the consequential changes in N and P concentrations will affect the allocation of nutrients to a plant's cellular components, life history strategies, and physiological functions [58]. This could partially explain our findings indicating that the shoot dry matter and shoot P were not significantly affected in the VW- and GM-treated plants, whereas shoot N was significantly higher in the VW-treated plants. Overall, there was considerable variation in the C:N:P stoichiometry of plants receiving the single and dual treatments, thus, indicating the possible effects of VW on plants. However, further detailed studies on nutrient acquisition mechanisms from different sources are required to clearly understand nutrient stoichiometry in plants.

Conclusions

Overall, the present study clearly indicates that the foliar spray of *C. assamicum* with VW and AMF inoculation in the soil improves plant growth and nutrient uptake under pot conditions. In addition, when combined with AMF treatment, the VW foliar spray significantly and differentially influences the growth and nutrient stoichiometry of plants. The influence of the VW treatment was more pronounced using the GM+VW treatment when compared to that of the RI+VW treatment. Interestingly, the VW treatment appeared to contribute more N to plants than that under the AMF treatment, thus, leading to changes in the C:N:P stoichiometry of the plant shoots. Furthermore, increased K dependency observed under the dual treatments suggests the significance of these treatments in improving crop growth under salt stress. These findings have important implications for selecting effective nutrient sources for crop growth. Future studies, by using advanced techniques such as continuous flow-isotope ratio mass spectrometry, are necessary for investigating the nutrient acquisition mechanisms of plants from different sources and for further precise understanding of the patterns observed in the current study.

Author Contributions

Conceived and designed the experiments: MKM VV LS MCK. Performed the experiments: MHK MKM RG. Analyzed the data: MHK MKM. Contributed reagents/materials/analysis tools: RG. Wrote the paper: MHK MKM.

References

1. Ansari AA, Sukhraj K (2010) Effect of vermiwash and vermicompost on soil parameters and productivity of okra (*Abelmoschus esculentus*) in Guyana. Afr J Agric Res 5: 1794–1798.

2. Meghvansi MK, Khan MH, Gupta R, Gogoi HK, Singh L (2012) Vegetative and yield attributes of okra and naga chilli as affected by foliar sprays of vermiwash on acidic soil. J Crop Improv 26: 520–531.

3. Naidu Y, Meon S, Siddiqui Y (2013) Foliar application of microbial-enriched compost tea enhances growth, yield and quality of muskmelon (*Cucumis melo* L.) cultivated under fertigation system. Sci Hort 159: 33–40.
4. Arancon NQ, Edwards CA, Dick R, Dick L (2007) Vermicompost tea production and plant growth impacts. Biocycle 48: 51–52.
5. Garcia M, Cruz Sosa IF, Saavedra AL, Hernandez MS (2002) Extraction of auxin-like substances from compost. Crop Res 24: 323–327.
6. Arthur GD, Jager AK, Van Staden J (2001) The release of cytokinin-like compounds from *Gingko biloba* leaf material during composting. Environ Exp Bot 45: 55–61.
7. Sensoy S, Demir S, Turkmen O, Erdinc C, Savur OB (2007) Responses of some different pepper (*Capsicum annuum* L.) genotypes to inoculation with two different arbuscular mycorrhizal fungi. Sci Hort 113: 92–95.
8. Meghvansi MK, Prasad K, Harwani D, Mahna SK (2008) Response of soybean cultivars toward inoculation with three arbuscular mycorrhizal fungi and *Bradyrhizobium japonicum* in the alluvial soil. Eur J Soil Biol 44: 316–323.
9. Willis A, Rodrigues BF, Harris PJC (2013) The ecology of arbuscular mycorrhizal fungi. Cr Rev Plant Sci 32: 1–20.
10. van der Heijden MGA, Boller T, Wiemken A, Sanders IR (1998) Different arbuscular mycorrhizal fungal species are potentially determinants of plant community structure. Ecology 79: 2082–2091.
11. Wilson GWT, Harnett DC (1998) Interspecific variation in plant responses to mycorrhizal colonization in tall grass prairie. Am J Bot 85: 1732–1738.
12. Zhang FS, Shen JB, Li L, Liu X (2004) An overview of rhizosphere processes related with plant nutrition in major cropping systems in China. Plant Soil 260: 89–99.
13. Siddiqui Y, Islam TM, Yuvarani N, Sariah M (2011) The conjunctive use of compost tea and inorganic fertiliser on the growth, yield and terpenoid content of *Centella asiatica* (L.) urban. Sci Hort 130: 289–295.
14. León-Anzueto E, Abud-Archila M, Dendooven L, Ventura-Canseco L, Gutiérrez-Miceli F (2011) Effect of vermicompost, worm-bed leachate and arbuscular mycorrhizal fungi on lemon grass (*Cymbopogon citratus*) growth and composition of its essential oils. Electronic J Biotechnol 14: 1–9.
15. Khorshidi M, Bahadori F, Behnamnia M (2013) The effects of arbuscular mycorrhizal fungi (*Glomus intraradices*) and vermicompost application on yield and nutrient uptake in garden thyme (*Thymus vulgaris* L.) under field conditions. Int J Agric Crop Sci 5: 1191–1194.
16. Sharma PK, Verma SP, Bhumbla DR (1980) Transformation of added phosphorus into inorganic phosphorus fractions in some acid soils of Himachal Pradesh. J Indian Soc Soil Sci 28: 450.
17. Vrede T, Dobberfuhl DR, Kooijman S, Elser JJ (2004) Functional connections among organism C:N:P stoichiometry, macromolecular composition and growth. Ecology 85: 1217–1229.
18. Elser JJ, Sterner RW, Gorokhova E, Fagan WF, Markow TA, et al. (2000) Biological stoichiometry from genes to ecosystems. Ecol Lett 3: 540–550.
19. Kyle M, Watts T, Schade J, Elser JJ (2003) A microfluorometric method for quantifying RNA and DNA in terrestrial insects. J Insect Sci 3: 1–7.
20. Agren GI (2008) Stoichiometry and nutrition of plant growth in natural communities. Annu Rev Ecol Evol Syst 39: 153–70.
21. Hessen DO, Jensen TC, Kyle M, Elser JJ (2007) RNA responses to N- and P-limitation; reciprocal regulation of stoichiometry and growth rate in *Brachionus*. Funct Ecol 21: 956–962.
22. Jordan NR, Zhang J, Huerd S (2000) Arbuscular-mycorrhizal fungi, potential roles in weed management. Weed Res 40: 397–410.
23. Gosling P, Hodge A, Goodlass G, Bending GD (2006) Arbuscular mycorrhizal fungi and organic farming. Agric Ecosyst Environ 113: 17–35.
24. Alloush GA, Clark RB (2001) Maize response to phosphate rock and arbuscular mycorrhizal fungi in acidic soil. Commun Soil Sci Plant Anal 32: 231–254.
25. Meghvansi MK, Siddiqui S, Khan MH, Gupta VK, Vairale MG, et al. (2010) Naga Chilli: A potential source of capsaicinoids with broad-spectrum ethnopharmacological applications. J Ethnopharmacol 132: 1–14.
26. Ismail SA (1997) Vermicology: The biology of earthworms. Chennai, India: Orient Longman.
27. Walkley AJ, Black IA (1934) Estimation of soil organic carbon by the chromic acid titration method. Soil Sci 37: 29–38.
28. Jackson ML (1973) Soil chemical analysis. New Delhi: Prentice Hall of India Ltd.
29. Bray RH, Kurtz LT (1945) Determination of total, organic and available form of phosphorus in soils. Soil Sci 59: 39–45.
30. Hanway JJ, Heidel H (1952) Soil analysis methods as used in Iowa state college soil testing laboratory. Iowa Agri 57: 1–31.
31. Peng S, Garcia F, Laza R, Kassman KG (1992) Leaf thickness affects the estimation of leaf N using a chlorophyll meter. Intl Rice Res Newsl 17: 19–20.
32. Tandon HLS (1993) Methods of analysis of soils, plants, waters and fertilizers. F.D.C.O, New Delhi.
33. Phillips JM, Hayman DS (1970) Improved procedures for clearing roots and staining parasitic and vesicular-arbuscular mycorrhizal fungi for rapid assessment of infection. Trans Br Mycol Soc 55: 158–161.
34. Giovanetti M, Mosse B (1980) An evaluation of techniques for measuring vesicular–arbuscular mycorrhizal infection in roots. New Pyhtol 84: 489–500.
35. Gerdemann JW, Nicolson TH (1963) Spores of mycorrhizal *Endogone* species extracted from soil by wet sieving and decanting. Trans Br Mycol Soc 46: 235–244.
36. van der Heijden MGA (2002) Arbuscular Mycorrhizal Fungi as a determinant of plant diversity: in search of underlying mechanisms and general principles. In: Van der Heijden MGA, Sanders I, editors. Mycorrhizal Ecology, Springer-Verlag. Berlin Heidelberg. pp. 243–265.
37. Porras-Soriano A, Soriano-Martín ML, Porras-Piedra A, Azcón R (2009) Arbuscular mycorrhizal fungi increased growth, nutrient uptake and tolerance to salinity in olive trees under nursery conditions. J Plant Physiol 166: 1350–1359.
38. Haytova D (2013) Review of foliar fertilization of some vegetables crops. Ann Rev Res Biol 3: 455–465.
39. Holloway PJ (1993) Structure and chemistry of plant cuticles. Pestic Sci 37: 203–232.
40. Schonherr J (2000) Calcium chloride penetrates plant cuticles via aqueous pores. Planta 212: 112–118.
41. Kuepper G (2003) Foliar fertilization current topic. ATTRA – National sustainable Agriculture Information service. NCAT Agriculture specialist.
42. Medina A, Azcon R (2010) Effectiveness of the application of arbuscular mycorrhiza fungi and organic amendments to improve soil quality and plant performance under stress conditions. J Soil Sci Plant Nutr 10: 354–372.
43. Medina A, Vassilev N, Azcon R (2010) The interactive effect of an AM fungus and an organic amendment with regard to improving inoculum potential and the growth and nutrition of *Trifolium repens* in Cd-contaminated soils. Appl Soil Ecol 44: 181–189.
44. Muthukumar T, Udaiyan K (2002) Growth and yield of cowpea as influenced by changes in arbuscular mycorrhiza in response to organic manuring. J Agron Crop Sci 188: 123–132.
45. Scheuerell S, Mahaffee W (2002) Compost tea: Principles and prospects for plant disease control. Compost Sci Util 10: 313–338.
46. Siddiqui Y, Sariah M, Razi I (2008) *Trichoderma*-fortified compost extracts for the control of *Choanephora* wet rot in okra production. Crop Prot 27: 385–390.
47. Plenchette C, Fortin JA, Furlan V (1983) Growth response of several plant species to mycorrhizae in a soil of moderate P-fertility. I. Mycorrhizal dependency under field conditions. Plant Soil 70: 199–209.
48. Doubkova P, Suda J, Sudova R (2012) The symbiosis with arbuscular mycorrhizal fungi contributes to plant tolerance to serpentine edaphic stress. Soil Biol Biochem 44:56–64.
49. Johnson NC (2010) Resource stoichiometry elucidates the structure and function of arbuscular mycorrhizas across scales. New Phytol 185: 631–647.
50. Dai O, Singh RK, Nimasow G (2011) Effect of arbuscular mycorrhizal (AM) inoculation on growth of chili plant in organic manure amended soil. Afr J Microbiol Res 5: 5004–5012.
51. Thomas JMG, Boote KJ, Allen LH, Gallo-Meagher M, Davis JM (2003) Seed physiology and metabolism: Elevated temperature and carbon dioxide effects on soybean seed composition and transcript abundance. Crop Sci 43: 1548–1557.
52. Sterner RW (1995) Elemental stoichiometry of species in ecosystems. In: Jones C, Lawton J, editors. Linking species and ecosystem. New York: Chapman and Hall. pp 240–252.
53. Sterner RW, Elser JJ (2002) Ecological stoichiometry: the biology of elements from molecules to the biosphere. Princeton University Press. Princeton and Oxford.
54. Elser JJ, Bracken MES, Cleland EE, Gruner DS, Harpole WS, et al. (2007) Global analysis of nitrogen and phosphorus limitation of primary producers in freshwater, marine and terrestrial ecosystems. Ecol Lett 10: 1135–1142.
55. Guo W, Zhao R, Zhao W, Fu R, Guo J, et al. (2013) Effects of arbuscular mycorrhizal fungi on maize (*Zea mays* L.) and sorghum (*Sorghum bicolor* L. Moench) grown in rare earth elements of mine tailings. Appl Soil Ecol 72: 85–92.
56. Milleret R, Le Bayon R-C, Gobat J-M (2009) Root, mycorrhiza and earthworm interactions: their effects on soil structuring processes, Plant Soil nutrient concentration and plant biomass. Plant Soil 316: 1–12.
57. Elser JJ, Acharya K, Kyle M, Cotner J, Makino W, et al. (2003) Growth rate-stoichiometry couplings in diverse biota. Ecol Lett 6: 936-943.
58. Chen MM, Yin HB, O'Conner P, Wang YS, Zhu YG (2010) C:N:P stoichiometry and specific growth rate of clover colonized by arbuscular mycorrhizal fungi. Plant Soil 326: 21–29.

4

Dynamics of Soil Bacterial Communities in Response to Repeated Application of Manure Containing Sulfadiazine

Guo-Chun Ding[1], Viviane Radl[2], Brigitte Schloter-Hai[2], Sven Jechalke[1], Holger Heuer[1], Kornelia Smalla[1]*, Michael Schloter[2]

1 Julius Kühn-Institut - Federal Research Centre for Cultivated Plants (JKI), Institute for Epidemiology and Pathogen Diagnostics, Braunschweig, Germany, **2** Helmholtz Zentrum München, German Research Center for Environmental Health, Research Unit for Environmental Genomics, Neuherberg, Germany

Abstract

Large amounts of manure have been applied to arable soils as fertilizer worldwide. Manure is often contaminated with veterinary antibiotics which enter the soil together with antibiotic resistant bacteria. However, little information is available regarding the main responders of bacterial communities in soil affected by repeated inputs of antibiotics via manure. In this study, a microcosm experiment was performed with two concentrations of the antibiotic sulfadiazine (SDZ) which were applied together with manure at three different time points over a period of 133 days. Samples were taken 3 and 60 days after each manure application. The effects of SDZ on soil bacterial communities were explored by barcoded pyrosequencing of 16S rRNA gene fragments amplified from total community DNA. Samples with high concentration of SDZ were analyzed on day 193 only. Repeated inputs of SDZ, especially at a high concentration, caused pronounced changes in bacterial community compositions. By comparison with the initial soil, we could observe an increase of the disturbance and a decrease of the stability of soil bacterial communities as a result of SDZ manure application compared to the manure treatment without SDZ. The number of taxa significantly affected by the presence of SDZ increased with the times of manure application and was highest during the treatment with high SDZ-concentration. Numerous taxa, known to harbor also human pathogens, such as *Devosia*, *Shinella*, *Stenotrophomonas*, *Clostridium*, *Peptostreptococcus*, *Leifsonia*, *Gemmatimonas*, were enriched in the soil when SDZ was present while the abundance of bacteria which typically contribute to high soil quality belonging to the genera *Pseudomonas* and *Lysobacter*, *Hydrogenophaga*, and *Adhaeribacter* decreased in response to the repeated application of manure and SDZ.

Editor: Gabriele Berg, Graz University of Technology (TU Graz), Austria

Funding: This work was funded by the Deutsche Forschungsgemeinschaft (DFG) in the framework of the Research Unit FOR 566 "Veterinary medicines in soil: Basic research for risk analysis" (SM59/5-3). The funders had no role in study design, data collection and analysis, decision to publish, or preparation of the manuscript.

Competing Interests: The authors have declared that no competing interests exist.

* E-mail: kornelia.smalla@jki.bund.de

Introduction

The use of animal manure for fertilization of agricultural soils has a long tradition in many parts of the world and is generally assumed to be ecologically more friendly and sustainable than mineral fertilizer. In particular organic and bio-dynamic farms depend on manure or compost as sources for fertilization. However, industrial husbandries with typically large numbers of animals sharing limited space depend on the prophylactic and therapeutic use of antibiotics. In addition, in several parts of the world antibiotics are still being used as growth promoters. Although antibiotics as growth promoters were banned in many European countries, considerable amounts of antibiotics such as tetracyclines, β-lactams and sulfonamides are still used in animal husbandries [1]. Depending on their physicochemical properties, many antibiotics such as sulfonamides are to a large extent excreted via urine or feces and are not or only to a low extent degraded during manure storage [2,3]. Thus, spreading manure on agricultural soils does not only introduce nutrients required for maintaining the soil fertility but also antibiotics, their metabolites and antibiotic resistant bacteria. Indeed, antibiotics have been detected in the environment due to the use of manure for soil fertilization or direct deposition via dung and urine of animals grazing on pastures [4,5].

The rapid sequestration of most antibiotics in soils, as e.g. observed for the sulfonamide antibiotic SDZ [6], leads usually to low concentrations of bioavailable SDZ, which are far below the minimal inhibitory concentrations after a single application of manure. Most data published indicated only short-term effects on the microbial community after a single application of manure spiked with antibiotics followed by a fast regeneration of the community structure [7,8] and its function [9–11]. However, agricultural management implies in most cases a repeated application of manure mainly during the vegetation period to keep the level of nutrients needed for best possible plant growth. Surprisingly, little is known of the effects of repeated application of manure containing antibiotics on microbial communities in soil which might cause cumulative effects. Despite a rapid dissipation and sequestration of SDZ in soil [3,6,12] an increased abundance of *sul1* and *sul2* resistance genes and their transferability was observed in soils treated with manure containing SDZ compared to control manure under microcosm, mesocosm and field conditions [13–15], suggesting that manure containing antibiotics enhanced the spreading of the antibiotic resistance genes in soils. In particular the repeated application of manure in combination with antibiotics may set the ground for an increased abundance of resistant bacteria as recently reported by Heuer *et al.* [16] and thus might stimulate the spreading of antibiotic resistance genes and mobile genetic elements in agricultural ecosystems. Antibiotic resistance genes localized on mobile genetic elements can be captured by human and veterinary pathogens and thus pose a threat to the treatment of bacterial diseases [17–19].

In the present microcosm experiment, the effect of repeated application of manure containing antibiotics on the soil bacterial community was investigated by 454 sequencing of 16S rRNA gene fragments amplified from total community (TC)-DNA. We have chosen the sulfonamide SDZ as a model compound as it is still frequently used in pig husbandry [4] and highly persistent in manure [20]. Manure unspiked and spiked with SDZ in two concentrations was applied three times, and the bacterial community composition was monitored over a period of 193 days. We hypothesized that repeated application of manure spiked with SDZ to soil would increase the degree of disturbance and reduce the resilience of soil bacterial communities compared to a single application of manure spiked with SDZ or soils treated with manure not containing antibiotics. Resilience was measured as the stability of bacterial diversity after changes in soil properties ([21]; here manure addition). To differentiate between short and long-term effects, the soil was sampled 3 and 60 days after each manure application, respectively.

Materials and Methods

Experimental design

The experimental design is described in detail by Heuer *et al.* [16]. In short, topsoil samples (Ap horizon) of a silt loam soil (Orthic Luvisol) with no history of manure application were used. Pots were filled with 100 g of 2 mm-sieved soils and treated with 4 g of manure from healthy pigs (approximately 30 m^3/ha) spiked with SDZ or not to achieve a SDZ concentration in soil of 10 mg/kg, 100 mg/kg, or no SDZ (Sigma-Aldrich, Germany), which corresponded to samples S10, S100, and S0, respectively. Soil samples without manure application served as controls (U). To avoid effects based on different manure composition at the different time points of application, manure was taken before the experiment and frozen in aliquots at −20°C. For each time point of application, an aliquot of the manure was thawn and the corresponding amount of SDZ was directly applied before mixing the manure into the soil. The SDZ concentration S10 was chosen in accordance to maximal inputs of sulfonamides detected in agricultural soils [3,22]. Manure loads were applied on days 0, 63 and 133. The loosely covered pots were incubated at 15°C in the dark. Water was added to the microcosms twice a week to compensate for weight losses and to maintain a soil moisture of about 55% of the soil's water-holding capacity. For each treatment and time point 5 replicates were prepared and treated individually for the subsequent analysis. Samples were taken every 3rd and 60th day after manure application (days 3, 60, 66, 123, 136 and 193) and kept at −80°C until further analysis. No specific permissions were required for collecting the manure and soil and the study did not involve endangered or protected species.

DNA extraction

Extraction of total community DNA from soil samples was performed according to Griffith *et al.* [23]. Briefly, 0.4 g soil per replicate were added to a lysing matrix tube (MP Biomedicals, Germany) and submitted to phenol:chloroform extraction starting with homogenization for 30 seconds at 5.5 m S-1 (Precellys, PeqLab, Germany). Extracted DNA was finally resolved in 50 µl DNase-free water. Quality and quantity of DNA extracts were checked using a spectrophotometer (Nanodrop, PeqLab, Germany). For the extraction of total community DNA from manure, 10 ml manure were centrifuged at about 6500 *g* for 10 min, the pellet was homogenized and DNA was extracted from 0.5 g manure pellet using the FastDNA SPIN kit for soil (MP Biomedicals, Heidelberg, Germany), followed by a purification step using the Geneclean spin kit (MP Biomedicals, Heidelberg, Germany), according to the manufacturer's instructions.

Preparation of the 16S rRNA gene amplicons, pyrosequencing and data processing

To generate the amplicon library for *Bacteria*, 16S primers 926-F (5′-AAACTYAAAKGAATTGACGG-3′), *Escherichia coli* position 907–926 and 630-R (5′-CAKAAAGGAGGTGATCC-3′), *E. coli* position 1528–1544 [24], were selected. Fusion primers were designed according to the guidelines of ROCHE (www.my454.com), by extending the specific primers with a 10 base multiplex identifier (MID), a 4 base key and the respective sequencing primers A or B for bidirectional sequencing. The optimal PCR conditions were determined by gradient PCRs with annealing temperatures ranging from 50°C to 60°C. Amplicons were finally generated using 50 ng of extracted DNA, an annealing temperature of 50°C and 22 PCR cycles. The PCR products were purified with AMPure Beads (Agencourt, Beckman Coulter, Krefeld, Germany) according to the Amplicon Library Preparation Method Manual (www.my454.com) and pooled in equimolar amounts.

Sequencing of the 16S rRNA genes was performed on a second-generation pyrosequencer (454 GS FLX Titanium, ROCHE, Germany) following the manufacturer's protocol for amplicon sequencing (www.my454.com). The number of replicates per treatment included in the final sequencing run can be found in Table S1 in File S1.

The automatic amplicon pipeline of the GS Run Processor (ROCHE) was used to perform an initial quality filtering of the pyrosequencing raw reads in order to remove failed reads, low quality reads and adaptor sequences. Sequence files were submitted to the NCBI Sequence Read Archive (www.ncbi.nlm.nih.gov/sra/) and are available with the study accession number SRP038712.

Sequence analysis

The two data sets acquired by the forward and reverse primers were analyzed separately to avoid the systematical bias of individual primers. The analyses were mainly performed according to Ding *et al.* [25]. Only those sequences with a length above 200 bp after removing the barcode, primer and unpaired regions were subjected to further analysis. The unpaired regions were truncated based on a standalone BlastN analysis against a bacterial database described by Pruesse *et al.* [26]. Sequences were grouped into operational taxonomic units (OTUs>97% sequence identity) using software package mothur (v1.14.0). The classification of sequences was performed using software RDP MultiClassifier at >80% confidence [27]. A taxonomic OTU report with each row representing one OTU containing taxonomic positions (from phylum to genus) and the number of sequences for each sample was constructed based on the OTU assignment and on the classification.

Analyses based on the taxonomic OTU report were done with R (version 2.14). To allow the comparison of samples with different number of reads, all algorithms selected rely on relative

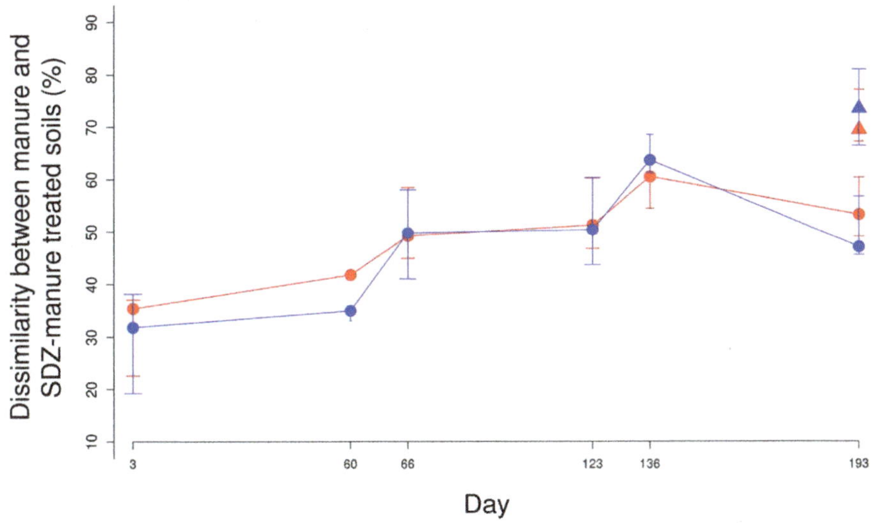

Figure 1. Dissimilarity between soils treated with manure (S0) and manure spiked with SDZ (S10 indicated by circles; S100 indicated by triangles) at different sampling times. Red symbols: results based on the data set acquired by the forward primer; blue symbols: results based on the data set acquired by the reverse primer. Error bars indicate the first and third quartiles.

abundance of taxonomic groups. To check whether the forward and reverse primers revealed a similar taxonomic composition or not, the percent of Bray-Curtis similarity of taxonomic composition between two data sets was calculated for each sample based on the percent of classified taxa at phylum, class, order, family, and genus level using the R add-on package "vegan". To evaluate the influence of repeated application of manure containing SDZ on the bacterial community compositions, the pairwise Pearson dissimilarities between S0 and S10 or between S0 and S100 for day 193 were calculated. To compare the microbial community composition between samples, non-metric multidimensional scaling (NMDS) analysis was carried out based on the Pearson dissimilarity matrix using R package MASS. The 3-dimensional plots of NMDS analysis were created with the R package "rgl" in conjunction with the Linux command "import". Tukey's honest significance tests under a generalized linear model via a logistic function for binomial data with the package multcomp [28] was applied to identify the discriminative taxa between treatments using a Bonferroni adjusted p value <0.05.

Results

Effects of the repeated application of manure containing SDZ on the composition of soil bacterial communities

In this study, bacterial 16S rRNA gene fragments amplified from TC-DNA extracted from 50 composite soil samples were sequenced using forward (926F) and reverse (630R) primers. A total of 132,278 sequences for 50 samples with 790–5685 high quality sequences per sample were obtained with the forward primer, and 162,446 sequences for 49 samples with 1105–7053 high quality sequences per sample with the reverse primer.

To study the effect of repeated applications of manure containing SDZ on the soil bacterial community composition, 16S rRNA gene pyrosequencing data sets acquired from TC-DNA of soils treated with SDZ manure or control manure were compared. Sequences obtained with the forward and reverse primers were treated independently and grouped into OTUs (>97% sequence identity). Both datasets revealed similar changes of bacterial community composition after manure application (Figure 1). In general, the Pearson dissimilarities of the bacterial community compositions between the S0 and S10 soils increased with repeated manure application (Figure 1). Shortly after each manure application the dissimilarity between the bacterial community composition of S0 and S10 soils was strongly increased. Sixty days after the first and second manure applications the dissimilarity remained at the same level as shortly after manure application, while 60 days after the third application a decrease in dissimilarity between S0 and S10 soil was observed. Compared with the S10 soils on day 193, the S100 soils were less similar to S0 soils in the bacterial community composition, indicating that a higher concentration of SDZ caused more pronounced changes of the soil bacterial communities (Figure 1). Non-metric multidimensional scaling based on the OTU reports acquired for the forward (Figure S1 in File S1) and reverse (Figure 2) data set confirmed the findings presented in Figure 1 as it showed that the bacterial community composition of S10 soils increasingly deviated from the S0 soils with time, and this difference was even more pronounced for S100 soils. Furthermore, the faint color symbols in Figure 2 representing the bacterial community composition shortly after manure application strongly deviated from the community composition of untreated soil suggesting a pronounced disturbance of the system. In contrast to S10 and S100 soils, the bacterial community composition of S0 soils (indicated by cubes) 60 days after the third manure application still grouped in the vicinity of the untreated soils (Figures 2 and S1) suggesting a stronger resilience of the soil bacterial community towards manure when no antibiotic was present.

The difference between bacterial communities of the initial soil (day 3 untreated) and of the treated soil was calculated based on Pearson dissimilarity (Figure 3). A pronounced increase in difference was observed for soils shortly after each manure application (S0, S10). The dissimilarity increased with the repeated applications of manure containing SDZ for samples collected 60 days after each manure application. The highest deviation was observed for soil treated three times with manure containing the high concentration (100 mg/kg) of SDZ. At the same sampling time, the least difference was found for the untreated control soils.

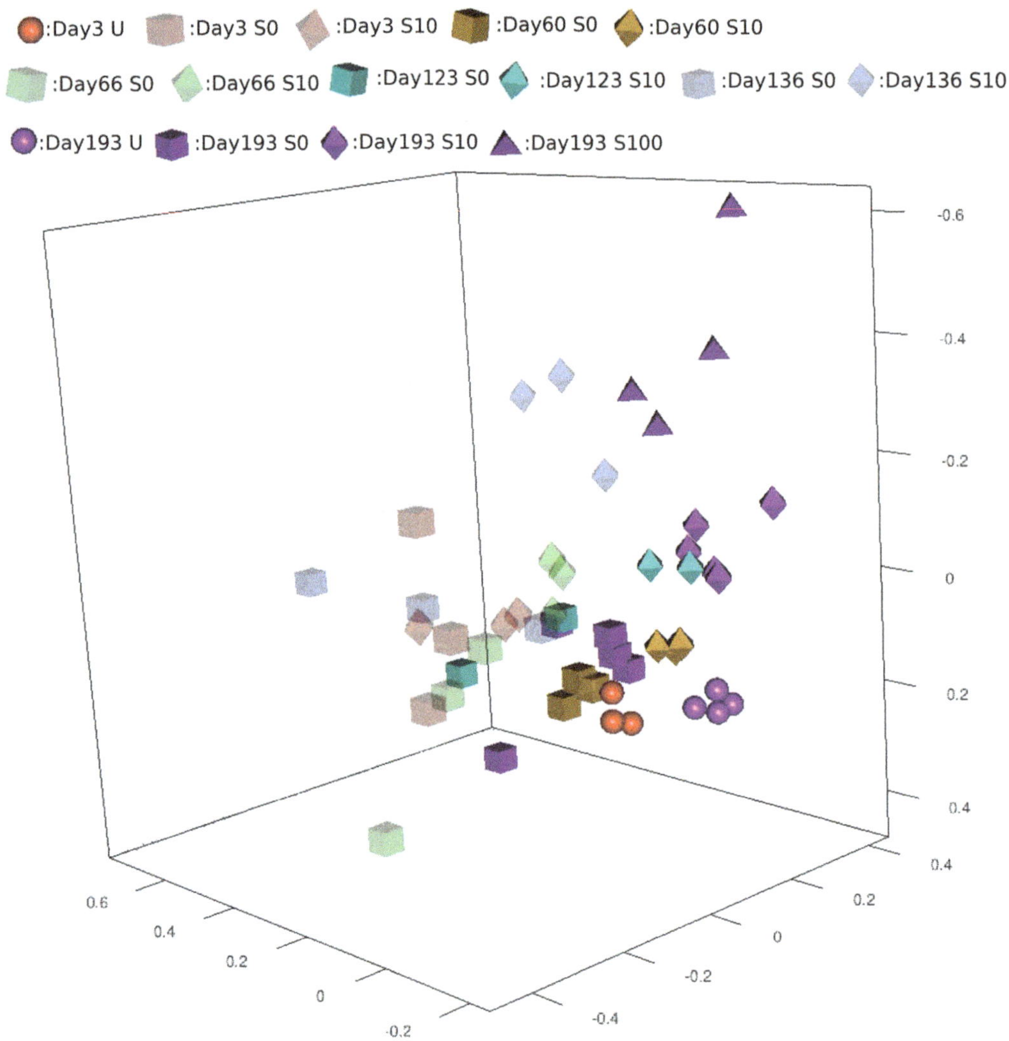

Figure 2. Non metric multidimensional scaling based on OTU reports acquired by the reverse primer sequencing. Balls: untreated soils; cubes: S0; octahedron: S10; pyramid: S100.

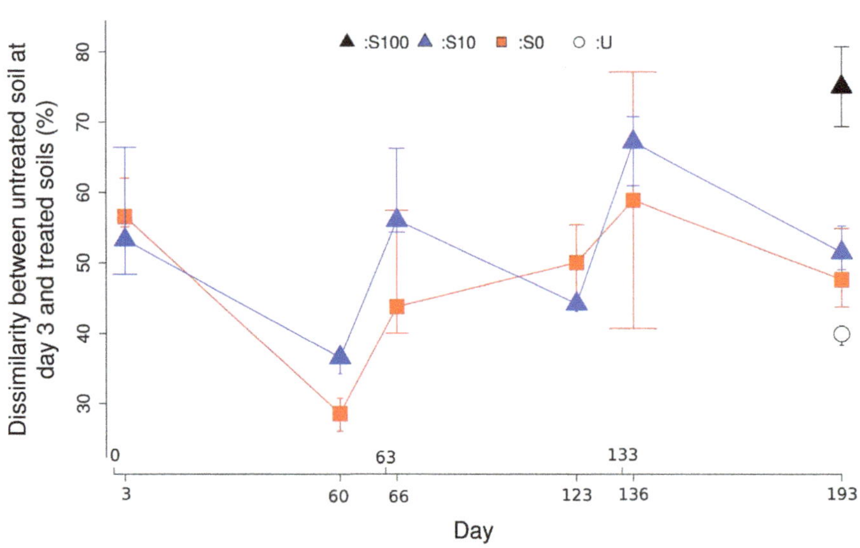

Figure 3. Dissimilarity between soils treated with manure (S0, S10 and S100) or not (U) and the untreated initial soils (Day3 U) at different sampling times. Dissimilarity was calculated from the data acquired by the reverse primer. Error bars indicate the first and third quartiles.

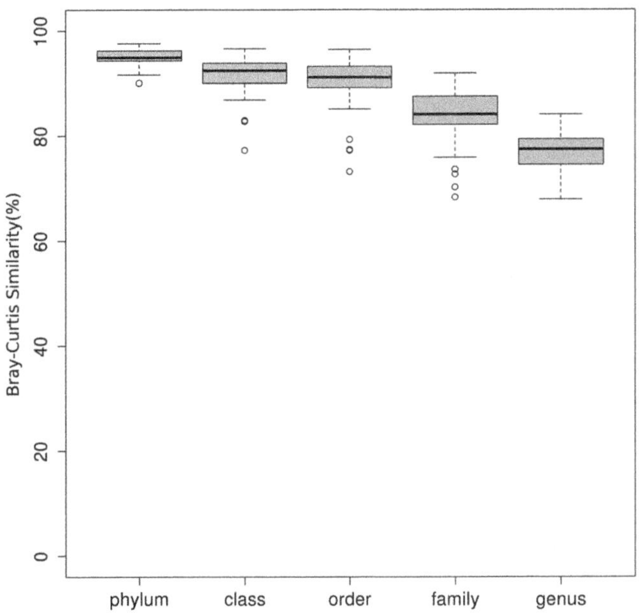

Figure 4. Box plot showing the similarity of taxonomic composition at different ranks between data acquired by the forward and reverse primers.

Taxa with significantly altered relative abundance in response to repeated application of manure spiked with SDZ or unspiked

About 86% of the sequences obtained with the forward primer could be assigned to 21 phyla while 90% of the reverse primer derived sequences were affiliated to 19 phyla. The percent Bray-Curtis similarities of the taxonomic composition between the data sets acquired with the forward and reverse primers, respectively, were calculated for each sample based on the percent of the classified taxa at different taxonomic ranks (from phylum to genus level). Although from the phylum to the genus level the similarity of the taxonomic composition decreased as shown in Figure 4, the overall taxonomic composition of both data sets was very similar.

To identify the taxa which significantly (Bonferroni adjusted p value<0.05) responded to the repeated application of manure, comparisons were made between soils treated with S0 or not (U). All taxa with significantly different relative abundance (Bonferroni adjusted p value<0.05) were considered. In soil samples taken three days after the addition of manure short-term responders were identified (summarized in Table S2 for the reverse primer derived sequence set and in Table S3 in File S1 for the sequence set obtained with the forward primer). Based on these data (Table S2 and S3) we could mainly identify genera belonging to the *Proteobacteria* such as *Delftia* (*Betaproteobacteria*), *Pseudomonas*, *Pseudoxanthomonas*, *Stenotrophomonas* (*Gammaproteobacteria*), *Bacillus*, *Lactobacillus*, *Streptocococcus*, *Clostridiales* (*Firmicutes*) and *Arthrobacter* (*Actinobacteria*) which were significantly increased in relative abundance in S0 soils compared to the untreated soil as short time responders to the manure application. The increase in the relative abundance of *Firmicutes* (Table S4 in File S1) is likely related to the high abundance of these organisms in the manure (data not shown). In turn the relative abundance of different genera of the *Acidobacteria* and *Planctomycetaceae* but also *Gemmatimonadetes*, the genus *Methylibium* (*Betaproteobacteria*) decreased in S0 soils compared to the untreated soils (Table S2 and S3 in File S1). An increased relative abundance of the genera *Devosia*, *Clostridium*, *Peptostreptococcus* and *Adheribacter* in response to the manure application was observed by comparing U and S0 soil on day 193 (Table S5 and S6 in File S1). In contrast, *Myxococcales* seemed to be negatively affected by manure. On day 193 the relative abundance of *Acidobacteria* was still significantly decreased in all manure treatments compared to untreated soil with the exception of GP10 which showed a higher relative abundance in S10. Some taxa, such as the family *Planctomycetaceae*, were not affected by manure application.

With repeated application of manure the number of taxa with significantly altered relative abundance in S10 compared to S0 soil increased (summarized in Table 1 for the reverse primer and in Table S7 in File S1 for the forward primer). An increased abundance in S10 soils was observed in particular after repeated manure application for the genera *Devosia*, *Sphingobium*, Gp10 and *Gemmatimonas*. Shortly after manure application a significantly increased abundance was observed in S10 compared to the S0 soils for the phylum *Chloroflexi*, the class of the *Deltaproteobacteria* and the orders of *Clostridales* and *Actinomycetales*. Acidobacterial genera Gp4, Gp6 and Gp10 with one exception showed always higher relative abundances in S10 but only in some cases these differences were found to be significant (Table 1). The relative abundances of *Pseudomonas*, *Lysobacter* and *Adheribacter* were lower in all S10 compared to S0 soils indicating that the presence of SDZ exerted a negative effect on these taxa.

Samples taken 60 days after the first, second and third manure applications mainly revealed the same taxa with increased or decreased relative abundance in S10 compared to S0 as in the samples taken shortly after manure application (shown in Table 2 for the reverse primer and in Table S8 in File S1 for the forward primer). With repeated application of manure (S0, S10) the number of taxa with significantly different relative abundance increased from 3 (60 days after the first manure application) to 11 (60 days after the third manure application). Although not significant at all time points the genera *Devosia*, *Gemmatimonas* and *Acidobacteria* Gp6 and Gp10 showed an increased relative abundance in S10 compared to S0 soils. In S100 soils analyzed only for the samples taken 60 days after the third manure application (day 193) the relative abundances of *Devosia* and *Gemmatimonas* were increased compared to their abundance in S10 and S0 soils clearly indicating that bacteria belonging to these genera take profit from high concentrations of SDZ. A higher relative abundance in S100 soils was also observed for *Anerolineaceae* (*Chloroflexi*), *Microbacteriaceae* and the genus *Clostridium*. Most striking was the selection of *Stenotrophomonas* in S100 soils as sequences of this genus were not detected in S0 and in very low relative abundance in S10 soils on day 193. Also acidobacterial genera were identified as responders to SDZ. However, the different genera of *Acidobacteria* responded differently to SDZ manure. The S10 soils had a higher abundance of *Acidobacteria* belonging to the genera Gp10 and Gp6 (except day 60) compared to S0. On day 193, the relative abundance of Gp4, Gp6 and Gp10 was lower in S100 than in S10 soil. The negative effect of SDZ on *Pseudomonas*, *Lysobacter* and *Adheribacter* was also detected in S100 but there was no significant difference in their relative abundance between S10 and S100.

Discussion

The importance of the soil microbiome for various ecosystem services such as nutrient cycling, soil fertility, degradation of pollutants, and plant growth promotion is well recognized [29]. However, antibiotics introduced into agricultural fields via manure might alter the ability of soil microbes to fulfill crucial ecosystem services, changing the diversity and activity of key functional

Table 1. Taxa with significantly different relative abundance between manured soils with or without SDZ treatments shortly after manure addition as revealed by barcoded pyrosequencing using the reverse primer (Bonferroni adjusted p value<0.05).

Phylum	Class	Order	Family	Genus	Day 3		Day 66		Day136	
					S0	**S10**	**S0**	**S10**	**S0**	**S10**
Proteobacteria	*Alphaproteobacteria*	*Rhizobiales*	*Hyphomicrobiaceae*	*Devosia*	0.2±0	0.3±0	0.5±0	**1.5±0**	1.1±0	**2.4±0**
		Sphingomonadales	*Sphingomonadaceae*	*Sphingobium*	0.1±0	0.1±0	0.1±0	0.3±0	0.1±0	**0.4±0**
	Betaproteobacteria	*Burkholderiales*	*Comamonadaceae*		**3.8±3**	1.6±0	3.3±1	2.1±0	2.6±1	3±1
			Oxalobacteraceae	*Herminiimonas*	0.1±0	0.1±0	0.5±0	0±0	**3±4**	0±0
	Deltaproteobacteria				1.1±0	1.2±0	1.4±0	1.3±0	3.1±1	**4.4±2**
	Gammaproteobacteria	*Pseudomonadales*	*Pseudomonadaceae*	*Pseudomonas*	**0.8±1**	0.2±0	**2.6±4**	0.2±0	0.6±1	0±0
		Xanthomonadales	*Xanthomonadaceae*	*Lysobacter*	**2.1±1**	1.2±0	**4.1±1**	0.8±1	**1.7±1**	0.2±0
				Stenotrophomonas	**1.6±2**	0.1±0	0±0	0±0	0.1±0	0.3±0
Acidobacteria	*Acidobacteria_Gp10*			*Gp10*	1±0	1.5±0	1.7±0	**3±1**	1±0	**2.1±0**
	Acidobacteria_Gp4			*Gp4*	3.3±0	**4.7±1**	4.8±1	5.3±0	4.2±2	4.1±2
	Acidobacteria_Gp6			*Gp6*	8.7±1	10.5±2	12.1±2	**16.1±1**	8.7±4	9.4±2
Firmicutes	*Bacilli*	*Lactobacillales*			5.4±1	5.3±2	3.9±1	**6.3±2**	2.8±1	3.8±1
	Clostridia	*Clostridiales*			5.9±2	6.4±1	6±1	8.2±1	7.5±1	**10.5±1**
Actinobacteria	*Actinobacteria*	*Actinomycetales*			7.1±1	5.8±2	2.9±1	3.3±1	4.7±5	**5.9±6**
Bacteroidetes	*Sphingobacteria*	*Sphingobacteriales*	*Cytophagaceae*	*Adhaeribacter*	**1.5±0**	0.7±0	**2.3±1**	0.6±0	**1.5±0**	0.4±0
Chloroflexi					3.3±1	3.6±0	1.9±1	3.8±1	4.5±0	**6.2±1**
Gemmatimonadetes	*Gemmatimonadetes*	*Gemmatimonadales*	*Gemmatimonadaceae*	*Gemmatimonas*	0.2±0	0.3±0	0.3±0	0.5±0	0.4±0	**1±1**

Bold numbers indicate taxa with significantly higher relative abundance in SDZ treated or non treated soils. Soils were collected three days after each treatment.

Table 2. Taxa with significantly different relative abundance between manured soils with or without SDZ treatments as revealed by barcoded pyrosequencing using the reverse primer (Bonferroni adjusted p value<0.05).

Phylum	Class	Order	Family	Genus	Day 60		Day 123		Day 193		
					S0	S10	S0	S10	S0	S10	S100
Proteobacteria	*Alphaproteobacteria*	*Rhizobiales*	*Hyphomicrobiaceae*	*Devosia*	0.4±0	**1.5±0**	1.8±1	2.2±0	1.5±0b	2.6±1a	3.5±1a
			Rhizobiaceae		0.4±0	0.3±0	0.3±0	0.5±0	0.6±0a	0.4±0ab	0.4±0b
		Rhodospirillales	*Rhodospirillaceae*	*Magnetospirillum*	0.4±0	0.4±0	0.3±0	0.6±0	0.8±0ab	0.9±0a	0.5±0b
		Sphingomonadales	*Sphingomonadaceae*		4±1	3.4±1	4±1	3.4±1	5.4±1a	3.6±0b	3.4±1b
	Betaproteobacteria	*Burkholderiales*	*Alcaligenaceae*		0.1±0	0.1±0	0.1±0	0.1±0	0.1±0b	0.1±0b	0.8±0a
			Burkholderiales_incertae_sedis	*Methylibium*	1.8±0	1.3±0	1.2±0	1.3±0	2±1a	1.8±1a	0.9±0b
			Comamonadaceae	*Hydrogenophaga*	0.1±0	0.1±0	0.6±0	0.2±0	0.5±0a	0.1±0b	0±0b
			Oxalobacteraceae	*Herminiimonas*	0±0	0±0	**1.2±0**	0±0	0.5±0a	0±0ab	0±0b
		Methylophilales	*Methylophilaceae*		0.1±0	0.1±0	0.3±0	0.1±0	0.3±0a	0.1±0ab	0±0b
		Rhodocyclales	*Rhodocyclaceae*	*Shinella*	0±0	0±0	0±0	0.1±0	0±0b	0.2±0b	1.2±1a
	Gammaproteobacteria	*Pseudomonadales*	*Pseudomonadaceae*	*Pseudomonas*	0.5±1	0.1±0	0.1±0	0±0	1.2±2a	0.2±0b	0±0b
		Xanthomonadales	*Xanthomonadaceae*	*Lysobacter*	**1.6±0**	0.7±0	**1.7±1**	0.6±0	0.9±0a	0.2±0b	0.1±0b
				Pseudoxanthomonas	0.7±1	0.4±0	**1.9±1**	0.4±0	0.7±0b	0.4±0b	2±2a
				Rhodanobacter	0±0	0±0	0.3±0	0.3±0	0.3±0b	0.9±1a	0.2±0b
				Stenotrophomonas	0±0	0±0	0.1±0	0.3±0	0±0b	0.2±0b	3.5±3a
	Deltaproteobacteria	*Myxococcales*	*Haliangiaceae*	*Haliangium*	0.5±0	0.1±0	**0.8±0**	0.1±0	0.8±0a	0.1±0b	0.1±0b
			Nannocystaceae		0.3±0	0±0	0.2±0	0±0	0.3±0a	0.2±0ab	0±0b
			Polyangiaceae		0.2±0	0.3±0	0.2±0	0.3±0	0.3±0b	0.2±0b	0.8±1a
Acidobacteria	*Acidobacteria_Gp10*			*Gp10*	2.8±0	4.3±1	2±0	**5.4±1**	1.6±0c	4.4±0a	2.5±1b
	Acidobacteria_Gp4			*Gp4*	7.9±1	8.6±1	4.4±0	5±1	7.5±2a	5.8±2a	4.5±1b
	Acidobacteria_Gp6			*Gp6*	18.2±2	**22.1±3**	14.2±2	**17.8±0**	13.6±3b	15.8±3a	11.1±2c
Firmicutes	*Bacilli*	*Lactobacillales*			0.1±0	0.1±0	0±0	0.4±0	0.1±0b	0.2±0ab	0.5±0a
	Clostridia	*Clostridiales*	*Clostridiaceae*	*Clostridium*	0.5±0	0.4±0	1.2±1	1.1±0	1.4±0b	1.7±0b	3.7±1a
	Clostridia		*Peptostreptococcaceae*	*Peptostreptococcus*	0.1±0	0.1±0	0.1±0	0.5±0	0.3±0b	0.4±0ab	0.8±0a
Bacteroidetes	*Sphingobacteria*	*Sphingobacteriales*	*Cytophagaceae*	*Adhaeribacter*	1.7±0	0.9±0	1.5±0	0.7±0	2.1±1a	0.6±0b	0.4±0b
Actinobacteria	*Actinobacteria*	*Actinomycetales*	*Microbacteriaceae*	*Leifsonia*	0±0	0.1±0	0±0	0.1±0	0±0b	0.2±0b	0.4±0a
Chloroflexi	*Anaerolineae*	*Anaerolineales*	*Anaerolineaceae*		1±0	1.1±0	1.9±0	1.9±0	1.5±0b	1.8±1ab	2.2±1a
Gemmatimonadetes	*Gemmatimonadetes*	*Gemmatimonadales*	*Gemmatimonadaceae*	*Gemmatimonas*	1.1±1	1.6±0	0.5±0	**1.7±2**	0.7±0c	1.6±1b	2.5±1a
Planctomycetes	*Planctomycetacia*	*Planctomycetales*	*Planctomycetaceae*		2.7±0	2.6±1	2.5±0	2.4±0	2.6±1a	2.5±0a	1.7±0b
Verrucomicrobia	*Subdivision3*			*Subdivision3_genera_incertae_sedis*	0.4±0	0.3±0	0.6±0	0.5±0	0.4±0a	0.5±0a	0.1±0b

Bold numbers indicate taxa with significantly higher relative abundance in SDZ treated or non treated soils; letters a, b and c indicate taxa with significant difference in relative abundance between S0, S10 and S100 treated soils collected at day 193. Soils were collected 60 days after each manure application.

groups by enhancing antibiotic resistant populations while decreasing the abundance of sensitive populations in soils [16,30]. In the present study, the effects of repeated application of SDZ manure on soil bacterial communities were explored by barcoded pyrosequencing of 16S rRNA gene amplicons from TC-DNA. More pronounced changes of bacterial communities were observed for soils that were repeatedly treated with SDZ manure. The strongest effects were observed for soils treated three times with manure containing a high concentration of SDZ (Figures 2 and 4). For the first time, taxonomic groups affected by the presence of SDZ were systematically identified, and numerous taxa were found with significantly altered relative abundance (Tables 1 and 2). These findings largely extended our understanding of the influence of SDZ introduced via manure on soil bacterial communities.

Although the bioavailable fraction of SDZ in soil rapidly decreased after each manure application [16], cumulative inputs of SDZ via manure still affected bacterial communities. Also in the study by Byrne-Bailey *et al.* [31], the proportion of sulfachloropyridazine resistant bacteria increased in soils after treatment with manure containing antibiotics. The repeated application of SDZ manure to grassland soil was found to affect nitrite oxidizing bacteria [32]. The influence of SDZ on soil bacterial communities is more likely to be detected when additional nutrients (root exudates, manure) are added to stimulate bacterial growth [33,34] as SDZ affects only growing bacteria and bacterial activities are generally low in oligotrophic environments such as soils.

The amendment of soil with SDZ manure resulted in a significantly increased relative abundance of numerous Gram-negative and Gram-positive taxa such as *Devosia*, *Shinella*, *Stenotrophomonas*, *Clostridium*, *Peptostreptococcus*, *Leifsonia*, *Gemmatimonas*, suggesting that the presence of SDZ provided a selective advantage for these taxa. Mainly the increase in bacteria affiliated to *Clostridia* detected shortly after the addition of manure likely resulted from manure-derived bacteria. In our study more than 60% of sequences in manure could be affiliated to *Clostridium* (data not shown), which confirms data from previous studies where the dominance of *Clostridium* in manure has been reported [35–37]. A steep decrease of *Clostridia* was observed 60 days after the application. Only in S100 soils the relative abundance of *Clostridia* was found to be significantly higher than in the other soils on day 193, suggesting that the high concentration of SDZ slowed down the decline of *Clostridium* populations. Whether the enhanced abundance of *Clostridium* in S100 soils was due to the selection of SDZ resistant populations deserves further research.

Although the resistance to SDZ was not reported yet for most of these taxa, it is likely that some of the populations enriched in the present study might be resistant to SDZ either due to the presence of resistance genes or due to intrinsic resistance. Isolates belonging to *Stenotrophomonas* and *Clostridium* were previously reported to be resistant to other sulfonamide antibiotics [38,39]. Based on the same TC-DNA, Heuer *et al.* [16] found an accumulation of *sul* gene carrying populations in the soil repeatedly treated with SDZ manure. *Sul* genes located on mobile genetic elements such as class 1 integrons or plasmids might have spread among different taxa [40,41]. The presence of antibiotics, even at very low concentrations could accelerate the genetic exchanges between bacteria [42,43]. The enhanced transferability of resistant genes was recently also reported by Jechalke *et al.* [15] in a field experiment using manure from SDZ-treated pigs. However, we cannot exclude that some of the taxa were intrinsically resistant to, or even able to mineralize SDZ as a SDZ degrader identified as *Microbacterium lacus* was recently isolated from the same Merzenhausen soil [44]. Interestingly, in the present study an OTU affiliated to *Leifsonia*, which is phylogenetically closely related to *Microbacterium* was found to be enriched in S100 soils.

In this study, the consensus sequences of an enriched OTU in soils treated with manure shared high similarity with the 16S rRNA gene of *S. maltophilia* which is also known as a human pathogen [38]. However, these enriched bacteria are not necessarily human pathogens as *Stenotrophomonas* was also found in different environments such as rhizosphere or soils [45,46]. Nevertheless, the enrichments of *Shinella*, *Stenotrophomonas*, *Clostridium* and *Peptostreptococcus* in soil which was continually treated with manure containing a high concentration of SDZ still urge for further investigation due to potential implications for public health. The enrichment of these bacteria, which are phylogenetically closely related to human pathogens, may improve the chance of transferring antibiotic resistance genes to human pathogens, since horizontal gene transfer is more prevalent between closely related organisms than between those distantly related (reviewed by Boto [47]). Soil particles carrying viable bacteria can be transported over long distances and might contribute to the spreading of antibiotic resistant bacteria over wide geographic ranges [48]. The ecological role of taxa such as *Devosia*, *Leifsonia* and *Gemmatimonas* has not been fully understood and thus the effect of their changed abundance on soil functions remains to be explored. Few studies suggested that some members of *Devosia* [49,50] and *Gemmatimonas* [51] might be associated with nitrogen cycling.

The relative abundances of several other taxa such as *Pseudomonas*, *Lysobacter*, *Hydrogenophaga*, *Haliangium*, and *Adhaeribacter* were found to be significantly lower in the soils treated by SDZ manure. Several members of *Lysobacter* or *Pseudomonas* are known as biological control agents against soilborne phytopathogens such as *Rhizoctonia solani*, *Thielaviopsis basicola* [52,53]. Several studies suggested that these beneficial bacteria belonging to *Pseudomonas* or *Lysobacter* probably play an important role for soil suppressiveness and plant growth [45,54]. In the present study, the consensus sequences of an OTU with decreased relative abundance in soil treated with SDZ manure shared 100% similarities with 16S rRNA gene of the strain of *P. brassicacearum* (NCBI accession number: NR_074834) which is a beneficial root-associated bacterium [55]. The decline of *Pseudomonas* and *Lysobacter* in arable soils might increase the susceptibility towards fungal pathogens. However, in general the application of manure to soils is one strategy to suppress soilborne diseases. Our data showed that the relative abundance of *Pseudomonas* was much higher in soils treated with manure than in untreated soils on days 3 and 193. The presence of SDZ in manure likely reduces the beneficial effects of manure on improving soil health as the abundance of *Pseudomonas* was reduced. However, resilience of these beneficial populations might still occur in the rhizosphere as decreased concentrations of SDZ were recently observed in the rhizosphere of maize and grass compared to bulk soil [6]. Little is known about the ecological roles of *Hydrogenophage*, *Haliangium*, *Adhaeribacter* in soil. *Hydrogenophaga* spp. were reported as main biphenyl degraders in the rhizosphere of horseradish (*Armoracia rusticana*) contaminated with polychlorinated biphenyls [56].

The present study is the first to explore the influence of veterinary antibiotics entering soil via manure on the diversity and abundance of *Acidobacteria*. The results revealed that Gp4 was negatively affected by repeated application of manure containing the high concentration of SDZ. Low concentrations of SDZ in manure seemed to favor Gp6 and Gp10, while high concentrations of SDZ could deprive the effects. *Acidobacteria* are known to be adapted to an oligotrophic lifestyle. Indeed, in the present study the relative abundance of *Acidobacteria* was significantly higher in the untreated soil than in the soil amended with manure. Due to their high abundance and ubiquitous distribution in soils, *Acidobacteria* might play an important role in terrestrial ecosystems. In combination with the analysis of three genomes of a*cidobacterial* isolates, Ward *et al.* [57] suggested that *Acidobacteria* might significantly contribute to the terrestrial carbon cycle.

Along with nutrients, a large amount of manure bacteria was also introduced into soil which could have also contributed to the dramatically altered bacterial community compositions observed in the present study shortly after each manure application. However, the majority of these introduced manure bacteria might not be very well adapted to soil, allowing the resilience of the indigenous soil bacteria community. In the present study we showed that repeated application of manure spiked with SDZ to soil increased the degree of disturbance and reduced the resilience of soil bacterial communities compared to a single application of manure spiked with SDZ or soils treated with manure without antibiotics. In a recent study by Poulsen *et al.* [58], the taxonomic composition of bacterial communities based on the abundance of phyla and proteobacterial classes were highly similar between soils with amendments of different organic wastes such as manure. In the present study, we observed clear effects of manure on the soil bacterial community composition as the NMDS analysis revealed separated groups for S0 and untreated soils on day 193 (Figure 3). In contrast to several subgroups of the *Acidobacteria*, *Myxococcales* (*Deltaproteobacteria*), numerous taxa belonging to different phyla (*Proteobacteria*, *Firmicutes* and *Bacteriodetes*) were observed with significantly higher relative abundance in the soils amended with manure than in the untreated control 60 days after the third application. Very likely the added nutrients favored taxa with a more copiotrophic lifestyle. In contrast to the field experiment analyzed by Poulsen *et al.* [58], the conditions in the present experiment were more controlled and the time period after the last manure amendment might have been shorter although this was not specified in the Poulsen study.

In summary, repeated input of SDZ into the Merzenhausen soil via manure, especially at a high concentration, caused pronounced changes in soil bacterial communities. The presence of SDZ provided a selective advantage for species affiliated to the genera *Devosia, Shinella, Stenotrophomonas, Clostridium, Peptostreptococcus, Leifsonia, Gemmatimonas*, while suppressing *Pseudomonas, Lysobacter, Hydrogenophage, Haliangium, Adhaeribacter*. We could show that SDZ-containing manure caused a more pronounced disturbance of the soil bacterial community compared to control manure and reduced resilience at least for the time period studied. However, whether the findings obtained can be generalized to other soil types with different soil properties remains an open question which needs further attention. Furthermore, mainly the long-term consequences of repeated manure application for soil ecosystem functions and human health demand further investigations.

Supporting Information

File S1 Supporting Files. Table S1. The number of replicates per treatment included in the final sequencing run using the reverse primer. The numbers in brackets indicate the number of samples sequenced using the forward primer. Table S2. Taxa with significantly different relative abundance between soils on day 3 using the reverse primer (Bonferroni adjusted p value<0.05). Table S3. Taxa with significantly different relative abundance between soils on day 3 using the forward primer (Bonferroni adjusted p value<0.05). Table S4. Relative abundance of detected phyla based on the average of the forward and reverse dataset. Table S5. Taxa with significantly different relative abundance between soils on day 193 using the reverse primer (Bonferroni adjusted p value<0.05). Table S6. Taxa with significantly different relative abundance between soils on day 193 using the forward primer (Bonferroni adjusted p value<0.05). Table S7. Taxa with significantly different relative abundance between manured soils with or without SDZ treatments as revealed by barcoded pyrosequencing using the forward primer (Bonferroni adjusted p value<0.05). Table S8. Taxa with significantly different relative abundance between manured soils with or without SDZ treatments as revealed by barcoded pyrosequencing using the forward primer (Bonferroni adjusted p value<0.05). Figure S1. Non metric multidimensional scaling based on OTU reports acquired by the forward primer sequencing. Balls: untreated soils; cubes: S0; octahedron: S10; pyramid: S100.

Author Contributions

Conceived and designed the experiments: HH KS MS. Performed the experiments: VR BSH. Analyzed the data: GCD. Contributed reagents/materials/analysis tools: GCD VR KS MS. Wrote the paper: GCD BSH HH KS MS SJ.

References

1. Kools SAE, Moltmann JF, Knacker T (2008) Estimating the use of veterinary medicines in the European union. Regulatory Toxicology and Pharmacology 50: 59–65.
2. Lamshöft M, Sukul P, Zühlke S, Spiteller M (2007) Metabolism of C-14-labelled and non-labelled sulfadiazine after administration to pigs. Anal Bioanal Chem 388: 1733–1745.
3. Heuer H, Focks A, Lamshöft M, Smalla K, Matthies M, et al. (2008) Fate of sulfadiazine administered to pigs and its quantitative effect on the dynamics of bacterial resistance genes in manure and manured soil. Soil Biol Biochem 40: 1892–1900.
4. Sarmah AK, Meyer MT, Boxall ABA (2006) A global perspective on the use, sales, exposure pathways, occurrence, fate and effects of veterinary antibiotics (VAs) in the environment. Chemosphere 65: 725–759.
5. Sittig S, Kasteel R, Groeneweg J, Vereecken H (2012) Long-term sorption and sequestration dynamics of the antibiotic sulfadiazine: a batch study. J Environ Qual 41: 1497–1506.
6. Rosendahl I, Siemens J, Groeneweg J, Linzbach E, Laabs V, et al. (2011) Dissipation and sequestration of the veterinary antibiotic sulfadiazine and its metabolites under field conditions. Environ Sci Technol 45: 5216–5222.
7. Hammesfahr U, Heuer H, Manzke B, Smalla K, Thiele-Bruhn S (2008) Impact of the antibiotic sulfadiazine and pig manure on the microbial community structure in agricultural soils. Soil Biol Biochem 40: 1583–1591.
8. Hammesfahr U, Bierl R, Thiele-Bruhn S (2011) Combined effects of the antibiotic sulfadiazine and liquid manure on the soil microbial-community structure and functions. J of Plant Nutr Soil Sci 174: 614–623.
9. Schauss K, Focks A, Heuer H, Kotzerke A, Schmitt H, et al. (2009) Analysis, fate and effects of the antibiotic sulfadiazine in soil ecosystems. Trac-Trends Anal Chem 28: 612–618.
10. Kotzerke A, Sharma S, Schauss K, Heuer H, Thiele-Bruhn S, et al. (2008) Alterations in soil microbial activity and N-transformation processes due to sulfadiazine loads in pig-manure. Environ Pollut 153: 315–322.
11. Kotzerke A, Hammesfahr U, Kleineidam K, Lamshoft M, Thiele-Bruhn S, et al. (2011) Influence of difloxacin-contaminated manure on microbial community structure and function in soils. Biol Fertil Soils 47: 177–186.
12. De Gelder L, Ponciano JM, Abdo Z, Joyce P, Forney LJ, et al. (2004) Combining mathematical models and statistical methods to understand and predict the dynamics of antibiotic-sensitive mutants in a population of resistant bacteria during experimental evolution. Genetics 168: 1131–1144.
13. Heuer H, Smalla K (2007) Manure and sulfadiazine synergistically increased bacterial antibiotic resistance in soil over at least two months. Environ Microbiol 9: 657–666.
14. Kopmann C, Jechalke S, Rosendahl I, Groeneweg J, Krögerrecklenfort E, et al. (2013) Abundance and transferability of antibiotic resistance as related to the fate of sulfadiazine in maize rhizosphere and bulk soil. FEMS Microbiol Ecol 83: 125–134.

15. Jechalke S, Kopmann C, Rosendahl I, Groeneweg J, Weichelt V, et al. (2013) Increased abundance and transferability of resistance genes after field application of manure from sulfadiazine-treated pigs. Appl Environ Microbiol 79: 1704–1711.
16. Heuer H, Solehati Q, Zimmerling U, Kleineidam K, Schloter M, et al. (2011) Accumulation of sulfonamide resistance genes in arable soils due to repeated application of manure containing sulfadiazine. Appl Environ Microbiol 77: 2527–2530.
17. Heuer H, Schmitt H, Smalla K (2011) Antibiotic resistance gene spread due to manure application on agricultural fields. Curr Opin Microbiol 14: 236–243.
18. Gaze WH, Krone SM, Larsson DGJ, Li X, Robinson JA, et al. (2013) Influence of humans on evolution and mobilization of environmental antibiotic resistome. Emerg Infect Dis 19: Article 871.
19. Forsberg KJ, Reyes A, Wang B, Selleck EM, Sommer MOA, et al. (2012) The shared antibiotic resistome of soil bacteria and human pathogens. Science 337: 1107–1111.
20. Sukul P, Lamshoft M, Zuhlke S, Spiteller M (2008) Sorption and desorption of sulfadiazine in soil and soil-manure systems. Chemosphere 73: 1344–1350.
21. Griffiths BS, Philippot L (2013) Insights into the resistance and resilience of the soil microbial community. FEMS Microbiol Rev 37: 112–129.
22. Lamshöft M, Sukul P, Zühlke S, Spiteller M (2010) Behaviour of C-14-sulfadiazine and C-14-difloxacin during manure storage. Sci Total Environ 408: 1563–1568.
23. Griffiths RI, Whiteley AS, O'Donnell AG, Bailey MJ (2000) Rapid method for coextraction of DNA and RNA from natural environments for analysis of ribosomal DNA- and rRNA-based microbial community composition. Appl Environ Microbiol 66: 5488–5491.
24. Timmers RA, Rothballer M, Strik D, Engel M, Schulz S, et al. (2012) Microbial community structure elucidates performance of *Glyceria maxima* plant microbial fuel cell. Appl Microbiol Biotechnol 94: 537–548.
25. Ding G-C, Heuer H, Smalla K (2012) Dynamics of bacterial communities in two unpolluted soils after spiking with phenanthrene: soil type specific and common responders. Front Microbiol 3: doi: 10.3389/fmicb.2012.00290.
26. Pruesse E, Quast C, Knittel K, Fuchs BM, Ludwig W, et al. (2007) SILVA: a comprehensive online resource for quality checked and aligned ribosomal RNA sequence data compatible with ARB. Nucleic Acids Res 35: 7188–7196.
27. Wang Q, Garrity GM, Tiedje JM, Cole JR (2007) Naïve bayesian classifier for rapid assignment of rRNA sequences into the new bacterial taxonomy. Appl Environ Microbiol 73: 5261–5267.
28. Hothorn T, Bretz F, Westfall P (2008) Simultaneous inference in general parametric models. Biometrical J 50: 346–363.
29. Torsvik V, Øvreås L (2002) Microbial diversity and function in soil: from genes to ecosystems. Curr Opin Microbiol 5: 240–245.
30. Zhu YG, Johnson TA, Su JQ, Qiao M, Guo GX, et al. (2013) Diverse and abundant antibiotic resistance genes in Chinese swine farms. Proc Natl Acad Sci U S A 110: 3435–3440.
31. Byrne-Bailey KG, Gaze WH, Kay P, Boxall ABA, Hawkey PM, et al. (2009) Prevalence of sulfonamide resistance genes in bacterial isolates from manured agricultural soils and pig slurry in the United Kingdom. Antimicrob Agents Chemother 53: 696–702.
32. Ollivier J, Schacht D, Kindler R, Groeneweg J, Engel M, et al. (2013) Effects of repeated application of sulfadiazine-contaminated pig manure on the abundance and diversity of ammonia and nitrite oxidizers in the root-rhizosphere complex of pasture plants under field conditions. Front Microbiol 4: Article 22.
33. Brandt KK, Sjøholm OR, Krogh KA, Halling-Sørensen B, Nybroe O (2009) Increased pollution-induced bacterial community tolerance to sulfadiazine in soil hotspots amended with artificial root exudates. Environ Sci Technol 43: 2963–2968.
34. Zielezny Y, Groeneweg J, Vereecken H, Tappe W (2006) Impact of sulfadiazine and chlorotetracycline on soil bacterial community structure and respiratory activity. Soil Biol Biochem 38: 2372–2380.
35. Doornbos RF, van Loon LC, Bakker P (2012) Impact of root exudates and plant defense signaling on bacterial communities in the rhizosphere. A review. Agron Sustain Dev 32: 227–243.
36. Peu P, Brugere H, Pourcher AM, Kerouredan M, Godon JJ, et al. (2006) Dynamics of a pig slurry microbial community during anaerobic storage and management. Appl Environ Microbiol 72: 3578–3585.
37. Yi J, Zheng R, Li FG, Chao Z, Deng CY, et al. (2012) Temporal and spatial distribution of *Bacillus* and *Clostridium histolyticum* in swine manure composting by fluorescent in situ hybridization (FISH). Appl Microbiol Biotechnol 93: 2625–2632.
38. Brooke JS (2012) Stenotrophomonas maltophilia: an emerging global opportunistic pathogen. Clin Microbiol Rev 25: 2–41.
39. Neikirk RL, Krieg NR (1964) Sulfonamide resistance of *Clostridium sporogenes*. J Bacteriol 87: 1526–1527.
40. Heuer H, Kopmann C, Binh CTT, Top EM, Smalla K (2009) Spreading antibiotic resistance through spread manure: characteristics of a novel plasmid type with low %G plus C content. Environ Microbiol 11: 937–949.
41. Heuer H, Smalla K (2012) Plasmids foster diversification and adaptation of bacterial populations in soil. FEMS Microbiol Rev 36: 1083–1104.
42. Davies J, Spiegelman GB, Yim G (2006) The world of subinhibitory antibiotic concentrations. Curr Opin Microbiol 9: 445–453.
43. Martinez JL (2009) Environmental pollution by antibiotics and by antibiotic resistance determinants. Environ Pollut 157: 2893–2902.
44. Tappe W, Herbst M, Hofmann D, Koeppchen S, Kummer S, et al. (2013) Degradation of sulfadiazine by *Microbacterium lacus* strain SDZm4, isolated from lysimeters previously manured with slurry from sulfadiazine-medicated pigs. Appl Environ Microbiol 79: 2572–2577.
45. Hayward AC, Fegan N, Fegan M, Stirling GR (2010) *Stenotrophomonas* and *Lysobacter*: ubiquitous plant-associated gamma-proteobacteria of developing significance in applied microbiology. J Appl Microbiol 108: 756–770.
46. Ryan RP, Monchy S, Cardinale M, Taghavi S, Crossman L, et al. (2009) The versatility and adaptation of bacteria from the genus *Stenotrophomonas*. Nat Rev Microbiol 7: 514–525.
47. Boto L (2010) Horizontal gene transfer in evolution: facts and challenges. Proc Biol Sci 277: 819–827.
48. Prospero JM, Blades E, Mathison G, Naidu R (2005) Interhemispheric transport of viable fungi and bacteria from Africa to the Caribbean with soil dust. Aerobiologia 21: 1–19.
49. Vanparys B, Heylen K, Lebbe L, De Vos P (2005) *Devosia limi* sp. nov., isolated from a nitrifying inoculum. Int J Syst Evol Microbiol 55: 1997–2000.
50. Rivas R, Velazquez E, Willems A, Vizcaino N, Subba-Rao NS, et al. (2002) A new species of *Devosia* that forms a unique nitrogen-fixing root-nodule symbiosis with the aquatic legume *Neptunia natans* (L.f.) druce. Appl Environ Microbiol 68: 5217–5222.
51. Yang Y, Wang X, Shi JF, Li JJ (2012) The influence of the discharging sewage on microbial diversity in sludge from Dongting Lake. World J Microb Biot 28: 421–430.
52. Sullivan RF, Holtman MA, Zylstra GJ, White JF, Kobayashi DY (2003) Taxonomic positioning of two biological control agents for plant diseases as *Lysobacter enzymogenes* based on phylogenetic analysis of 16S rDNA, fatty acid composition and phenotypic characteristics. J Appl Microbiol 94: 1079–1086.
53. Weller DM (2007) Pseudomonas biocontrol agents of soilborne pathogens: looking back over 30 years. Phytopathology 97: 250–256.
54. Mendes R, Kruijt M, de Bruijn I, Dekkers E, van der Voort M, et al. (2011) Deciphering the rhizosphere microbiome for disease-suppressive bacteria. Science 332: 1097–1100.
55. Ortet P, Barakat M, Lalaouna D, Fochesato S, Barbe V, et al. (2011) Complete genome sequence of a beneficial plant root-associated bacterium, *Pseudomonas brassicacearum*. J Bacteriol 193: 3146.
56. Uhlik O, Jecna K, Mackova M, Vlcek C, Hroudova M, et al. (2009) Biphenyl-metabolizing bacteria in the rhizosphere of horseradish and bulk soil contaminated by polychlorinated biphenyls as revealed by stable isotope probing. Appl Environ Microbiol 75: 6471–6477.
57. Ward NL, Challacombe JF, Janssen PH, Henrissat B, Coutinho PM, et al. (2009) Three genomes from the phylum *Acidobacteria* provide insight into the lifestyles of these microorganisms in soils. Appl Environ Microbiol 75: 2046–2056.
58. Poulsen PHB, Abu Al-Soud W, Bergmark L, Magid J, Hansen LH, et al. (2013) Effects of fertilization with urban and agricultural organic wastes in a field trial - prokaryotic diversity investigated by pyrosequencing. Soil Biol Biochem 57: 784–793.

5

Spatial Variability and Stocks of Soil Organic Carbon in the Gobi Desert of Northwestern China

Pingping Zhang[1], **Ming'an Shao**[1,2]*

[1] State Key Laboratory of Soil Erosion and Dryland Farming on the Loess Plateau, Northwest A & F University, Yangling, China, [2] Key Laboratory of Ecosystem Network Observation and Modeling, Institute of Geographical Science and Natural Resources, Chinese Academy of Sciences, Beijing, China

Abstract

Soil organic carbon (SOC) plays an important role in improving soil properties and the C global cycle. Limited attention, though, has been given to assessing the spatial patterns and stocks of SOC in desert ecosystems. In this study, we quantitatively evaluated the spatial variability of SOC and its influencing factors and estimated SOC storage in a region (40 km^2) of the Gobi desert. SOC exhibited a log-normal depth distribution with means of 1.6, 1.5, 1.4, and 1.4 g kg^{-1} for the 0–10, 10–20, 20–30, and 30–40 cm layers, respectively, and was moderately variable according to the coefficients of variation (37–42%). Variability of SOC increased as the sampling area expanded and could be well parameterized as a power function of the sampling area. Significant correlations were detected between SOC and soil physical properties, i.e. stone, sand, silt, and clay contents and soil bulk density. The relatively coarse fractions, i.e. sand, silt, and stone contents, had the largest effects on SOC variability. Experimental semivariograms of SOC were best fitted by exponential models. Nugget-to-sill ratios indicated a strong spatial dependence for SOC concentrations at all depths in the study area. The surface layer (0–10 cm) had the largest spatial dependency compared with the other layers. The mapping revealed a decreasing trend of SOC concentrations from south to north across this region of the Gobi desert, with higher levels close to an oasis and lower levels surrounded by mountains and near the desert. SOC density to depths of 20 and 40 cm for this 40 km^2 area was estimated at 0.42 and 0.68 kg C m^{-2}, respectively. This study provides an important contribution to understanding the role of the Gobi desert in the global carbon cycle.

Editor: Ben Bond-Lamberty, DOE Pacific Northwest National Laboratory, United States of America

Funding: This work was supported by grants from the Major Program of the National Natural Science Foundation of China No. 91025018. The funders had no role in study design, data collection and analysis, decision to publish, or preparation of the manuscript.

Competing Interests: The authors have declared that no competing interests exist.

* E-mail: mashao@ms.iswc.ac.cn

Introduction

Soil organic carbon (SOC) has an important influence on the physical, chemical, and biological properties of soil and is critical for improving soil fertility and quality, increasing the water holding capacity of soil, reducing soil erosion, and enhancing crop productivity [1,2]. With climate change and environmental issues dominating global concerns, SOC has received increasing attention worldwide because of its important role in the global C cycle and its potential feedback on the global warming [3–6]. As one of the largest and most dynamic component in the global C cycle, the SOC stock is at least two times the amount of C stored in the vegetation and atmosphere [7]. Thus, a small loss of SOC pool due to changes in fertilization, cropping system, farming practices, and soil erosion could significantly increase the atmospheric CO_2 [8–11]. On the other hand, soils can increase the existing SOC pool by sequestration of C from the atmosphere [12–15], the processes of which are an active area of study. Reliable assessment of the spatial patterns and stocks of SOC at one timeline as a baseline is essential for understanding the potential of soils to sequester C, for quantifying the SOC sink or source capacity of soils in changing environments, and for developing the strategies necessary to mitigate the effects of global warming [16,17].

In recent years, extensive work has been conducted toward estimating the SOC stocks and distribution patterns at the global, continental, country, and regional scales [11,18–24]. For example, the global SOC stock has been estimated to be about 2400 Pg C in the top 2 m [4]. However, these estimations are highly uncertain because of the gaps in spatial coverage for many regions that causes difficulties to develop a harmonized SOC baseline [22–24]. In addition, the selection of the type of SOC database, the land use and/or soil map, the mapping resolution, reference depth, bulk density or other information can also have a great effect on the final SOC stock estimation [25]. Similarly, due to inconsistent estimation methods and limited data, the SOC stock estimations in China are also uncertain and has varied greatly, from 50 to180 Pg, and SOC density from 54.6 to 190.5 t C/ha [22]. The accuracy of these large-scale SOC stock estimations largely depends on the data availability from site-based or small-scale measurements [6,24]. To reduce the uncertainty of SOC stocks estimation and better understand the role of SOC in the global C cycle, reliable baseline datasets providing information on SOC stocks in all types of sites and ecosystems are necessary.

Desertification is one of the most severe types of land degradation in arid and semiarid areas of the world [26]. Due to the harsh natural conditions and the fragile ecological environment, desert ecosystems are more sensitive to climate change, leading to the emission of CO_2 to the atmosphere and a reduction in the pool of SOC [27,28]. In contrast, it is possible to increase SOC concentrations in desert soils through the adoption of restorative measures such as the establishment of plants [14,29,30] and the prohibition of grazing [31]. [32] indicated that the control of desertification could globally sequester 0.9–1.9 Pg C yr^{-1} over a period of 25–50 years.

China is also seriously threatened by desertification [33,34]. [27] estimated that desertified land in China potentially covers 158 Mha, comprising 81 Mha of slight, 61 Mha of moderate, and 35 Mha of severe desertification. The widely distributed desertified lands in China thus likely have a considerable effect on the regional terrestrial C balance and the feedbacks that affect climate change [35]. Although some studies have been conducted on assessment of SOC concentrations/stocks in desert area, many questions still remain open. So far, most studies on SOC stock estimates from desert ecosystems have been carried out in sandy desert [36–39], and only few from the Gobi desert are available [40]. Due to the difficulties and associated costs of soil sampling, most studies estimating SOC stocks rely on information taken from a relatively small number of representative sampling points or profiles [41]. This decreases the estimation accuracy of SOC stocks in desert ecosystems and limits our capability to evaluate the C budget, to assess its contribution to the increasing global concentrations of CO_2, and to propose measures to increase the sequestration of organic C in soils. Therefore, to better understand the SOC pool in desert environments, it is necessary to conduct more intensive site and local scale estimates of the variance in SOC, and the possible spatial controls on this variance structure.

Gansu province is one of the main desertified areas induced by wind and one of the source regions of sandstorms in northern China. In this paper, a typical fenced region of the Gobi desert was chosen as a study case. The objectives of this study were: (1) to estimate SOC concentrations and determine the spatial distribution of SOC, (2) to analyse the relationships between SOC concentrations and environmental factors, and (3) to estimate SOC density and storage in the study area.

Materials and Methods

Ethics statement

The study area belong to Linze Inland River Basin Research Station (39°21′ N and 100°07′E,1389 m), a department of the Cold and Arid Regions Environmental and Engineering Research Institute, Chinese Academy of Sciences. The study was approved by the Cold and Arid Regions Environmental and Engineering Research Institute, Chinese Academy of Sciences.

Study area

The study area, occupying approximately 40 km^2 (5 km×8 km), is located in the Gobi desert in the middle of Gansu province (the central reaches of the Heihe River Basin) of Northwestern China, between latitudes 39°24′ and 39°28′N and longitudes 100°08′ and 100°11′E (Fig. 1*a,b*). The region is a relatively flat alluvial plain (elevation ranging from 1390 to 1470 m) bordered by a young oasis to the southwest, the remnants of the Qilian Mountains to the north, and an extension of the Badain Jaran Desert to the southeast (Fig. 1*c*). The area is characterized by low and seasonal variability in rainfall and is classified as a typical temperate desert. The mean annual precipitation and air temperature are 117 mm and 7.6°C, respectively. Rainfall in brief summer showers contributes 65% of the annual total precipitation. The mean annual pan-evaporation is approximately 2390 mm, twenty times greater than the annual precipitation. The average annual wind speed is 3.2 m s^{-1}, with the resultant wind coming from the northwest, and the dominant windy days and wind storms occur between March and May [42]. The zonal soil is classified as gray-brown desert soil, derived from gravelly diluvial-alluvial materials of the denuded monadnock [43]. Stones are present in a significant proportion of the surface and sub-soil horizons. The aboveground plant cover is discontinuous and can be described as patches of sub-shrubs surrounded by bare areas. The study area has been fenced and is protected from grazing for the purpose of revegetation and reclamation. The dominant plant species are *Nitraria sphaerocarpa* Maxim. and *Reaumuria soongorica* (Pall.) Maxim., and the accompanying plant species are mainly *Kalidium gracile* Fenzl., *Allium mongolicum* Rgl., *Bassia dasyphylla* (Fisch. and Mey.) Kuntze. and *Halogeton arachnoideus* Moq.

Soil sampling and laboratory analysis

Soil samples were collected from a total of 187 locations on a regular grid of 500 m×500 m throughout the study area from August to September in 2011. A portable GPS receiver (Garmin GPSmap 62 s) was used to locate the sampling site, as displayed in Figure 1*c*. At each location, soil was collected from four depths (0–10, 10–20, 20–30, and 30–40 cm) at five randomly selected sampling points within a radius of approximately 20 cm. The five sub-samples were then combined to produce one representative sample. In total, 748 soil samples were collected. The samples were all air dried, weighed, and sieved to 2 mm to separate the coarse (>2 mm) and fine (<2 mm) fractions. The former was reweighed to determine the stone content. The latter was separated into two parts: one was subject to further particle-size analyses by laser diffraction using a Mastersizer 2000 (Malvern Instruments, Malvern, England), and the other was ground to pass through a 0.25-mm mesh for SOC concentration analysis. The SOC concentration (g kg^{-1}) was measured using the potassium dichromate-wet combustion method [44]. Undisturbed soil samples cores (100 cm^3) were also collected for determining soil bulk density (BD) in each layer. To reduce the influences of stones on BD measurement, we averaged five replicate measurements for each location and layer.

Re-sampling method

Estimating SOC variability at different scales is important for effective soil survey sampling design and SOC change prediction [2]. In this study, to detect the change tendency of SOC variability with the size of the sampling area, a series of sampling point allocations with different areas were done through re-sampling using all sampling points (n = 187) in the study area (5 km×8 km) [45,46]. For convenience, the west-east and south-north distances of the re-sampling area were set integral multiples of 1 km to generate five re-sampling options for the west-east distance and eight options for the south-north distance. The random combination of these two sets of options thus produced 40 potential re-sampling methods. Some of these methods, though, were the same within the study area (e.g., the re-sampling methods of 2 km×6 km, 3 km×4 km, and 4 km×3 km yielded the same area), so finally a total of 24 kinds of re-sampling areas were obtained (Table 1). SOC variability at a certain area was computed by averaging the CV values of all the possible re-sampling scenarios with the same area.

Calculation of SOC density and stock

The SOC density of a single layer was estimated based on Eq. (1):

$$SOCD_i = \frac{SOC_i \times BD_i \times d_i \times (1 - CF_i/100)}{100} \quad (1)$$

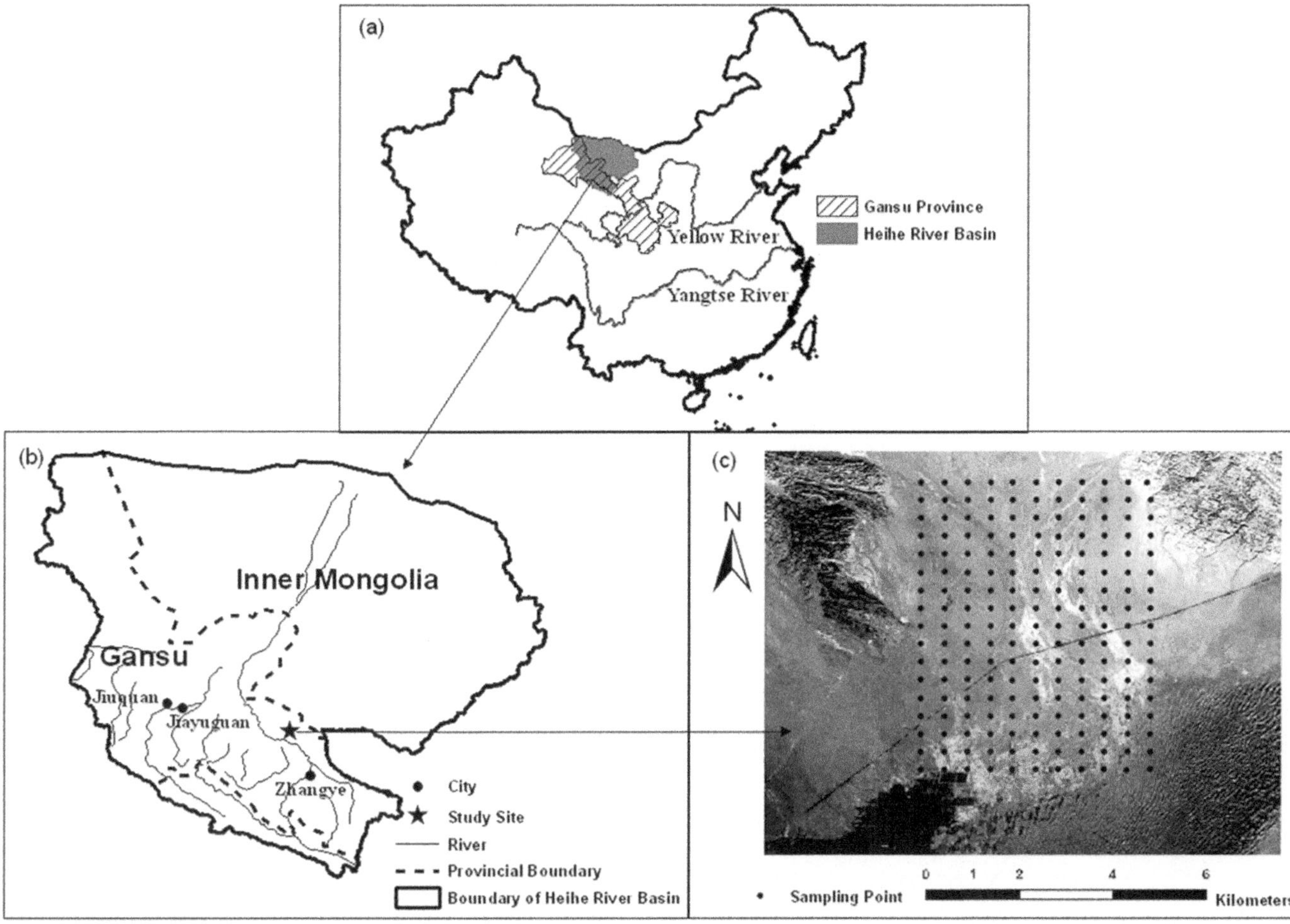

Figure 1. The location of Gansu Province and the Heihe River Basin, China (a), the study site in the Heihe River Basin (b), and the soil-sampling points in the study area (c).

Table 1. Details of the re-sampling areas and sampling methods.

Re-sampling area (km^2)	Sampling method	Re-sampling area (km^2)	Sampling method
1	1×1*	15	3×5; 5×3
2	1×2; 2×1	16	2×8; 4×4
3	1×3; 3×1	18	3×6
4	1×4; 4×1; 2×2	20	4×5; 5×4
5	1×5; 5×1	21	3×7
6	1×6; 2×3; 3×2	24	3×8; 4×6
7	1×7	25	5×5
8	1×8; 2×4; 4×2	28	4×7
9	3×8	30	5×6
10	2×5; 5×2	32	4×8
12	2×6; 3×4; 4×3	35	5×7
14	2×7	40	5×8

*The digit before the multiplication sign represents the west-east sampling distance (km), and the digit after it represents the south-north sampling distance (km).

Table 2. Selected soil physical properties at different soil depths.

Variables	Statistical parameter	Soil depth			
		0–10 cm	10–20 cm	20–30 cm	30–40 cm
Stones	Mean (%)	12.1 a	13.3 ab	15.2 b	69.1 c
	CV (%)	100.5	103.4	101.3	24.5
Sand	Mean (%)	70.9 a	72.4 a	71.1 a	26.0 a
	CV (%)	20.1	19.7	21	22.1
Silt	Mean (%)	7.5 a	6.5 a	6.5 a	2.4 a
	CV (%)	85.5	97.5	107.5	111.9
Clay	Mean (%)	9.5 a	7.8 ab	7.2 b	2.6 b
	CV (%)	84.4	102.4	111.3	123
BD	Mean (g cm^{-3})	1.4 a	1.4 a	1.4 a	0.5 a
	CV (%)	8.8	10.4	9.7	10.1

BD, bulk density; CV, coefficient of variation; Stones, >2 mm; Sand, 2–0.2 mm; Silt, 0.2–0.002 mm; Clay, <0.002 mm. Mean values followed by the same letter are not significantly different ($\alpha = 0.05$) at different soil depths using the LSD method.

where $SOCD_i$ and SOC_i are SOC density (kg C m^{-2}) and concentration (g kg^{-1}) of the ith layer, BD_i is the bulk density of the ith layer (g cm^{-3}), d_i is the depth of the ith layer (cm), and CF_i is the fraction (%) of coarse fragments >2 mm in the ith layer.

For an individual profile with a depth D (cm), SOC densities were then calculated by summarising the SOC density of each soil layer i:

$$SOCD_D = \sum_{i=1}^{n} SOCD_i \qquad (2)$$

The total SOC storage is calculated as:

$$SOC_{stock} = \sum_{j=1}^{m} S_j \times SOCD_{D,j} \qquad (3)$$

where S_j is the total area (m^2) of a given land-use type j and $SOCD_{D,j}$ is the average SOC density of the jth land-use type (kg m^{-2}).

Statistical analysis

A descriptive statistical analysis was first used to illustrate the central trend and the overall variation of the variables. This analysis included descriptions of the minimum, maximum, mean, median, skewness, Kurtosis, standard deviation (SD), and coefficients of variation (CVs). A one-sample Kolmogorov-Smirnov (K-S) test was used to examine the normality of the data, and natural logarithmic transformations were performed where necessary to meet the normality requirement of geostatistical analysis. The means of different layers were compared by a one-way analysis of variance (ANOVA). Correlation analysis and stepwise linear regression analysis were performed to understand relationships between SOC concentrations and the environmental factors. All statistical analyses used the programme SPSS v. 16.0 (SPSS Inc., Chicago, IL, USA).

Geostatistical methods such as semivariogram calculation, cross-validation, kriging, and mapping have been widely applied in the study of SOC spatial distribution [47,48]. Semivariograms were used to determine the degree of spatial dependence. Before semivariogram calculation, a preliminary semivariogram surface analysis was performed to detect any zonal effect or trend in direction [49]. The experimental semivariogram, $\gamma(h)$, is half the

Table 3. Summary of statistical parameters for soil organic carbon (SOC) concentrations at different soil depths.

Variables	SOC (g kg^{-1})			
	0–10 cm	10–20 cm	20–30 cm	30–40 cm
Maximum	4.5	3.9	4.3	3.6
Minimum	0.4	0.5	0.5	0.5
Mean	1.6 a	1.5 ab	1.4 b	1.4 b
SD	0.6	0.6	0.6	0.6
CV	36.8	38.5	41.0	42.1
Skewness	1.83	1.26	1.64	1.15
Kurtosis	5.95	2.40	4.66	1.04
P of K-S test	0.001	0.012	0.009	0.006

SD, standard deviation; CV, coefficient of variation. Mean values followed by the same letter are not significantly different ($\alpha = 0.05$) at different soil depths using the LSD method.

expected squared difference between paired data separated by a distance h and is expressed as:

$$\gamma(h)=\frac{1}{2N(h)}\sum_{i=1}^{N(h)}[Z(x_i)-Z(x_i+h)]^2 \quad (4)$$

where $Z(x_i+h)$ and $Z(x_i)$ are observations at positions x_i+h and x_i, respectively, and $N(h)$ denotes the number of data pairs separated by the lag distance, h. A semivariogram model contains three important parameters which interpret the spatial structure of soil properties: nugget (C_0), sill ($C+C_0$), and range (A). Nugget represents the undetectable measurement error, inherent variability or the variation within the minimum sampling distance. Sill is the upper limit of the semivariogram model, representing the total variation. The separation distance at which the sill is reached is the range of spatial dependence. Samples separated by distances smaller than the range are spatially related, whereas samples separated by larger distances are not spatially related. The nugget ratio (C_0/C_0+C) can be regarded as a criterion for classifying the spatial dependence of soil properties. A variable is considered to have strong, moderate, or weak spatial dependence if the ratio is less than 0.25; between 0.25 and 0.75; and over 0.75, respectively [50].

In this study, all semivariograms were estimated with fixed distance intervals of 500 m, and the maximum distance was set to 4700 m. Because 4700 m is less than half the maximum distance between sampling sites, it coincides with the requirement of geostatistical analysis [51]. A total of nine sets of class intervals were generated. There are several commonly used semivariogram models (such as spherical, exponential, Gaussian, and linear models). The best model was based on two criteria: small residual sum of squares (RSS) and high coefficient of determination (R^2). The best-fit parameters were subsequently estimated using weighted least squares regression method.

After selecting the best-fit semivariogram models, ordinary kriging was used as an interpolation method to predict values for SOC concentrations. The prediction was calculated as the linear sum:

$$Z^*(x_0)=\sum_{i=1}^{n}\lambda_i Z(x_i) \quad (5)$$

where $Z^*(x_0)$is the value to be predicted at the location x_0, $Z(x_i)$ is the known value at the sampling location x_i, n is the number of locations within the search neighbourhood used for the prediction, and λ_i is the kriging weight assigned to $Z(x_i)$.

The predicted map quality of the SOC was tested by cross-validation with replacement. Three indices were calculated to assess the effectiveness of kriging: mean error (ME), root mean square error (RMSE), and root mean square standardised error (RMSSE) [52].

The geostatistical analysis was performed with GS+ v. 7.0 (Gamma Design Software, Plainwell, Michigan, USA), and contour maps through ordinary kriging were produced with GIS software ArcView v. 3.3 and its extension module of Spatial Analysis v. 2 (ESRI Inc., Redlands, California, USA).

Results and Discussion

Soil physical properties

Due to the lack of mineral weathering and to long-term wind and water erosion, this natural desert exhibited a high content of stones, with an average of 27.42% stones in the top 40 cm of soil. The stone content increased slowly from 12.2% in the 0–10 cm layer to 15.2% in the 20–30 cm layer, and then abruptly increased

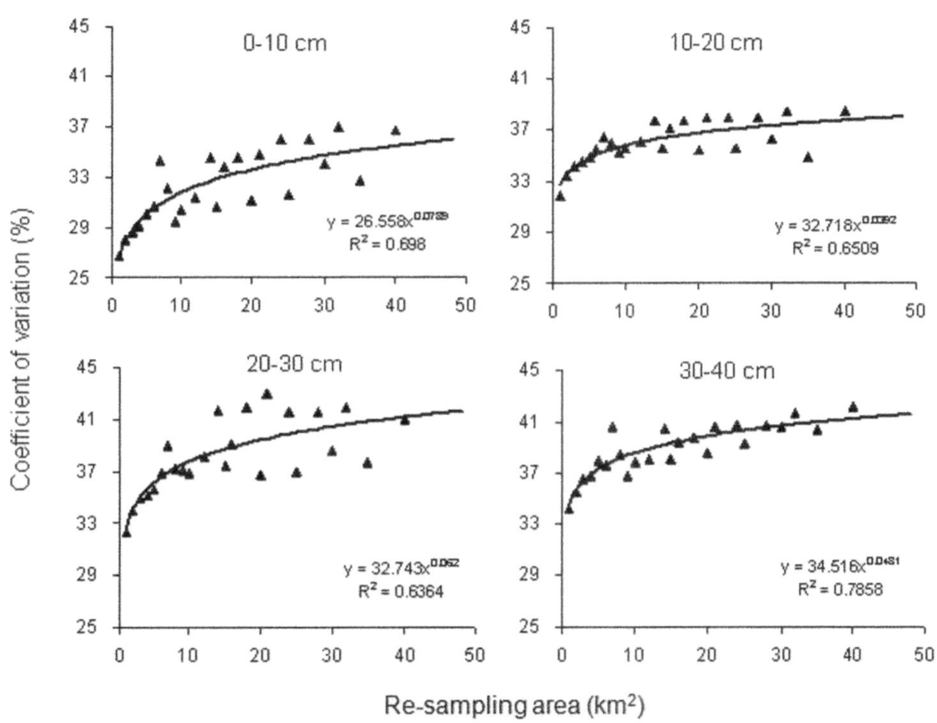

Figure 2. Coefficient of variation of soil organic carbon (SOC) concentrations at different re-sampling areas.

Table 4. Correlation analysis between soil organic carbon (SOC) concentrations and soil physical properties at different soil depths.

Variables	SOC (g kg^{-1})			
	0–10 cm	10–20 cm	20–30 cm	30–40 cm
Stones (%)	−0.196**	−0.201**	−0.201**	−0.448**
Sand (%)	−0.679**	−0.605**	−0.572**	−0.564**
Silt (%)	0.676**	0.606**	0.572**	0.567**
Clay (%)	0.663**	0.593**	0.561**	0.552**
BD (g cm^{-3})	−0.413**	−0.387**	−0.432**	−0.398**

BD, bulk density; ** denotes significance of correlation at $P<0.01$.

to 69.1% in the 30–40 cm layer (Table 2). All soil layers, except for 30–40 cm, contained more than 70% sand and much less silt and clay. Surface layer (0–10 cm) tended to have a higher content of clay ($P<0.01$), perhaps due to erosion. Clay contents tend to increase at shallower depths as the amount of soil erosion increases [53]. Beyond that, other soil properties (sand, silt, and BD) remained constant throughout the entire soil profile. Compared with the lower variability of sand (CVs of 19.6–22.1%), stone (CVs of 24.5–100.5%), silt (CVs of 85.5–111.9%) and clay (CVs of 84.4–123.0%) showed moderate or strong variability. The CV of the BD at each depth was ≤10.5%, indicating that BD was not very variable throughout the study area. The CVs indicated that the variabilities of all constituent contents, except for stone content, generally increased with soil depth.

SOC concentration and its variability

The calculation of variation function generally should be in accord with normal distribution, otherwise it may cause the proportional effect, raising the sill or nugget values [54]. As shown in Table 3, the raw SOC exhibited a positively skewed distribution in this study area. Thus, we used the logarithmic transformation to reduce the data skewness. The Ln-transformed data for all four layers passed the Kolmogorov-Smirnov test at a significance level higher than 0.05 (not shown) and consequently could be used in the analysis of the geostatistical variation function.

SOC concentrations were generally variable, ranging between 0.4–4.5 g kg^{-1}, 0.5–3.9 g kg^{-1}, 0.5–4.3 g kg^{-1}, and 0.5–3.6 g kg^{-1} for the four descending layers, respectively (Table 3). According to the soil-nutrient classification standards from the

Table 5. Stepwise multiple linear regression of soil organic carbon (SOC) concentrations with selected soil variables at different soil depths.

Soil depth	Independent variables	Coefficient	Explained Variance (%)
0–10 cm	Constant	3.589 **	
	Sand	−2.617 **	39.78
	Stones	0.006 **	7.30
	Adjusted R^2	0.470	
	MSE	0.417	
10–20 cm	Constant	0.989 **	
	Silt	5.200 **	31.18
	Stones	0.006 **	6.52
	Adjusted R^2	0.377	
	MSE	0.444	
20–30 cm	Constant	0.950 **	
	Silt	4.599 **	27.20
	Stones	0.006 **	6.80
	Adjusted R^2	0.340	
	MSE	0.466	
30–40 cm	Constant	1.613 **	
	Silt	3.206 **	24.20
	Stones	−0.007 **	10.00
	Adjusted R^2	0.342	
	MSE	0.477	

MSE, mean squared error; R^2, coefficient of determination; ** Denotes significance of correlation at $P<0.01$.

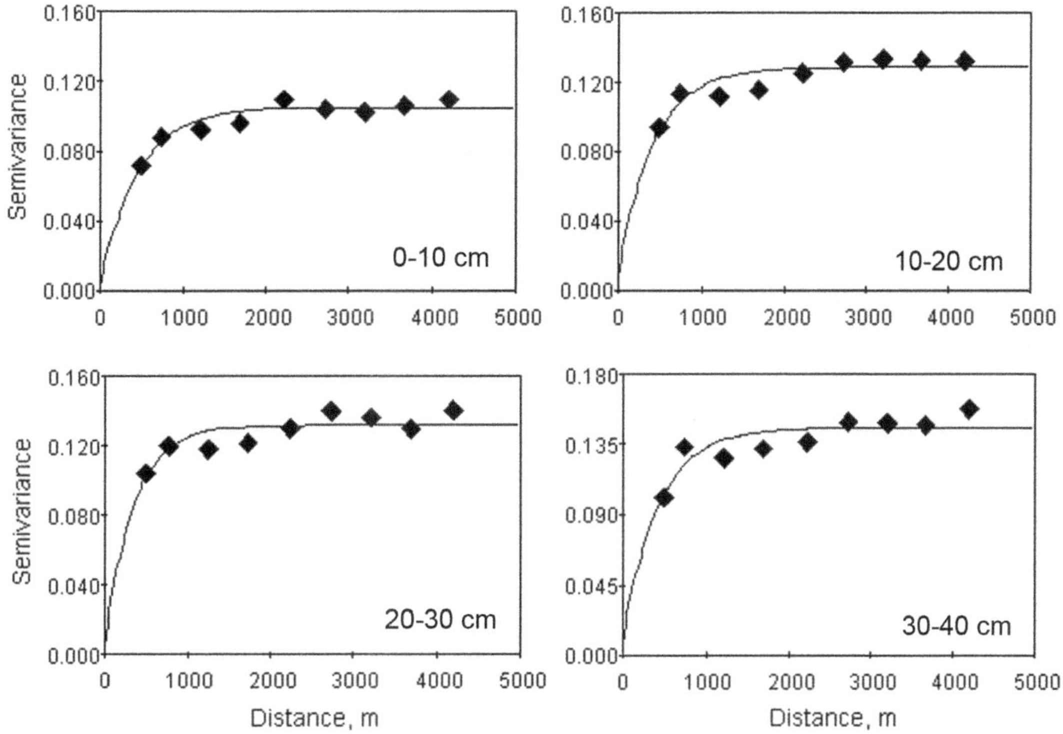

Figure 3. Isotropic semivariograms of soil organic carbon (SOC) concentrations at different soil depths.

Second National Soil Survey in China [55], the SOC level in this area was very low; only seven of the 748 soil samples were above the lowest classification standard (3.5 g kg^{-1}). The mean were 1.6, 1.5, 1.4, and 1.4 g kg^{-1} for the 0–10, 10–20, 20–30, and 30–40 cm layers, respectively, and progressively decreased with soil depth. The surface layer (0–10 cm) had the highest SOC concentrations, due to inputs of organic material that accumulated from organic litter and root residues [25,56], and higher soil aeration enabled higher soil enzyme activities in the surface layer than in the deeper layers [29]. Some researchers have also reported that SOC content decreased with depth in other natural ecosystems [6,10], but this trend is unlikely at sites with substantial human intervention such as orchards and tree nurseries [6]. In these sites, the topsoil may not have SOC contents much higher than the underlying layers, as a result of the management practices, such as heavy tillage, that could have enhanced SOC runoff and soil respiration [6].

The CVs of SOC concentrations varied from 36.8% to 42.1%, which are considered as moderate variation [57]. A moderate variability of SOC has also been reported in other studies at multiple scales [11,25,41,58]. [11,25,58] reported a decreasing trend of SOC variability with increasing soil depth. Our study, however, found the opposite trend. This can probably be explained by the different land-use types and vegetation characteristics. In their studies, croplands, grasslands, and forestland are the main land-use types. SOC is thus greatly affected by human activities, such as grazing and deforestation, and the agricultural managements of plowing, fertilisation, harvesting, and crop rotation which can have a greater impact on the surface soil layer than on the deeper layers [2,6,31,59]. Moreover, the high vegetation coverage in these land-use types indicates a high biomass of plant roots and litter, and an active soil microbial and enzyme activities, which mainly occur in the surface soil layer [29,56]. By contrast, the natural desert region in our study is rarely disturbed by human activities and the vegetation growth is limited by inferior soil and water conditions. The surface layer is more susceptible to processes, such as sedimentation and erosion, which can homogenise the distribution of SOC [60]. Therefore, SOC in the deeper layer tend to have greater spatial variability. Additionally, the preferential transport of SOC via cracks during

Table 6. Parameters of the semivariogram models estimated for soil organic carbon (SOC) concentrations at different soil depths.

Soil depth	Model	Nugget C_0	Still (C_0+C)	C_0/(C_0+C)	Range (m)	R^2
0–10 cm	exponential	0.0019	0.1048	0.018	1347	0.884
10–20 cm	exponential	0.0069	0.1288	0.054	1251	0.770
20–30 cm	exponential	0.0047	0.1324	0.035	1047	0.681
30–40 cm	exponential	0.0069	0.1458	0.047	1254	0.724

R^2, coefficient of determination.

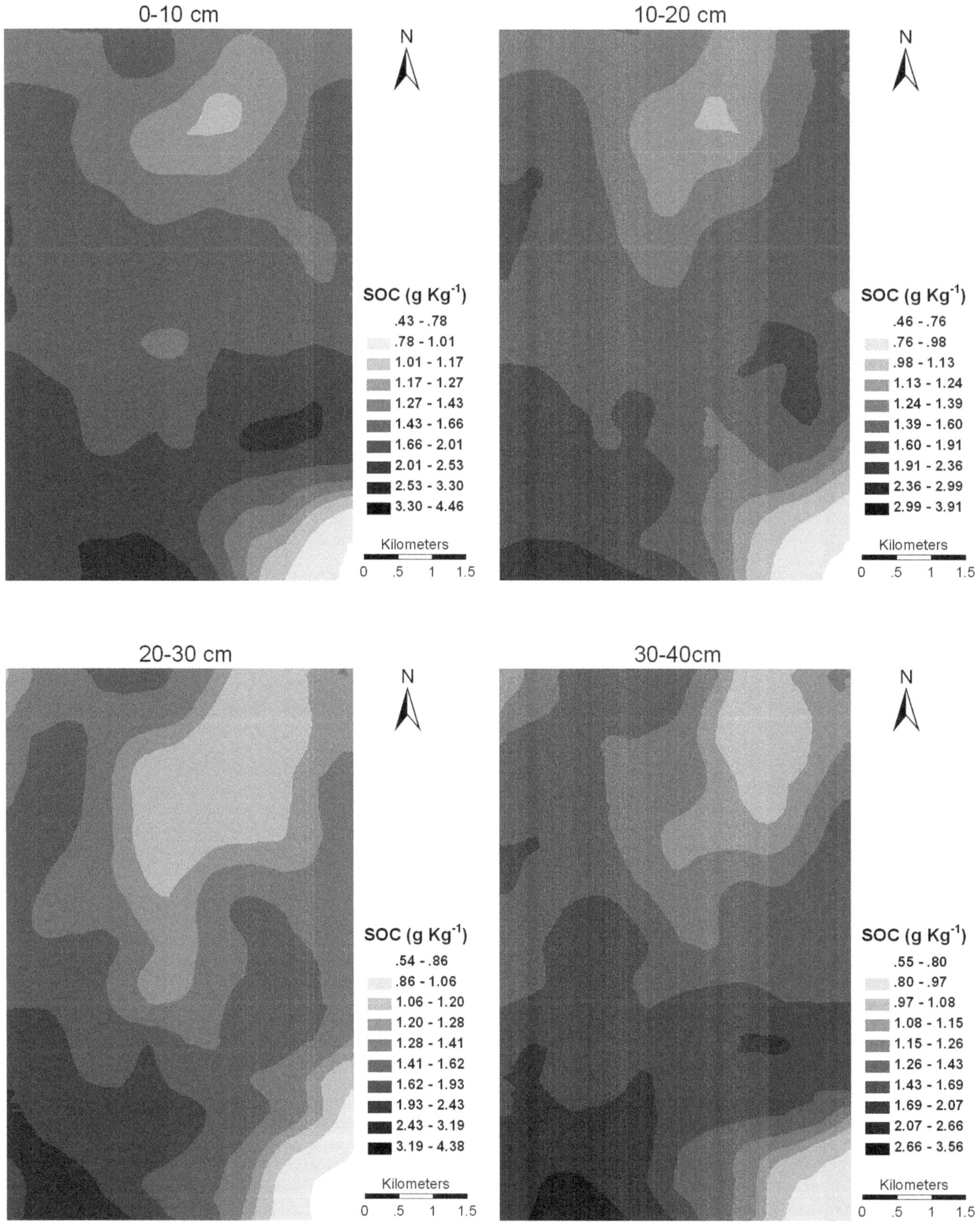

Figure 4. Distribution of soil organic carbon (SOC) concentrations across the study area.

Table 7. Verification of interpolation reliability for soil organic carbon (SOC) concentrations at different soil depths.

Soil depth	ME	RMSE	RMSSE
0–10 cm	−0.00821	0.535	1.105
10–20 cm	−0.00083	0.544	1.056
20–30 cm	0.00173	0.504	1.017
30–40 cm	−0.00143	0.570	1.104

ME, mean error; RMSE, root mean square error; RMSSE, root mean square standardized error.

dry periods could further increase the heterogeneity of SOC in the deeper soil horizons [10].

Response of SOC variability to the expansion of area

Fig. 2 demonstrated that CV of SOC concentration increased with the expansion of area and could be well parameterized as a power function of the sampling area for all the four soil layers. The factors influencing SOC concentration variability are scale dependent. For example, parent material, precipitation and geological history are of major importance to affect SOC at large scales. However, microtopography (such as the run-off gullies) and vegetation may be the dominant factors of SOC variability at small scales [61]. As sampling area increases, the origin of SOC variations above may get increasingly complex and heterogeneous, contributing to greater variability [2]. Moreover, the value of the fractal power parameter in the surface layer (0.0789) was much larger than that in the other three layers (0.0392–0.0620), indicating SOC concentration variability in the surface layer was more sensitive to the expansion of area. If increasing the same area, the SOC variability will increase more in the surface layer than that in the deeper layers.

The function between CV of SOC concentration and sampling area can also be used to estimate the variability of SOC concentration at a desired area and, consequently, the number of required samples (NRS). In order to obtain the mean value of SOC concentration with an accuracy level of Δ, at a confidence level of 1-α, the sample size should reach the requirement of NSR $= \lambda_\alpha^2 (CV\%/\Delta)^2$($\lambda_\alpha$ is the value of the Student's t-distribution at the confidence level of 1-α) [48]. Take the surface layer (0–10 cm) for example, NRS is found to be equal to 110, 227 and 326 for an area of 1, 100 and 1000 km^2, respectively, by assuming an accuracy level of 5% and a confidence level of 95%.

Pearson correlation and stepwise linear regression analyses

Under the extremely arid climate, the vegetational cover in this region of the Gobi desert is very low, and the chemical and biotic influences on soil development are relatively minor. Climate, soil type, and terrain can be disregarded as variables considering the small size and flatness of the study area. The physical properties of the soil were thus most likely responsible for the SOC variability in our study area.

Table 4 shows the correlation analyses between the soil physical properties and SOC concentrations. The four soil layers had similar patterns. SOC concentrations were positively correlated with both silt and clay contents and were negatively correlated with stone and sand contents and BD. The positive effects of clay and silt contents on SOC concentrations are likely due to the ability of clay and silt particles to adsorb organic matter. Finer particles are better than larger particles in protecting bound organic matter for longer times [62]. Moreover, the distribution of soil particle size and BD can indirectly impact SOC dynamics by affecting the physical structure, drainage, and aeration of soils [63].

The stepwise linear regression analysis was further performed to delineate the effect of different factors on SOC and to find the best predictive variables for SOC. A summary of these linear models is shown in Table 5. For the surface layer (0–10 cm), the regression model explained 47.0% of the overall SOC variability in which most of the variability was attributable to sand (39.7%) and stone contents (7.3%). For the other three layers, however, silt and stone contents together accounted for approximately 35% of the total variance of SOC concentrations. These results indicated that the relatively coarse fractions (stones, sand, and silt) were more important for the explanation of variability in SOC concentrations in this study area

Semivariogram and parameters

Spatial structure was not significantly associated with direction in the study area. Only isotropic semivariograms were thus plotted for SOC concentrations by using the model best fitted by the least squares regression method. Exponential models were theoretically optimal for all four soil layers.

The semivariogram models and best-fitted model parameters of SOC concentration at different soil depths are given in Fig. 3 and Table 6. The semivariogram of the SOC concentrations indicated a slightly smaller nugget effect (C_0) in the surface soil layer (0–10 cm) than in the other three layers, implying that the deeper soil layers had higher undetectable experimental error, short-range variability, and random and inherent variability of SOC concentration than the surface soil layer [3]. The sill values, representing total variation, showed an increasing trend from the surface soil layer to the deepest layer, which further validated the results obtained by conventional statistical methods. The nugget ratio (C_0/C_0+C) ranged from 0.018 to 0.054, indicating a strong spatial dependence for SOC concentrations for all four soil layers in our study [50]. Strong spatial dependency of soil properties can usually be attributed to intrinsic factors. In this natural desert, fenced enclosures for natural restoration have been implemented for many years and are rarely disturbed by human activities and grazing, so we suggest that the variability of SOC concentrations in this region may be highly dependent on the mineralogical composition of the parental material and on the weathering processes that have led to its formation [64–66].

The range also changed significantly with soil depth, which is likely due to the control of the distribution pattern of SOC by different soil processes. The range for the 0–10 cm layer was the largest, indicating a larger spatial autocorrelation in the surface soil layer than in the deeper soil layers. The surface soil layer is most sensitive to erosion, which could homogenise the distribution of SOC and increase the spatial autocorrelation distance [60]. [67]

Table 8. Comparison of SOC concentrations and SOC stocks in other desert areas in China.

Region	Environment	Mean annual precipitation (mm)	Area (km^2)	Reference depth (cm)	SOC concentration (g kg^{-1})	SOC stock (kg C m^{-2})	Reference
Erdos, Inner Mongolia	Temperate shrub desert	170 (161–209)	Profile measurement	0–300	0.42		[75]
Aershan, Inner Mongolia	Temperate desert	102 (61–121)	Profile measurement	0–300	0.25		[75]
Alxa league, Inner Mongolia	Desert steppe	134	Point measurement (sites with different grazing degrees)	0–40	1.73–2.05		[74]
Horqin Sand Land, Inner Mongolia	Desert grassland	350–450	Point measurement (sites with different desertification types and degrees)	0–30	0.37–4.35	0.19–1.76	[38]
Alxa Left County, Inner Mongolia	Desert shrubland	60–160	2.5	0–30		0.21–2.80	[39]
Horqin region, Inner Mongolia	Sand Land (sand dunes)	360	Point measurement (grazed and restored sites)	0–20	1.31–2.03		[37]
Horqin region, Inner Mongolia	Sand Land (mobile dunes)	360	Point measurement (sites with different restoration processes of dune vegetation)	0–20	0.68–1.29		[77]
Hunshandake, eastern Inner Mongolia	Sandy Land	300 (165–572)	21 400	variable	0.44–4.37		[76]
Central Gansu province	Gobi desert	117	40	0–40	1.45 (0.43–4.46)	0.68	This study

also showed that wind erosion significantly changed the spatial distribution patterns of SOC over two or three windy seasons. With the increasing of soil depth, the influence of erosion on SOC distribution was weakened, and the SOC tended to be much more independent and was characterized by a stochastic pattern. As shown in Table 6, the range decreased from 1347 to 1047 m from 0–10 cm to 20–30 cm. The range in the 30–40 cm layer, however, increased slightly, reflecting the influence of parental materials on the spatial structure of SOC in deeper layers. In deeper soil layers, the SOC was probably inherited from the soil parent materials because the soils are young with little weathering or anthropogenic impacts [65]. Since the parental materials are distributed quite uniformly across the study area [66], it leads to a better distribution of spatial continuity. Because the range was larger than our sampling interval (500 m), our sampling system was sufficiently robust to detect spatial relationships on the scale of the landscape.

Kriging of spatial variation of SOC concentrations

The semivariogram models were used as input to ordinary kriging, and the resulting distribution maps are shown in Fig. 4. The interpolation cross-validation was carried out to test the effectiveness of the prediction maps, and the associated prediction errors for each map are shown in Table 7. ME determined the degree of bias in the estimates and should be close to zero. RMSE quantified the average differences between prediction and observation and should be as small as possible. If the model accurately described the data, RMSSE should be close to unity. Based on the above criteria, the predicted maps of SOC concentration for the study area were reliable.

The character of SOC distribution is clearly similar for each layer, and the entire study area is characterised by low concentrations of SOC and high variation. SOC concentrations generally decreased gradually from south to north. The southwest part of the study area, close to the oasis, tended to have higher concentrations of SOC. The shelter belts in front of the oasis functioned as natural barriers to reduce wind velocity and intercept the fine material, which tends to be rich in SOC [68]. The increased fine fractions also improved the soil properties, encouraging the colonisation of some annuals, e.g. *B. dasyphylla* and *H. arachnoideus*. Their rapid growth and death provided an important influx of SOC. In contrast, relatively low concentrations of SOC were mainly distributed in the north and the southwest. The northern part of the site was surrounded by mountains and was characterised by coarse soils and little vegetations, facilitating the drifting of fine particles and a noticeable decline in SOC. On the other hand, the southwestern region was linked with the Badain Juran desert and the longtime encroachment of drifted sand from the desert produced a more sandy texture in this area.

SOC density and stocks

SOC density is indispensable for the assessment of SOC stocks and is the required measurement of account for the Clean Development Mechanism of the United Nations Framework Convention on Climate Change [69]. SOC density was 0.22, 0.20, 0.19, and 0.07 kg C m^{-2} for the 0–10, 10–20, 20–30, and 30–40 cm layers, respectively, and dropped sharply with increasing depth due to the high stone content. The low value in the 30–40 cm layer indicates that the SOC density would be even lower below 40 cm because of the increasing number of stones. We can thus reasonably conclude that SOC in this region of the Gobi desert was mainly stored in the upper 30 cm of soil. The overall SOC stocks in the upper 20 cm was 0.42 kg C m^{-2} and to a depth of 40 cm 0.68 kg C m^{-2}. When compared to the reported values for other regions in China [11,25,70,71,72], due to natural drought, large number of stones, and intense soil erosion, SOC stocks in the Gobi desert region was very low. However, since the Gobi desert (568,980 km^2) accounts for about 5.9% of China's total territory [73] and is sensitive to climate change [17], it is likely to have a considerable effect on the terrestrial C balance in China.

Table 8 summarises the results of SOC studies (including SOC concentrations and densities) in other desert regions in China. The SOC concentrations and densities in our study area were generally in the same range as those in other regions, except that SOC concentrations were a little higher than those in the Erdos and Aershan regions. Different reference soil depths, different sampling methods, and the higher patchiness of our study area may account for our higher SOC concentrations. Even though the soils in our study area had higher fractions of stones, patches with more fertile soil still allowed the growth of vegetation, which thus increased the SOC concentrations. Compared with some of the other regions mentioned in Table 8, SOC in our study site is less affected by grazing, which can give rise to a considerable decrease in ground coverage and primary productivity, and thus accelerate soil erosion by wind and result in loss of SOC [74]. However, the improvement of soil quality through the adoption of grazing prohibition can increase SOC concentrations [36]. Moreover, higher SOC concentrations in our study site can also be attributed to the reduced mineralization rate of SOC, because of the lack of water in this region [66]. Comparisons among the studies listed in Table 8, however, remain limited due to differences in sampling methods of SOC measurement and reference soil depths. Identifying the dynamics of SOC in changing desert environments is difficult. We should thus develop more site inventories of SOC in desert environments or use a comparable approach to better understand the potential changes of SOC in desert environments, which will lay the groundwork for developing more effective strategies to combat soil desertification and reduce the risk of desertification in the future.

Conclusions

We have provided estimates of spatial SOC concentrations and stocks for this region of the Gobi desert that are more accurate than previous estimates. Classical statistics indicated that SOC concentrations decreased with increasing soil depth and were moderately variable in the study area. The deepest soil layer (30–40 cm) had the highest amount of variation in SOC concentrations. Significant correlations were detected between SOC and selected physical properties of the soil, especially the stone, sand, and silt contents. The composition of the parental material (such as the distribution of soil particle size) and the weathering (such as erosion and sedimentation) that led to its formation may be responsible for the strong spatial dependence of SOC. This dependence implies that SOC in desert ecosystem is sensitive to climate change and thus represents an important dynamic pool of C in the global C cycle. The kriging interpolated maps indicated a decreasing trend of SOC concentrations from south to north across the study area, which was apparently related to the location of the study area. This study contributes to our understanding of the role of Gobi desert ecosystem in the global C cycle and incorporation of small-scale spatial variations of SOC into large-scale spatiotemporal models.

Acknowledgments

Special thanks go to the staff of the Linze Inland River Basin Research Station of the Institute of Cold and Arid Regions Environmental and Engineering Research of CAS.

Author Contributions

Conceived and designed the experiments: PPZ MAS. Performed the experiments: PPZ. Analyzed the data: PPZ. Wrote the paper: PPZ.

References

1. Rossi J, Govaerts A, De Vos B, Verbist B, Vervoort A, et al. (2009) Spatial structures of soil organic carbon in tropical forests-A case study of Southeastern Tanzania. Catena 77: 19–27.
2. Wang DD, Shi XZ, Lu XX, Wang HJ, Yu DS, et al. (2010) Response of soil organic carbon spatial variability to the expansion of scale in the uplands of Northeast China. Geoderma 154: 302–310.
3. Schlesinger WH, Raikes JA, Hartley AE, Cross AF (1996) On the spatial pattern of soil nutrients in desert ecosystems. Ecology 77: 364–374.
4. Amundson R (2001) The carbon budget in soils. Annual Review of Earth and Planetary Sciences 29: 535–562.
5. Davidson EA, Janssens IA (2006) Temperature sensitivity of soil carbon decomposition and feedbacks to climate change. Nature 440: 165–173.
6. Su ZY, Xiong YM, Zhu JY, Ye YC, Ye M (2006) Soil organic carbon content and distribution in a small landscape of Dongguan, South China. Pedosphere 16: 10–17.
7. IPCC (Intergovernmental Panel on Climate Change) (2000) Land use, land use change and forestry, a special report of the IPCC. In: Watson RT, Noble IR, Bolin B, Ravondranath NH, Verardo DJ, Dokken DJ, editors. Cambridge: Cambridge University Press. pp. 1–51.
8. Li H, Parent LE, Karam A, Tremblay C (2004) Potential of Sphagnum peat for improving soil organic matter, water holding capacity, bulk density and potato yield in a sandy soil. Plant and soil 265: 355–365.
9. Zhang C, McGrath D (2004) Geostatistical and GIS analyses on soil organic carbon concentrations in grassland of southeastern Ireland from two different periods. Geoderma 119: 261–275.
10. Don A, Schumacher J, Scherer-Lorenzen M, Scholten T, Schulze ED (2007) Spatial and vertical variation of soil carbon at two grassland sites-implications for measuring soil carbon stocks. Geoderma 141: 272–282.
11. Liu ZP, Shao MA, Wang YQ (2011) Effect of environmental factors on regional soil organic carbon stocks across the Loess Plateau region, China. Agriculture, Ecosystems and Environment 142: 184–194.
12. Batjes NH (1999) Management options for reducing CO_2-concentrations in the atmosphere by increasing carbon sequestration in the soil. Dutch National Research Programme on Global Air Pollution and Climate Change. Wageningen: ISRIC. Technical Paper 30, Report 410-200-031.
13. Marland G, Schlamadinger B (1999) The Kyoto Protocol could make a difference for the optimal forest-based CO_2 mitigation strategy: some results from GORCAM. Environmental Science and Policy 2: 111–124.
14. Schlesinger WH (1999) Carbon sequestration in soils. Science 284: 2095.
15. Lal R (2004) Carbon sequestration in dryland ecosystems. Environmental management 33: 528–544.
16. Venteris E, McCarty G, Ritchie J, Gish T (2004) Influence of management history and landscape variables on soil organic carbon and soil redistribution. Soil science 169: 787–795.
17. Hoffmann U, Yair A, Hikel H, Kuhn NJ (2012) Soil organic carbon in the rocky desert of northern Negev (Israel). Journal of Soils and Sediments 12:811–825.
18. Eswaran H, Van Den Berg E, Reich P (1993) Organic carbon in soils of the world. Soil Science Society of America Journal 57: 192–194.
19. Batjes NH (1996) Total carbon and nitrogen in the soils of the world. European Journal of Soil Science 47: 151–163.
20. Batjes N, Dijkshoorn J (1999) Carbon and nitrogen stocks in the soils of the Amazon Region. Geoderma 89: 273–286.
21. Arrouays D, Deslais W, Badeau V (2001) The carbon content of topsoil and its geographical distribution in France. Soil use and Management 17: 7–11.
22. Yu D, Shi X, Wang H, Sun W, Chen J, et al. (2007) Regional patterns of soil organic carbon stocks in China. Journal of Environmental Management 85: 680–689.
23. Grimm R, Behrens T, Märker M, Elsenbeer H (2008) Soil organic carbon concentrations and stocks on Barro Colorado Island-digital soil mapping using Random Forests analysis. Geoderma 146: 102–113.
24. Baritz R, Seufert G, Montanarella L, Van Ranst E (2010) Carbon concentrations and stocks in forest soils of Europe. Forest Ecology and Management 260: 262–277.
25. Wang YF, Fu B, Lü Y, Song C, Luan Y (2010) Local-scale spatial variability of soil organic carbon and its stock in the hilly area of the Loess Plateau, China. Quaternary Research 73: 70–76.

26. Murdock LW, Frye WW (1983) Erosion: its effect on soil properties, productivity and profit. publication AGR-102. College of Agriculture, University of Kentucky.
27. Duan ZH, Xiao HL, Dong ZB, He XD, Wang G (2001) Estimate of total CO_2 output from desertified sandy land in China. Atmospheric Environment 35: 5915–5921.
28. Li X, Wang Y, Liu L, Luo G, Li Y, et al. (2013) Effect of land use history and pattern on soil carbon storage in arid region of Central Asia. PloS one 8: e68372.
29. Cao CY, Jiang SY, Ying Z, Zhang FX, Han XS (2011) Spatial variability of soil nutrients and microbiological properties after the establishment of leguminous shrub Caragana microphylla Lam. plantation on sand dune in the Horqin Sandy Land of Northeast China. Ecological Engineering 37: 1467–1475.
30. Deng L, Shangguan ZP, Sweeney S (2013) Changes in soil carbon and nitrogen following land abandonment of farmland on the Loess Plateau, China. PloS one 8: e71923.
31. Zhou ZY, Li FR, Chen SK, Zhang HR, Li G (2011) Dynamics of vegetation and soil carbon and nitrogen accumulation over 26 years under controlled grazing in a desert shrubland. Plant and Soil 341: 257–268.
32. Lal R (2001) Potential of desertification control to sequester carbon and mitigate the greenhouse effect. Climatic Change 51: 35–72.
33. Zhu ZD, Chen GT (1994) Sandy desertification in China. Science Press, Beijing, pp. 20–33.
34. Feng Q, Endo K, Guodong C (2002) Soil carbon in desertified land in relation to site characteristics. Geoderma 106: 21–43.
35. Qi F, Guoduong C, Masao M (2001) The carbon cycle of sandy lands in China and its global significance. Climatic Change 48: 535–549.
36. Zhou RL, Li YQ, Zhao HL, Drake S (2008) Desertification effects on C and N content of sandy soils under grassland in Horqin, northern China. Geoderma 145: 370–375.
37. Zuo XA, Zhao HL, Zhao XY, Zhang TH, Guo YR, et al. (2008) Spatial pattern and heterogeneity of soil properties in sand dunes under grazing and restoration in Horqin Sandy Land, Northern China. Soil and Tillage Research 99: 202–212.
38. Zhao HL, He YH, Zhou RL, Su YZ, Li YQ, et al. (2009) Effects of desertification on soil organic C and N content in sandy farmland and grassland of Inner Mongolia. Catena 77: 187–191.
39. Zhou ZY, Li FR, Chen SK, Zhang HR, Li G (2011) Dynamics of vegetation and soil carbon and nitrogen accumulation over 26 years under controlled grazing in a desert shrubland. Plant and soil 341: 257–268.
40. Lioubimtseva E, Simon B, Faure H, Faure-Denard L, Adams J (1998) Impacts of climatic change on carbon storage in the Sahara–Gobi desert belt since the Last Glacial Maximum. Global and Planetary Change 16: 95–105.
41. Grüneberg E, Schöning I, Kalko EK, Weisser WW (2010) Regional organic carbon stock variability: A comparison between depth increments and soil horizons. Geoderma 155: 426–433.
42. Li QY, He ZB, Zhao WZ, Li QS (2004) Spatial pattern of Nitraria spaerocarpa population and dynamics in different habitats. Journal of Desert Research 24: 484–488 (in Chinese).
43. FAO/UNESCO (1988) 'Soil map of the world, revised legend.' (FAO/UNESCO: Rome)
44. Nelson DW, Sommers LE (Eds.) (1982) Total carbon, organic carbon and organic matter. In: Page AL, Miller RH, Keeney DR, editors. Methods of Soil 591 Analysis. Part 2. Agronomy Monograph, 2nded. ASA and SSSA, Madison, WI. pp 534–580.
45. Hu W, Shao MA, Wang QJ (2005) Scale-dependency of spatial variability of soil moisture on a degraded slope-land on the Loss Plateau. Transactions of the CSAE 21: 11–16 (in Chinese).
46. Zhang FS, Liu ZX, Zhang Y, Miao YG, Qu W, et al. (2009) Scaling effect on spatial variability of soil organic matter in crop land. Journal of the Graduate School of the Chinese Academy of Sciences 26: 350–356 (in Chinese).
47. Bond-Lamberty B, Brown KM, Goranson C, Gower ST (2006) Spatial dynamics of soil moisture and temperature in a black spruce boreal chronosequence. Canadian Journal of Forest Research-Revue Canadienne De Recherche Forestiere 36: 2794–2802.
48. Loescher HW, Ayres E, Duffy P, Luo H, Brunke M (2014) Spatial variation in soil properties among North American ecosystems and guidelines for sampling designs. Plos one 9: e83216.
49. Trangmar BB, Yost RS, Uehara G (1985) Application of geostatistics to spatial studies of soil properties. Advances in agronomy 38: 45–94.
50. Cambardella C, Moorman T, Parkin T, Karlen D, Novak J, et al. (1994) Field-scale variability of soil properties in central Iowa soils. Soil Science Society of America Journal 58: 1501–1511.
51. Wei JB, Xiao DN, Zhang XY, Li XZ, Li XY (2006) Spatial variability of soil organic carbon in relation to environmental factors of a typical small watershed in the black soil region, northeast China. Environmental Monitoring and Assessment 121: 597–613.
52. Marchetti A, Piccini C, Francaviglia R, Mabit L (2012) Spatial distribution of soil organic matter using geostatistics: A key indicator to assess soil degradation status in central Italy. Pedosphere 22: 230–242.
53. Arriaga FJ, Lowery B (2005) Spatial distribution of carbon over an eroded landscape in southwest Wisconsin. Soil and Tillage Research 81: 155–162.
54. Wang SQ, Zhou CH (1999) Estimating soil carbon reservoir of terrestrial ecosystem in China. Geographical Research 18: 349–356.
55. Office for the Second National Soil Survey of China (1993) Second National Soil Survey of China. Beijing: China Agricultural Press.
56. Liu ZP, Shao MA, Wang YQ (2012) Large-scale spatial variability and distribution of soil organic carbon across the entire Loess Plateau, China. Soil Research 50: 114–124.
57. Lei ZD, Yang SX, Xu ZR. (1985) Research spatial variability of soil properties. Journal of Hydraulic Engineering 9: 10–21 (in Chinese).
58. Fang X, Xue ZJ, Li BC, An SS (2012) Soil organic carbon distribution in relation to land use and its storage in a small watershed of the Loess Plateau, China. Catena, 88: 6–13.
59. Zibilske LM, Bradford JM, Smart JR (2002) Conservation tillage induced changes in organic carbon, total nitrogen and available phosphorus in a semi-arid alkaline subtropical soil. Soil and Tillage Research 66: 153–163.
60. Li J, Okin GS, Alvarez L, Epstein H (2008) Effects of wind erosion on the spatial heterogeneity of soil nutrients in two desert grassland communities. Biogeochemistry 88: 73–88.
61. Wang J, Fu B, Qiu Y, Chen L (2001) Soil nutrients in relation to land use and landscape position in the semi-arid small catchment on the loess plateau in China. Journal of Arid Environments 48: 537–550.
62. Six J, Conant RT, Paul EA, Paustian K (2002) Stabilization mechanisms of soil organic matter: Implications for C-saturation of soils. Plant and Soil 241: 155–176.
63. Saxton KE, Rawls WJ (2006) Soil water characteristic estimates by texture and organic matter for hydrologic solutions. Soil Science Society of America Journal 70: 1569–1578.
64. Tack FMG, Verloo MG, Vanmechelen L, Van Ranst E (1997) Baseline concentration levels of trace elements as a function of clay and organic carbon contents in soils in Flanders (Belgium). Science of the Total Environment 201: 113–123.
65. Zhang XP, Deng W, Yang XM (2002) The background concentrations of 13 soil trace elements and their relationships to parent materials and vegetation in Xizang (Tibet), China. Journal of Asian Earth Sciences 21: 167–174.
66. Su YZ, Yang R (2008) Background concentrations of elements in surface soils and their changes as affected by agriculture use in the desert-oasis ecotone in the middle of Heihe River Basin, North-west China. Journal of Geochemical Exploration 98: 57–64.
67. Li J, Okin GS, Alvarez L, Epstein H (2007) Quantitative effects of vegetation cover on wind erosion and soil nutrient loss in a desert grassland of southern New Mexico, USA. Biogeochemistry 85: 317–332.
68. Yan H, Wang S, Wang C, Zhang G, Patel N (2005) Losses of soil organic carbon under wind erosion in China. Global Change Biology 11: 828–840.
69. United Nations Framework Convention on Climate Change (2007) Tool for the Demonstration and Assessment of Additionality in A/R CDM Project Activities, Annex 17, CDM Executive Board 35, Bonn, Germany. Available: http://cdm.unfccc.int/methodologies/ARmethodologies/tools /ar-am-tool-01-v2.pdf
70. Li Z, Jiang X, Pan XZ, Zhao QG (2001) Organic carbon storage in soils of tropical and subtropical China. Water, Air, and Soil Pollution 129: 45–60.
71. Fan Y, Liu S, Zhang S, Deng L (2006) Background organic carbon storage of topsoil and whole profile of soils from Tibet District and their spatial distribution. Acta Ecologica Sinica 20: 2834–2846 (in Chinese).
72. Zhang Y, Zhao YC, Shi XZ, Lu XX, Yu DS, et al. (2008) Variation of soil organic carbon estimates in mountain regions: a case study form Southwest China. Geoderma 146: 449–456.
73. Feng YZ, Yang GH (2003) Countermeasures of sustainable development of land resources in the Northwest region. Gansu Science and Technology Press, Lanzhou, pp. 102.
74. Pei SF, Fu H, Wan CG (2008) Changes in soil properties and vegetation following exclosure and grazing in degraded Alxa desert steppe of Inner Mongolia, China. Agriculture, ecosystems and environment 124: 33–39.
75. Wang Y, Li Y, Ye X, Chu Y, Wang X (2010) Profile storage of organic/inorganic carbon in soil: From forest to desert. Science of the Total Environment 408: 1925–1931.
76. Yang X, Zhu B, Wang X, Li C, Zhou Z, et al. (2008) Late Quaternary environmental changes and organic carbon density in the Hunshandake Sandy Land, eastern Inner Mongolia, China. Global and Planetary Change 61, 70–78.
77. Zuo XA, Zhao XY, Zhao HL, Zhang TH, Guo YR, et al. (2009) Spatial heterogeneity of soil properties and vegetation-soil relationships following vegetation restoration of mobile dunes in Horqin Sandy Land, Northern China. Plant and soil 318, 153–167.

6

Tree Species Composition Influences Enzyme Activities and Microbial Biomass in the Rhizosphere: A Rhizobox Approach

Shengzuo Fang[1]*, Dong Liu[1], Ye Tian[1], Shiping Deng[1,2], Xulan Shang[1]

1 College of Forest Resources and Environment, Nanjing Forestry University, Nanjing, People's Republic of China, **2** Department of Plant and Soil Sciences, Oklahoma State University, Stillwater, Oklahoma, United States of America

Abstract

Monoculture causes nutrient losses and leads to declines in soil fertility and biomass production over successive cultivation. The rhizosphere, a zone of usually high microbial activities and clearly distinct from bulk soil, is defined as the volume of soil around living roots and influenced by root activities. Here we investigated enzyme activities and microbial biomass in the rhizosphere under different tree compositions. Six treatments with poplar, willow, and alder mono- or mixed seedlings were grown in rhizoboxes. Enzyme activities associated with nitrogen cycling and microbial biomass were measured in all rhizosphere and bulk soils. Both enzyme activities and microbial biomass in the rhizosphere differed significantly tree compositions. Microbial biomass contents were more sensitive to the changes of the rhizosphere environment than enzyme activities. Tree species coexistence did not consistently increase tested enzyme activities and microbial biomass, but varied depending on the complementarities of species traits. In general, impacts of tree species and coexistence were more pronounced on microbial composition than total biomass, evidenced by differences in microbial biomass C/N ratios stratified across the rhizosphere soils. Compared to poplar clone monoculture, other tree species addition obviously increased rhizosphere urease activity, but greatly reduced rhizosphere L-asparaginase activity. Poplar growth was enhanced only when coexisted with alder. Our results suggested that a highly productive or keystone plant species in a community had greater influence over soil functions than the contribution of diversity.

Editor: Han Y.H. Chen, Lakehead University, Canada

Funding: This work was supported by the National Basic Research Program of China (973 program, 2012CB416904) and the Knowledge Innovation Group Programs of Jiangsu Province, as well as the Doctorate Fellowship Foundation of Nanjing Forestry University. The funders had no role in study design, data collection and analysis, decision to publish, or preparation of the manuscript.

Competing Interests: The authors have declared that no competing interests exist.

* E-mail: fangsz@njfu.edu.cn

Introduction

Nutrient availability in the rhizosphere is governed by soil properties, plant nutrient uptake characteristics, and interactions of plants, microorganisms, and their surrounding soil environment. Growing evidence suggests that plant species have significant impacts not only on soil physicochemical properties and organic matter quality but also on the abundance and composition of soil microbial community [1–3]. It is well established that microbial abundance, composition and diversity can fundamentally alter soil processes that lead to changes in soil nutrient availability [4,5].

The rhizosphere is known to be a hot spot of plant-microbial interactions and a driving force of soil processes. Plant species could affect quantity and quality of carbon resources in the rhizosphere, which would influence the composition and diversity of microbial community in these environments [6–8]. Different plant species can promote proliferation of different microbial communities by releasing different amount and types of root exudates [9,10]. Coexistence of multiple plant species may enhance the complexity of soil microorganisms by increasing the heterogeneity of root exudates and carbon that are contributed from roots and decomposing litter [11,12]. Moreover, plants may directly or indirectly impact soil nutrient availability by influencing soil enzyme activities through releasing extracellular enzymes and/or altering microbial community that is known to be major contributors of enzyme activities in soil [13–15]. It has been reported that plants have a major influence in shaping rhizosphere microbial communities when they are grown in monoculture [6,16–18]. However, the effects of plant species coexistence on soil microbes and soil enzyme activities are less known. Some studies reported herbaceous plant species coexistence did not significantly affect soil microbial biomass C, but increased microbial functional group diversity index and enzyme activity under contaminated soils [15,19]. Mao et al. [20] suggested that mixing N-fixing shrub species into poplar stands can improve metabolic quotient and soil fertility and increase productivity in a long run. However, few studies have been conducted to elucidate effects of woody plant coexistence on rhizosphere microbial community and enzyme activities and the relationships between below-ground soil biology and above-ground plant growth characteristics.

Microbial characterization of tree rhizosphere provides vital information relating to evaluating soil nutrient status [18,21]. However, rhizosphere studies have been challenged by the lack of a satisfying method for obtaining sufficient soil samples for subsequent laboratorial analysis. Several procedures based on

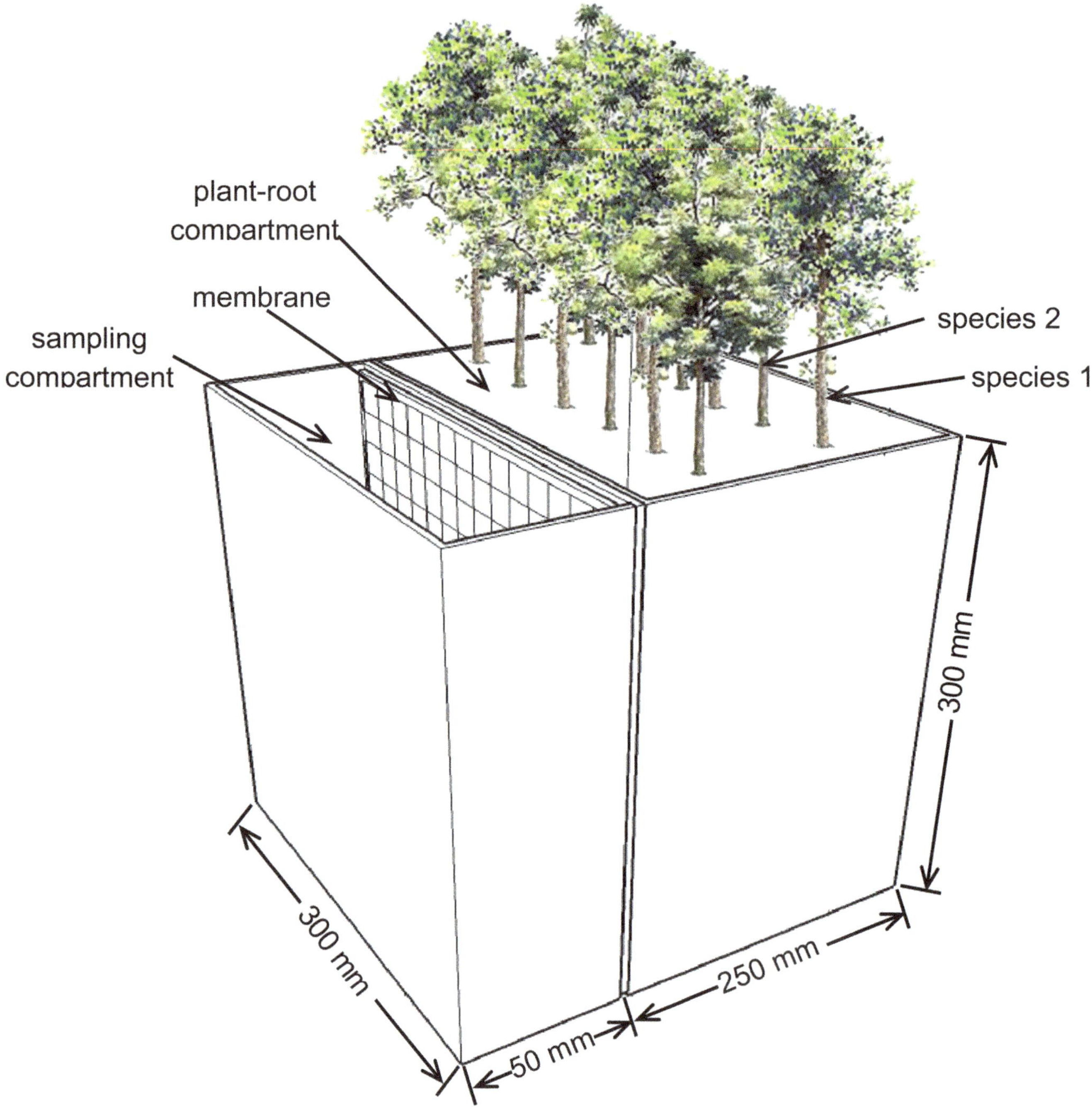

Figure 1. Schematics of the rhizobox used in the experiment. The rhizobox is 300×300 mm (depth ×width) with an internal volume of 22.5 L in the plant-root compartment and 4.5 L in the sampling compartment.

shaking or washing-off soil particles adhering to roots have been proposed for the separation of rhizosphere soil from bulk soil [22–24]. However, it is known that soil texture and moisture strongly influence the amount of soil adhering to the root system, and root induced changes on some soil variables have been observed up to about 7 mm from the surface of an active root segment or a root mat [25,26]. Therefore, rhizosphere studies often involve sampling procedures that permit evaluation of this gradient variation, making it challenging for data interpretation and comparison of results from different experiments.

Fast-growing plantations, such as poplar, are playing an increasingly important role in ameliorating environments and facilitating socio-economic sustainable development [27,28]. However, monoculture plantations lead to increasing challenges in pest control and declines in soil fertility and plant biomass productivity over successive rotations [20,29]. Therefore, mixed plantations have been recommended to combat insect damages and soil degradation while maintaining or increasing stand productivity [30,31].

Much of the past research efforts were devoted to understanding crop production ecosystems, with less known on the impact of tree compositions and species coexistence on soil nutrient cycling and

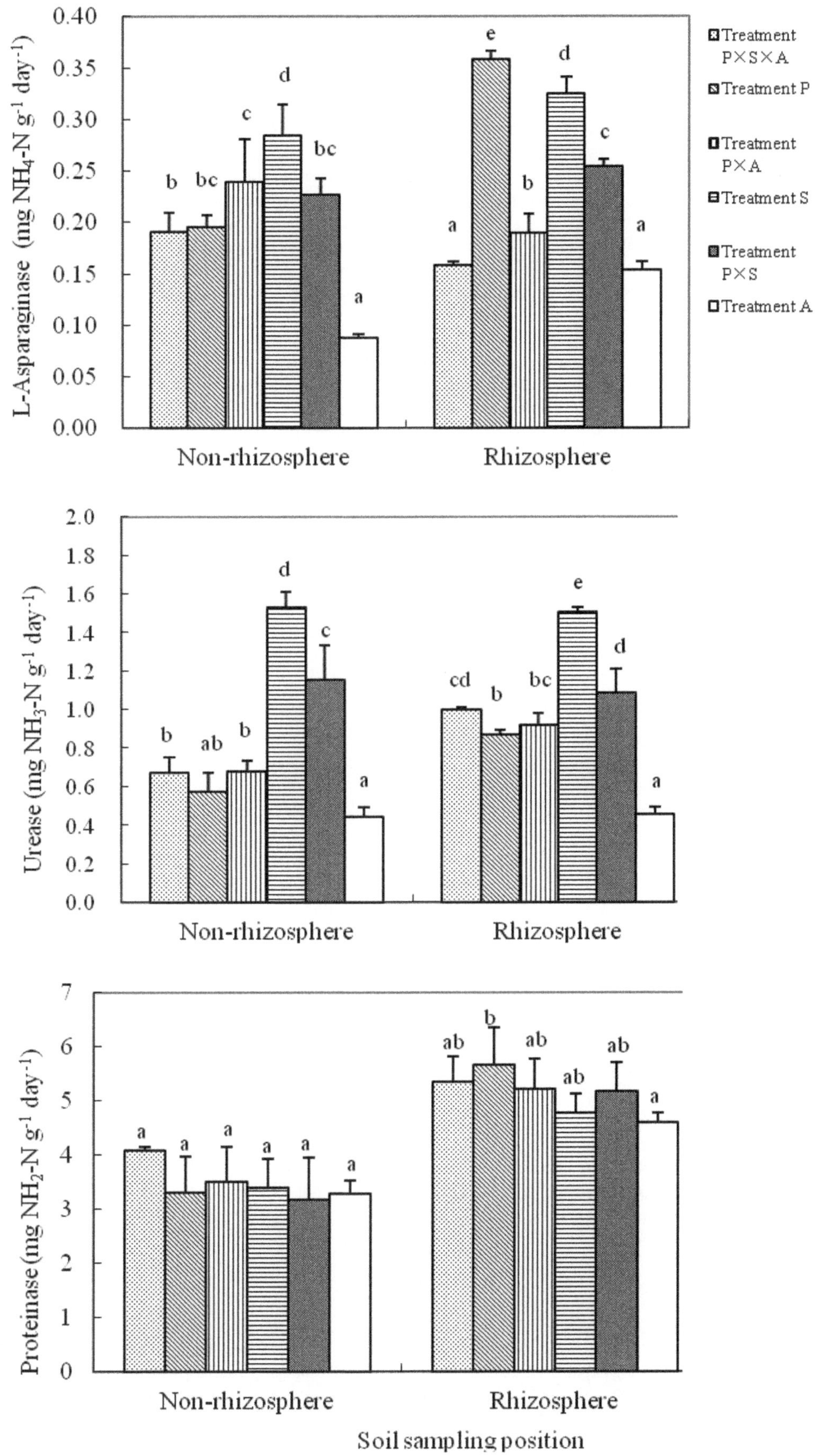

L-Asparaginase (mg NH_4-N g^{-1} day^{-1})
0.40
0.35
0.30
0.25
0.20
0.15
0.10
0.05
0.00
b bc c d bc a
a e b d c a
Non-rhizosphere
Rhizosphere
Treatment P×S×A
Treatment P
Treatment P×A
Treatment S
Treatment P×S
Treatment A
Urease (mg NH_3-N g^{-1} day^{-1})
2.0
1.8
1.6
1.4
1.2
1.0
0.8
0.6
0.4
0.2
0.0
b ab b d c a
cd b bc e d a
Non-rhizosphere
Rhizosphere
Proteinase (mg NH_2-N g^{-1} day^{-1})
7
6
5
4
3
2
1
0
a a a a a a
ab b ab ab ab a
Non-rhizosphere
Rhizosphere
Soil sampling position

Figure 2. Activities (mean ± standard deviation) of L-asparaginase, urease and proteinase in the rhizosphere (0 to 10 mm to the root mats) and non-rhizosphere soils (>10 mm to the root mats) under six different tree species compositions. Bars indicate standard deviations. Treatments P, S, A, P×S, P×A and P×S×A refer to poplar, willow, alder, poplar-willow, poplar-alder, and poplar-willow-alder, respectively. Different letters indicate significantly different means within the same sampling position at P<0.05 according to Duncan's multiple range test.

seedling growth [5]. In this study, we conducted a greenhouse experiment to evaluate the effect of monoculture and mixed plantations of three tree species on rhizosphere enzyme activities and microbial biomass content. The specific objectives were to: (1) assess the effects of tree species compositions on enzyme activities and microbial biomass in rhizospheric and non-rhizospheric soils and (2) investigate the relationships between enzyme activities and nitrogen (N) cycling and microbial biomass in rhizospheric soils.

Materials and Methods

Ethics Statement

No specific permissions were required for sampling the soil in the study area because the location is not privately-owned or protected in any way and there are no endangered or protected species in the soil sampled.

Rhizobox Design and the Planting Experiment

The rhizoboxes were made of two top-open black-plexiglas compartments separated by a polyamide membrane that has 30 µm pore-diameter and is 75 µm thick (Fig. 1). The plant-root compartment was 250 mm wide and had an internal volume of 22.5 L. The sampling compartment, where rhizosperic and bulk soils were sampled, was 50 mm wide and had an internal volume of 4.5 L. The soil used was taken from 5–15 cm depth of flooded area along Yangtze River near Maanshan City in Anhui Province (N31°41′, 118°26′E). This soil has pH of 7.2 and organic carbon of 19.0 g kg^{-1}, total nitrogen of 1.93 g kg^{-1}, total phosphorus of 0.40 g kg^{-1}, and total potassium of 6.96 g kg^{-1}. The two compartments of the rhizoboxes were packed with the soil at bulk density of 1.05 g cm^{-3} after the soil was sieved through a 2 mm mesh.

Three tree species that are commonly found in seasonal flooding areas and six compositions of these species were evaluated. The species included poplar (*Populus* × *euramericana* cv. 'Nanlin-895'), willow (*Salix* × *jiangsuensis* cv. J799), and alder (*Alnus trabeculosa*). The six tested treatments (planting patterns) included poplar (P), willow (S), alder (A), poplar-willow (P×S, 1:1), poplar-alder mixture (P×A, 1:1), and poplar-willow-alder mixture (P×S×A, 1:1:1). The unrooted cuttings of poplar and willow, and 1-year-old bare seedlings of alder were planted in March 2010. The cutting length for the two species was 15–18 cm, while the cutting diameters for poplar and willow were 1.0–1.2 cm and 0.6–0.8 cm, respectively. During the planting, heights of alder seedlings ranged from 30 to 35 cm and basal diameter was from 0.3 to 0.5 cm, with intact root systems. In the plant-root compartment of each rhizobox, 12 trees were planted in two rows that were spaced 5 cm × 10 cm between plants and 2.5 cm or 7.5 cm from the compartment edges. For treatments employed more than one species, different tree species were randomly placed in alternative spots within rows (see Fig. 1). Each treatment was replicated three times. Plants were grown in a greenhouse with night and day temperatures of 10 and 35°C and soil water content maintained at 60% water holding capacity. During the 8-month growing period, roots penetrated soil in the plant-root compartments, and root mats were formed along the membranes, which were in contact with soil in the rhizosphere compartments.

Soil Sampling and Preparation

After eight months of growth in the rhizobox (November, 2010), heights and basal diameters of all planted trees were measured; the two compartments of each rhizobox were separated carefully. In the sampling compartments, soils in contact with root mats were sliced vertically into three sequential sections of approximately 2 mm, 2 mm, and 6 mm width and sampled as rhizosphere soils, while the soils at the distance of 10–50 mm to the root mats were sampled as bulk soils (non-rhizosphere soils). Each sample was mixed thoroughly, sieved through a 2-mm mesh, and then divided into two portions. One was air dried for chemical analysis, and the other was stored at 4°C before subsequent analysis of microbial biomass and enzymatic activities.

Enzymatic Activity Assays

Enzyme activities associated with N cycling, including L-asparaginase, urease, and proteinase, were evaluated. L-asparaginase activity was determined as described by Frankenberger and Tabatabai [32] and was expressed as mg NH_4^+–N released g^{-1} soil day^{-1}. Briefly, 5 g of soil was incubated with buffered (0.1 M THAM, pH 10) L-asparagine at 37°C for 0.5 h, followed by the quantification of NH_4^+ released. Urease and proteinase activities were determined in triplicates at optimal pH for each enzyme according to methods described by Liu et al. [18]. Urease activity was expressed as mg NH_3^+–N released g^{-1} soil day^{-1}, while proteinase activity was expressed as mg tyrosine released g^{-1} soil day^{-1}.

Microbial Biomass Measurements

Fumigation and microbial biomass was measured following the procedure described by Joergensen and Brookes [33]. Each fresh soil (5 g on an oven-dry basis) was fumigated with ethanol-free $CHCl_3$ for 24 h. After adding 0.5 M K_2SO_4 (4:1 solution to soil ratio), the soil was shaken for 30 min on an oscillating shaker. The suspension was centrifuged, and then the supernatant was filtered through a Whatman polyethersulfone syringe filter (pore size: 0.45 µm). Non-fumigated soil was used as a control. All results are expressed on an oven-dry soil basis (105°C, 24 h) and are means of triplicate laboratory analyses of samples from three replicate treatments.

Microbial organic C (MOC) in the K_2SO_4 soil extracts was measured by an automated uv-persulphate oxidation method [34]. Microbial biomass C (MBC) was calculated from MBC = 2.22 MOC, where MOC = [(C extracted from fumigated soil)-(C extracted from non-fumigated soil)] [34]. Ninhydrin–N in the K_2SO_4 soil extracts was measured according to the method described by Joergensen and Brookes [33] and was used to calculate microbial biomass N (MBN) (MBN = 5.0 ninhydrin–N) [33,35].

Statistical Analyses

Statistical analyses were carried out using SPSS 13.0 (SPSS, Chicago, IL, USA). Both one-way and two-way ANOVA were conducted to compare enzymatic activities and microbial biomass in the rhizosphere and non-rhizosphere soils among tree composition treatments and distances to root mats. A Duncan's multiple range test was used to detect differences between treatment means. All statistical analyses were carried out at

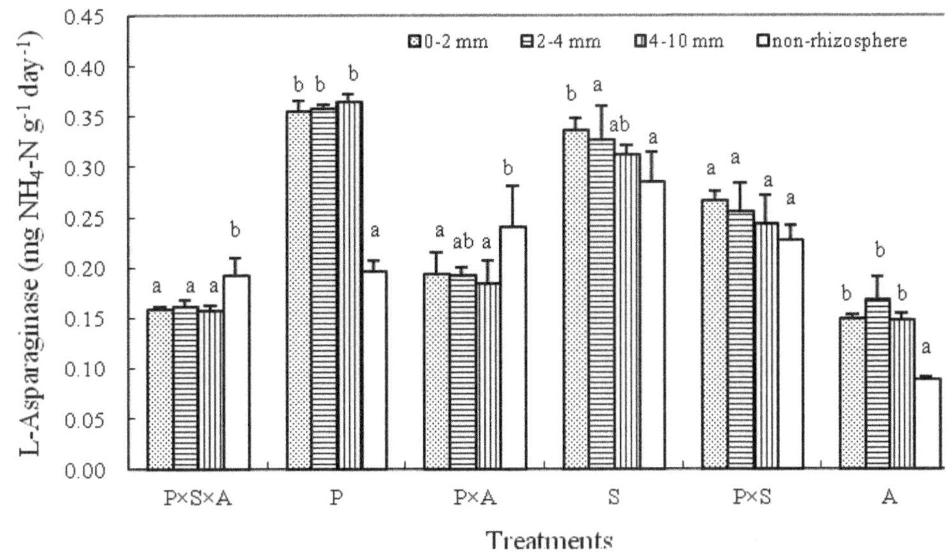
L-Asparaginase (mg NH4-N g-1 day-1)
0-2 mm
2-4 mm
4-10 mm
non-rhizosphere
P×S×A
P
P×A
S
P×S
A
Treatments

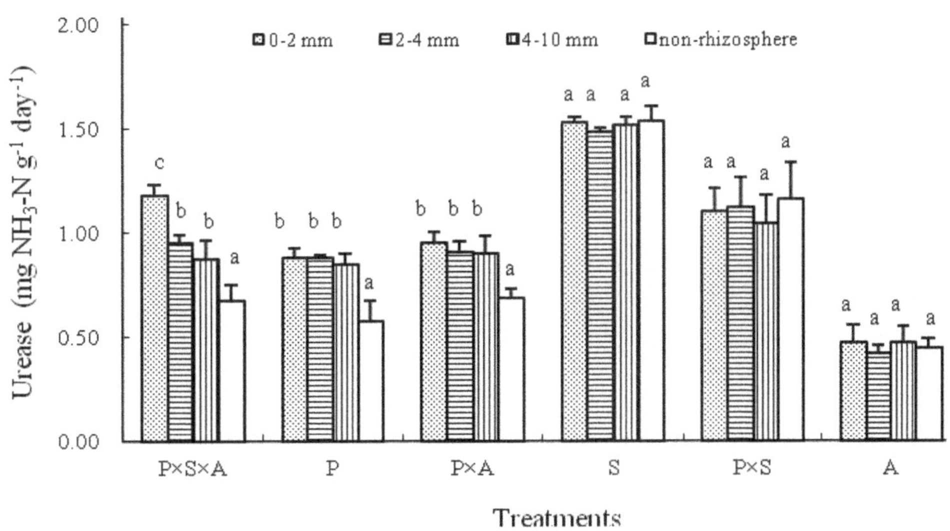
Urease (mg NH3-N g-1 day-1)
0-2 mm
2-4 mm
4-10 mm
non-rhizosphere
P×S×A
P
P×A
S
P×S
A
Treatments

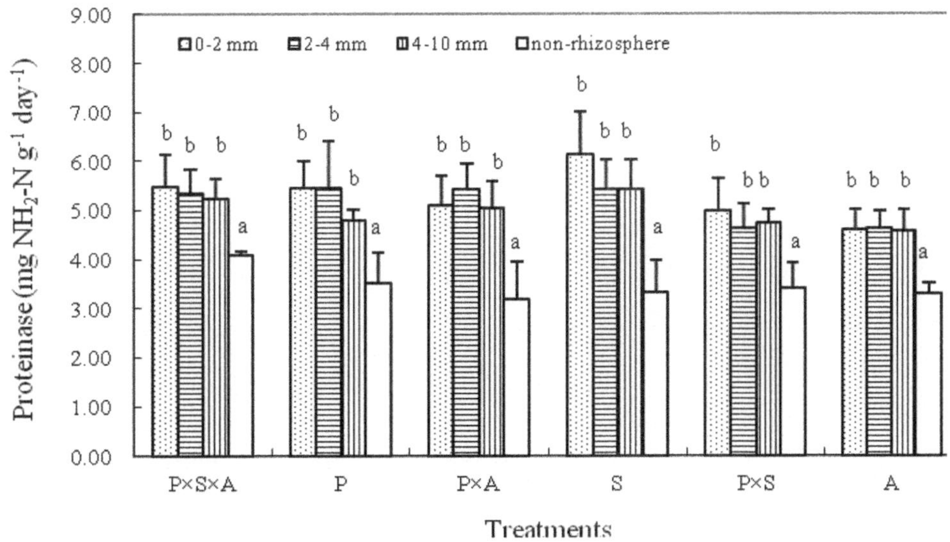
Proteinase (mg NH2-N g-1 day-1)
0-2 mm
2-4 mm
4-10 mm
non-rhizosphere
P×S×A
P
P×A
S
P×S
A
Treatments

Figure 3. Activities (mean ± standard deviation) of L-asparaginase, urease and proteinase in soils at different distances to the root mats under different tree species compositions. Bars indicate standard deviations. Treatments P, S, A, P×S, P×A and P×S×A refer to poplar, willow, alder, poplar-willow, poplar-alder, and poplar-willow-alder, respectively. Different letters indicate significantly different means within the same treatment at P<0.05 according to Duncan's multiple range test.

$p<0.05$. Relationships between enzyme activities and microbial biomass were evaluated using Pearson's correlation analysis.

Results

Enzymatic Activities

L-asparaginase urease, and proteinase in both rhizosphere and non-rhizosphere soils were significantly affected by tree species compositions ($p<0.05$), with the exception of proteinase activity in non-rhizospere soils (Fig. 2). In the rhizosphere, L-asparaginase varied from 0.15 to 0.36 mg NH_4^+–N g^{-1} day^{-1}, with the highest activity in treatment P (poplar monoculture), followed by treatment S (willow monoculture), and the lowest in treatment A (alder monoculture). L-asparaginase activity in the monoculture of poplar or willow was significantly higher than that in the mixed culture of poplar and willow (Fig. 2). Compared to poplar monoculture, L-asparaginase activity in treatment P×S×A (poplar-willow-alder coexistence), treatment P×A (poplar-alder coexistence) and treatment P×S (poplar-willow coexistence) was reduced by 126.1%, 98.1% and 51.6%, respectively. The trend of L-asparaginase activity in the non-rhizosphere soils of different planting patterns was somewhat different from that found in the rhizophere soils. The highest L-asparaginase activity in the non-rhizosphere soils was observed in treatment S, followed by patterns 3, 5, 2, and then treatment P×S×A, with the lowest found in treatment A (Fig. 2).

The rhizosphere urease activity ranged from 0.45 to 1.51 mg NH_3^+–N g^{-1} day^{-1}, with the highest in treatment S, followed by

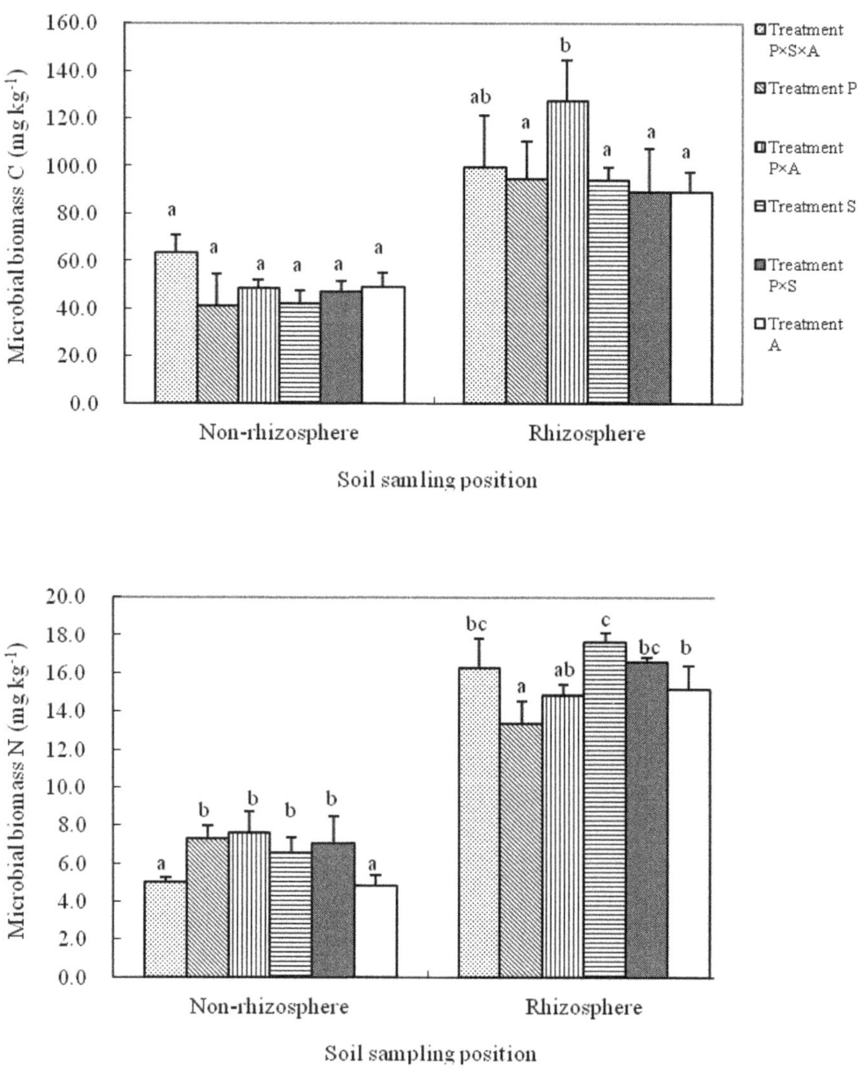

Figure 4. Content of microbial biomass C and N in the rhizosphere and non-rhizosphere soils under different tree species compositions (means ± standard deviation). Treatments P, S, A, P×S, P×A and P×S×A refer to poplar, willow, alder, poplar-willow, poplar-alder, and poplar-willow-alder, respectively. Different letters indicate significantly different means within the same sampling position at P<0.05 according to Duncan's multiple range test.

Table 1. Gradient variation of microbial biomass nitrogen (N) and carbon (C) content (mg kg^{-1}) in the rhizosphere of different tree species compositions.†

Index	Tree species composition (Planting patterns)	Microbial biomass N or C content in soils at distance to the root mats (mm)			
		0–2	2–4	4–10	10–50
Microbial biomass N	Poplar (treatment P)	17.0±1.6c	12.3±1.4b	10.9±1.5b	7.3±0.7a
	Willow (treatment S)	16.0±0.3b	16.5±1.1b	20.5±0.5c	6.6±0.8a
	Alder (treatment A)	17.5±0.8c	13.9±1.5b	14.2±1.6b	4.8±0.6a
	Poplar-willow (treatment P×S)	15.6±0.9b	17.9±0.8c	16.4±0.9bc	7.1±1.4a
	Poplar-alder (treatment P×A)	15.7±1.2b	15.1±0.9b	13.8±0.9b	7.6±1.1a
	Poplar-willow-alder (treatment P×S×A)	17.1±1.5c	18.7±1.5c	13.1±1.8b	5.0±0.2a
Microbial biomass C	Poplar	107.3±21.6b	99.3±11.5b	77.1±13.7b	41.5±12.9a
	Willow	100.3±3.4b	92.5±1.1b	89.3±9.9b	42.1±5.4a
	Alder	98.3±3.4c	82.3±9.1b	86.7±5.3bc	49.4±5.7a
	Poplar-willow	108.2±0.41c	87.1±5.9b	72.9±15.7b	47.0±4.6a
	Poplar-alder	137.9±17.5c	136.3±8.9c	107.7±12.6b	48.8±3.1a
	Poplar-willow-alder	123.0±21.0c	95.6±9.4b	81.1±7.5b	53.6±7.3a

†Data represent means ± standard deviation. Different lower case letters indicate significantly different means within the row at $P<0.05$ according to Duncan's multiple range test.

treatment P×S, and the lowest in treatment A (Fig. 2). Compared to poplar monoculture, other species addition significantly increased rhizosphere urease activity, where the urease activity in treatment P×S×A, treatment P×A and treatment P×S was 15.1%, 5.8% and 25.3% higher than the treatment P, respectively. Similar trend was observed in the non-rhizosphere soils (Fig. 2).

Trend for proteinase activities was somewhat different from those for activity of urease or L-asparaginase. Activity of proteinase was significantly higher in the rhizosphere soils than in non-rhizosphere soils, and planting patterns showed limited impact on its activities (Fig. 2). Proteinase activity ranged from 4.60 to 5.66 mg NH_2–N g^{-1} day^{-1} in the rhizosphere soils and 3.18 to 4.09 mg NH_2–N g^{-1} day^{-1} in the non-rhizosphere soils (Fig. 2). In the rhizosphere soils, the highest proteinase activity was found in treatment P, and the lowest in treatment A. However, proteinase activity in the non-rhizosphere soils was not significantly different among different planting patterns.

The spatial variation of enzyme activities in the rhizosphere is showed in Fig. 3. In most cases, no significant difference was detected for tested enzymes within 10 mm to the root mats in the rhizosphere soils. However, the rhizosphere L-asparaginase activity in treatment S and the urease activity in treatment P×S×A were significantly different among the three distances to the root mats within the rhizoshpere, showing an increasing tendency as approaching the root mats (Fig. 3).

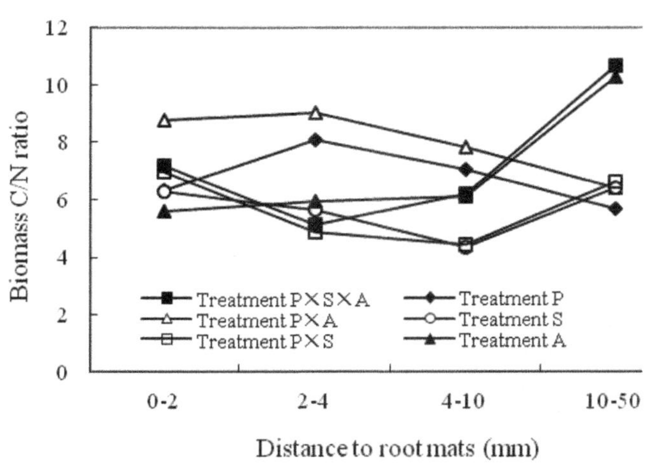

Figure 5. Ratios of microbial biomass C and N content in soils at different distances to the root mats under different tree species compositions. Treatments P, S, A, P×S, P×A and P×S×A refer to poplar, willow, alder, poplar-willow, poplar-alder, and poplar-willow-alder, respectively.

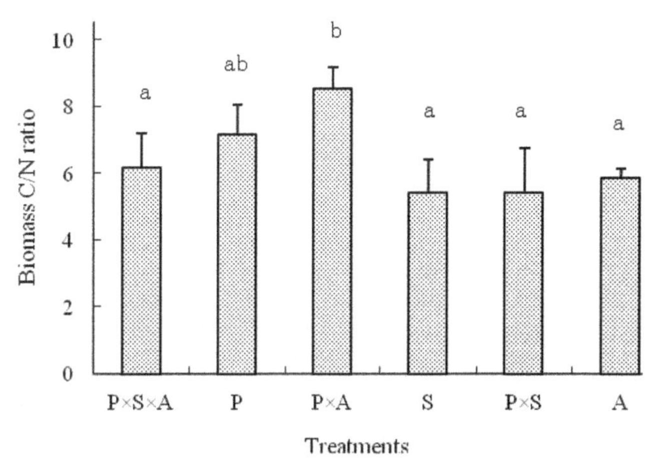

Figure 6. Ratios of microbial biomass C and N content in the rhizosphere soils under different tree species compositions (means ± standard deviation). Treatments P, S, A, P×S, P×A and P×S×A refer to poplar, willow, alder, poplar-willow, poplar-alder, and poplar-willow-alder, respectively. Means with the same letter are not significantly different at $P<0.05$ according to Duncan's multiple range test.

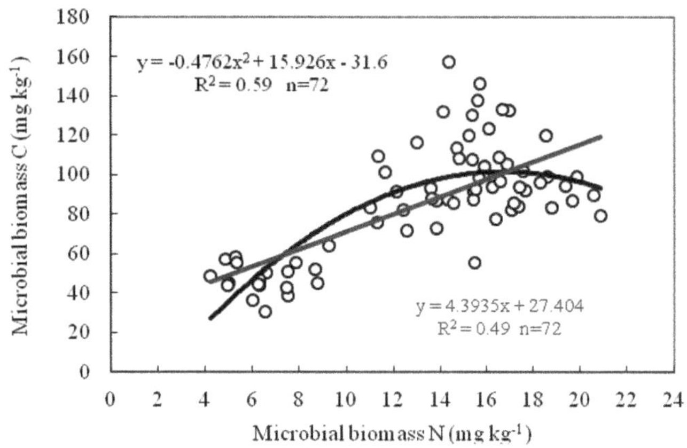

Figure 7. Relationships between microbial biomass C and N content in all soils tested (n = 72).

Microbial Biomass

Soil MBC and MBN varied significantly in the rhizosphere soils among different planting patterns (Fig. 4). The highest MBC was observed in treatment P×A, reaching 127.3 mg kg^{-1}. The MBC value in treatment P×A was significantly greater than the values in patterns 2, 4, 5 and 6, but was slightly higher than the value in treatment P×S×A. Compared to poplar monoculture, alder addition increased rhizosphere MBC in poplar mixtures, e.g. increasing 5.6% in treatment P×S×A and 34.6% in treatment P×A, respectively. However, the greatest MBN was measured in treatment S, followed by treatment P×S and the lowest was found in treatment P (Fig. 4). Compared to poplar monoculture, other species addition significantly increased rhizosphere MBN, where the MBN in treatment P×S×A, treatment P×A and treatment P×S was 21.6%, 11.2% and 23.9% higher than the treatment P, respectively.

Tree species compositions did not significantly affect MBC in the non-rhizosphere soils (Fig. 4). However, a significant difference in MBN among the planting patterns was observed, with treatment P×S×A and 6 being significantly lower than those found in other tested patterns (Fig. 4).

Comparing to MBC at 4–10 mm distance, its content at 0–2 mm was about 1.3–1.5 fold in mixed planting patterns, and 1.1–1.4 fold in monoculture. However, the MBN values at 4–10 mm distance in willow monoculture and poplar-willow mixture) was 28.1% and 5.1% higher than at 0–2 mm distance respectively, though the MBN at 0–2 mm distance increased by a factor of 1.1–1.6 in the other planting patterns compared to the distance of 4–10 mm.

The spatial variation of MBN and MBC in both rhizosphere and non-rhizosphere soils is showed in Table 1. For all planting patterns, MBN and MBC contents were significantly higher in rhizosphere than non-rhizosphere soils. In most cases, a significant difference in MBN and MBC was detected among the three distances to the root mats in the rhizosphere soils, except for MBN in treatment P×A and MBC in treatment P and treatment S (see Table 1). The mean MBN in the rhizosphere, where treatments with six planting patterns averaged within distance classes, was 16.5±0.8, 15.7±2.4 and 14.8±3.3 mg kg^{-1} at distances of 0–2, 2–4 and 4–10 mm to the root mats, respectively, while the mean MBC was 112.5^{b}±15.2, 98.9^{ab}±19.3 and 85.8^{a}±12.3 mg kg^{-1} (where different superscript letters indicate significant difference at $p<0.05$).

Table 3. Correlation coefficients between microbial biomass contents and enzyme activities in the soils (n = 24).

Index	Microbial biomass C	MBN	Urease	Proteinase
Microbial biomass N (MBN)	0.754**			
Urease	0.162	0.326		
Proteinase	0.822**	0.765**	0.154	
L-asparaginase	0.011	0.118	0.590**	0.228

**Significant at P<0.01.

Microbial C/N Ratio

The microbial C/N ratio in the rhizosphere ranged 5.6–8.6, 4.9–9.0, 4.4–7.8 at the distances of 0–2, 2–4 and 4–10 mm to the root mats, respectively, while in the non-rhizosphere was from 5.7 to 10.7 (Fig. 5). The mean C/N ratios across six planting patterns within distance classes, was 6.9±1.1, 6.4±1.7, 6.0±1.4 and 7.7±2.2 at 0–2, 2–4, 4–10 and 10–50 mm (non-rhizosphere soil) to the root mats respectively, and the observed variations between the distance to the root mats were not statistically significant ($p = 0.365$). However, the results demonstrated that the microbial C/N ratio in the rhizosphere was significantly influenced by the planting patterns (Fig. 6). The highest C/N ratio was achieved in

Table 2. Tree growth in eight months under various planting patterns.[†]

Planting patterns	Net height increment (cm)			Net basal diameter increment (cm)		
	poplar	willow	alder	poplar	willow	alder
Poplar-willow-alder	82.9±13.5a	170.8±20.3a	44.1±3.8a	0.65±0.22a	0.70±0.13ab	0.20±0.06a
Poplar-willow	83.1±10.5a	178.1±1.4a	/	0.58±0.08a	0.71±0.02b	/
Poplar-alder	141.9±6.8b	/	58.9±7.8b	0.96±0.05b	/	0.26±0.05a
Poplar	120.0±14.8b	/	/	0.91±0.11b	/	/
Willow	/	164.1±6.1a	/	/	0.53±0.06a	/
Alder	/	/	60.3±2.9b	/	/	0.33±0.06a

[†]Data represent means ± standard deviation. Different letters indicate significantly different means among the planting patterns at P<0.05 according to Duncan's multiple range test.

treatment P×A, which is 38.3, 19.3, 57.5, 57.4 and 45.1% greater than those in patterns 1, 2, 4, 5 and 6, respectively.

Furthermore, a significant correlation existed between soil microbial biomass C and N (Fig. 7), and R^2 reached 0.59 and 0.49 for Quadratic and linear equation, respectively ($n = 72$, $p<0.001$).

Seedling Growth

Seedling height and basal diameter growth of three tree species were affected by planting patterns (Table 2). At the end of the 8-month growing period, the highest increment of height and basal diameter for poplar and willow were not found in monoculture but mixed planting patterns. Poplar grew best in mixed culture with alder (treatment P×A), while willow in both mixed planting patterns. Alder grow best in monoculture, but its seedling height and basal diameters were not significantly higher than those in poplar-alder mixed planting patterns. Although willow grew best when planted with other two tree species, its presence led to low growth increments for poplar and alder. Compared to poplar monoculture (treatment P), annual increment of height in treatment P×A increased by 18.3%, whereas decreased by 30.9% and 30.8% in patterns 1 and 5, respectively. Similar trend was found in basal diameter increment.

Discussion

Tree Species Coexistence and Microbial Activity

In natural ecosystems, plant diversity influences the composition and biomass of microbial communities [36,37] because rhizodeposits or organic compounds released by plants can be highly specific for a given plant species or even a particular cultivar [17,38]. However, interactions between plant species and soil organisms have been reported to be context dependent [39] or even lack such relationships [17,40,41]. Results from the present study indicated that rhizosphere microbial activity was significantly affected by the tree species compositions, but the highest L-asparaginase, urease and proteinase as well as the greatest MBC value were observed in monoculture except that the highest MBN was achieved in the poplar-alder mixture (Fig. 2 and Fig. 4). Our results support the idea that a highly productive or keystone plant species present in the community results in greater influence over soil functions than the contribution of additional diversity per se [2,42,43], but in contrast to the idea that a higher diversity will have higher functioning by complementarity of species traits [42,44].

Data from this study indicated that composition and complementarity of species traits are also important factors influencing rhizosphere microbial community and functionality. This influence is, in part, evidenced by differences in microbial biomass C/N ratios at different distances from the root mats or among different planting patterns. Different microbial biomass C/N ratios signify changes in microbial community composition. Marilley and Aragno [45] revealed that members of the Cytophaga-Flexibacter-Bacteriodes phylum were ubiquitous in the bulk and rhizosphere soils of *Trifolium repens* and *Lolium perenne*. In fact, most soil microbes appear to be ubiquitous [46]. Based on PCR-DGGE detection, Kandeler et al. [47] showed that about 10% of the detected bacteria in the rhizosphere of maize were rhizosphere-specific. Nevertheless, the impact of plants to rhizosphere microbial community was detectable. Soil microbial C/N ratios often range between 5 to 8 [48]. In this study, microbial biomass C/N ratios were as high as about 11 under poplar-willow-alder or alder monoculture, suggesting that these rhizospheres promoted proliferation of actinomycetes and fungal community. All the high microbial biomass C/N ratios detected in this study were under alder mono- or mixed culture. It seems that alder could be playing a vital role inducing the observed higher microbial biomass C/N ratios. This coincided to certain degree with trend observed on seedling growth. Alder grew best in monoculture, while poplar grew best in mixed culture with alder. Further studies are needed to elucidate factors that induced the observed changes of microbial biomass to significantly higher C/N ratios than commonly observed in soil and relationship of these change to seedling growth. However, when included all data in the analysis, microbial biomass C/N ratio was about 4.4, which clearly indicate dominance of bacteria in the soil environment.

Relationship between Enzyme Activity and Microbial Biomass

A positive correlation between tested enzyme activity and microbial biomass was observed in this study (Table 3), but correlation coefficients were very different for the specific enzyme activity. Proteinase activity was significantly and positively correlated with MBC ($r = 0.822$, $p<0.01$) and MBN ($r = 0.765$, $p<0.01$), indicating that proteinase activity was associated with active microorganisms in the soil that are the major source of soil enzymes. However, urease activity was weakly correlated with MBC and MBN, and almost no correlation was found between L-asparaginase activity and soil microbial biomass (see Table 3), supporting the idea that plants also liberate enzymes to the soil through root exudates or after the death and rupture of the cells [49]. The correlations between microbial biomass and enzyme activity are influenced by many factors [50], and results reported are also inconsistent with different researchers. For example, Böhme et al. reported that a strong correlation between the activity of proteases and soil microbial biomass C was observed in the Bad Lauchstädt soil, while little correlation found for the soil from Keszthely [51]. Okur et al. indicated MBC was significantly and positively correlated with dehydrogenase, protease, urease, and alkaline phosphatase [52]. Highly positive correlations were observed between MBC and MBN as well as between L-asparaginase and urease activities in the present study ($p<0.01$), in coincidence with the results from Stark et al. [50].

Stratification of Enzyme Activity and Microbial Biomass in the Rhizosphere

Rhizoboxes have been widely used to study the rhizosphere of annual crops and young plants, but have been rarely applied to tree species [24,53]. In the present study we assessed enzyme activity and microbial biomass gradients in the rhizosphere (defined the soil within 10 mm to the root mats) associated with different tree species compositions. Our results indicated that in most cases no significant differences were detected for tested enzymes among the three distances (0–2, 2–4 and 4–10 mm) to the root mats in rhizosphere soils (Fig. 3), and urease activity at the distance of 0–2 mm increased by a factor of about 1.3 in treatment P×S×A compared to the distances of 2–4 and 4–10 mm. The results from this study are not similar to the reported result by Kandeler et al. [47] where enzyme activities increased by a factor of approximately 2 in the first 1 mm of a maize (*Zea mays*) rhizosphere, but similar to the results *in situ* rhizobox reported by Avianil et al. [53] where the gradients of amidase activity and hydrolytic activity in fruit tree rhizosphere (within 5 mm to the root mats) were barely observed. The observed differences among different studies could be due to difference in root systems of plant species under evaluation. However, the mean enzyme activities across all planting patterns were significantly higher within 10 mm than 10–50 mm to the root mats. This suggests that root effect on enzyme activities may extend beyond 10 mm. This is evidenced by the significantly different activities of L-asparaginase and urease in the non-rhizosphere soils (Fig. 3).

On the contrary, significant differences in MBN and MBC were observed in most cases among the three distances to the root mats in the rhizosphere soils (see Table 1). Regardless of planting patterns, content of MBN and MBC decreased with increasing distance to root mats. The higher microbial biomass in the rhizosphere than bulk soil was, in part, due to migration of microbes toward growing seedlings [54] or translocated by mass flow due to moisture gradient induced by plant transpiration. The microscale variability reflected changes in substrate quantity, quality, and composition in the rhizosphere as well as sensitivity of microbial community to these changes. This is consistent with observations that rhizosphere bacterial community is predominated by r-strategists that are capable of proliferating rapidly in response to changes of substrate [47]. Generally, the values in rhizosphere gradients of microbial biomass in the present study were less than the reported results from annual crops, where microbial biomass increased by a factor of 2–4 in ryegrass [55] or up to 10 in a model rhizosphere [56] for a 1 mm distance from the root surface.

Enzyme activity and microbial biomass along rhizosphere gradient are very important information for sampling procedures of rhizosphere soil and results from different experiments should be compared carefully. Our results indicated that mean activity of 3 tested enzymes and mean MBN in the rhizosphere, where treatments with six planting patterns averaged within distance classes, were not significantly different at distances of 0–2, 2–4 and 4–10 mm to the root mats, while a significant difference in mean MBC was observed between 0–2 mm and 4–10 mm distances from root mats. Results from this study suggest that defining the soil at 0–4 mm distance from the root mats as rhizosphere soil should be more reliable for this specific rhizobox design and experiment, though sampling procedures for rhizosphere soil should depend on soil texture and structure, plant species and observed parameter, in agreement with the results from an *in situ* rhizobox approach [18].

Conclusions

Composition and complementarity of species traits are important factors influencing rhizosphere microbial community and functionality. Both enzyme activity and microbial biomass in the rhizosphere were significantly influenced by tree species composition. However, impact of tree species coexistence on enzyme activities and microbial biomass varied, depending on complementarity of species traits. Compared to poplar monoculture, other tree species addition did not significantly affect rhizosphere proteinase activity, but led to significantly lower L-asparaginase and higher urease activity in the rhizosphere. Microbial biomass contents are more sensitive to changes in the rhizosphere environment than enzyme activities. The growth of poplar was enhanced only when coexisted with alder. In general, a highly productive or keystone plant species in a community has greater influence over soil functions than the contribution of diversity. Incorporate willow in a plantation may lead to marked increase in the capacity of soil to cycle N through activities of L-asparaginase and urease.

Acknowledgments

We thank Honghui Qu, Qing Chen, Qiongqiong Li and Wenwen Liu for their excellent assistance in soil collection and sample analysis. We are also grateful to Prof. Xizeng Xu for his valuable comments and useful suggestions on the research.

Author Contributions

Conceived and designed the experiments: SF YT. Performed the experiments: DL. Analyzed the data: DL. Contributed reagents/materials/analysis tools: DL YT XS. Wrote the paper: SF SD DL.

References

1. Ushio M, Wagai R, Balser TC, Kitayama K (2008) Variations in the soil microbial community composition of a tropical montane forest ecosystem: does tree species matter? Soil Biol Biochem 40: 2699–2702.
2. Ladygina N, Hedlund K (2010) Plant species influence microbial diversity and carbon allocation in the rhizosphere. Soil Biol Biochem 42: 162–168.
3. Zhang C, Liu G, Xue S, Song Z (2011) Rhizosphere soil microbial activity under different vegetation types on the Loess Plateau, China. Geoderma 161: 115–125.
4. Balser TC, Firestone MK (2005) Linking microbial community composition and soil processes in a California annual grassland and mixed-conifer forest. Biogeochemistry 73: 395–415.
5. Ushio M, Kitayama K, Balser TC (2010) Tree species effects on soil enzyme activities through effects on soil physicochemical and microbial properties in a tropical montane forest on Mt. Kinabalu, Borneo. Pedobiologia 53: 227–233.
6. Grayston SJ, Wang SQ, Campbell CD, Edwards AC (1998) Selective influence of plant species on microbial diversity in the rhizosphere. Soil Biol Biochem 30: 369–378.
7. Bardgett RD, Mawdsley JL, Edwards S, Hobbs PJ, Rodwell JS, et al. (1999) Plant species and nitrogen effects on soil biological properties of temperate upland grasslands. Funct Ecol 13: 650–660.
8. Berg G, Smalla K (2009) Plant species and soil type cooperatively shape the structure and function of microbial communities in the rhizosphere. FEMS Microbiol Ecol 68: 1–13.
9. Chen X, Tang JJ, Fang ZG, Hu S (2002) Phosphate-solubilizing microbes in rhizosphere soils of 19 weeds in southeastern China. J Zhejiang Univ Sci 3: 355–361.
10. Nguyen C (2003) Rhizodeposition of organic C by plants: mechanisms and controls. Agronomie 23: 375–396.
11. Broughton LC, Gross KL (2000) Patterns of diversity in plant and soil microbial communities along a productivity gradient in a Michigan old field. Oecologia 125: 420–427.
12. Stephan A, Meyer AH, Schmid B (2000) Plant diversity affects culturable soil bacteria in experimental grassland communities. J Ecol 88: 988–998.
13. Martens DA, Johanson JB, Frankerberger WT Jr (1992) Production and persistence of soil enzymes with repeated addition of organic residues. Soil Sci 153: 53–61.
14. Tabatabai MA (1994) Soil enzymes. In: Methods of Soil Analysis: Part 2. Microbiological and Biochemical Properties of Soils (Eds.: R.W. Weaver, G.S. Angle, P.S. Bottomley, D. Bezdicek, S. Smith, M.A. Tabatabai, A. Wollum). Soil Science Society of America, Madison, Wisconsin, 775–833.
15. Yang R, Tang J, Chen X, Hu S (2007) Effects of coexisting plant species on soil microbes and soil enzymes in metal lead contaminated soils. Appl Soil Ecol 37: 240–246.
16. Steer J, Harris JA (2000) Shifts in the microbial community in rhizosphere and non-rhizosphere soils during the growth of *Agrostis stolonifera*. Soil Biol Biochem 32: 869–878.
17. Singh BK, Munro S, Potts JM, Millard P (2007) Influence of grass species and soil type on rhizosphere microbial community structure in grassland soils. Appl Soil Ecol 36: 147–155.
18. Liu D, Fang S, Tian Y, Dun X (2012) Variation in rhizosphere soil microbial index of tree species on seasonal flooding land: An in situ rhizobox approach. Appl Soil Ecol 59: 1–11.
19. Gao Y, Zhou P, Mao L, Zhi Y, Zhang C, et al. (2010) Effects of plant species coexistence on soil enzyme activities and soil microbial community structure under Cd and Pb combined pollution. J Environ Sci 22: 1040–1048.
20. Mao R, Zeng DH, Ai GY, Yang D, Li LJ, et al. (2010) Soil microbiological and chemical effects of a nitrogen-fixing shrub in poplar plantations in semi-arid region of Northeast China. Eur J Soil Biol 46: 325–329.
21. Sinha S, Masto RE, Ram LC, Selvi VA, Srivastava NK, et al. (2009) Rhizosphere soil microbial index of tree species in a coal mining ecosys-tem. Soil Biol Biochem 41: 1824–1832.
22. Luster J, Göttlein A, Nowack B, Sarret G (2009) Sampling, defining, characterising and modeling the rhizosphere–the soil science tool box. Plant Soil 321: 457–482.
23. Naim MS (1965) Development of rhizosphere and rhizoplane microflora of *Artistida coerulescens* in the Lybian desert. Arch Mikrobiol 50: 321–325.
24. Yanai RD, Majdi H, Park BB (2003) Measured and modeled differences in nutrient concentrations between rhizosphere and bulk soil in a Norway spruce stand. Plant Soil 257: 133–142.
25. Jungk A, Claassen N (1997) Ion diffusion in the soil–root system. Adv Agron 61: 53–110.

26. Jones DL, Dennis PG, Owen G, Hees PW (2003) Organic acid behavior in soil-misconceptions and knowledge gaps. Plant Soil 248: 31–41.
27. Fang S, Liu J, Liu D, Xie B (2010) Enzymatic activity and nutrient availability in the rhizosphere of poplar plantations treated with fresh grass mulch. Soil Sci Plant Nutr 56: 483–491.
28. Fang S, Xue J, Tang L (2007) Biomass production and carbon sequestration potential in poplar plantations with different management patterns. J Environ Manag 85: 672–679.
29. Kelty MJ (2006) The role of species mixtures in plantation forestry. For Ecol Manag 233: 195–204.
30. Rothe A, Binkley D (2001) Nutritional interactions in mixed species forests: a synthesis. Can J For Res 31: 1855–1870.
31. Forrester DI, Bauhus J, Cowie AL, Vanclay JK (2006) Mixed-species plantations of Eucalyptus with nitrogen-fixing trees: a review. For Ecol Manag 233: 211–230.
32. Frankenberger WT Jr, Tabatabaim MA (1991) L-Asparaginase activity of soils. Biol Fertil Soils 11: 6–12.
33. Joergensen RG, Brookes PC (1990) Ninhydrin-reactive nitrogen measurements of microbial biomass in 0.5 M soil extracts. Soil Biol Biochem 22: 1023–1027.
34. Wu J, Joergensen RG, Pommerening B, Chaussod R, Brookes PC (1990) Measurement of soil microbial biomass C by fumigation-extraction-an automated procedure. Soil Biol Biochem 22: 1167–1169.
35. Inubushi K, Brookes PC, Jenkinson DS (1991) Soil microbial biomass C, N and ninhydrin-N in aerobic and anaerobic soils measured by the fumigation extraction method. Soil Biol Biochem 23: 737–741.
36. Wardle DA, Yeates GW, Williamson W, Bonner KI (2003) The response of a three trophic level soil food web to the identity and diversity of plant species and functional groups. Oikos 102: 45–56.
37. Prieto LH, Bertiller MB, Carrera AL, Olivera NL (2011) Soil enzyme and microbial activities in a grazing ecosystem of Patagonian Monte, Argentina. Geoderma 162: 281–287.
38. Jones DL, Hodge A, Kuzyakov Y (2004) Plant and mycorrhizal regulation of rhizodeposition. New Phytol 163: 459–480.
39. Bezemer MT, Lawsons CS, Hedlund K, Edwards AR, Brook AJ, et al. (2006) Plant species and functional group effects on abiotic and microbial soil properties and plant-soil feedback responses in two grasslands. J Ecol 94: 893–904.
40. Nunan N, Daniell TJ, Singh BK, Papert A, McNicol JW, et al. (2005) Links between plant and rhizoplane bacterial communities in grassland soils characterized using molecular techniques. Appl Environ Microbiol 71: 6784–6792.
41. Kielak A, Pijl AS, van Veen JA, Kowalchuk GA (2008) Differences in vegetation composition and plant species identity lead to only minor changes in soil-born microbial communities in a former arable field. FEMS Microbiol Ecol 63: 372–382.
42. Tilman D, Knops J, Wedin D, Reich R, Ritchie M, et al. (1997) The influence of functional diversity and composition on ecosystem processes. Science 277: 1300–1302.
43. Lepš J, Brown VK, Diaz TA, Gormsen D, Hedlund K, et al. (2001) Separating the chance effect from diversity per se effect in functioning of plant communities. Oikos 92: 123–134.
44. Loreau M (1998) Biodiversity and ecosystem functioning: a mechanistic model. Proc Natl Acad Sci USA 95: 5632–5636.
45. Marilley J, Aragno M (1999) Phylogenetic diversity of bacterial communities differing in degree of proximity of Lolium perenne and Trifolium repens roots. Appl Soil Ecol 13: 127–136.
46. Marschner P, Yang CH, Lieberei R, Crowley DE (2001) Soil and plant specific effects on bacterial community composition in the rhizosphere. Soil Biol Biochem 33: 1437–1445.
47. Kandeler E, Marscher P, Tscherko D, Gahoonia TS, Nielsen NE (2002) Microbial community composition and functional diversity in the rhizosphere of maize. Plant Soil 238: 301–312.
48. Sylvia DM, Fuhrmann A, Hartel PG, Zuberer DA (2005) Principles and Applications of Soil Microbiology. Pearson Education, Inc.: Upper Saddle River, New jersey.
49. Buée M, Boer WD, Martin F, van Overbeek L, Jurkevitch E (2009) The rhizosphere zoo: An overview of plant-associated communities of microorganisms, including phages, bacteria, archaea, and fungi, and of some of their structuring factors. Plant Soil 321: 189–212.
50. Stark C, Condron LM, Stewart A, Di HJ, O'Callaghan M (2007) Effects of past and current crop management on soil microbial biomass and activity. Biol Fertil Soils 43: 531–540.
51. Böhme L, Langer U, Böhme F (2005) Microbial biomass, enzyme activities and microbial community structure in two European long-term field experiments. Agric Ecosyst Environ 109: 141–152.
52. Okur N, Altindışl A, Cengel M, Gocmez S, Kayikcioğlu HH (2009) Microbial biomass and enzyme activity in vineyard soils under organic and conventional farming systems. Turk J Agric For 33: 413–423.
53. Avianil I, Laor Y, Raviv M (2006) Limitations and potential of in situ rhizobox sampling for assessing microbial activity in fruit tree rhizosphere. Plant Soil 279: 327–332.
54. Bashan Y, Holguin G (1997) Azospirillum-plant relationships: Environmental and physiological advances (1990–1996). Can J Microbiol 43: 103–121.
55. De Neergaard A, Magid J (2001) Influence of the rhizosphere on microbial biomass and recently formed organic matter. Eur J Soil Sci 52: 377–384.
56. Yeates G, Darrah PR (1991) Microbial changes in a model rhizosphere. Soil Biol Biochem 23: 963–971.

7

Rice-Straw Mulch Reduces the Green Peach Aphid, *Myzus persicae* (Hemiptera: Aphididae) Populations on Kale, *Brassica oleracea* var. *acephala* (Brassicaceae) Plants

Reinildes Silva-Filho[1], Ricardo Henrique Silva Santos[2], Wagner de Souza Tavares[2], Germano Leão Demolin Leite[3], Carlos Frederico Wilcken[4], José Eduardo Serrão[5], José Cola Zanuncio[1]*

1 Departamento de Biologia Animal, Universidade Federal de Viçosa, Viçosa, Brazil, **2** Departamento de Fitotecnia, Universidade Federal de Viçosa, Viçosa, Brazil, **3** Instituto de Ciências Agrárias, Universidade Federal de Minas Gerais, Montes Claros, Brazil, **4** Departamento de Proteção Vegetal, Universidade Estadual Paulista, UNESP, campus of Botucatu, Botucatu, Brazil, **5** Departamento de Biologia Geral, Universidade Federal de Viçosa, Viçosa, Brazil

Abstract

Organic mulches, like peel and rice-straw, besides other materials affect the UV and temperature, which cause a reduction in the aphid arrival. The aim was to evaluate the effect of covering the soil with straw on the populations of the green peach aphid, *Myzus persicae* on the kale, *Brassica oleracea* var. *acephala* plants. The first experiment evaluated the direct effect of the rice-straw mulch and the second its indirect effect on aphid immigration, testing the plant characteristics that could lead to the landing preference of this insect. The third experiment evaluated the direct effect of the mulch on the aphid population. In the second and third experiments, four plants, each in a 14 L polyethylene pot with holes at the bottom, were used in areas with and without soil mulching. These pots were changed between areas, after seven days, to evaluate the effects of this change on the arrival of the winged aphids to the plants. Each plant was covered with anti-aphid gauze and inoculated with one winged *M. persicae*. Winged and apterous adults of this insect were counted per plant after 15 days. The temperature increased in the mulched plots to a maximum of 21–36°C and to 18–32°C in the plots with or without soil covering, respectively. Plant growth reduced the numbers of the winged aphids landing before and after they were moved to the bare soil plots. The nutrient content was similar in plants in both the mulched and no mulched plots. The population growth of *M. persicae* was higher in the control than in the mulched plots. This was partially due to temperatures close to 30°C in these plots and changes in the plant physiology. The soil mulching with rice-straw decreased the *M. persicae* landing, increased the plot temperatures and improved the vegetative growth of the kale plants.

Editor: Martin Heil, Centro de Investigación y de Estudios Avanzados, Mexico

Funding: Funding was received from the Conselho Nacional de Desenvolvimento Científico e Tecnológico (CNPq), Coordenação de Aperfeiçoamento de Pessoal de Nível Superior (CAPES) and Fundação de Amparo à Pesquisa do Estado de Minas Gerais (FAPEMIG). The funders had no role in study design, data collection and analysis, decision to publish, or preparation of the manuscript.

Competing Interests: The authors have declared that no competing interests exist.

* E-mail: zanuncio@ufv.br

Introduction

The chemical control of pests on the *Brassica* spp. L. (Brassicaceae) plants increased the production costs and environment contamination [1,2]. However, reflective mulches, including rice-straw, *Oryza sativa* L. (Poaceae) is a suitable alternative to reduce it [3–5]. Mulch with synthetic materials and rice peels or rice-straw can reduce the aphid infestation [3,6,7]. The organic and synthetic mulches can repel the winged aphid due to theirs color, heat reflection and increase in the UV and air temperature near to the mulch, although synthetic ones are more reflective [3,8].

The peel and rice-straw reflect UV, which reduced the virus transmitted by the cabbage aphid, *Brevicoryne brassicae* L., 1758 (Hemiptera: Aphididae) to potato, *Solanum tuberosum* L. (Solanaceae) and kale, *Brassica oleracea* var. *acephala* (Brassicaceae) plants [4,9]. This covering is more significant than the synthetic ones, as it is more economical and does not require removal [10–12], improves the soil by incorporating organic material and does not contaminate the environment [13–15]. Mulch can also directly affect the aphid immigration and their populations indirectly by changes in the plant physiology which could lead to the lower landing preference and population growth of these insects [16–18].

The green peach aphid, *Myzus persicae* Sulzer, 1776 (Hemiptera: Aphididae) is one of the most important agricultural pests in the world, which reduces the yield production in more than 40 different plant families [19,20]. This peach aphid is an efficient vector of more than 100 plant viruses [21,22] and resistant to the major insecticides being used [23,24].

This work studied the effects of the rice-straw mulch on the *M. persicae* populations on kale plants, to test the following hypotheses: (a) the mulch reduces the aphid landing on the kale plants; (b) this effect is due to the higher air temperature in the plant environment, (c) plant nutrient content changes, and (d) the factors that increase aphid population.

Materials and Methods

The Direct Effect of the Mulch on Aphid Immigration

This experiment evaluates the direct effect of the rice-straw mulch on the *M. persicae* immigration to the kale plants in the field at the "Universidade Federal de Viçosa (UFV)" in Viçosa, Brazil. The kale clone cultivar (*B. oleraceae* var. *acephala*) was obtained from the germplasm bank of the UFV and the rice-straw mulch from the agro-ecological sector of the UFV. The experiment was conducted with five randomized blocks and two treatments (with or without the mulch). Each plot had 16 plants spaced at 0.80 m×0.80 m and 1.20 m apart.

The kale seedlings were then transplanted and the plants were thinned on October 16 and 17, 2012 leaving them with five leaves each. A total of 2.5 L of chicken manure and 200 g of 4-14-8 NPK were applied per hole during the transplantation of the kale plants and they were daily sprinkler-watered over the next 15 days.

Winged and apterous aphids were removed from the kale leaves and 1.056 Kg of dry matter.m^{-2} of rice-straw per hectare was placed on the soil. The winged aphids landing on the four central plants per plot were counted and removed daily at 5:00 P.M. between October 18 and November 1, 2012. The four central plants were chosen to avoid edge effects on the lateral ones. The daily maximum temperature was recorded with a max-min thermometer in the center of the plots at plant canopy height, opposed to direct sunlight. The maximum temperature was recorded because it was positively correlated to the aphid flying period [25,26].

The data were evaluated by covariance analysis, with Poisson error and $\chi 2$ of Pearson to correct over dispersion, following F test at 5% probability. The average winged *M. persicae* which landed on the four kale plants per plot were obtained over 15 days. The maximum average temperature was the first exploratory variable (x_1), the mulched and no mulched treatments, the second one (x_2), and the mean number of the winged aphids the response one (y) on plots sampled. "Sistema para Análises Estatísticas e Genéticas (SAEG)" version 9.1 [27] (Supplier: UFV) was the software used for data analysis.

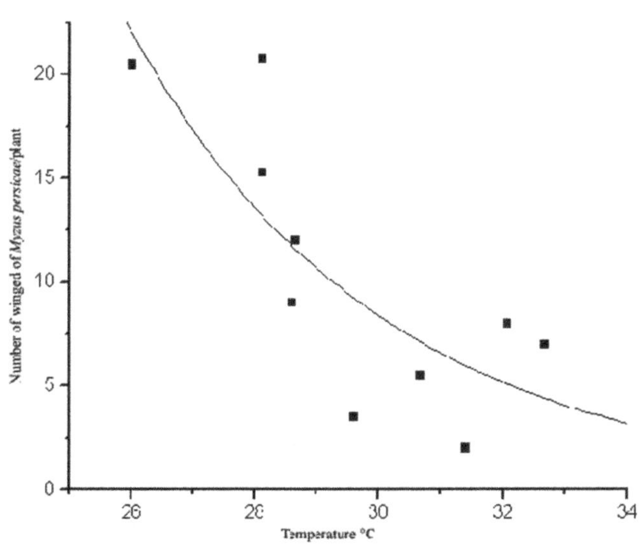

Figure 1. Number of winged green peach aphid, *Myzus persicae* (Hemiptera: Aphididae) landed on kale, *Brassica oleracea* var. *acephala* (Brassicaceae) plants as related to maximum temperature in the plots, with $F_{1,9}=39.03$, $p\leq 0.008$, $r^2=0.60$. Each dot corresponds to the mean number of aphids per plant in 10 plots with a Poisson error and $\chi 2$ of Pearson to correct over dispersion, followed F test at 5% probability.

The Indirect Effect of the Mulch on Aphid Immigration

The second experiment evaluated the indirect effect of the mulch on aphid immigration to test if the effect on the nutrient content of the kale plants was correlated with the aphid landing. These plants were cultivated in pots in bare soil or in mulch plots and their positions were changed after seven days. The change in the positions of the pots was made after the evaluations: those in the plots with soil covering were transferred to plots without covering and vice-versa. Counting and removal of the winged aphids continued for more seven days. The lower degree of immigration of this insect to the bare soil plots after changing the pots positions would imply that the plants carried over some characteristics from the previous environment. The kale seedlings were transplanted on April 30, 2013 to 14 L polyethylene pots with soil fertilized with 50 g of NPK (formulation 4-14-8) and 2.0 L of chicken manure which was chosen due to its richness in nutrients and availability [28]. Poultry manure may have some additional effects on aphid population along with effects of rice-straw mulch such as increase plant vigor and water retention [29]. All the aphids were eliminated and the pots were placed into holes at ground level in the field on May 30, 2013. The plot location and design, leaf thinning, sampling, aphid removal, rice-straw layer and irrigation were conducted similar to the first experiment.

The aphids were counted from May 30 to June 5, 2012 ('beginning' period) and from June 6 to June 12, 2012 (latter period) after the pots had their positions changed. Two leaves were taken from a kale plant per plot in June 5, 2012 to determine their macronutrient content. The material sampled was washed with deionized water, dried until constant weight, grounded and the NO_3^-, NH_4^+, P, K, Ca, Mg and S content were determined based on the dry matter [30,31].

Data were evaluated by multiple regression analysis, with Poisson error, and $\chi 2$ Pearson to correct over dispersion followed by the F test at 5% probability. The average daily number of winged *M. persicae* on the plants per plot was obtained during the seven day period. The exploratory variables were $x_1=NO_3^-$, $x_2=NH_4^+$, $x_3=P$, $x_4=K$, $x_5=Ca$, $x_6=Mg$, $x_7=S$ and $x_8=$ mulch treatments and the mean number of winged aphids was the result (y) on the plots sampled. The nutrient content of the kale plants on the mulched and bare soils was compared with the analysis of variance ($p\leq 0.05$).

The Effect of Mulch on the Aphid Population Growth

One winged *M. persicae* was inoculated per kale leaf to evaluate the mulch effect on the growth of the aphid colony. These leaves were covered with anti-aphid gauze to prevent the inoculated aphids from flying out, other aphids from landing and to avoid predation. The aphids were inoculated on June 12, 2012 and all the nymphs and winged adults of *M. persicae* were counted on June 27, 2012. This experiment had five replications, and the plot design and plant manipulation were similar to the second experiment.

The data were evaluated using the variance analysis with Poison error and F test at 5% probability. The treatments (with or without mulch) were considered the exploratory variable (x_1) and the number of nymphs and adult winged aphids were the response variables (y).

No specific permits are required for the plantation of kale and rearing *M. persicae* in Brazil. The field and laboratory studies did not involve endangered or protected species.

Results

The number of winged *M. persicae* was lower on the kale plants of the mulched plots than on those in the bare soil ($F = 39.03$; d.f. = 1.8; $p \leq 0.008$), 83 and 327 individuals, respectively, during the monitoring period. The mulch reduced the number of the winged *M. persicae* on the kale plants during the 15-day period (Fig. 1) and increased the maximum temperature at canopy height (Fig. 2). The maximum temperature ranged between 21–36°C and 18–32°C on the mulched and no mulched plots, respectively, which implies that these factors had reduced the arrival of the winged aphids.

As in the first experiment, the mulch decreased the *M. persicae* immigration ($F = 26.52$; d.f. = 1.7; $p \leq 0.0001$). The number of individuals of this aphid was 1 and 61 and 520 and 351, at the beginning and after the change in the positions of the pots, in the mulched and no mulched plots, respectively. The number of aphids which landed was higher in the no mulched plants moved to the mulched plots ($F = 26.52$; d.f. = 1.7; $p \leq 0.0001$), compared with the 'beginning' period. On the other hand, plants from the mulched plots had a lower number of winged aphids when moved to the no mulched plots (Fig. 3). The macronutrient content of the kale leaves was similar in the mulched and no mulched plots ($p \geq 0.05$), with means of: $NO_3^- = 1.30$ dag.Kg^{-1}, $NH_4^+ = 1.36$ dag.Kg^{-1}, Ca = 1.88 dag.Kg^{-1}, S = 0.10 dag.Kg^{-1}, K = 2.41 dag.Kg^{-1}, Mg = 0.43 dag.Kg^{-1}, and P = 0.46 dag.Kg^{-1}. The nutrient content boron no relation to the number of aphids on the kale plants ($p \geq 0.05$).

The mulch reduced the population growth of *M. persicae* with a lesser number of nymphs, adults ($F = 35.62$; d.f. = 1.8; $p \leq 0.0001$) and winged individuals ($F = 8.12$; d.f = 1.8; $p \leq 0.03$) in the mulched plots 15 days after the inoculation with the winged insect (Fig. 4). In all, 31 adults and nymphs and 1 winged individual were found per colony of *M. persicae* 15 days after the inoculation, with a temperature of 33.8°C in the mulched plots. This value was 318 adults and nymphs and 3 winged adults per colony over the same 15 days in the no mulched plots which had a temperature of 28.9°C.

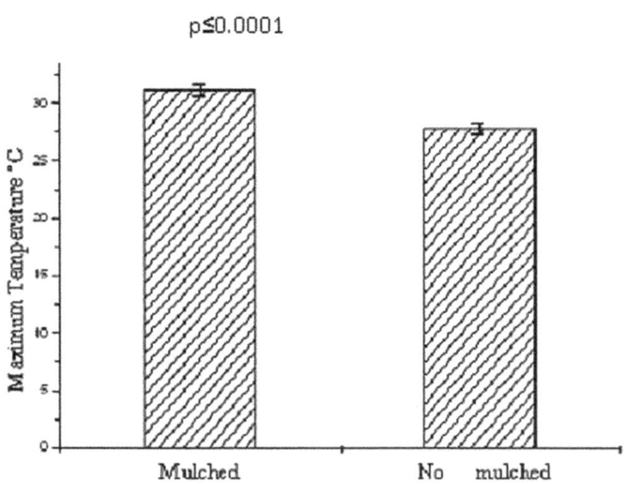

Figure 2. Maximum temperature in the plots with kale, *Brassica oleracea* var. *acephala* (Brassicaceae) plants on mulched and no mulched plots (average of 15 days) with Poisson error and $\chi 2$ of Pearson to correct over dispersion followed by an F test at 5% probability.

Discussion

The temperature increasing in the plots by the mulch may explain the reduced number of arriving winged *M. persicae* on the kale plants as reported for the lower populations of *Aphis* sp. (Hemiptera: Aphididae), *Myzus* sp. and silverleaf whitefly, *Bemisia argentifolii* Bellows & Perring, 1994 (Hemiptera: Aleyrodidae) on watermelon, *Citrullus lanatus* (Thunb.) Matsum. & Nakai (Cucurbitaceae) and corn, *Zea mays* L. (Poaceae) with the polyethylene mulch which was attributed to raising the soil temperature and plant growth besides light reflectance [32–34]. This confirms that the mulch can decrease the aphid immigration due to the temperature increase with a negative impact on the aphid populations by slowing down the development and reducing the fecundity [19,33]. Mulch obtained from a crop of sunn hemp, *Crotalaria juncea* L. (Fabaceae) or weeds (original weed cover) acts as physical barrier to lesser cornstalk borer, *Elasmopalpus lignosellus* Zeller, 1848 (Lepidoptera: Pyralidae) around host plants of bush bean, *Phaseolus vulgaris* L. (Fabaceae) in Florida, USA [35]. High temperatures can be harmful to the developing embryos and thus, reduce the population growth in subsequent generations [36,37].

The indirect effect of the mulch on the *M. persicae* immigration to the kale plants may have induced this aphid to move to the no mulched plots. This result suggests that the differences in the plant physiology in each environment were maintained for at least seven days. On the other hand, the effectiveness of these characteristics (higher plants and better leaf development) to repel the winged aphids is lower than the mulch effect. Mulching can improve the physical and chemical properties of the soil and favors plant growth [38,39]. The maintenance of adequate soil conditions is important to assure satisfactory crop growth and high yields [8,40] with a good impact on the insect pests [41,42]. Plants cultivated on the mulched plots were more vigorous and with higher biomass which could allow more effective defense mechanisms. This probably includes the production of semiochemicals which repel the aphids, thus resulting in lower *M. persicae* populations [4,43,44]. This is due to the humidity maintenance, higher soil temperature, better weed control and mineral nutrient increase in the plants in the mulched plots [45,46]. The use of cover crops and mulches are convenient non-chemical methods for managing some insect pests and weeds. The numbers of Cicadellidae (Hemiptera),

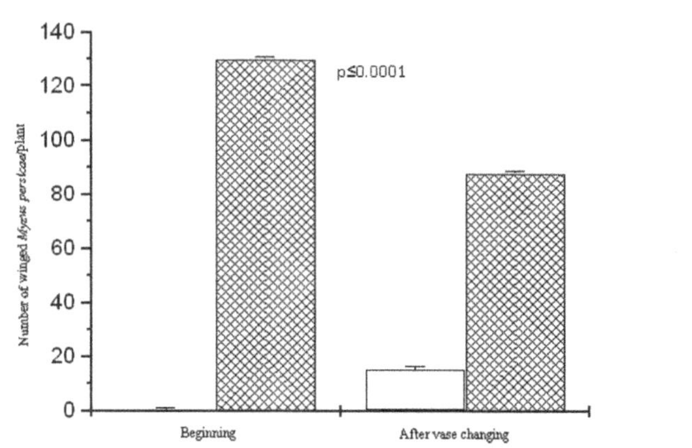

Figure 3. Mean number of winged green peach aphid, *Myzus persicae* (Hemiptera: Aphididae) landed on kale, *Brassica oleracea* var. *acephala* (Brassicaceae) plants on mulched and no mulched plots (hatched bars) at the beginning (days 1-7) and after vase position was changed (days 8-14). Poisson error and $\chi 2$ of Pearson to correct over dispersion followed by an F test at 5% probability.

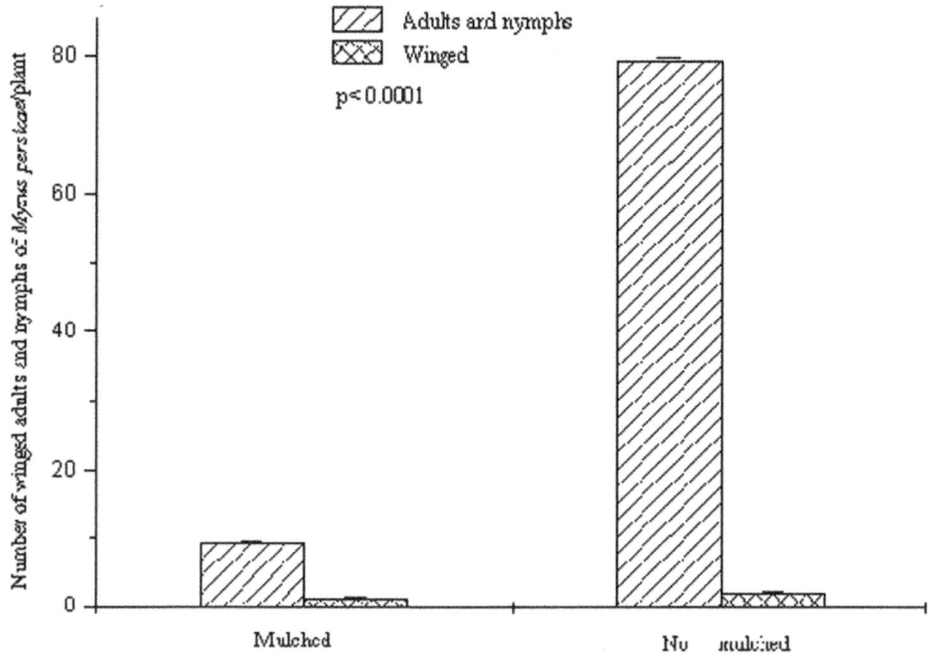

Figure 4. Number of winged adults+nymphs of green peach aphid, *Myzus persicae* (Hemiptera: Aphididae) on kale, *Brassica oleracea* var. *acephala* (Brassicaceae) leaves on mulch and no mulched plots after 15 days of infestation. Poisson error and $\chi 2$ of Pearson to correct over dispersion followed by an F test at 5% probability.

Formicidae (Hymenoptera), Orthoptera, and small-bodied plant-feeders (aphids, Aphidoidea; thrips, Thysanoptera, and whiteflies, Aleyrodidae; Hemiptera) were higher in control (no mulched) or cowpea, *Vigna unguiculata* (L.) Walp. (Fabaceae) plots than several of the other mulch treatment plots (sunn hemp mulch; sorghum-sudangrass mulch, *Sorghum* × *drummondii* (Nees ex. Steud.) Millsp. & Chase (Poaceae), and pine bark nuggets, *Pinus* sp., Pinaceae) in Florida, USA, possibly because weed levels were higher in control and cowpea plots [47]. Fire ants, *Solenopsis* spp. (Hymenoptera: Formicidae) were more abundant where there was mulched cover and are important predators of weed seed and pest insects in killed cover crop plots and that cover crop mulches in summer pepper (*Capsicum annuum* L. cv. 'Camelot', Solanaceae) and fall collard (*Brassica oleracea* L. cv. 'Champion', Brassicaceae) production are potentially viable alternatives to black plastic mulch and soil fumigation in South Carolina, USA [48].

The indirect effect of the mulch on *M. persicae* immigration was not related to the macronutrient content of the kale leaves and no negative relationship between the aphid landing and expected K content of the plant was found. Nutrients make these plant tissues more resistant and with thicker cellular walls which could affect the feeding preference and increase the resistance to the insects [49,50]. The UV radiation could increase the tannin accumulation in the lumen and the phenols in the epidermal cells. These factors make the epidermal cells thicker, although with a smaller mesophyll area, suggesting that more carbon is allocated as a protective mechanism with the lower photosynthetic rate [51]. This indicates that the occurrence of these factors and the reduced number of winged aphids on the plants grown or maintained in the mulched plots may be caused by an indirect effect of the UV light reflectance [4,52]. The inorganic mulch reflects the short-wave UV light [53], which confuses and repel the incoming winged aphids, thus reducing their incidence on the plants [11]. Although the impact of the short-wave light to repel the whiteflies was not conclusive, these insects may respond similarly to those of the winged *M. persicae* approaching the reflective (Al) surfaces [54]. The mulch can influence the concentrations of the carbohydrates and soluble sugars in the kale plants [55], which may have reduced the number of aphids probably due the production of semiochemicals.

The reduction in the *M. persicae* population by the mulch helps to explain the effect of light reflectance. This parameter was not evaluated. However, the UV was responsible as seen by the lesser number of aphids landing on the kale plants [56] and the temperature increased in the cantaloupe, *Cucumis melo* var. *cantalupensis* Naudin (Cucurbitaceae) and tulip, *Tulipa* sp. (Liliaceae) plants on the mulched soil [54]. Temperatures between 25–30°C increased the population growth of the cotton aphid, *Aphis gossypii* Glover, 1887; the turnip aphid, *Lipaphis erysimi* Kaltenbach, 1843 (Hemiptera: Aphididae) and the *M. persicae* on the cotton, *Gossypium hirsutum* L. (Malvaceae) and citrus, *Citrus* sp. (Rutaceae) [57,58], but temperatures above 30°C reduced their populations [59]. This shows that the higher temperatures in the mulched plots, probably, may have reduced the population growth of *M. persicae* during the 15-day period. Besides, the temperature influences the synthesis of the acyclic polyol (mannitol), common in most animals, but registered only for a few insects [60]. The healthy polyols also protect the proteins against denaturation at high temperatures [61]. The aphids may adapt to the high temperatures by accumulating the polyhydric alcohols in their hemolymph, as observed for mannitol in the greenflies (Hemiptera: Aphidoidea) [62]. The synthesis of these compounds is stimulated by high temperatures. They are not excreted by the aphids; they accumulate in the hemolymph and thus act as a thermo protective mechanism [62]. However, the aphids may not synthesize the mannitol due to the high energy expense and for this reason show a preference for plants with lower temperatures.

The rice-straw mulch decreased the number of winged aphids landing on the kale plants and the growth of this insect colony, which concurs with the lower aphid populations and the reduced incidence of the aphid-borne viruses on the wheat, *Triticum aestivum* L. (Poaceae) and rice plants on mulched soil [4,63,64]. The incidence of the *Cucumber mosaic virus* (CMV) *Cucumovirus* (Bromoviridae) on the narrow-leafed lupins, *Lupinus angustifolius* L. (Fabaceae) and the *Bean yellow mosaic virus* (BYMV) *Potyvirus* (Potyviridae) on cereal straw was lower than in the bare soil in Australia. This was explained by the lower occurrence of the virus with a decreasing incidence of the winged aphids landing [65]. The increasing UV reflectance of the backgrounds decreased *Aphis* spp. and *M. persicae* landing on the *B. brassica* plants with a negative correlation of this parameter with the aphid numbers [11]. Straw mulch, therefore, reduced the populations and the numbers of this aphid on the green traps want confirms results obtained in potato [66] and faba beans, *Vicia faba* L. (Fabaceae) [67].

Conclusions

The lower number of winged *M. persicae* landing and the better vegetative development of the kale plants may be explained by the higher temperature in the mulched plots, as well as to the physiological changes of the kale plants. Besides, the soil covering may reduce the winged *M. persicae* to find the kale plants. The strategy using rice-straw mulch presents a potential to be used in integrated aphid management in the kale plantations.

Acknowledgments

Global Edico Services of India edited and proofread this manuscript.

Author Contributions

Conceived and designed the experiments: RS-F RHSS GLDL JES JCZ. Performed the experiments: RS-F WST CFW. Analyzed the data: RS-F RHSS GLDL. Contributed reagents/materials/analysis tools: RS-F RHSS GLDL JES JCZ. Wrote the paper: RS-F RHSS WST GLDL CFW JES JCZ.

References

1. Trdan S, Žnidarčič D, Kač M, Vidrih M (2008) Yield of early white cabbage grown under mulch and non-mulch conditions with low populations of onion thrips (*Thrips tabaci* Lindeman). Int J Pest Manage 54: 309–318.
2. Žnidarčič D, Valič N, Trdan S (2008) Epicuticular wax content in the leaves of cabbage (*Brassica oleracea* L. var. *capitata*) as a mechanical barrier against three insect pests. Acta Agric Slov 91: 361–370.
3. Sarangi SK, Saikia US, Lama TD (2010) Effect of rice (*Oryza sativa*) straw mulching on the performance of rapeseed (*Brassica campestris*) varieties in rice-rapeseed cropping system. Indian J Agr Sci 80: 603–605.
4. Summers CG, Mitchell JP, Stapleton JJ (2004) Management of aphid-borne viruses and *Bemisia argentifolii* (Homoptera: Aleyrodidae) in zucchini squash by using UV reflective plastic and wheat straw mulches. Environ Entomol 33: 1447–1457.
5. Broad ST, Schellhorn NA, Lisson SN, Mendham NJ, Corkrey R (2008) Host location and parasitism of *Brevicoryne brassicae* in diversified broccoli cropping systems. Entomol Exp Appl 129: 166–171.
6. Neto SED, Ferreira RLF, Pontes FST (2009) Profitability of lettuce cultivars organic production in greenhouse and field with different tillage soil. Cienc Rural 39: 1362–1368.
7. Kasirajan S, Ngouajio M (2012) Polyethylene and biodegradable mulches for agricultural applications: a review. Agron Sustain Dev 32: 501–529.
8. Greer L, Dole JM (2003) Aluminum foil, aluminium-painted, plastic, and degradable mulches increase yields and decrease insect-vectored viral diseases of vegetables. Horttechnology 13: 276–284.
9. Nyoike TW, Liburd OE (2010) Effect of living (buckwheat) and UV reflective mulches with and without imidacloprid on whiteflies, aphids and marketable yields of zucchini squash. Int J Pest Manage 56: 31–39.
10. Schmidt NP, O'Neal ME, Singer JW (2007) Alfalfa living mulch advances biological control of soybean aphid. Environ Entomol 36: 416–424.
11. Döring TF, Kirchner SM, Kühne S, Saucke H (2004) Response of alate aphids to green targets on coloured backgrounds. Entomol Exp Appl 113: 53–61.
12. Chakraborty D, Garg RN, Tomar RK, Singh R, Sharma SK, et al. (2010) Synthetic and organic mulching and nitrogen effect on winter wheat (*Triticum aestivum* L.) in a semi-arid environment. Agr Water Manage 97: 738–748.
13. Reddy N, Yang YQ (2006) Properties of high-quality long natural cellulose fibers from rice straw. J Agr Food Chem 54: 8077–8081.
14. Gill HK, McSorley R, Branham M (2011) Effect of organic mulches on soil surface insects and arthropods. Fla Entomol 94: 226–232.
15. Homma SK, Tokeshi H, Mendes LW, Tsai SM (2012) Long-term application of biomass and reduced use of chemicals alleviate soil compaction and improve soil quality. Soil Till Res 120: 147–153.
16. Costello MJ (1995) Spectral reflectance from a broccoli crop with vegetation or soil as background - influence on immigration by *Brevicoryne brassicae* and *Myzus persicae*. Entomol Exp Appl 75: 109–118.
17. Zanic K, Ban D, Ban SG, Ciljac TG, Dumicic G (2009) Response of alate aphid species to mulch colour in watermelon. J Food Agric Environ 7: 496–502.
18. Zanic K, Ban D, Culjak TG, Ban SG, Dumicic G, et al. (2013) Aphid populations (Hemiptera:Aphidoidea) depend of mulching in watermelon production in the Mediterranean region of Croatia. Span J Agric Res 11: 1120–1128.
19. Davis JA, Radcliffe EB, Ragsdale DW (2006) Effects of high and fluctuating temperatures on *Myzus persicae* (Hemiptera: Aphididae). Environ Entomol 35: 1461–1468.
20. Mirmohammadi S, Allahyari H, Nematollahi MR, Saboori A (2009) Effect of host plant on biology and life table parameters of *Brevicoryne brassicae* (Hemiptera: Aphididae). Ann Entomol Soc Am 102: 450–455.
21. Moreno-Delafuente A, Garzo E, Moreno A, Fereres A (2013) A plant virus manipulates the behavior of its whitefly vector to enhance its transmission efficiency and spread. PLoS ONE 8: e61543.
22. Rajabaskar D, Wu Y, Bosque-Perez NA, Eigenbrode SD (2013) Dynamics of *Myzus persicae* arrestment by volatiles from Potato leafroll virus-infected potato plants during disease progression. Entomol Exp Appl 148: 172–181.
23. van Toor RF, Drayton GM, Lister RA, Teulon DAJ (2009) Targeted insecticide regimes perform as well as a calendar regime for control of aphids that vector viruses in seed potatoes in New Zealand. Crop Prot 28: 599–607.
24. Hansen LM, Nielsen SL (2012) Efficacy of mineral oil combined with insecticides for the control of aphid virus vectors to reduce potato virus Y infections in seed potatoes (*Solanum tuberosum*). Acta Agr Scand B-S P 62: 132–137.
25. Slosser JE, Pinchak WE, Rummel DR (1998) Biotic and abiotic regulation of *Aphis gossypii* Glover in west Texas dryland cotton. Southwest Entomol 23: 31–65.
26. Culjak TG, Grubisic D, Jelovcan S (2008) Seasonal abundance of *Rhopalosiphum padi* (L.) (Homoptera: Aphididae) in wheat and its role as barley yellow dwarf virus vectors. Cereal Res Commun 36: 27–30.
27. (2007) SAEG Sistema para Análises Estatísticas, Versão 9.1. Fundação Arthur Bernardes - UFV - Viçosa.
28. Silva EE, De-Polli H, Guerra JGM, Aguiar-Menezes EL, Resende ALS, et al. (2011) Organic crop succession of maize and collard greens intercropped with legumes in no-tillage system. Hortic Bras 29: 57–62.
29. Birkhofer K, Bezemer TM, Bloem J, Bonkowski M, Christensen S, et al. (2008) Long-term organic farming fosters below and aboveground biota: Implications for soil quality, biological control and productivity. Soil Biol Biochem 40: 2297–2308.
30. Avalhães CC, Prado RD, Romualdo LM, Rosane DE, Correia MAR (2009) Omission of macronutrients of the growth and nutritional status of plants of cabbage grown in nutrient solution. Biosci J 25: 21–28.
31. Castoldi R, Charlo HCD, Vargas PF, Braz LT (2009) Growth, nutrients accumulation and crop productivity of cauliflower. Hortic Bras 27: 438–446.
32. Prasifka JR, Schmidt NP, Kohler KA, O'Neal ME, Hellmich RL, et al. (2006) Effects of living mulches on predator abundance and sentinel prey in a corn-soybean-forage rotation. Environ Entomol 35: 1423–1431.
33. Ban D, Zanic K, Dumicic G, Culjak TG, Ban SG (2009) The type of polyethylene mulch impacts vegetative growth, yield, and aphid populations in watermelon production. J Food Agric Environ 7: 543–550.
34. Simmons AM, Kousik CS, Levi A (2010) Combining reflective mulch and host plant resistance for sweetpotato whitefly (Hemiptera: Aleyrodidae) management in watermelon. Crop Prot 29: 898–902.
35. Gill HK, McSorley R, Goyal G, Webb SE (2010) Mulch as a potential management strategy for lesser cornstalk borer, *Elasmopalpus lignosellus* (Insecta: Lepidoptera: Pyralidae), in bush bean (*Phaseolus vulgaris*). Fla Entomol 93: 183–190.
36. Chiu MC, Chen YH, Kuo MH (2012) The effect of experimental warming on a low-latitude aphid, *Myzus varians*. Entomol Exp Appl 142: 216–222.
37. Chen CY, Chiu MC, Kuo MH (2013) Effect of warming with temperature oscillations on a low-latitude aphid, *Aphis craccivora*. B Entomol Res 103: 406–413.
38. Kamara AY, Akobundu IO, Sanginga N, Jutzi SC (2000) Effect of mulch from selected multipurpose trees (MPTs) on growth, nitrogen nutrition and yield of maize (*Zea mays* L.). J Agron Crop Sci 184: 73–80.

39. Gencsoylu I, Yalcin I (2004) Advantages of different tillage systems and their effects on the economically important pests, *Thrips tabaci* Lind. and *Aphis gossypii* Glov. in cotton fields. J Agron Crop Sci 190: 381–388.
40. Karamanos AJ, Bilalis D, Sidiras N (2004) Effects of reduced tillage and fertilization practices on soil characteristics, plant water status, growth and yield of upland cotton. J Agron Crop Sci 190: 262–276.
41. Palumbo JC, Toscano NC, Blua MJ, Yoshida HA (2000) Impact of *Bemisia* whiteflies (Homoptera: Aleyrodidae) on alfalfa growth, forage yield, and quality. J Econ Entomol 93: 1688–1694.
42. Hayes RC, Li GD, Dear BS, Humphries AW, Tidd JR (2009) Persistence, productivity, nutrient composition, and aphid tolerance of *Cullen* spp. Crop Pasture Sci 60: 1184–1192.
43. da Costa JG, Pires EV, Riffel A, Birkett MA, Bleicher E, et al. (2011) Differential preference of *Capsicum* spp. cultivars by *Aphis gossypii* is conferred by variation in volatile semiochemistry. Euphytica 177: 299–307.
44. Fernandez-Grandon GM, Woodcock CM, Poppy GM (2013) Do asexual morphs of the peach-potato aphid, *Myzus persicae*, utilise the aphid sex pheromone? Behavioural and electrophysiological responses of *M. persicae* virginoparae to (4aS,7S,7aR)-nepetalactone and its effect on aphid performance. B Entomol Res 103: 466–472.
45. Cividanes FJ, dos Santos-Cividanes TM (2010) Occurrence of alate aphids and their relationship with meteorological factors and host plants. Pesqui Agropecu Bras 45: 7–15.
46. Durak R, Borowiak-Sobkowiak B (2013) Influence of temperature on the biological parameters of the anholocyclic species *Cinara tujafilina* (Hemiptera: Aphidoidea). Cent Eur J Biol 8: 570–577.
47. Gill HK, McSorley R (2011) Impact of different organic mulches on the soil surface arthropod community and weeds. Int J Pest Manage 58: 33–40.
48. Pullaro TC, Marino PC, Jackson DM, Harrison HF, Keinath AP (2006) Effects of killed cover crop mulch on weeds, weed seed, and herbivores. Agric Ecosyst Environ. 115: 97–104.
49. Leite GLD, Picanço M, Guedes RNC, Skowronski L (1999) Effect of fertilization levels, age and canopy height of *Lycopersicon hirsutum* on the resistance to *Myzus persicae*. Entomol Exp Appl 91: 267–273.
50. Ranger CM, Singh AP, Frantz JM, Canas L, Locke JC, et al. (2009) Influence of silicon on resistance of *Zinnia elegans* to *Myzus persicae* (Hemiptera: Aphididae). Environ Entomol 38: 129–136.
51. Laakso K, Sullivan JH, Huttunen S (2000) The effects of UV-B radiation on epidermal anatomy in loblolly pine (*Pinnus taeda* L.) and Scots pine (*Pinnus sylvetris* L.). Plant Cell Environ 23: 461–472.
52. Hu ZQ, Zhao HY, Thieme T (2013) Probing behaviors of *Sitobion avenae* (Hemiptera: Aphididae) on enhanced UV-B irradiated plants. Arch Biol Sci 65: 247–254.
53. Reitz SR, Yearby EL, Funderburk JE, Stavisky J, Momol MT, et al. (2003) Integrated management tactics for *Frankliniella* thrips (Thysanoptera: Thripidae) in field-grown pepper. J Econ Entomol 96: 1201–1214.
54. Summers CG, Stapleton JJ (2002) Use of UV reflective mulch of to delay the colonization and reduce the severity of *Bemisia argentifolii* (Homoptera: Aleyrodidae) infestations in cucurbits. Crop Prot 21: 921–928.
55. Ibarra L, Flores J, Díaz-Pérez JC (2001) Growth and yield of muskmelon in response to plastic mulch and row covers. Sci Hortic 87: 139–145.
56. Legarrea S, Diaz BM, Plaza M, Barrios L, Morales I, et al. (2012) Diminished UV radiation reduces the spread and population density of *Macrosiphum euphorbiae* (Thomas) [Hemiptera: Aphididae] in lettuce crops. Hortic Sci 39: 74–80.
57. Zamani AA, Talebi AA, Fathipour Y, Baniameri V (2006) Effect of temperature on biology and population growth parameters of *Aphis gossypii* Glover (Hom., Aphididae) on greenhouse cucumber. J Appl Entomol 130: 453–460.
58. Parajulle MN (2007) Influence of constant temperatures on life history parameters of the cotton aphid, *Aphis gossypii*, infesting cotton. Environ Entomol 36: 666–672.
59. Satar S, Kersting U, Uygun N (2008) Effect of temperature on population parameters of *Aphis gossypii* Glover and *Myzus persicae* (Sulzer) (Homoptera: Aphididae) on pepper. J Plant Dis Protect 115: 69–74.
60. Saeidi F, Moharramipour S, Barzegar M (2012) Seasonal patterns of cold hardiness and cryoprotectant profiles in *Brevicoryne brassicae* (Hemiptera: Aphididae). Environ Entomol 41: 1638–1643.
61. Haque I, Islam A, Singh R, Moosavi-Movahedi AA, Ahmad F (2006) Stability of proteins in the presence of polyols estimated from their guanidinium chloride-induced transition curves at different pH values and 25 degrees C. Biophys Chem 119: 224–233.
62. de Munoz FGG, Lanz-Mendoza H, Hernandez-Hernandez FC (2007) Free radical generation during the activation of hemolymph prepared from the homopteran *Dactylopius coccus*. Arch Insect Biochem 65: 20–28.
63. Magar SJ, Nirmal DD (2008) Effect of different mulching on the incidence of yellow vein mosaic disease of okra (*Abelmoschus esculentus*). Indian J Agr Sci 78: 726–729.
64. Olinik JR, Oliveira A, Kepp MA, Reghin MY (2011) Yield of italian hibrid squash on different soil coverings. Hortic Bras 29: 130–134.
65. Latham LJ, Jones RAC, McKirdy J (2001) Cucumber mosaic cucumovirus infection of cool-season crop, annual pasture, and forage legumes: susceptibility, sensitivity, and seed transmission. Aust J Agri Res 52: 683–697.
66. Saucke H, Döring TF (2004) Potato virus Y reduction by straw mulch in organic potatoes. Ann App Biol 144: 347–355.
67. Heimbach U, Eggeres C, Thieme T (2002) Weniger Blattläuse durch Mulchen? Gesunde Pflanz 54: 119–125.

8

Environmental Fate of Double-Stranded RNA in Agricultural Soils

Samuel Dubelman*, Joshua Fischer, Fatima Zapata, Kristin Huizinga, Changjian Jiang, Joshua Uffman, Steven Levine, David Carson

Regulatory Division, Monsanto Company, St. Louis, Missouri, United States of America

Abstract

A laboratory soil degradation study was conducted to determine the biodegradation potential of a DvSnf7 dsRNA transcript derived from a Monsanto genetically modified (GM) maize product that confers resistance to corn rootworm (CRW; *Diabrotica spp.*). This study provides new information to improve the environmental assessment of dsRNAs that become pesticidal through an RNAi process. Three agricultural soils differing in their physicochemical characteristics were obtained from the U.S., Illinois (IL; silt loam), Missouri (MO; loamy sand) and North Dakota (ND; clay loam), and exposed to the target dsRNA by incorporating insect-protected maize biomass and purified (*in vitro*-transcribed) DvSnf7 RNA into soil. The GM and control (non-GM maize) materials were added to each soil and incubated at ca. 22°C for 48 hours (h). Samples were collected at 12 time intervals during the incubation period, extracted, and analyzed using QuantiGene molecular analysis and insect bioassay methods. The DT_{50} (half-life) values for DvSnf7 RNA in IL, MO, and ND soils were 19, 28, and 15 h based on QuantiGene, and 18, 29, and 14 h based on insect bioassay, respectively. Furthermore, the DT_{90} (time to 90% degradation) values for DvSnf7 RNA in all three soils were <35 h. These results indicate that DvSnf7 RNA was degraded and biological activity was undetectable within approximately 2 days after application to soil, regardless of texture, pH, clay content and other soil differences. Furthermore, soil-incorporated DvSnf7 RNA was non-detectable in soil after 48 h, as measured by QuantiGene, at levels ranging more than two orders of magnitude (0.3, 1.5, 7.5 and 37.5 µg RNA/g soil). Results from this study indicate that the DvSnf7 dsRNA is unlikely to persist or accumulate in the environment. Furthermore, the rapid degradation of DvSnf7 dsRNA provides a basis to define relevant exposure scenarios for future RNA-based agricultural products.

Editor: Guy Smagghe, Ghent University, Belgium

Funding: Monsanto Company funded this study. The funder provided support in the form of salaries for all authors, but did not have any additional role in the study design, data collection and analysis, decision to publish, or preparation of the manuscript. The specific roles of these authors are articulated in the 'author contributions' section.

Competing Interests: All authors are employees of Monsanto Company. This does not alter the authors' adherence to PLOS ONE policies on sharing data and materials.

* E-mail: samuel.dubelman@monsanto.com

Introduction

Genetically modified (GM) crops expressing traits that confer protection from insect pests and facilitate weed management provide significant value in agriculture and require novel modes of action for sustainable efficacy. Comprehensive science-based ecological risk assessments (ERA) are required to ensure that products with benefits for agriculture are also safe for the environment. Ecological risk from the cultivation of a GM crop will be a function of both potential harm (hazard) and environmental exposure [1]. Determining the fate of a transgenic trait in soil, whether it be an insecticidal protein or dsRNA, is a component of the ERA for a GM crop, and defines realistic routes and levels of exposure to non-target organisms in the environment.

With the introduction of GM crops, attention has focused on the environmental fate of transgenic-derived DNA and proteins. The environmental exposure of Cry proteins derived from *Bacillus thuringiensis (Bt)* is minimal due to the lack of persistence of *Bt* proteins in soil [2,3]. For example, laboratory studies measuring the degradation rate of the Cry1Ac protein expressed in *Bt* cotton showed that Cry1Ac has a half-life (DT_{50}) of 16 days (d) when incorporated into soil microcosms [4]. Comparable results were reported for Cry1Ab from laboratory microcosm studies with *Bt* maize [5,6]. Results obtained from laboratory soil degradation studies were confirmed in multi-year monitoring studies of large agricultural fields with continuous cultivation of the *Bt* crops, which showed no detection, persistence, or accumulation of Cry1Ac in *Bt*-cotton fields [7] or of Cry1Ab in *Bt*-maize fields [8,9]. Based on the results of these laboratory and field studies, EPA has concluded that *Bt* proteins do not accumulate as a result of continuous cultivation [3].

Similar to the case of *Bt* proteins, the rate of nucleic acid degradation can be influenced by the biophysical characteristics of soil and sediments in addition to other abiotic and biotic factors [10–13]. DNA degradation kinetics have been defined in several instances using plant tissue or extracellular DNA applied to microbiologically active soil and sediment. For example, genomic DNA (gDNA) from tomato leaves exhibits a half-life of about 1.5 d when introduced to viable soil [14]. Similarly, transgenic and conventional soybean DNA from freeze-dried leaf tissue exhibits a half-life of about 1.4 d [15]. Once DNA is released from plant tissues or applied directly to soil, more rapid and extensive degradation occurs. Purified GM and non-GM maize DNA exposed to leachate water at ambient room temperature exhibits a half-life of <2 h while GM and non-GM soybean DNA exposed to similar conditions has a half-life of about 4 h [16].

Degradation of DNA in the soil environment is likely due to processing by DNases. High molecular weight gDNA molecules may first be cleaved into oligonucleotides of approximately 400-bp by microbial restriction endonuclease I activity. These oligonucleotide substrates can then undergo further degradation through the activity of microbial-produced DNases. Bacterial populations may also increase by approximately one order of magnitude in soil exhibiting DNA-degrading activity [17].

Less information is available on the degradation kinetics of RNA in soil, however data collected to date provides no evidence that RNA persists any longer than DNA [18,19]. In Keown *et al.*, soil samples amended with either dried DNA or RNA were analyzed for released nitrogen and CO_2 evolution as a measure of nucleic acid degradation. The results of this study found that RNA degraded as quickly as DNA in different soils [19].

Crops that express double-stranded RNA (dsRNA) molecules are being developed which take advantage of the endogenous RNAi machinery of target insects and can produce highly specific insecticidal oligonucleotides (siRNA) for agricultural pest control. DvSnf7 dsRNA expressed in GM maize confers protection against the Western Corn Rootworm (WCR) [20]. Ingestion of DvSnf7-expressing tissue by a susceptible insect host results in the uptake of the dsRNA leading to mRNA and protein suppression of a targeted essential gene in WCR, systemic spread, and eventual mortality [20,21]. Because of the potential for increased development of products with RNA-based modes of action, more information is needed to understand the fate of dsRNA in the soil environment [22,23]. Results reported here assessing the environmental fate of DvSnf7 dsRNA from an insect-protected maize product indicate that the insecticidal dsRNA degrades readily and is unlikely to persist in the soil environment.

Materials and Methods

Synthesis of IVT DvSnf7

In vitro T7 RNA Polymerase-transcribed 968 nucleotide (nt) DvSnf7 RNA (IVT DvSnf7 RNA) was used to treat soil samples and serve as the reference standard for the QuantiGene 2.0 microplate assay [24]. This 968-nt transcript is produced in transgenic maize containing a DvSnf7 suppression cassette and was determined by cDNA sequencing to include a 240-nt inverted repeat region and adjacent 5′- and 3′ sequences. The 968-nt DvSnf7 fragment was amplified by PCR from the original plant transformation vector and cloned downstream of a synthetic T7 promoter sequence (TAATACGACTCACTATAGGG) in pUC19 plasmid. The identity of the recombinant pUC plasmid template used to produce the IVT DvSnf7 RNA was verified by DNA sequencing. For IVT RNA synthesis, the pUC plasmid was linearized by BglII restriction digestion and incubated with nucleoside triphosphates (8 mM each, Sigma) and T7 RNA Polymerase in transcription buffer overnight at 37°C. The reaction was then treated with DNase I (100 units per 1 mL, Ambion) and extracted with phenol:chloroform (1:1 volume:volume). The size of the IVT DvSnf7 RNA was confirmed by agarose gel electrophoresis and its concentration was determined using a NanoDrop 8000 (Thermo Scientific, Wilmington, DE) according to the manufacturer's instruction. Aliquots of the IVT DvSnf7 RNA sample were stored in a −80°C freezer.

Soil Sample Preparation and Dosing

Three soil types were collected from the top 15 cm of fields in Illinois (IL), Missouri (MO) and North Dakota (ND) and subsequently characterized for percent sand, silt and clay; USDA textural class; bulk density; percent organic matter; pH; cation exchange capacity (CEC); field moisture capacity (FMC) at 1/3 Bar and at 15 Bar; and a number of chemical elements (AGVISE Laboratories, Northwoods, ND). The three soils were sieved at AGVISE Laboratories through a 2 mm screen to remove non-soil debris prior to analysis. Upon receipt, each soil was spread on large trays to air dry and then repacked and stored at ambient temperature under aerobic conditions. At the start of the study, 1 g soil samples were weighed into 50 ml conical tubes. Samples containing maize tissue were amended with a mixture of DvSnf7-expressing lyophilized root and shoot tissue (20:80 w/w) derived from V8-V10 growth stage plants at a rate of 20 mg tissue per g soil. Due to relatively low expression of DvSnf7 in maize tissue, a solution of IVT DvSnf7 was diluted in 1× PBST (1 mM KH_2PO_4, 10 mM Na_2HPO_4, 137 mM NaCl, 2.7 mM KCl, pH 7.0; Roche Diagnostics, Indianapolis, IN) to the desired concentrations. Given the mass ratio of plant tissue to soil in the samples, the QuantiGene and insect bioassay methods were not sufficiently sensitive to generate quantifiable measurements of the endogenously expressed DvSnf7 RNA over the entire experimental range (90% degradation). Samples were therefore treated with an IVT DvSnf7 RNA solution of sufficient concentration to detect dsRNA after 90% of the initial DvSnf7 RNA had degraded. The IVT DvSnf7 solution was diluted such that ~400 µl solution was required for dosing each sample. Samples were incubated at ambient temperature for the required interval and then placed in −80°C storage until all samples were collected and prepared for extraction.

Sample Extraction

All samples were extracted prior to QuantiGene analysis and insect bioassay. Briefly, 25 steel ball bearings were added to each sample in 50 ml conical tubes. For Illinois and Missouri soil samples, sufficient 1× PBST pH 7 was added to achieve a 25:1 buffer:soil ratio. For high clay soil samples (e.g. ND), the pH 7 buffer was substituted with 1× PBST pH 12 to improve the extraction efficiency of clay-bound RNA. Samples were then agitated in a paint shaker (Harbil), for 10 minutes and passed through a 0.22 µm vacuum filter. Samples containing high clay ND soil were centrifuged for 10 minutes at 7500× *g* prior to filtration to avoid clogging the filter with fine clay particles. Filtered sample extracts were aliquoted and frozen at −80°C until analyses were performed.

QuantiGene Analysis

Prior to sample analysis by a DvSnf7-specific QuantiGene assay (Affymetrix, Inc., Santa Clara, CA), aliquots of sample extracts were purified to remove potential hybridization inhibitors (e.g. humic acid, polyphenols). Briefly, 1 ml aliquots of the sample extracts were mixed with an equal volume of citrate-buffered phenol pH 4.3 (Sigma-Aldrich, Saint Louis, MO) and chloroform (Sigma-Aldrich, Saint Louis, MO). Emulsified samples were separated via centrifugation and the RNA was precipitated from the aqueous phase with isopropanol. Following precipitation, samples were centrifuged, air-dried, and resuspended in DEPC-treated water (Sigma-Aldrich, Saint Louis, MO). The concentration of DvSnf7 RNA in purified samples was determined with a QuantiGene 2.0 assay following the manufacturer's instructions. A custom QuantiGene probe set was designed by the manufacturer to hybridize to the sequence within DvSnf7 and IVT DvSnf7

RNA that confers insecticidal activity. The IVT DvSnf7 RNA and the sequence-specific recognition by the QuantiGene probe set has previously been described in detail [24]. Each purified sample was diluted 1:100 in assay diluent and analyzed in triplicate. The DvSnf7 concentration was calculated using a four-parameter curve fit [25] and the mean concentration of each sample was calculated from triplicate analysis.

Determination of DvSnf7 RNA Concentration by SCR Mortality Assay

Southern Corn Rootworm (SCR) was selected for these bioassays because it was more amenable than Western Corn Rootworm (WCR) to laboratory bioassays and displayed comparable sensitivity to the DvSnf7 RNA [20]. Four separate soil samples were collected at each of 12 collection times for each of the three soil types and extracted prior to assay. Three replicates of each of these extracts were used for insect bioassays. Each bioassay determination included a concentration response curve with several sequential dilutions of *IVT* DvSnf7 RNA that served as a standard curve to quantify DvSNf7 RNA in soil extracts. For each bioassay, extract solutions were prepared by aliquoting each test and control soil sample extract into three replicates, diluting with purified water, incorporating the dilution into an agar-based SCR diet (Bio-Serv, #F9800B) and then vortex-mixing the solution until visually homogeneous. These diet mixtures were then dispensed in 0.25 ml aliquots into 48-well plates (Becton Dickson Labware, Franklin Lakes, NJ; Prod. # 3078). SCR larvae (≤30 h after the first observation of hatching) were individually placed on these diets with a target number of 24 insects tested for each sample replicate. The insects were incubated for a period of 12 d at 27°C, 70% relative humidity and in total darkness. Assays were evaluated for survival at the end of the incubation period and results were subjected to a correction to account for control mortality. Standard curves for SCR bioassays were analyzed by probit analysis under PROC PROBIT in SAS, the OPTC function was used to correct for control mortality, and DvSnf7 soil concentrations were estimated by inverse prediction.

Environmental Fate Modeling

A standard 3-parameter logistic equation [26], run under PROC NLMIXED within SAS (Release 9.3), was used to model the time-dependent decrease in DvSNf7 RNA concentration in soil for the QuantiGene and insect bioassay data, and to estimate DT_{50} and DT_{90} values, the time periods required for 50% and 90% degradation of the initial DvSnf7 RNA concentration in soil.

Results

The environmental fate of DvSnf7 dsRNA was determined in three different soils that have physical characteristics representative of world regions where maize is grown, including a silt loam (IL), loamy sand (MO), and clay loam (ND) soil. Prior to the study, the three soils were characterized by determining percent sand, silt and clay, USDA textural class, bulk density, percent organic matter, pH, cation exchange capacity (CEC), and field moisture capacity (FMC) at 1/3 Bar and 15 Bar (Table 1).

Because DvSnf7 dsRNA is expressed in GM maize, soil samples were amended with a mixture of lyophilized root and shoot maize tissue to simulate post-harvest field environmental conditions. Given that the *in planta* expression of DvSnf7 RNA is low, tissue-amended soil samples were also supplemented with a solution of IVT DvSnf7 dsRNA to ensure that a robust quantitation could be obtained at all time intervals of an experimental range covering at least 90% degradation of the initial DvSnf7 RNA concentration.

The concentration of DvSnf7 in soil samples collected at 12 time intervals during the 48 h incubation was determined using a QuantiGene assay. The QuantiGene assay detects both *in planta*-expressed DvSnf7 and *in vitro*-transcribed DvSnf7 via hybridization with target sequence-specific probes. After an initial induction period of a few hours, DvSnf7 RNA degraded rapidly in all three soils and was no longer detected by QuantiGene after 24–36 h (Figure 1). As a control, a solution of DvSnf7 incubated at ambient temperature in the absence of soil or maize tissue was stable after 48 h and showed no signs of degradation (data not shown). The biological activity of DvSnf7 RNA in these samples was also measured by a mortality bioassay using Southern Corn Rootworm (SCR) as a sensitive target species. It has previously been reported that the dsRNA produces an insect mortality response when the double stranded segments are ≥60 base pairs [20] and this result is consistent with the findings of Miller *et al.* [27]. Results indicate that the biological activity (insect mortality) of DvSnf7 RNA declined rapidly in all three soils similar to results obtained using the QuantiGene assay. While the QuantiGene assay has the potential to detect partially degraded dsRNA species in addition to the parent compound (see reference 24), the high degree of correlation observed between the QuantiGene and the insect-bioassay results suggests that the dsRNA is being degraded to less than 60 bp at minimum. Insect bioassay results exhibited a similar initial induction period, followed by a rapid degradation phase, and ultimately declining to a level indistinguishable from the background mortality of the control insect group (Figure 1).

Degradation kinetics of DvSnf7 RNA in soil amended with maize tissue were calculated using a 3-parameter logistic equation. The estimated DT_{50} values ranged from about 14 h (ND soil) to about 29 h (MO soil) using results obtained from QuantiGene and SCR bioassay (Table 2). The estimated DT_{90} values ranged from 21 to 35 h for the three different soils. Furthermore, these results demonstrated high consistency between two analytical methods that rely on uniquely different end-points (molecular hybridization versus insect mortality). These results indicate that the DvSnf7 RNA was degraded and devoid of any detectable biological activity within 2 days of incorporation into soil.

Since DvSnf7 RNA was observed to degrade rapidly in soils amended with maize tissue, we tested whether dsRNA would be degraded with similar kinetics when applied to soil alone. The three representative soils were treated with a solution of IVT DvSnf7 dsRNA and incubated at ambient temperature for up to 48 h. Analysis by QuantiGene demonstrated that regardless of soil class, DvSnf7 RNA degraded rapidly in soil alone, with similar kinetics to soil amended with maize tissue (Figure 2). These results indicate that dsRNA degrades rapidly in soil, in the presence or absence of maize biomass.

To determine whether the observed DvSnf7 degradation kinetics are dependent on the amount of dsRNA applied to the soil microcosms, the relationship between degradation kinetics and the initial dsRNA concentration was investigated. Tissue-amended silt loam soil (IL) was treated with four different DvSnf7 RNA concentrations ranging from to 0.3 to 37.5 µg per gram of soil and degradation kinetics of the dsRNA was measured at each treatment level. The lowest treatment level (0.3 µg/g) was the lowest initial dsRNA concentration that would allow quantifiable QuantiGene measurements after at least 90% degradation of the applied dsRNA. Despite varying the amount of DvSnf7 added to soil by more than two orders of magnitude, there was no significant observable change in dsRNA degradation kinetics (Figure 3). These results suggest that the rate of dsRNA degradation is largely independent of the dsRNA concentration applied to soil.

Table 1. Physicochemical characteristics of three soils used to determine the environmental fate of DvSnf7 RNA.

	Soil Characterization Results		
Parameter	**Illinois**	**Missouri**	**North Dakota**
USDA Textural Class:	Silt Loam	Loamy Sand	Clay Loam
Particle Size Distribution			
% Sand	23	86	26
% Silt	60	11	38
% Clay	17	3	36
Bulk Density (g/cm^3)	0.92	1.3	0.92
% Organic Matter	3.5	1.7	5.1
pH	5.8	5.5	7.1
Cation Exchange Capacity (meq/100 g)	8.5	6.0	24.4
Field Moisture Capacity (ml/100 g) at 1/3 Bar	28.9	6.9	37.1
at 15 Bar	9.6	3.6	25.1

Discussion

This study demonstrates that the DvSnf7 dsRNA rapidly degrades in the soil environment with a half-life of less than 30 h. This degradation is relatively rapid when compared to *Bt* proteins that exhibit half-lives typically ranging from approximately one to several days [4,5,8,28–30]. The biodegradation of dsRNA was established using laboratory microcosms containing soil alone or amended with maize tissue to simulate the composition of a maize field following harvest. Biodegradation rate data obtained from laboratory soil microcosms has proven to be a reliable indicator for predicting potential persistence of a given organic substrate in more complex environmental or field conditions [4,31,32], and the lack of persistence or accumulation of the Cry1Ab [8,9] and Cry1Ac [7] proteins has been demonstrated in field dissipation studies. Following an initial lag period, DvSnf7 dsRNA degrades rapidly as measured by quantitative molecular hybridization and by assessing the loss of biological activity (mortality) using a sensitive insect bioassay. The observed lag period (<10 h) is likely due to acclimation of the resident microbial community to the amended maize tissue or dsRNA and is a common feature of organic substrate biodegradation [33].

Viable resident microbial communities have been shown to be necessary for the decomposition of leaf litter and plant genomic DNA [14]. Degradation of the dsRNA in acclimated soil microcosms could result from direct metabolism by the resident microbial community or from plant-encoded nucleases contained within the amended maize tissue [34]. To test this, DvSnf7 RNA was incorporated directly into soil in the absence of maize tissue. Similar to soil microcosms amended with maize tissue, there was an initial acclimation phase followed by rapid biodegradation. The estimated half-life of DvSnf7 RNA was less than 24 h in soil alone regardless of soil type, which indicates that the activity of the resident soil microbial community is sufficient to degrade dsRNA. This finding has particular implications for exposure assessments of potential nucleic acids formulated for applications in which an exposure scenario may not necessarily include the presence of plant detritus. Consistent with our findings for dsRNA, Southern hybridization of phage P22 DNA recovered from soil microcosms

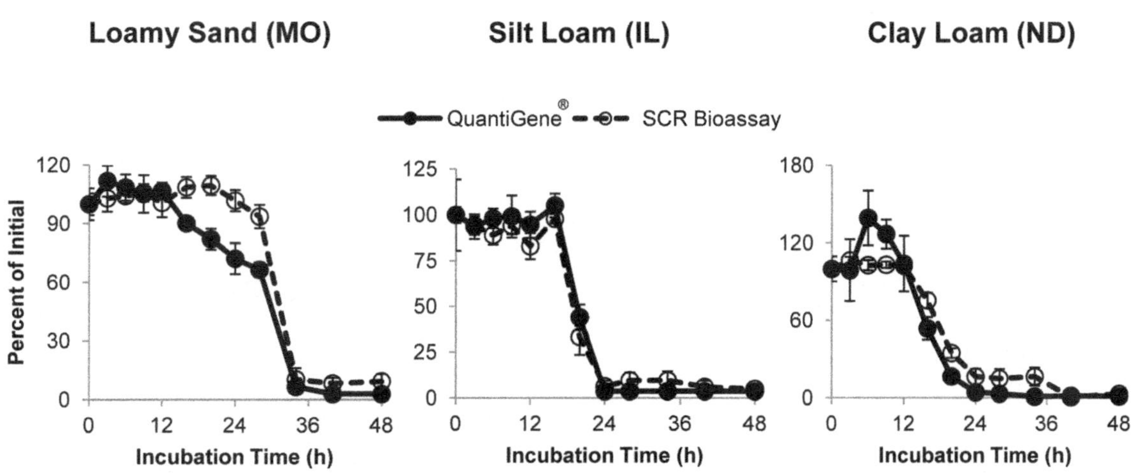

Figure 1. DvSnf7 RNA degrades rapidly when applied to soil. Soil samples were amended with lyophilized maize tissue and fortified with 7.5 µg DvSnf7 RNA per gram of soil. DvSnf7 RNA concentration in soil was measured with a QuantiGene assay. Biological activity against Southern Corn Rootworm was determined by assessing insect mortality in a 12-day bioassay. Error bars represent one standard error of the mean (n = 4).

Table 2. QuantiGene and insect bioassay methods provide comparable degradation rate estimates.

		QuantiGene			Insect Mortality		
Soil	**Rate Estimate**	**Time (±SE)**	**LCI**	**UCI**	**Time (±SE)**	**LCI**	**UCI**
Loamy Sand (MO)	DT_{50}	27.8 (0.6)	26.6	29.1	29.0 (1.2)	26.4	31.5
	DT_{90}	34.5 (0.5)	33.4	35.6	34.3 (0.8)	32.5	36.1
Silt Loam (IL)	DT_{50}	19.4 (0.4)	18.6	20.3	17.7 (1.4)	14.6	20.9
	DT_{90}	23.0 (0.3)	22.3	23.6	23.1 (1.1)	20.5	25.6
Clay Loam (ND)	DT_{50}	15.1 (0.7)	13.7	16.4	13.7 (0.8)	11.9	15.6
	DT_{90}	21.6 (0.6)	20.5	22.7	20.5 (0.8)	18.8	22.2

The degradation rate estimates (DT_{50} and DT_{90}), in hours, were calculated for DvSnf7 RNA in each soil type as measured by QuantiGene and insect bioassay. Numbers in parenthesis represent the standard error of the mean (SE) in hours. LCI and UCI represent the lower and upper 95% confidence intervals respectively, in hours.

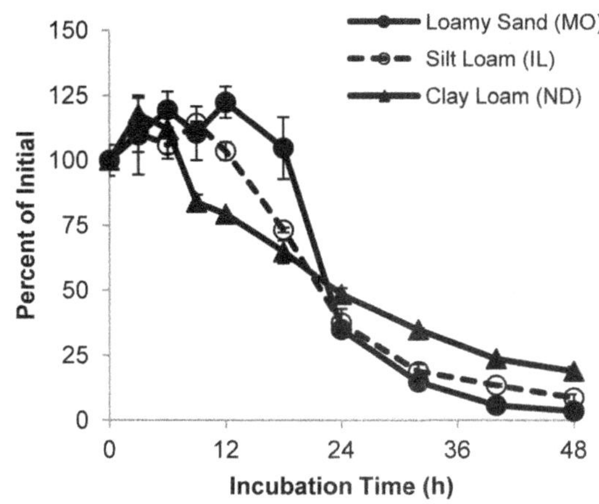

Figure 2. DvSnf7 RNA degrades in soil alone. Soil samples were fortified with 7.5 μg DvSnf7 RNA per gram of soil in the absence of maize tissue. DvSnf7 RNA concentration was determined with a QuantiGene assay. Error bars represent one standard error of the mean (n = 2).
doi:10.1371/journal.pone.0093155.g002

indicates that DNA undergoes rapid degradation within 5–24 h of application. Furthermore, an increase in observable counts of soil microorganisms temporally coincides with degradation of the applied DNA. Evidence suggests this process is mediated significantly by the soil bacterial communities as degradation can be impeded by adding rifampicin and chloramphenicol, but not cyclohexamide [17].

The biodegradation of dsRNA is also likely mediated by microbial-produced ribonucleases [12,17]. Gulden and colleagues demonstrated that the half-life of DNA sequences decreased with increasing temperature implicating enzymatic processes are responsible for degrading free nucleic acid [12,16]. For RNA, various ribonucleases (RNases) produced by pro- and eukaryotes may play critical roles in its processing and degradation. In particular, RNase III is an endoribonuclease that is responsible for

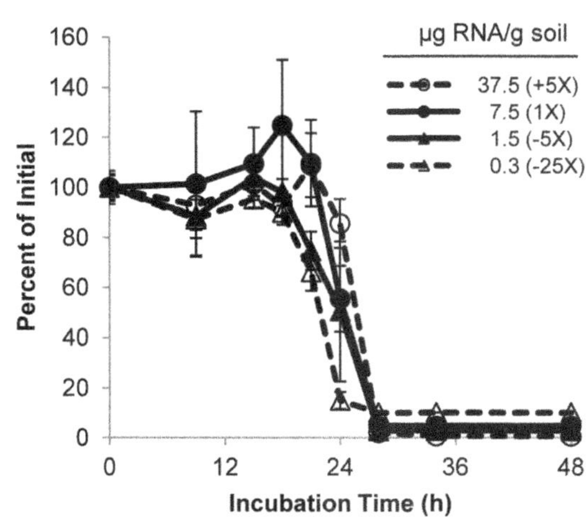

Figure 3. Degradation kinetics for DvSnf7 RNA are independent of initial dsRNA concentration. Soil samples were amended with lyophilized maize tissue and fortified with the indicated amounts of DvSnf7 RNA (μg RNA/g soil). Soil concentration of DvSnf7 RNA was determined with a QuantiGene assay. Numbers in parenthesis indicate the relative DvSnf7 RNA treatment. Error bars represent one standard error of the mean (n = 3).

the cleavage of dsRNA substrates [35]. Recently, the conserved Eri1 family of RNases was shown to degrade double-stranded silencing RNAs [36,37] such as those that may be used in GM agricultural products.

When RNA is released into the environment, the same RNases responsible for intra-cellular degradation are active in an extra-cellular capacity. Single-stranded RNA (ssRNA) from viral genomes exhibit a half-life on the order of days in seawater [38]. Similarly, RNA from human norovirus added to seawater and incubated under simulated winter-like conditions ($8\pm0.5°C$ and UV radiation at 1 mW/cm^2) exhibited an exponential decrease in concentration with a DT_{90} of approximately 141 h and no detectable amount of RNA found after 14 d [39]. In sand, free extracellular microbial RNA labeled with ^{14}C and ^{3}H exhibited up to 62% degradation after 14 d [40].

In comparison to *Bt* proteins, the expression of DvSnf7 *in planta* is relatively low such that addition of a DvSnf7 RNA solution was required to produce a full and quantitative degradation profile in soil microcosms. Therefore, we tested whether the degradation kinetics were dependent on the concentration of dsRNA applied to soil. Despite varying the amount of dsRNA in soil microcosms by over two orders of magnitude, there was little observed change in the half-life of DvSnf7. Furthermore, there was no discernible correlation between the applied amount of the dsRNA and efficiency of analytical recovery from soil suggesting the soil microcosms were not saturated with the applied nucleic acid (data not shown). Consistent with these results, saturation was not observed when DNA was applied to microcosms over a range of concentrations up to 28 μg of DNA per gram of soil [17]. Overall, results of these experiments suggest that the dsRNA expressed *in planta* by this GM maize is unlikely to persist in the soil environment.

In summary, we have demonstrated the environmental degradation of dsRNA in multiple soil environments. The degradation of dsRNA was characterized by an initial acclimation phase followed by a rapid loss of DvSnf7 RNA and its associated biological activity in three different soils, exhibiting half-lives ranging from about 14 to 30 h. Rapid degradation of the dsRNA was observed in soil alone or soil amended with maize tissue and the rate of degradation was largely independent of the RNA treatment level. Taken together these results demonstrate that the DvSnf7 dsRNA from a CRW-protected GM maize is unlikely to persist or accumulate in the soil environment. In a broader context, these results provide fundamental information to define environmental exposure scenarios for ecological risk assessment of future RNA-based agricultural products.

Acknowledgments

The authors wish to thank Bonnie Ayden for assistance in executing soil microcosm experiments and Geoffrey Mueller, Joanne Warren, Jennifer Fridley and John Lake for assistance with performing insect bioassays.

Author Contributions

Conceived and designed the experiments: SD JF SL. Performed the experiments: SD JF FZ JU. Analyzed the data: SD JF JU CJ. Wrote the paper: SD JF KH SL DC.

References

1. US EPA (1998) Guidelines for ecological risk assessment, Federal Register 63(93):26846–26924, EPA/630/R-95/002F. U.S. Environmental Protection Agency, Washington, D.C.
2. Icoz I, Stotzky G (2008) Fate and effects of insect-resistant *Bt* crops in soil ecosystems. Soil Biology and Biochemistry 40: 559–586.
3. Kough JL, Edelstein R (2012) EPA regulatory requirements for plant-incorporated protectants. Regulation of Agricultural Biotechnology: The United States and Canada: 163–174.
4. Sims SR, Ream JE (1997) Soil inactivation of the *Bacillus thuringiensis* subsp *kurstaki* CryIIA insecticidal protein within transgenic cotton tissue: Laboratory microcosm and field studies. Journal of Agricultural and Food Chemistry 45: 1502–1505.
5. Sims SR, Holden LR (1996) Insect bioassay for determining soil degradation of *Bacillus thuringiensis* subsp. *kurstaki* Cry1A(b) protein in corn tissue. Environmental Entomology 25: 659–664.
6. Badea EM, Chelu F, Lăcătuşu A (2010) Results regarding the levels of Cry1Ab protein in transgenic corn tissue (MON810) and the fate of Bt protein in three soil types. Romanian Biotechnological Letters 15: 55–62.
7. Head G, Surber JB, Watson JA, Martin JW, Duan JJ (2002) No detection of Cry1Ac protein in soil after multiple years of transgenic *Bt* cotton (Bollgard) use. Environmental Entomology 31: 30–36.
8. Dubelman S, Ayden BR, Bader BM, Brown CR, Jiang CJ, et al. (2005) Cry1Ab protein does not persist in soil after 3 years of sustained *Bt* corn use. Environmental Entomology 34: 915–921.
9. Gruber H, Paul V, Meyer HHD, Muller M (2012) Determination of insecticidal Cry1Ab protein in soil collected in the final growing seasons of a nine-year field trial of Bt-maize MON810. Transgenic Research 21: 77–88.
10. England LS, Holmes SB, Trevors JT (1998) Persistence of viruses and DNA in soil. World Journal of Microbiology and Biotechnology 14: 163–169.
11. England LS, Trevors JT (2003) The microbial DNA cycle in soil. Rivista Di Biologia/Biology Forum 96: 317–326.
12. Levy-Booth DJ, Campbell RG, Gulden RH, Hart MM, Powell JR, et al. (2007) Cycling of extracellular DNA in the soil environment. Soil Biology and Biochemistry 39: 2977–2991.
13. Pietramellara G, Ascher J, Borgogni F, Ceccherini MT, Guerri G, et al. (2009) Extracellular DNA in soil and sediment: fate and ecological relevance. Biology and Fertility of Soils 45: 219–235.
14. Poté J, Rossé P, Rosselli W, Van VT, Wildi W (2005) Kinetics of mass and DNA decomposition in tomato leaves. Chemosphere 61: 677–684.
15. Levy-Booth DJ, Campbell RG, Gulden RH, Hart MM, Powell JR, et al. (2008) Real-time polymerase chain reaction monitoring of recombinant DNA entry into soil from decomposing Roundup Ready leaf biomass. Journal of Agricultural and Food Chemistry 56: 6339–6347.
16. Gulden RH, Lerat S, Hart MM, Powell JR, Trevors JT, et al. (2005) Quantitation of transgenic plant DNA in leachate water: Real-time polymerase chain reaction analysis. Journal of Agricultural and Food Chemistry 53: 5858–5865.
17. Blum SAE, Lorenz MG, Wackernagel W (1997) Mechanism of retarded DNA degradation and prokaryotic origin of DNases in nonsterile soils. Systematic and Applied Microbiology 20: 513–521.
18. Greaves MP, Wilson MJ (1970) The degradation of nucleic-acids and montmorillonite nucleic-acid complexes by soil microorganisms. Soil Biology and Biochemistry 2: 257–268.
19. Keown H, O'Callaghan M, Greenfield LG (2004) Decomposition of nucleic acids in soil. New Zealand Natural Sciences 29: 13–19.
20. Bolognesi R, Ramaseshadri P, Anderson J, Bachman P, Clinton W, et al. (2012) Characterizing the mechanism of action of double-stranded RNA activity against western corn rootworm (*Diabrotica virgifera virgifera* LeConte). PLoS ONE 7: e47534.
21. Ramaseshadri P, Segers G, Flannagan R, Wiggins E, Clinton W, et al. (2013) Physiological and Cellular Responses Caused by RNAi- Mediated Suppression of Snf7 Orthologue in Western Corn Rootworm (Diabrotica virgifera virgifera) Larvae. PLoS ONE 8.
22. Auer C, Frederick R (2009) Crop improvement using small RNAs: Applications and predictive ecological risk assessments. Trends in Biotechnology 27: 644–651.
23. Lundgren JG, Duan JJ (2013) RNAi-Based Insecticidal Crops: Potential Effects on Nontarget Species. Bioscience 63: 657–665.
24. Armstrong TA, Chen H, Ziegler TE, Iyadurai KR, Gao A-G, et al. (2013) Quantification of transgene-derived double-stranded RNA in plants using the QuantiGene nucleic acid detection platform. Journal of Agricultural and Food Chemistry 61: 12557–12564.
25. De Lean A, Munson PJ, Rodbard D (1978) Simultaneous analysis of families of sigmoidal curves: application to bioassay, radioligand assay, and physiological dose-response curves. Am J Physiol 235: E97–E102.
26. Van Ewijk PH, Hoekstra JA (1993) Calculation of the EC50 and its confidence interval when subtoxic stimulus is present. Ecotoxicology and Environmental Safety 25: 25–32.
27. Miller SC, Miyata K, Brown SJ, Tomoyasu Y (2012) Dissecting systemic RNA interference in the red flour beetle *Tribolium castaneum*: Parameters affecting the efficiency of RNAi. PLoS ONE 7: e47431.
28. Palm CJ, Schaller DL, Donegan KK, Seidler RJ (1996) Persistence in soil of transgenic plant produced *Bacillus thuringiensis* var. *kurstaki* d-endotoxin. Canadian Journal of Microbiology 42: 1258–1262.

29. Herman RA, Wolt JD, Halliday WR (2002) Rapid degradation of the Cry1F insecticidal crystal protein in soil. Journal of Agricultural and Food Chemistry 50: 7076–7078.
30. Donegan KK, Palm CJ, Fieland VJ, Porteous LA, Ganio LM, et al. (1995) Change in levels, species, and DNA fingerprints of soil microorganisms associated with cotton expressing the *Bacillus thurgiensis* var *kurstaki* endotoxin. Applied Soil Ecology 2: 111–124.
31. Gledhill WE, Trehy ML, Carson DB (1991) Comparative biodegradability of anionic surfactants in synthetic and natural test systems. Chemosphere 22: 873–880.
32. Carson DB, Saeger VW, Gledhill WE (1990) Use of Microcosms versus Convenstional biodegradation testing for estimating chemical persistence. In: Landis WG, van der Schalie WH, editors. Aquatic Toxicology and Risk Assessment: Thirtheenth Volume: ASTM STP 1096. pp. 48–59.
33. Cripe CR, Pritchard PH (1990) Aquatic test systems for studying the fate of xenobiotic compounds. In: Landis WG, van der Schalie WH, editors. Aquatic Toxicology and Risk Assessment: Thirtheenth Volume: ASTM STP 1096. pp. 29–47.
34. Green PJ (1994) The ribonucleases of higher plants. Annual Review of Plant Physiology and Plant Molecular Biology 45: 421–445.
35. Arraiano CM, Andrade JM, Domingues S, Guinote IB, Malecki M, et al. (2010) The critical role of RNA processing and degradation in the control of gene expression. FEMS Microbiology Reviews 34: 883–923.
36. Kennedy S, Wang D, Ruvkun G (2004) A conserved siRNA-degrading RNase negatively regulates RNA interference in *C. elegans*. Nature 427: 645–649.
37. Iida T, Kawaguchi R, Nakayama J-i (2006) Conserved ribonuclease, Eri1, negatively regulates heterochromatin assembly in fission yeast. Current Biology 16: 1459–1464.
38. Tsai YL, Tran B, Palmer CJ (1995) Analysis of viral RNA persistence in seawater by reverse transcriptase-PCR. Applied and Environmental Microbiology 61: 363–366.
39. Dancer D, Rangdale RE, Lowther JA, Lees DN (2010) Human norovirus RNA persists in seawater under simulated winter conditions but does not bioaccumulate efficiently in Pacific oysters (*Crassostrea gigas*). Journal of Food Protection 73: 2123–2127.
40. Novitsky JA (1986) Degradation of dead microbial biomass in a marine sediment. Applied and Environmental Microbiology 52: 504–509.

Multiple Exposure and Effects Assessment of Heavy Metals in the Population near Mining Area in South China

Ping Zhuang[1], Huanping Lu[1], Zhian Li[1]*, Bi Zou[1], Murray B. McBride[2]

1 Key Laboratory of Vegetation Restoration and Management of Degraded Ecosystems, South China Botanical Garden, Chinese Academy of Sciences, Guangzhou, China,
2 Department of Crop and Soil Sciences, Cornell University, Ithaca, New York, United States of America

Abstract

The objective of this study was to investigate the levels of Cd, Pb, Cu and Zn in the environment and several important food sources grown and consumed in the vicinity of Dabaoshan mine in Southern China, and evaluate potential health risks among local residents. The Cd, Pb, Cu and Zn concentrations of arable soils and well water near the mines exceeded the quality standard values. The concentrations of Cd and Pb in some food crops (rice grain, vegetable and soybean) samples were significantly higher than the maximum permissible level. The Cd and Pb concentrations in half of the chicken and fish meat samples were higher than the national standard. The residents living near Dabaoshan mine had higher Cd and Pb levels in hair than those of a non-exposed population. The intake of rice was identified as a major contributor to the estimated daily intake of these metals by the residents. The hazard index values for adults and children were 10.25 and 11.11, respectively, with most of the estimated risks coming from the intake of home-grown rice and vegetables. This study highlights the importance of multiple pathways in studying health risk assessment of heavy metal exposure in China.

Editor: Aimin Chen, University of Cincinnati, United States of America

Funding: This research work was financially supported by the National Natural Science Foundation of China (No. 40871221 and No. 41301571) and the Research Fund Program of Guangdong Provincial Key Laboratory of Environmental Pollution Control and Remediation Technology (No. 2013K0008). The funders had no role in study design, data collection and analysis, decision to publish, or preparation of themanuscript.

Competing Interests: The authors have declared that no competing interests exist.

* E-mail: lizan@scbg.ac.cn

Introduction

Mining and smelting activities have had an important role for local and national economies; however, mining-related industries have commonly been performed in an uncontrolled way, giving rise to severe soil erosion and environmental problems, especially heavy metal pollution [1,2]. The discharge of acidic mine drainage (AMD), with elevated levels of heavy metals, can contaminate the downstream water, agricultural soils, food crops and biota and pose a health risk to residents near the mining areas [3,4]. Heavy metal contamination by mining is a major environmental concern on a global scale, particularly in developing countries. Within the global mineral resources industry, China has been one of the largest producers and consumers of several metalliferous and nonmetallic mineral commodities for many years. In China, there are about 8,000 state-owned mining enterprises and 230,000 collectively owned mines that produce hundreds of millions of tons of mining wastes annually [5]. As a result, health related incidents caused by heavy metal pollution in China have risen sharply since 2005, with major accidents attracting nationwide attention [6,7]. It was reported that high Cd in rice on the Chinese market was mainly the result of contaminated fields affected by AMD. Health risks in mine areas affect not only workers, but the whole population living around the areas, in particular children, so that millions of people in the world are estimated to be exposed to metals in mine areas [8–10].

There exist multiple exposure pathways for residents living close to mining or mineral-processing sites, including direct ingestion of soil and water, dermal contact by contaminated soil and water, inhalation of dusts, and consumption of food crops and animals. Various studies have been conducted to evaluate human health risks due to heavy metal exposure through soil [11], water [12], rice [13,14], vegetables [15,16], and even dust [17] from metalliferous mining areas throughout the world, such as Romania, Poland, Korea, China, and France. In general, dietary intake has been recognized as the main route of exposure for most populations, although inhalation can play an important role in very contaminated sites [18,19]. Simple media or pathway-specific approaches to risk assessment may fail to ensure public safety, so it is necessary to apply multi-pathway risk analysis involving all relevant environmental media to identify the dominant pathway of potential concern [9,20].

For metal toxicity monitoring and human health risk assessment, human hair has been widely used in biomonitoring of heavy metals in recent years to estimate environmental exposure levels and assess nutritional status [21]. As a metabolically inactive tissue, hair has become well established, especially for investigating levels of and changes in many heavy metals that accumulate in the body [22]. Human hair analysis has the advantages over other tissues of being less invasive to sample, more convenient to store and transport, and less hazardous to handle. Furthermore, hair levels of metals are less sensitive to immediate intake and could therefore also be a useful biological indicator in characterizing long-term exposure to the measured metal contaminant.

The aims of this study are: (1) to determine the levels of heavy metal concentrations in the environment and foodstuffs near the mining area, (2) to provide a better understanding of each exposure pathway and evaluate the potential health impacts of these metals on the general population. The results of this assessment will aid development of management options and health intervention policies for the affected areas near metal mining and smelting in China and around the world.

Materials and Methods

Ethics Statement

This research was approved by South China Botanical Garden, Chinese Academy of Sciences. The permit for each location was obtained by the authority of Shaxi town. The specific location of field study was provided in the text. The vegetables were purchased from the farmer. All necessary permits were obtained for the described field study. The field studies did not involve endangered or protected species. All participants were informed about the objectives and methods of the study before the investigation. And written consent was obtained from all participants. Data will be made available upon request.

Study Area

Dabaoshan mine (24°31′37″ N; 113°42′49″ E), the largest mine in South China, is located in eastern Shaoguan city, Guangdong, and has been in full-scale operation as a large-scale and integrative quarrying mine since the 1970s (Figure 1). The climate in this area is characterized by a humid subtropical climate, with an annual average temperature of 20.3°C and rainfall of 1762 mm. The Hengshi river (Southward) originates at the mine site and is the main drainage pathway for effluent from the Dabaoshan mine, with the Chuandu river (Northward) being another drainage pathway. The rivers thus deliver significant quantities of heavy metals to numerous villages in this region. After about 40 years of exposure to several heavy metals, some local residents in the mining area have begun to acquire upper gastrointestinal diseases; specifically, oesophageal and stomach cancer are reported to be prevalent. Thus, certain villages around Dabaoshan mine have been termed endemic cancer villages, with a mortality rate approaching 56% in humans [23]. Previous studies in this area have reported that arable soil and foodstuffs in the vicinity of the mine were severely polluted by heavy metals [24] and that there was an increased risk of behavioral problems in school-aged children associated with metal exposures [25]. Wang et al [26] also reported elevated human cancer mortality rates among the metal-exposed populations in Dabaoshan mine area. Six sampling sites near the mine were selected as the study areas (See Supporting Information S1). The agricultural soil in this region was repeatedly irrigated with polluted water from the Hengshi river and Chuandu river. There are about 60–200 households in each village, and they have similar population structures, living conditions and lifestyle.

Sample Collection

Several different environmental media samples, including 122 soils, 54 rice grains, 320 vegetables, 30 soybeans, 12 sediments of fishponds, 150 fishes, 20 well waters, 48 chickens, and 64 hair samples were collected from different sites around Dabaoshan mine (Table S1). Rice grains and tens of species of home-grown vegetables (Table S2), and their corresponding soil samples were collected from SX, DS, FD, LQ, SB, and XJ villages during November 2007 – May 2008. There are 3–5 replications for each vegetable. Soil samples were collected from the top 0–10 cm layer. Soybean was sampled from DS and FD villages in August 2009. At each site of sampling, three to five subsamples were collected to form a composite sample which was stored in a polyethylene zip-bag, and immediately transported to the laboratory. During April - August 2009, a total of 24 wells were selected and sampled from

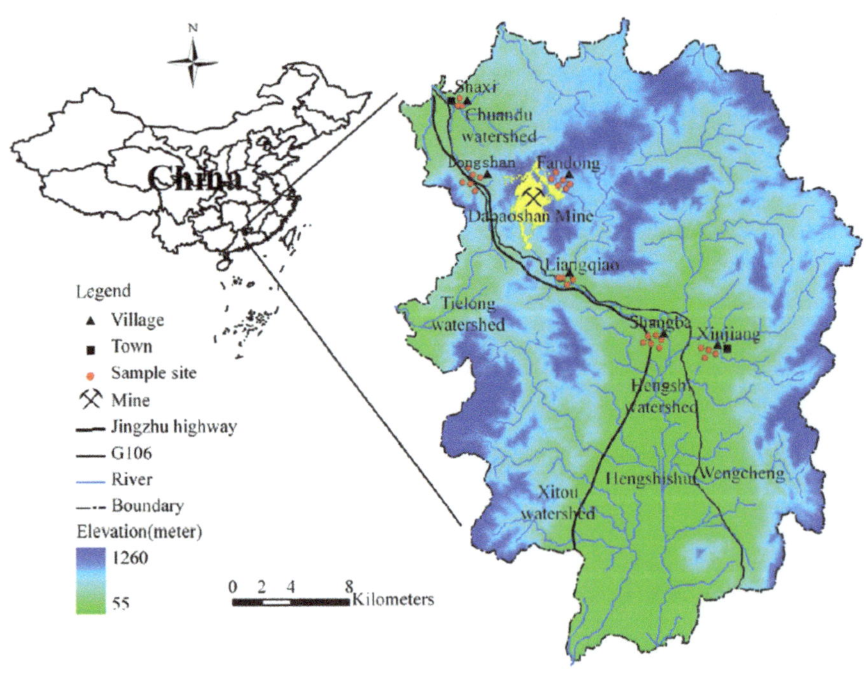

Figure 1. The location of sampling sites in the vicinity of Dabaoshan mine in Guangdong province in South China. The six sampled villages are Shaxi, Dongshan, Fandong, Liangqiao, Shangba and Xinjiang.

DS, FD and SB villages because these well waters were consumed by the local residents.

Sediment samples from 3 fishponds at FD village were collected from the surface down to a depth of 10 cm at five different locations, and these samples were pooled together. Six species of fish and sediment samples were sealed in polyethylene bags and kept cold on ice during transportation to the laboratory. The number of fish, total length, fresh weight (fw) and habitat of fish samples from three fishponds are shown in Table S3.

An experimental group of chicken was fed with rice grain (grown in FD contaminated soil), and the control group was fed with chickenfeed (bought from the market). The concentrations of Pb, Cd, Zn and Cu in the metal-enriched rice and chickenfeed fed with chickens are shown in Table S4. After feeding for 6 months, pectoral muscle of all chickens were separately dissected from the body of the specimens, frozen in liquid nitrogen, and freeze-dried. To characterize the exposure level and metal accumulation in human body, hair samples were taken from 24 inhabitants of both DS and FD village (n = 48). Sixteen further hair samples were collected from non-exposed populations. All the study participants gave their permission to be included in the study.

The soil samples were air-dried at room temperature, then pulverized and sieved through a 1 mm stainless-steel mesh. All home-grown food crops were washed thoroughly with Milli-Q water, and the fresh weight of the samples were recorded. All fish and sediment samples were sealed in polyethylene bags and kept cold on ice during transportation to the laboratory. Sediment samples were air-dried, crushed, sieved through a 2 mm screen, then pulverized and passed through a 0.2 mm mesh sieve.

Analysis of Samples

Four heavy metals (Cd, Pb, Cu and Zn) were analyzed in all the samples. Soil and sediment samples were digested in preparation for total metal analysis using a concentrated acid mixture (HNO_3, $HClO_4$ and HF). For vegetables and soybeans, dried samples were digested with HNO_3 and $HClO_4$ in a 5:1 ratio until a transparent solution was obtained [27]. Chicken muscle, fish muscle and hair samples were digested in HNO_3 (16 mol/L) and H_2O_2 (30%) by microwave. Each hair sample was cut into smaller sections, thoroughly mixed, and ample portions were washed using the method described by Altshul et al [28]. The concentrations of heavy metals in soil and foodstuff samples were analyzed using an atomic absorption spectrophotometer (AAS, GBC932AA), with the concentrations of Cd and Pb in rice grain, vegetables, fish and chicken being determined using graphite furnace atomic absorption spectrophotometer (GFAAS, GBC932AA). The hair digestion procedure was the same as the method described by Wang et al. [21]. The heavy metal concentrations of hair samples were analyzed with an inductively coupled plasma mass spectrometer (ICP-MS) (Agilent 7700x, Agilent Scientific Technology Ltd., USA).

Data Calculation

The translocation capability of heavy metals from the soil to the edible part of crops can be described using a bioaccumulation factor (BAF). The BAFs of Cd, Pb, Cu and Zn were calculated as follows:

$$BAF = \frac{Cplant}{Csoil}$$

where Cplant and Csoil on dry weight basis represent the heavy metal concentration in edible part of food crops and soils, respectively.

The average estimated daily intake (EDI) of heavy metals by the human subjects was calculated using the following equation, which is recommended by the US EPA [29].

$$EDI = \frac{C \times IR \times EF \times ED}{BW \times AT}$$

where EDI is the average daily intake or dose through ingestion (μg/kg bw/day); C is the heavy metal concentration in the exposure medium (mg/L or mg/kg); IR is the ingestion rate (L/day, or kg/day); EF is the exposure frequency (365 days/year); ED is the exposure duration (70 years, equivalent to the average lifespan); BW is the body weight (kg). Average adult and child body weights were considered to be 60 and 30 kg, respectively and AT is the time period over which the dose is averaged (365 days/year times number of exposure years, assumed to be 70 years in this study). A questionnaire-based survey was conducted in the studied villages to determine key risk factors such as dietary behaviors, daily activities and lifestyle of local people. We invited 50 local residents in each village to participate in the survey. In this study area, most foodstuffs are self-produced, whereas pork is purchased from the market. The average considered daily intakes of adults and children is shown in Table S5, according to the survey and reports by Ma et al. [30] and Zhai et al. [31].

The human health risk posed by heavy metal exposure are usually characterized by the target hazard quotient (THQ) [29], the ratio of the average estimated daily intake resulting from exposure to site media compared to the reference dose (RfD) for an individual pathway and chemical. Oral reference dose obtained from the Integrated Risk Information System [32], is an estimation of maximum permissible risk to a human population through daily exposure when taking into consideration a sensitive group during a lifetime. The applied RfD for Cd, Pb, Cu and Zn was 1.0, 4.0, 40, 300 μg/kg/d, respectively. The THQ based on non-cancer toxic risk is determined by

$$THQ = \frac{EDI}{RfD}$$

If the value of THQ is less than 1, the risk of non-carcinogenic toxic effects is assumed to be low. When it exceeds 1, there may be concerns for potential health risks associated with overexposure.

To assess the overall potential risk of adverse health effects posed by more than one metal, the THQs can be summed across contaminants to generate a hazard index (HI) to estimate the risk of a mixture of contaminants. The HI refers to the sum of more than one THQ for multiple substances and/or multiple exposure pathways. In the present study, the HI was used as a screening value to identify whether there is significant risk caused by heavy metals through average dietary consumption for the residents living near the Dabaoshan mine.

Quality Control

Appropriate quality assurance procedures and precautions were carried out to ensure reliability of the results for the different environmental media under investigation. Double distilled deionised water was used throughout the study. Glassware was properly cleaned, and the reagents were of analytical grade. Reagent blank determinations were used to correct the instrument readings. For validation of analytical procedures, a recovery study was carried out by spiking and homogenizing several already analyzed samples with varying amounts of standard solutions of the metals. Several standard reference materials (SRM) were obtained from the National Research Center for CRMs (Table S6) and used for

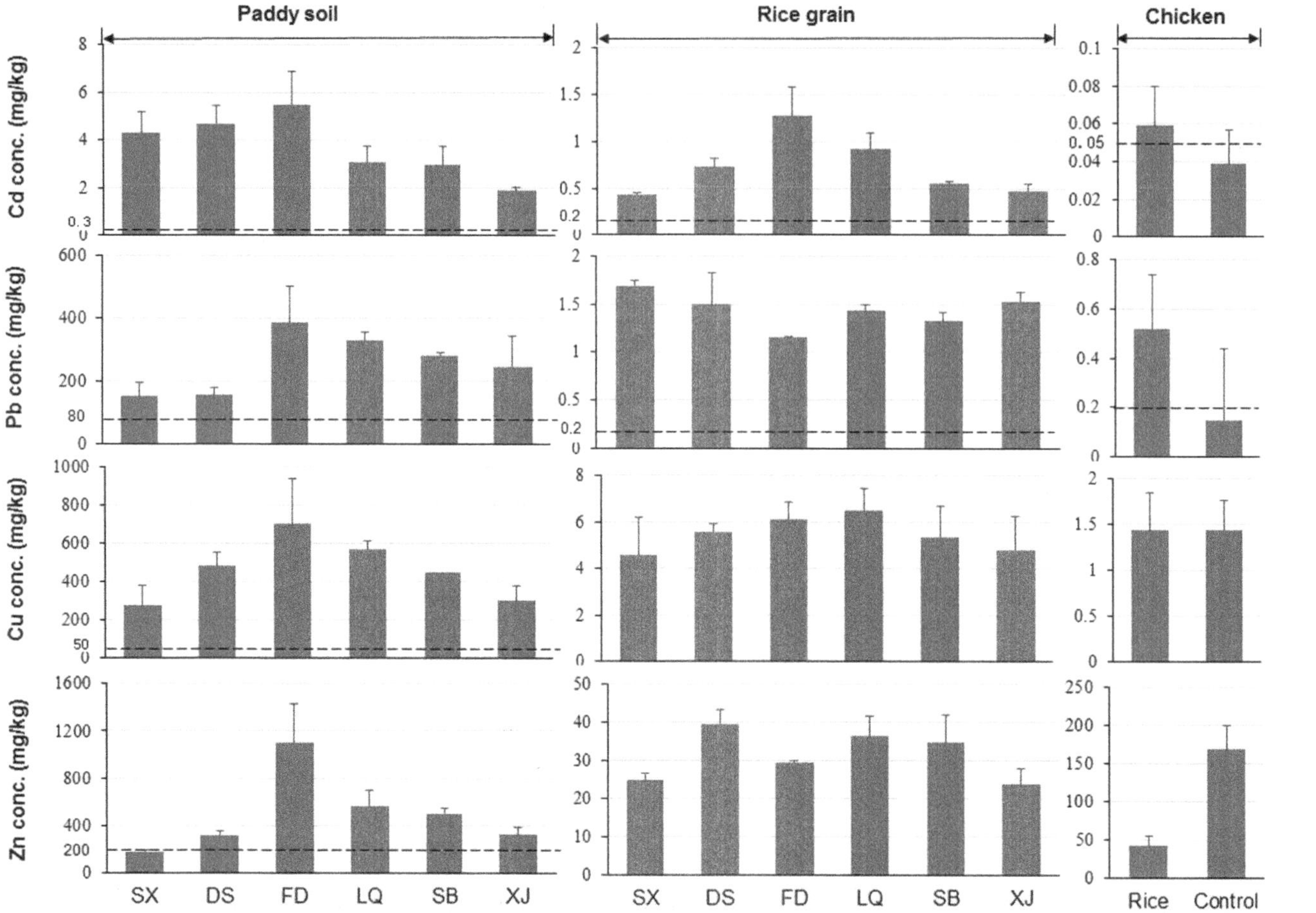

Figure 2. Comparison of concentrations (mg/kg, mean ± SD) of Cd, Pb, Cu and Zn and their respective Chinese national quality value (dotted line) in paddy soil, rice grain and chicken muscle samples collected at different localities in the vicinity of Dabaoshan mine. The number of samples is 9 for respective paddy soil, 8–10 for rice grain, 24 for each group chicken. For soil and rice samples, they were collected from six specific locations, including Shixi (SX), Dongshan (DS), Fandong (FD), Liangqiao (LQ), Shangba (SB) and Xinjiang (XJ). The chicken was fed with rice grain collected from Fandong village.

validation of the analytical procedure. Blank and drift standards were run after every twenty determinations to maintain instrument calibration. The coefficient of variation of replicate analyses was determined for the measurements to calculate analytical precision.

Results and Discussion

Heavy Metals in Different Environmental Media and Food Chain

The environmental sample analysis results showed widespread heavy metal (Cd, Pb, Cu and Zn) contamination in the different exposure media around the Dabaoshan mine (Figures 2–3). Among the six sites in the study area, the paddy soil samples collected from FD showed the highest concentrations of Cd (5.5 mg/kg), Pb (386 mg/kg), Cu (703 mg/kg), and Zn (1100 mg/kg), presumably because FD is located on the mountaintop of the Dabaoshan mine (Figure 2). In contrast, the soil samples collected from XJ, which is located far from Dabaoshan mines (>15 km), showed the lowest heavy metal concentrations. These results suggest that elevated heavy metal concentrations in soils are associated with the mining activities, and indicate that this area is unsuitable for agricultural use. The highest Cd concentration (4.9 mg/kg) in garden soils was found in DS village, whereas soil Pb (297 mg/kg) from LQ was significantly higher compared with the other sites. Concerning co-located arable soils vegetated by various food crops, generally the paddy soil contained higher heavy metal concentrations than the garden soil. This may be due to the fact that the paddy soils were irrigated with highly heavy metal-contaminated stream water whereas the water sources of the garden soils were mainly derived from well water or rainfall.

In comparison to Chinese soil quality guideline values, the soil concentrations of Cd, Pb, Cu and Zn were frequently exceeded (Figures 2–3). Distance from the mine was a key determinant of soil pollution, as there are significant spatial differences among these six studied villages. This finding was consistent with the previous studies [3]. The heavy metal concentrations in soil samples around Dabaoshan mine were remarkably high in general, being comparable with those recorded near a Pb–Zn mine of Spain [33], and higher than those reported for the Songcheon Au–Ag mine in Korea [34], an old mining area in Romania [15] and an abandoned mine in Thailand [35]. From Figures 2–3, it is clear that the characteristics of soil significantly affect the soil level at which the heavy metals will exceed recommended limits in food crops and pose risks in terms of food safety and animal health.

The heavy metal concentrations in rice grain from the six sites decreased in the order Zn>Cu>Pb>Cd (Figure 2). The maximum concentrations of Cd (1.27 mg/kg) at FD and Pb

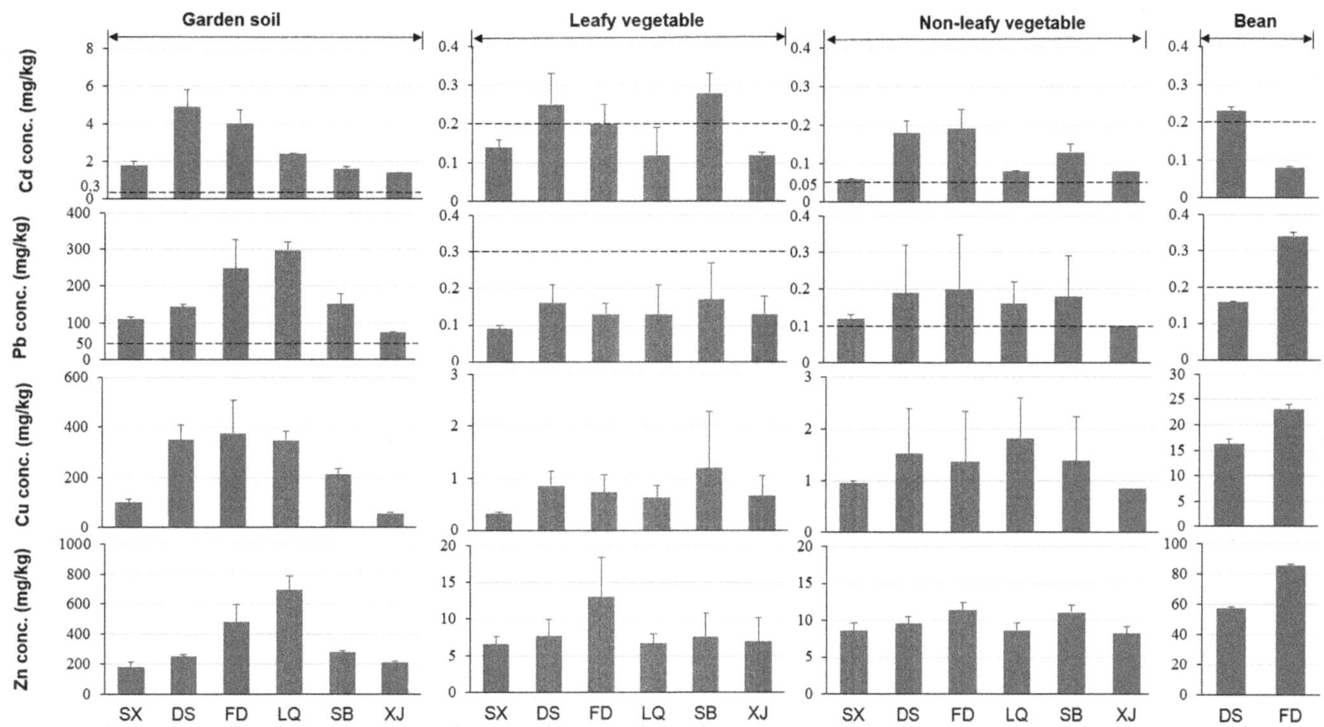

Figure 3. Comparison of concentrations (mg/kg, mean ± SD) of Cd, Pb, Cu and Zn and their respective Chinese national quality value (dotted line) in garden soil, leafy vegetables, non-leafy vegetables and bean samples from different locations. The number of samples is 9 for respective garden soil, 3–5 for each vegetable, 15 for respective soybean. For soil and vegetable samples, they were collected from six specific locations, including Shixi (SX), Dongshan (DS), Fandong (FD), Liangqiao (LQ), Shangba (SB) and Xinjiang (XJ). The bean samples were collected from DS and FD villages.

(1.69 mg/kg) at SX were approximately 6–8 folds greater than the maximum permissible level for both metals of 0.2 mg/kg of rice [36]. The relatively high heavy metal concentrations measured in the self-produced food crops were generally consistent with the elevated heavy metal concentrations found in the paddy and garden soils, although the consistently elevated Pb in rice grain did not follow the site-to-site variation of soil Pb concentration (Figure 2). In comparison with other mining areas, the highest Cd concentration (1.27 mg/kg) in rice grain collected from the present study area was 6–8 times higher than concentrations reported by Lee (0.15 mg/kg, in the Au-Ag-Pb-Zn mining area) [37] and by Ji (0.224 mg/kg, near abandoned metal mine) [38] from Korea, and about 5 times lower than those recorded in the Chenzhou Pb/Zn mining area from China [13], i.e. 6.99 mg/kg. The Pb concentrations in rice grain were higher than those reported for a Pb/Zn mining area from China (0.8 mg/kg) [13] and those reported for an abandoned metal mine from Korea [38].

The concentrations of heavy metals in garden soils and self-produced vegetables (e.g., mustard, Chinese cabbage, lettuce, spinach, garden pea, and tuber of sweet potato, etc. see Table S2) are presented in Figure 3. The leafy and non-leafy vegetables contained higher heavy metal concentrations, especially of Cd, in villages DS, FD and SB. Comparison of the results with the food quality guidelines as set by the Chinese government indicated that the Cd limit was exceeded in 60% of the vegetable samples. For soybean, the Cd concentration in DS and Pb concentration in FD (Figure 3) exceeded the maximum permissible level of 0.2 mg/kg [36]. The concentrations of heavy metals in homegrown vegetable samples around Dabaoshan mine were comparable with those reported in an old mining area of Romania [15], and higher than what was recorded in an abandoned metal mine area of Korea [38]. Of all the vegetables tested, leafy vegetables always contained higher Cd than the non-leafy vegetables (Figure 3).

Contamination of local wetlands, ponds, and rivers by sluicing waste and AMD could provide a pathway for heavy metals into

Table 1. Heavy metal concentrations (mg/L, mean ± SD) in well water sampled from three sites near Dabaoshan mine.

	SB	DS	FD	National standard[a]
Cd	0.0095±0.000	0.01±0.001	0.015±0.001	0.005
Pb	0.013±0.001	0.008±0.000	0.017±0.008	0.01
Cu	1.42±0.012	1.29±0.031	1.58±0.091	1.0
Zn	3.63±0.82	3.47±0.18	4.81±1.05	1.0

[a]Standards for drinking water quality (GB5749-2006) set by the Ministry of Health of the People's Republic of China.

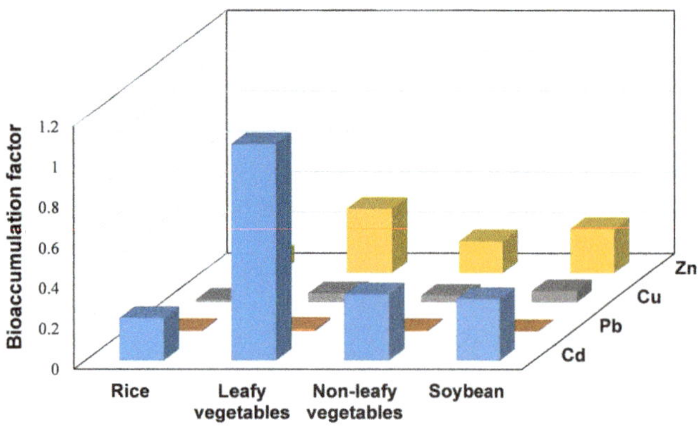

Figure 4. Bioaccumulation factor (BAF), a ratio of heavy metals concentration in the edible part of rice, leafy vegetables, non-leafy vegetables and soybean to that in the corresponding soil.

the aquatic food chain [39]. The Hengshi and Chuandu Rivers are the source of drinking and irrigation water for the residents near Dabaoshan mine, and groundwater also may be contaminated by the two rivers. The concentrations of heavy metals in the well water collected from DS, FD and SB villages exceeded the national standard (Table 1), indicating the drinking water in the local private wells were contaminated by acidic mine water from the Dabaoshan mine, which agreed with previous results [40]. In view of these results, a public water supply system was installed and the community was advised not to use private wells.

Heavy metal analyses in sediments taken from three different sites at FD village show that the levels of Cd (0.32–8.91 mg/kg), Pb (262–327 mg/kg), Cu (239–1477 mg/kg) and Zn (386–4524 mg/kg) were markedly higher than normal background levels (Figure S1). The heavy metal levels of fodder (Ryegrass) cultivated around the fishponds and fed to fish were 1.65, 3.95, 18.02 and 88.01 mg/kg fw for Cd, Pb, Cu and Zn, respectively. It is a positive result that average Cd, Cu and Zn concentrations in fish muscle tissue were within maximum permissible levels. However, the concentrations of Pb in more than 60% of the fish fresh samples from site 1 were above the maximum permissible level (Figure S1). These results suggested the high translocation of metals in site 1 were likely due to the low sediment pH values (5.22), which might cause the heavy metals to be more soluble and bioavailable [41]. Elevated concentrations of heavy metals in the water environment and its human health risk as a consequence of historical mining have been reported elsewhere all over the world [42,43]. Therefore, mining effluents increase metal levels in the aquatic system, and consumption by fish of contaminated sediments and water originating from mining operations is an important exposure route that can result in metal accumulation in fish.

Soil-to-plant transfer is one of the key components of human exposure to metals through the food chain. As shown in Figure 4, large variations in BAFs were observed among different food crops and metals. The results showed that BAFs of all food crops for these tested metals were in the order: Cd>Zn>Cu>Pb, consistent with a report by Li et al. [44]. The BAF values for Cd varied from 0.21 to 1.07, with the highest level in leafy vegetables. Leafy vegetables usually grow quickly and have high transpiration rate, which was in agreement with previous study [45]. The high BAFs of Cd and Zn for respective environmental media were similar to the results for some crops from Chenzhou Pb/Zn mine reported by Liu et al. [13]. Generally, the BAF of heavy metals is controlled by the chemical speciation of heavy metals in soil, soil properties, such as pH and salinity, and plant physiological features [46,47]. In the present study, the lower pH in the sandy soil (Table S7) can increase the solubility of heavy metal and may transfer into the crop tissues.

In the third food chain exposure route investigated (soil-rice-chicken), it is notable that Cd and Pb concentrations in chicken muscle fed with metal-contaminated rice grain (grown in contaminated soil at FD) were higher than those of the control (Figure 2). The levels of Cd and Pb in muscle of chicken fed with contaminated rice exceeded the maximum permissible levels for Cd and Pb in meat prescribed by China. For livestock, the ingestion of rice or grass is the most relevant pathway for the intake of heavy metals [48]. From the data produced in the present study, it is clear that cultivation of food crops on contaminated soil for human or livestock consumption can potentially lead to the uptake and accumulation of trace metals in the edible plant parts, with a resulting risk to human and animal health in the Dabaoshan mining area.

Exposure Assessment

Human exposure to heavy metals in soils near the Dabaoshan mine might occur directly through the ingestion of soil or indirectly through consumption of locally grown vegetables and dairy and meat products from locally raised farm animals. Table 2 summarizes the results of the actual exposure assessment for adults and children exposed to heavy metals through five exposure pathways around Dabaoshan mine. For adults, the total EDI were 5.45, 10.59, 45.14 and 278.38 μg/kg bw/d for Cd, Pb, Cu and Zn, respectively, which were much higher than those reported for an abandoned metal mine from Korea [38]. The EDIs of Cd, Pb and Cu for adults through these five exposure pathways were 5.45, 2.64 and 1.13 fold higher than the RfD of these metals, respectively. It should be cautioned that the EDIs of heavy metals for children were slightly higher than those for adults. The biggest contribution to the intake of heavy metals came from rice, with tens of times as much as intake through others ingestion pathway, as shown in Table 2.

Health risks to residents in the study area through the consumption of agricultural products, and through inadvertent ingestion of soil were assessed by estimating target hazard quotients (THQ). A THQ value greater than 1 would indicate that a potential health risk may exist. The THQs of heavy metals from multiple consumption pathways is in decreasing order Cd>Pb>Cu>Zn (Table 2). The THQs of all heavy metal in rice were the highest among all the crops, which higher than those grown on reclaimed tidal flat soil in the Pearl River Estuary [44]. Considering that rice is a staple food in the diet of the people in the area, intake of heavy metals through rice consumption is likely to be a main source of heavy metal intake among residents in the area. The HI values (the sum of all THQs) through diet and soil for adults and children in the Dabaoshan mine area were 10.25 and 11.11, respectively; both are higher than reported in other areas [16,49]. These results suggest that the adults and children living around the Dabaoshan mine may experience adverse health effects. Considering THQ determination for the different metals and pathways for children, besides rice, the highest risk was for ingestion of vegetables for Cd (0.71), followed in descending order by ingestion of soil for Pb (0.32) and ingestion of bean for Cu (0.26). Concern has focused particularly on children who become exposed to heavy metals to a greater extent than adults, which may harm brain and nervous system development [50]. Despite Zn having the highest concentration among the metals, its THQ value

Table 2. Estimated daily intake (EDI, ug/kg bw/d) and target hazard quotient (THQ) for adult and children exposed to contaminants in the vicinity of Dabaoshan mine.

Metal	Pathways	EDI		THQ	
		Adult	Child	Adult	Child
Cd	Ingestion of rice	4.54	4.83	4.54	4.83
	Ingestion of vegetables				
	Leafy vegetables	0.54	0.42	0.54	0.42
	Non-leafy vegetables	0.26	0.21	0.26	0.21
	Bean	0.04	0.08	0.04	0.08
	Ingestion of meat			0.03	0.04
	Fish	0.03	0.04		
	Chicken	0.02	0.02	0.02	0.02
	Ingestion of soil	0.00	0.02	0.00	0.02
	Ingestion of water	0.02	0.02	0.04	0.04
	Total	5.45	5.63	5.47	5.65
Pb	Ingestion of rice	8.93	9.50	2.23	2.38
	Ingestion of vegetables				
	Leafy vegetables	0.39	0.31	0.10	0.08
	Non-leafy vegetables	0.24	0.19	0.06	0.05
	Bean	0.07	0.13	0.02	0.03
	Ingestion of meat				
	Fish	0.56	0.57	0.14	0.14
	Chicken	0.16	0.21	0.04	0.05
	Ingestion of soil	0.22	1.30	0.05	0.32
	Ingestion of water	0.02	0.02	0.01	0.00
	Total	10.59	12.23	2.65	3.06
Cu	Ingestion of rice	34.11	36.31	0.85	0.91
	Ingestion of vegetables				
	Leafy vegetables	2.13	1.67	0.05	0.04
	Non-leafy vegetables	1.67	1.34	0.04	0.03
	Bean	5.57	10.48	0.14	0.26
	Ingestion of meat				
	Fish	0.85	0.86	0.02	0.02
	Chicken	0.43	0.57	0.01	0.01
	Ingestion of soil	0.39	2.32	0.01	0.06
	Ingestion of water	2.69	2.15	0.07	0.05
	Total	45.14	53.55	1.20	1.39
Zn	Ingestion of rice	194.86	207.43	0.65	0.69
	Ingestion of vegetables				
	Leafy vegetables	23.45	18.33	0.08	0.06
	Non-leafy vegetables	14.62	11.70	0.05	0.04
	Bean	20.23	38.08	0.07	0.13
	Ingestion of meat				
	Fish	4.74	3.39	0.02	0.01
	Chicken	12.60	16.80	0.04	0.06
	Ingestion of soil	0.41	2.49	0.00	0.01
	Ingestion of water	7.46	5.96	0.02	0.02
	Total	278.38	304.17	0.93	1.02
Hazard index				10.25	11.11

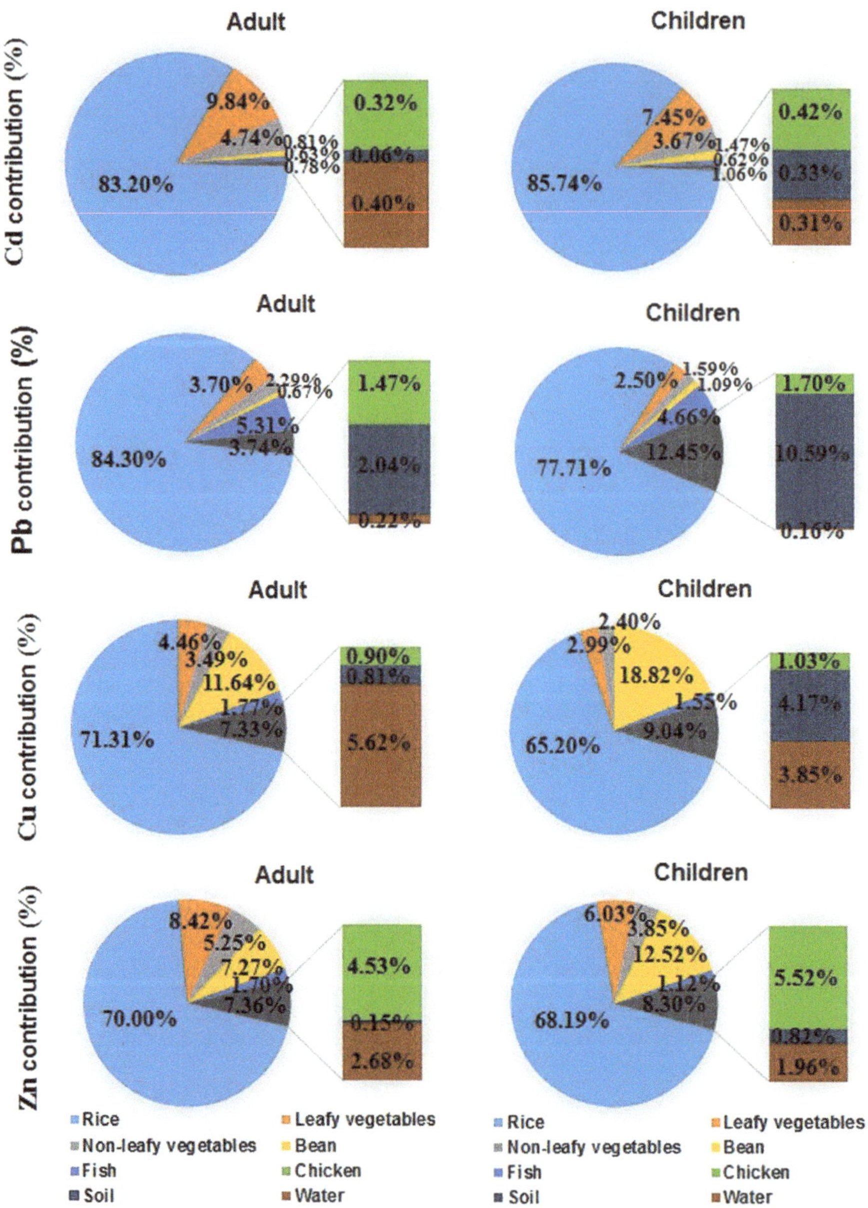

Figure 5. Relative contribution profile of daily human intake. Relative contribution was derived by dividing the daily intake estimate from a given source by the total daily intake estimate from all of the sources under investigation.

was below 1, indicating that Zn does not pose a health risk to the local residents.

Mining sites are usually characterized by soil contamination and a higher exposure to heavy metals for the population residing in the area [51,52]. Exposure level and risk assessment in the study area are compared with those of other studies [9,15,35,38,49,53–56] from mining and smelting sites around the world as shown in Table 3. The health risk for residents living in the vicinity of Dabaoshan mine associated with the consumption of local foodstuffs was markedly higher than that reported for other metal mines all over the world. Some evidences also revealed a large part of the population in the vicinity of Dabaoshan mine had contacted or ingested contaminated soil [24,57], contaminated water [23,40], consumed local agricultural products, such as foodcrops [24], fish [58] and chicken, over a long period of time. These concerns were further supported by the finding of behavioral problems for school-aged children [25] and significantly elevated blood levels of metals (38.9 and 24.1 μg/L for Pb and Cd in high exposed area vs 4.46 and 1.87 μg/L for Pb and Cd in low exposed area) [26] in local residents from the same mining area. Liu et al. [23] and Wang et al. [26] reported that an increased risk of mortality from all cancer (e.g. enteron tumors) was probably associated with long-term environmental exposure to both Cd and Pb.

Source Allocation for Different Metals

When the profiles of relative contribution of soil and foodstuffs to daily heavy metal intake estimates from all the sources being investigated (e.g., rice grain, vegetables, beans, fish, chicken, water and soil) were compared between adults and children (Figure 5), intake of rice was the major source of Cd, Pb, Cu and Zn exposure, accounting for 65.2–85.7% of the estimated total daily

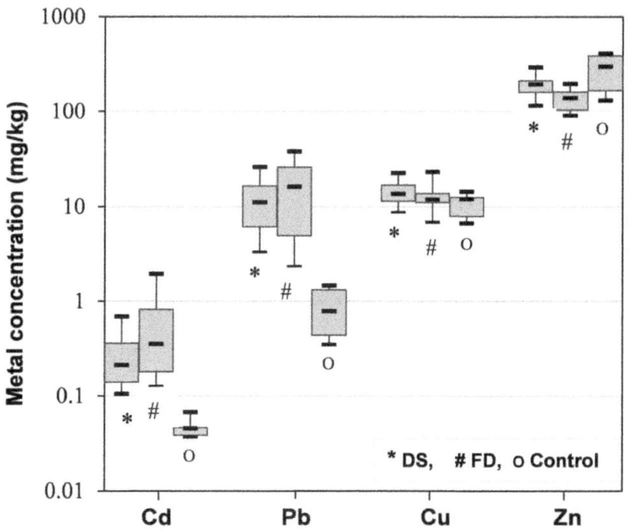

Figure 6. Box-plot diagram of four elements in hair samples from two villages (Dongshan, DS, Fandong, FD) in the vicinity of Dabaoshan mine and the control area (non-exposed site). Y-axis is presented in logarithmic scale. The central mark on each box is the median with the edges of the 25th and 75th percentiles.

intake in the study area. Interestingly, soil ingestion was a second important factor for daily Pb intake, accounting for 10.6% of children's exposure. Among all the main exposure pathways, the most important exposure pathway for heavy metals appeared to be the ingestion of locally grown rice, with about several times as much as intake through vegetables. The allocations of Cd and Pb from dietary exposure were comparable with that reported for Pb by Dong and Hu [20], but higher than the contribution from diet of 70% and 50% for Cd and Pb, respectively, estimated by Plant et al. [18].

The results of this study suggested that the contribution to total Cd and Pb exposure from drinking water is even lower than 1%, which was consistent with the finding by Dong and Hu [20]. It is clear that the allocation of dietary exposure is the highest in the study area, and the contributions by the soil and water exposure pathways are relatively low. Specifically for children, a health risk from Pb was indicated, with the second most important exposure pathway being the direct ingestion of soil. The source allocation of Pb for soil accounted for 10.6% of children's exposure, which was similar to that reported for an abandoned metal mine area of Korea (11.1–16.8%) [38]. Staessen et al. [19] also found that 2–4% of the variance in the long-term body burden of Cd was directly related to consumption of vegetables, implying that the ingestion of locally grown vegetables is an important source for human Cd exposure.

Compared to other countries, the relatively high source contribution from the diet in China is due to the high concentrations of heavy metals in the food. The high heavy metal concentrations in grains and vegetables were mainly due to enrichment from the soil [59]. The substantial contribution of rice and soil to the intake of four heavy metals indicate that the status of metal contamination of soil and rice should receive further attention in the metal mine areas throughout the country. In fact, the potential health risk from heavy metals can be for some individuals much higher than our calculations based on average ingestion of water, food and soil. Some local inhabitants, who consume more contaminated locally grown food crops, breathe contaminated air, or smoke, might be exposed to a health risk from dietary heavy metals well above the calculated risk. As a result, great effort is required to control concentrations of heavy metals in soil to reduce dietary metal exposure, and ultimately to eliminate heavy metal exposure in China.

Heavy Metal Levels in Hair and Global Health Implications

Human hair can serve as a useful direct biomonitoring tool to assess the extent of heavy metal exposure to residents in metal-polluted areas. The heavy metal concentrations in the hair samples are shown in Figure 6. The hair levels of four heavy metals were in the order of Zn>Pb, Cu>Cd. The hair samples from the residents

Table 3. Comparison of estimated dietary intake (EDI) and hazard index (HI) of heavy metals in different areas around the world.

Sites	EDI (ug/kg bw/d) or HI	Cd	Pb	Cu	Zn	References
Dabaoshan mine, Shaoguan, China	EDI	5.45–5.63	10.6–12.2	50.4–59.1	292–320	In this study
	HI	5.45–5.63	2.65–3.06	1.26–1.48	0.97–1.07	
Abandoned metal mine, Kanchanaburi, Thailand	EDI	0.25–5.34	0.71–1.46			[35]
	HI	0.25–5.34	0.18–0.36			
Abandoned metal mine, Katowice, Poland	HI	2.6–2.7		0.025–0.041	0.11–0.18	[56]
Old mining area, Banat, Romania	HI	<1	>1	<1	<1	[15]
Lead-zinc mining area, Jiangsu, China	HI	1.81–3.32	3.01–16.2	0.164–0.189	0.28–0.30	[9]
Huludao Zn plant, Liaoning, China	EDI	0.70	1.36	45.5	204	[49]
	HI	0.749	0.364	1.22	0.731	
Pb/Sb smelter, Nanning, China	HI	7.42	1.35	0.146	0.323	[53]
Chatian mercury mining deposit Hunan, China	EDI	0.05–1.66	0.359–1.6			[55]
	HI	0.05–1.66	2.67–2.99			
Songcheon Au-Ag mine, Gangneung-si, Korea	HI	0.75		0.078	0.26	[54]
Abandoned metal mine, Korea	EDI	0–0.195	0.001–0.233	0.009–13.4	0.208–66.6	[38]
	HI	0.179	0.077	0.027	0.063	

living in the Dabaoshan mine area (DS and FD villages) contained markedly higher average Cd, Pb and Cu concentrations than those of non-exposed populations, with Cd around 5–11 times higher (0.27–0.57 mg/kg) and Pb 15–28 times higher (12.5–23.9 mg/kg) than the non-exposed levels, which might be a consequence of long term exposure of the local residents to the mining activities. Some subjects in FD showed even higher levels of Cd and Pb, with the maxima estimated at 1.94 and 38 mg/kg, respectively. The average levels of hair Pb, Cd, Cu and Zn in the study area were lower than the levels in the scalp hair of adults from S. Domingos mine, Portugal [60], but higher than those recorded in a lead-zinc mining area from China [9]. Interestingly, the Zn concentration in hair of a non-exposed population was higher than that of subjects living near the Dabaoshan mine, which might be due to the higher concentration of toxic metals like Cd and Pb in the vicinity, which can interact with Zn and replace it in the heme enzymes and metallothioneins [61]. Consistent with the higher THQ value of the residents in FD village, their hair samples exhibited higher than those in DS village. This finding agreed well to the conclusions of previous studies [9]. Therefore, the elevated hair Cd and Pb levels provide an independent indicator that residents in the Dabaoshan mine region might be at high risk of toxic metal exposure due to the elevated levels of heavy metals in food crops, fish or livestock.

Increasing evidence shows heavy metal pollution of mined areas to cause health damage to the local inhabitants [4,26]. For example, itai-itai disease was caused by Cd poisoning in Japan due to the Kamioka mine releasing this metal into river water that was then used to irrigate rice paddy soils [62]. Moreover, tens of thousands of Kabwe's residents (Zambia) suffered from severe lead poisoning, which resulted from artisanal re-mining of and exposures to wastes from historical lead-zinc mining and smelting [63]. In northern Nigeria, confirmed deaths of at least 400 children were the result of artisanal processing of lead-rich gold ores [64]. There is some evidence to suggest that local inhabitants may develop a natural tolerance to the high soil metal concentrations encountered in their environment, since non-local people who have supposedly suffered from heavy metal poisoning caused by the consumption of home produced vegetables grown in gardens reclaimed from former mine dumps [61,65]. Thus, by understanding human exposure pathways for toxic metals, the routes by which heavy metals may enter the body and cause health effects, scientists can help identify other mining areas that may pose the highest risk for heavy metal poisoning and need for medical surveillance and intervention.

Supporting Information

Figure S1 Comparison of concentrations (mg/kg) of Cd, Pb, Cu and Zn and their respective Chinese national quality value (dotted line) in muscle samples from three different fishponds at Fandong village. Y-axis of Pb, Cu and Zn concentrations are presented in logarithmic scale.

Table S1 Number of soil, rice, vegetable, soybean, water, sediment, fish and chicken samples collected from different sampling sites in the present study.

Table S2 The edible part and water content of vegetables.

Table S3 The number of fish, total length, fresh weight (fw) and habitat of fish samples.

Table S4 Heavy metal concentrations (mg/kg, mean ± SD) in the metal-enriched rice grain and chickenfeed for both groups of chicken.

Table S5 Average considered daily intakes of adults and children in the present study.

Table S6 Standard reference materials for respective sample under investigation.

Table S7 The pH and organic matter (OM) of paddy and garden soils from Dabaoshan mine.

Supporting Information S1 Description of study area.

Acknowledgments

We thank Prof. Aimin Chen (University of Cincinnati, USA), the editor and two reviewers (Prof. Iosif Gergen and an anonymous reviewer) for their comments on an early version of this paper.

Author Contributions

Conceived and designed the experiments: PZ ZAL. Performed the experiments: PZ HPL. Analyzed the data: PZ. Contributed reagents/materials/analysis tools: PZ BZ. Wrote the paper: PZ MBM.

References

1. Rybicka EH (1996) Impact of mining and metallurgical industries on the environment in Poland. Appl Geochem 11: 3–9.
2. Dudka S, Adriano DC (1997) Environmental impacts of metal ore mining and processing: a review. J Environ Qual 26: 590–602.
3. Benin AL, Sargent JD, Kalton M, Roda S (1999) High concentrations of heavy metals in neighborhoods near ore smelters in northern Mexico. Environ Health Perspect 107: 279–184.
4. Plumlee GS, Morman SA (2011) Mine wastes and human health. Element 7: 399–404.
5. Shu WS, Zhang ZQ, Lan CY (2000) Strategies for restoration of mining wastelands in China. Ecol Sci 19: 24–29.
6. Gao Y, Xia J (2011) Chromium contamination accident in China: viewing environment policy of China. Environ Sci Technol 45: 8065–8056.
7. Zhang XW, Yang LS, Li YH, Li HR, Wang WY, et al. (2012) Impacts of lead-zinc mining and smelting on the environment and human health in China. Environ Monit Assess 184: 2261–2273.
8. Plumlee GS, Durant JT, Morman SA, Neri A, Wolf RE, et al. (2013) Linking geological and health sciences to assess childhood lead poisoning from Artisanal gold mining in Nigeria. Environ Health Perspect 121: 744–750.
9. Qu CS, Ma ZW, Yang J, Liu Y, Bi J, et al. (2012) Human exposure pathways of heavy metals in a Lead-Zinc mining area, Jiangsu Province, China. PLoS ONE 7: e46793.
10. Wcisło E, Ioven D, Kucharski R, Szdzuj J (2002) Human health risk assessment case study: an abandoned metal smelter site in Poland. Chemosphere 47: 507–515.
11. Kapusta P, Szarek-Łukaszewska G, Stefanowicz AM (2011) Direct and indirect effects of metal contamination on soil biota in a Zn-Pb post-mining and smelting area (S Poland). Environ Pollut 159: 1516–1522.
12. Park JH, Choi KK (2013) Risk assessment of soil, water and crops in abandoned Geumryeong mine in South Korea. J Geochem Explor 128: 117–123.
13. Liu HY, Probst A, Liao BH (2005) Metal contamination of soils and crops affected by the Chenzhou lead/zinc mine spill (Hunan, China). Sci Total Environ 339: 153–166.
14. Zhu YG, Sun GX, Lei M, Teng M, Liu YX, et al. (2008) High percentage inorganic arsenic content of mining impacted and nonimpacted Chinese rice. Environ Sci Technol 42: 5008–5013.
15. Harmanescu M, Alda LM, Bordean DM, Gogoasa I, Gergen I (2011) Heavy metals health risk assessment for population via consumption of vegetables

grown in old mining area; a case study: Banat County, Romania. Chem Cent J 5: 64.
16. Wang X, Sato T, Xing B, Tao S (2005) Health risk of heavy metals to the general public in Tianjin, China via consumption of vegetables and fish. Sci Total Environ 350: 28–37.
17. Kerin EJ, Lin HK (2010) Fugitive dust and human exposure to heavy metals around the Red Dog Mine. Rev Environ Contam Toxicol 206: 49–63.
18. Plant J, Smith D, Smith B, Williams L (2001) Environmental geochemistry at the global scale. Appl Geochem 16: 1291–1308.
19. Staessen JA, Vyncke G, Lauwerys RR, Roels HA, Celis HG, et al. (1992) Transfer of cadmium from a sandy acidic soil to man: a population study. Environ Res 58: 24–34.
20. Dong ZM, Hu JY (2012) Development of lead source-specific exposure standards based on aggregate exposure assessment: Bayesian inversion from biomonitoring information to multipathway exposure. Environ Sci Technol 46: 1144–1152.
21. Wang T, Fu JJ, Wang YW, Liao CY, Tao YQ, et al. (2009) Use of scalp hair as indicator of human exposure to heavy metals in an electronic waste recycling area. Environ Pollut 157: 2445–2451.
22. Druyan ME, Bass D, Puchyr R (1998) Determination of reference ranges for elements in human scalp hair. Biol Trace Elem Res 62: 183–187.
23. Liu YS, Gao Y, Wang KW, Mai XH, Chen GD, et al. (2005) Etiologic study on alimentary tract malignant tumor in villages of high occurrence. China Trop med 5: 1139–1141. (in Chinese).
24. Zhuang P, McBride MB, Xia HP, Li NY, Li ZA (2009) Health risk from heavy metals via consumption of food crops in the vicinity of Dabaoshan mine, South China. Sci Total Environ 407: 1551–1561.
25. Bao QS, Lu CY, Song H, Wang M, Ling WH, et al. (2009) Behavioural development of school-aged children who live around a multi-metal sulphide mine in Guangdong province, China: a cross-sectional study. BMC Publ Health 9: 217.
26. Wang M, Song H, Chen WQ, Lu CY, Hu QS, et al. (2011) Cancer mortality in a Chinese population surrounding a multi-metal sulphide mine in Guangdong province: an ecologic study. BMC Publ Health 11: 319.
27. Allen SE, Grimshaw HM, Rowland AP (1986) Chemical analysis, in: Moore, P.D., Chapman, S.B. (Eds.), Methods in Plant Ecology. Oxford, London: Blackwell Scientific Publication.
28. Altshul L, Covaci A, Hauser R (2004) The relationship between levels of PCSs and pesticides in human hair and blood: preliminary results. Environ Health Perspect 112: 1193–1199.
29. US EPA (United States Environmental Protection Agency) (1997) Exposure factors handbook. EPA/600/P-95/002F. Washington, DC.
30. Ma WJ, Deng F, Xu YJ, Xu HF, Nie DP (2005) The study on dietary intake and nutritional status of residents in Guangdong, South China. J Prev Med 31: 1–5. (in Chinese).
31. Zhai FY, He YN, Ma GS, Li YP, Wang ZH, et al. (2005) Study on the current status and trend of food consumption among Chinese population. Chin J Epidemiol 26(7): 485–488. (in Chinese).
32. US EPA (United States Environmental Protection Agency) (2007) Integrated Risk Information System-database. Philadelphia PA; Washington, DC.
33. Rodríguez L, Ruiz E, Alonso-Azcárate J, Rincón J (2009) Heavy metal distribution and chemical speciation in tailings and soils around a Pb–Zn mine in Spain. J Environ Manage 90: 1106–1116.
34. Lee JS, Chon HT, Kim KW (2005) Human risk assessment of As, Cd, Cu and Zn in the abandoned metal mine site. Environ Geochem Health 27: 185–191.
35. Nobuntou W, Parkpian P, Kim Oanh NT, Noomhorm A, Delaune RD, et al. (2010) Lead distribution and its potential risk to the environment: Lesson learned from environmental monitoring of abandon mine. J Environ Sci Health A Tox/Hazard Subst Environ Eng 45: 1702–1714.
36. MHPRC (Ministry of Health of the People's Republic of China) (2012) The limits of pollutants in foods (GB 2762–2012) (in Chinese). Beijing, China: MHPRC.
37. Lee CG, Chon HT, Jung MC (2001) Heavy metal contamination in the vicinity of the Daduk Au-Ag-Pb-Zn mine in Korea. Appl Geochem 16: 1377–1386.
38. Ji K, Kim J, Lee M, Park S, Kwon HJ, et al. (2013) Assessment of exposure to heavy metals and health risks among residents near abandoned metal mines in Goseong, Korea. Environ Pollut 178: 322–328.
39. Moiseenko TI, Kudryavtseva LP (2001) Trace metal accumulation and fish pathologies in areas affected by mining and metallurgical enterprises in the Kola Region, Russia. Environ Pollut 114: 285–297.
40. Chen A, Lin C, Lu W, Wu Y, Ma Y, et al. (2007) Well water contaminated by acidic mine water from the Dabaoshan Mine, South China: Chemistry and toxicity. Chemosphere 70: 248–255.
41. Adhikari S, Ghosh L, Ayyappan S (2006) Combined effects of water pH and alkalinity on the accumulation of lead, cadmium and chromium to *Labeo rohita* (Hamilton). Int J Env Sci Tech 3: 289–296.
42. Borgmann U, Couillard Y, Grapentine LC (2007) Relative contribution of food and water to 27 metals and metalloids accumulated by caged *Hyalella azteca* in two rivers affected by metal mining. Environ Pollut 145: 753–765.
43. Proctor PD (1984) Heavy metal additions to the environment near mines, mills, and smelters, Southeast Missouri. In: Nriagu, J.O. (Ed.), Environmental Impacts of Smelters. Wiley, New York; 89–115.
44. Li QS, Chen Y, Fu HB, Cui ZH, Shi L, et al. (2012) Health risk of heavy metals in food crops grown on reclaimed tidal flat soil in the Pearl River Estuary, China. J Hazard Mat 227–228: 148–154.
45. Albering HJ, van Leusen SM, Moonen EJC, Hoogewerff JA, Kleinjans JCS (1999) Human health risk assessment: A case study involving heavy metal soil contamination after the flooding of the river Meuse during the winter of 1993–1994. Environ Health Perspect 107: 37–43.
46. Golia EE, Dimirkou A, Mitsios IK (2008) Influence of some parameters on heavy metals accumulation by vegetables grown in agricultural soils of different soil orders. B Environ Contam Toxicol 81: 80–84.
47. McLaughlin MJ, Whatmuff M, Warne M, Heemsbergen D, Barry G, et al. (2006) A field investigation of solubility and food chain accumulation of biosolid-cadmium across diverse soil types. Environ Chem 3: 428–432.
48. Rodrigues SM, Pereira ME, Duarte AC, Römkens PF (2012) Soil-plant-animal transfer models to improve soil protection guidelines: a case study from Portugal. Environ Int 39: 27–37.
49. Zheng N, Wang QC, Zhang XW, Zheng DM, Zhang ZS, et al. (2007) Population health risk due to dietary intake of heavy metals in the industrial area of Huludao city, China. Sci Total Environ 383: 81–89.
50. Rieuwert JS, Farago ME, Cikrt M, Bencko V (2000) Differences in lead bioavailability between a smelting and a mining area. Water Air Soil Pollut 122: 203–229.
51. Alonso E, Cambra K, Martinez T (2001) Lead and cadmium exposure from contaminated soil among residents of a farm area near an industrial site. Arch Environ Health 56: 278–282.
52. Paoliello M, De Cpitani E, Cunha F, Matsuno T, Carvalho M (2002) Exposure of children to lead and cadmium from a mining area of Brazil. Environ Res 88: 120–128.
53. Cui YJ, Zhu YG, Zhai RH, Chen DY, Huang ZH, et al. (2004) Transfer of metals from soil to vegetables in an area near a smelter in Nanning, China. Environ Int 30: 785–791.
54. Lim HS, Lee JS, Chon HT, Sager M. (2008) Heavy metal contamination and health risk assessment in the vicinity of the abandoned Songcheon Au-Ag mine in Korea. J Geochem Explor 96: 223–230.
55. Sun HF, Li YH, Ji YF, Yang LS, Wang WY, et al. (2010) Environmental contamination and health hazard of lead and cadmium around Chatian mercury mining deposit in western Hunan Province, China. Trans Nonferrous Met Soc China 20: 308–314.
56. Wcisło E, Ioven D, Kucharski R, Szdzuj J. (2002) Human health risk assessment case study: an abandoned metal smelter site in Poland. Chemosphere 47(5): 507–515.
57. Zhou JM, Dang Z, Cai MF, Liu CQ (2007) Soil heavy metal pollution around the Dabaoshan mine, Guangdong province, China. Pedosphere 17: 588–94.
58. Zhuang P, Li ZA, McBride MB, Zou B (2013) Health risk assessment for consumption of fish originating from ponds near Dabaoshan mine, South China. Environ Sci Pollut Res 20(8): 5844–5854.
59. Hough RL, Breward N, Young SD, Crout NM, Tye AM, et al. (2004) Assessing potential risk of heavy metal exposure from consumption of home-produced vegetables by urban populations. Environ Health Perspect 112: 215–221.
60. Pereira R, Ribeiro R, Goncalves F (2004) Scalp hair analysis as a tool in assessing human exposure to heavy metals (S. Domingos mine, Portugal). Sci Total Environ 327: 81–92.
61. Waalkes MP (2003) Cadmium carcinogenesis. Mutat Res 533: 107–120.
62. Abrahams PW (2002) Soils: their implications to human health. Sci Total Environ 291: 1–32.
63. Branan N (2008) Mining leaves nasty legacy in Zambia. Geotimes website. Available: http://www.geotimes.org/jan08/article.html?id = nn_zambia.html. Accessed 2008 Jan.
64. UNEP/OCHA (2010) Lead Pollution and Poisoning Crisis, Environmental Emergency Response Mission, Zamfara State, Nigeria, United Nations Environment Program, September/October.
65. Thomas R (1980) Arsenic pollution arising from mining activities in south-west England. Inorganic Pollution and Agriculture. London: HMSO, 142–158.

Influence of 20-Year Organic and Inorganic Fertilization on Organic Carbon Accumulation and Microbial Community Structure of Aggregates in an Intensively Cultivated Sandy Loam Soil

Huanjun Zhang[1], Weixin Ding[1]*, Xinhua He[2,3], Hongyan Yu[1], Jianling Fan[1], Deyan Liu[1]

1 State Key Laboratory of Soil and Sustainable Agriculture, Institute of Soil Science, Chinese Academy of Sciences, Nanjing, China, **2** Forests NSW, NSW Department of Primary Industries, West Pennant Hills, New South Wales, Australia, **3** School of Plant Biology, University of Western Australia, Crawley, Western Australia, Australia

Abstract

To evaluate the long-term effect of compost (CM) and inorganic fertilizer (NPK) application on microbial community structure and organic carbon (OC) accumulation at aggregate scale, soils from plots amended with CM, NPK and no fertilizer (control) for 20 years (1989–2009) were collected. Soil was separated into large macroaggregate (>2,000 μm), small macroaggregate (250–2,000 μm), microaggregate (53–250 μm), silt (2–53 μm) and clay fraction (<2 μm) by wet-sieving, and their OC concentration and phospholipid fatty acids (PLFA) were measured. The 20-year application of compost significantly ($P<0.05$) increased OC by 123–134% and accelerated the formation of macroaggregates, but decreased soil oxygen diffusion coefficient. NPK mainly increased OC in macroaggregates and displayed weaker influence on aggregation. Bacteria distributed in all aggregates, while fungi and actinobacteria were mainly in macroaggregates and microaggregates. The ratio of monounsaturated to branched (M/B) PLFAs, as an indicator for the ratio of aerobic to anaerobic microorganisms, increased inversely with aggregate size. Both NPK and especially CM significantly ($P<0.05$) decreased M/B ratios in all aggregates except the silt fraction compared with the control. The increased organic C in aggregates significantly ($P<0.05$) negatively correlated with M/B ratios under CM and NPK. Our study suggested that more efficient OC accumulations in aggregates under CM-treated than under NPK-treated soil was not only due to a more effective decrease of actinobacteria, but also a decrease of monounsaturated PLFAs and an increase of branched PLFAs. Aggregations under CM appear to alter micro-habitats to those more suitable for anaerobes, which in turn boosts OC accumulation.

Editor: Dafeng Hui, Tennessee State University, United States of America

Funding: This work was supported by the National Basic Research Program of China (2011CB100503), the Chinese Academy of Sciences (KSCX2-EW-N-08), and National Science Foundation of China (41171190). The funders had no role in study design, data collection and analysis, decision to publish, or preparation of the manuscript.

Competing Interests: The authors have declared that no competing interests exist.

* E-mail: wxding@issas.ac.cn

Introduction

Sequestration of carbon (C) in soils has gained increasing recognition in recent years for its role in global environmental change [1]. Soil aggregation affects soil properties such as aeration and water infiltration and alters micro-habitats and activities of microorganisms [2]. A growing body of work has revealed a close relationship between aggregation and soil organic C (OC) concentrations [3,4] as OC stabilization in aggregates is the principal mechanism controlling OC turnover and sequestration [5,6,7].

The potential of soils to sequester C depends on soil type, management practices, and soil aggregate structure [5,8,9]. Previous studies have suggested that OC is more effectively protected and more stable in microaggregates, whereas macroaggregates provide a niche for the storage of labile C [3,5,10]. Fertilization, a common agricultural practice, has been found to significantly influence aggregate formation and OC distribution in aggregates [11,12]. For example, increases in OC following organic manure application were observed in aggregates of all sizes, especially in macroaggregates [11,13]. Kong *et al.* [14] found that the majority of OC derived from organic material was preferentially sequestered in microaggregates within small macroaggregates. In contrast, Sarkar *et al.* [15] and Bhattacharyya *et al.* [16] found that the highest rate of increase in OC following repeated inorganic fertilizer applications primarily occurred in the silt and clay fractions of OC-poor soils.

The turnover of OC in soil aggregates is not only determined by the physical protection offered by that fraction, but also by the abundance and community of microorganisms, especially Gram-positive (G^+) bacteria and actinobacteria [17,18,19]. Aggregates provide spatially heterogeneous habitats for microorganisms characterized by different substrates, oxygen concentrations and water contents [20]. In turn, microorganisms provide further ecosystem functions through their effects on soil structure by binding soil particles and organic matter to create aggregates [9,21]. Shifts in microbial community and function in response to different agricultural management practices can markedly alter rates of organic C loss from soils. Determining how fertilization affect the distribution of microbial functional groups among aggregates can lead to a better understanding of the differences in soil OC turnover under different management practices [2,22]. A previous study showed that different fertilizer treatments such as nitrogen fertilizer, green manure and sewage sludge did not dramatically alter the distribution pattern of bacteria in aggregates <200 μm [20].

The abundance and community structure of microorganisms are very responsive and can provide immediate and precise information on changes occurring in soil. In contrast, OC concentrations in soil and aggregates change at a slower rate [23]. To date, limited information has been available on the relationship between OC turnover and microbial community structure in soil aggregates. A long-term field experiment has been established in the Northern China plain to monitor changes in soil organic C under organic and inorganic fertilizer applications. In the present study, the abundance and community structure of microorganisms in aggregates were measured. The objectives of this study were to: (1) identify how repeat inorganic and organic fertilizer applications affect the distribution and community structure of microorganisms in aggregates; and (2) evaluate the relationship between OC accumulation and microbial community structure in aggregates.

Materials and Methods

Field Experiment and Soil Sampling

A long-term field experiment was set up in September 1989 to monitor the dynamic variation in OC following applications of compost and inorganic fertilizer to an intensively cultivated fluvo-aquic soil at the Fengqiu Agro-ecological Experimental Station, Chinese Academy of Sciences, Henan Province, China (35°00′N, 114°24′E). The soil was developed from alluvial sediments of the Yellow River and classified as Aquept [24]. Prior to the experiment, the field had been cultivated under a similar agricultural regime for at least 50 years, so the heterogeneity of soil fertility was minimal. The characteristics of the surface soil (0–20 cm) in September 1989 were listed in Table 1.

The experiment was based on a random design including three fertilization treatments with four replicate plots (9.5 m×5 m each): (1) no-fertilization control (control), (2) inorganic fertilizer NPK (NPK) and (3) compost (CM). Winter wheat (*Triticum aestivum*) was annually rotated with summer maize (*Zea mays*). A total of 150 kg N ha^{-1} for each crop was applied in the NPK and CM treatments. In the NPK treatment, urea was used at 60 or 90 kg N ha^{-1} as basal fertilizer in early June or early October for maize or wheat, respectively, and at 90 or 60 kg N ha^{-1} as supplement fertilizer in late July or late February for maize or wheat, respectively. In the NPK treatment, 75 kg P_2O_5 ha^{-1} (calcium superphosphate) and 150 kg K_2O ha^{-1} (potassium sulfate) were also added to each crop. In the CM treatment, a total of 2.76 tons ha^{-1} compost (made from wheat straw, oil cake and cotton cake, with a C/N ratio of approximately 8), which contained 1,164 kg C, 150 kg N, 51 kg P_2O_5 and 65 kg K_2O, was applied as basal fertilizer in early June for maize and early October for wheat. To match the same amounts of P and K with the NPK treatment, 24 kg P_2O_5 ha^{-1} (calcium superphosphate) and 85 kg K_2O ha^{-1} (potassium sulfate) were added to the compost prior to the application. No fertilizer was applied in the control treatment. Fertilizers and composts were broadcasted onto the soil surface by hand and the surface soil (0–20 cm) was immediately tilled. Seeds of 195 kg wheat ha^{-1} and 37.5 kg maize ha^{-1} were immediately sown into the soil by hand after the basal fertilization. Seeding rows were 70 cm wide for maize and 15 cm wide for wheat cultivation. Over the entire 20-year period, field management followed local practices.

Ten soil cores (0–20 cm) from each replicate plot were collected with a 2.5 cm diameter auger and then mixed to form one composite sample on 7 June 2009 after wheat harvest. Fresh samples were stored at 4°C and immediately transported to the laboratory for wet-sieving and phospholipid fatty acids (PLFA) analyses, while air-dried and sieved soils were used to determine the OC content by the wet oxidation-redox titration method [25]. A further eight undisturbed cylinder soil samples (100 cm^3) were taken from each plot to establish the soil water retention curve and bulk density.

Soil Water Retention Curve and the Effective Diffusion Coefficient of Oxygen

Water retention curves were determined with a ceramic pressure plate assembly at equilibrium matric potentials of −0.1, −0.2, −1, −3.5, −6, −10, −33, −50, −100, −200, −500 and −1,500 kPa in a pressure chamber. The obtained data were used to calculate the soil water retention curves and to derive the van Genuchten parameters by RETC (RETention Curve) software [26] using the following equation:

$$\theta(\psi) = \theta_r + \frac{\theta_s - \theta_r}{[1+(\alpha\psi)^n]^m}$$

where θ is the soil water content (cm^3 cm^{-3}), θ_s and θ_r are the saturated and residual water contents (cm^3 cm^{-3}), respectively, ψ is the matric potential (kPa) as indicated by pressure head, and the parameters of α, n and m ($m = 1-1/n$) are dimensionless.

According to the capillary rise theory, the pore size can be calculated using the following equation when the soil is hydrophilic at 20°C of water [27]:

$$d = \frac{3000}{|\psi|}$$

where d is the diameter of pore (μm), and $|\psi|$ is the absolute matric potential (kPa).

The effective diffusion coefficient (D) of oxygen through the pore space of the soil (m^2 s^{-1}) was calculated as follows [28]:

$$D = {}^1/_{N^2} \times [(D_{a0} \times Q_a{}^P) + (KH \times D_{w0} \times Q_w{}^P)]$$

where N is soil porosity, D_{a0} is the free diffusion coefficient in air

Table 1. Characteristics of the soil sampled in September 1989.

Texture (Sand, Silt, Clay)	pH_{H2O}	Organic C	Total N	Total P	Total K	Inorganic N
(%)	(extract 1:5, w/v)	(g C kg^{-1})	(g N kg^{-1})	(g P kg^{-1})	(g K kg^{-1})	(mg N kg^{-1})
52,33,15	8.65	4.48	0.43	0.5	18.6	9.51

(1.8×10^{-5} m^2 s^{-1} at 20°C), D_{w0} is the free diffusion coefficient in water (2.2×10^{-9} m^2 s^{-1} at 20°C), Q_a and Q_w are the proportion of soil porosity occupied by air and water, respectively, i.e. Q_a+ Q_w = 1, and p is a power constant (p = 3.4).

Q_w was calculated as follows:

$$Q_w = \rho \times \theta_m$$

where ρ is the soil bulk density, and θ_m is the soil moisture content.

Soil Fractionation

The water–stable aggregates of moist soils were separated using the wet–sieving protocol [29]. One hundred grams of moist soil samples (on an oven–dried basis) were submerged in deionized water for 5 min at room temperature on top of a 2,000–μm sieve. The sieve was manually moved up and down 3 cm, 50 times over a 2–min period. The fraction remaining on the 2,000–μm sieve was collected in a pre–weighed aluminum pan. Water plus the filtered soil was poured through a 250–μm sieve, and the sieving procedure repeated. Water plus the <250 μm fraction of soil was poured through a 53–μm sieve, and the sieving procedure repeated. To separate the silt (2–53 μm) from the clay (<2 μm) fraction, the remaining suspension was poured into centrifuge bottles and centrifuged at approximately 1,000 rpm for 3 min at 15°C. To obtain the clay fraction, all supernatants collected in several centrifuge bottles were centrifuged at 4,500 rpm for 30 min at 15°C. Large macroaggregate (>2,000 μm), small macroaggregate (250–2,000 μm), microaggregate (53–250 μm), silt (2–53 μm) and clay (<2 μm) fractions were obtained from the tested soils. An aliquot of each fraction was used to determine moisture content. Another subsample of each fraction was dried at 50°C to determine the OC content. The remaining sample was used to measure the microbial biomass and community structure using PLFA method.

PLFA Measurement

PLFAs were extracted using a modified Bligh–Dyer technique as described by Brant *et al.* [30]. In brief, three grams of soil sample (on an oven–dried basis) were incubated in a 2:1:0.8 solution of methanol, chloroform, and phosphate buffer. Soil extracts were centrifuged and the chloroform phases collected. Phospholipids were separated from glycolipids and neutral lipids using silicic acid bonded solid–phase–extraction columns. The phospholipid fractions were dried under a stream of nitrogen at 37°C. The fatty acid methyl esters (FAME) were produced through mild alkaline methanolysis, and the FAME were dried under nitrogen at 37°C. Finally, the FAME were dissolved in hexane, which contained a 19:0 (methyl nonadecanoate fatty acid) FAME standard. Samples were analyzed on a Shimadzu GC–MS QP 2010 PLUS (Shimadzu, Kyoto, Japan). Peaks were identified based on comparing retention times with known standards. Concentrations of each PLFA were obtained by comparing peak areas with the 19:0 FAME standards.

PLFAs have been used as biomarkers for various groups of microorganisms [31]. The PLFAs 16:3ω3, 18:1ω9, 18:2ω6,9 18:2ω9,12 and 20:1ω9 are used as biomarkers for fungi [19,32]; 16:1ω5 is generally attributed to arbuscular mycorrhizal fungi [33]; 10Me17:0, 10Me18:0, and 10Me20:0 for actinobacteria [34,35]; 3OH–15:0, 14:0, 15:0, 16:0, 17:0, 18:0, 19:0 and 20:0 for general bacteria; a15:0, i15:0, a16:0, i16:0, a17:0, i17:0 and i19:0 for G$^+$ bacteria [34]; and cy17:0, cy19:0, 16:1ω7c, 16:1ω7t, 18:1ω7c and 18:1ω7t for Gram–negative (G$^-$) bacteria [34,35]. According to Bossio *et al.* [36], 16:1ω7c, 16:1ω7t, 18:1ω7c, 18:1ω7t, 18:1ω9 and 16:1ω5 are monounsaturated PLFAs, while 3OH–15:0, a15:0, i15:0, a16:0, i16:0, a17:0, i17:0, i19:0, 10Me17:0, 10Me18:0 and 10Me20:0 are branched PLFAs. The ratio of monounsaturated to branched PLFAs is regarded as an indicator for the ratio of aerobic to anaerobic microorganisms and the development of aerobic conditions [36,37].

Statistical Analysis

The recovery rate of microbial PLFAs in all aggregates was calculated as follows:

$$\text{Re covery } rate(\%) = \sum (C_{aggregate} \times M_{aggregate}) \Big/ C_{soil} \times 100$$

where $C_{\text{aggregate}}$ is the concentration of microbial PLFAs in aggregates (nmol g^{-1} aggregate), $M_{\text{aggregate}}$ is the proportion of aggregates in whole soil by mass (kg kg^{-1} soil), and C_{soil} is the concentration of microbial PLFAs in soil (nmol g^{-1} soil).

Statistically significant differences among the treatments were determined using analysis of variance (one–way ANOVA) and Least Significant Difference (LSD) calculations at the 5% level with SPSS 18 for windows. Redundancy analysis (RDA) and regression analyses were used to test relationships between the increased organic C concentration in the CM and NPK treatments compared with the control and the abundance of microorganisms in aggregates in the CM and NPK treatments.

Results

Aggregate Mass Distribution and Aggregate–associated OC Concentration

Compost applications significantly (P<0.05) increased the mass proportion of large and small macroaggregates by 175% and 44%, respectively, at the expense of reduction in the silt fraction or microaggregate compared with the control. Inorganic fertilizer NPK only significantly (P<0.05) increased the proportion of small macroaggregates by approximately 30% (Fig. 1). The highest OC concentration was observed in small macroaggregates in all treatments, while the lowest occurred in the clay fraction (Table 2). Compost application more efficiently increased organic C in all aggregates than fertilizer NPK, and significantly (P<0.05) increased the values by 123–134% compared with the control. NPK mainly increased organic C in large and small macroaggregates, and to a lesser extent those in microaggregate, silt and clay fractions.

Pore Size and Effective Diffusion Coefficient of Oxygen

The proportion of pores with a neck diameter <4 μm in the CM treatment amounted to 65.12%, which was significantly (P< 0.05) higher than in the control and NPK treatments, the latter being 55.87 and 57.28%, respectively (Fig. 2a). In contrast, the proportion of pores with a neck diameter of 15–60 μm showed a decrease in the order: control>NPK>CM and there was a significant difference between treatments. Pores greater than 300 μm were significantly (P<0.05) increased by fertilization, but there was no obvious difference between CM and NPK treatments. The highest effective diffusion coefficient of oxygen was observed in the control treatment, peaking at 5.19×10^{-6} m^2 s^{-1}, while the lowest occurred in the CM treatment, measuring only 1.30×10^{-6} m^2 s^{-1} (Fig. 2b).

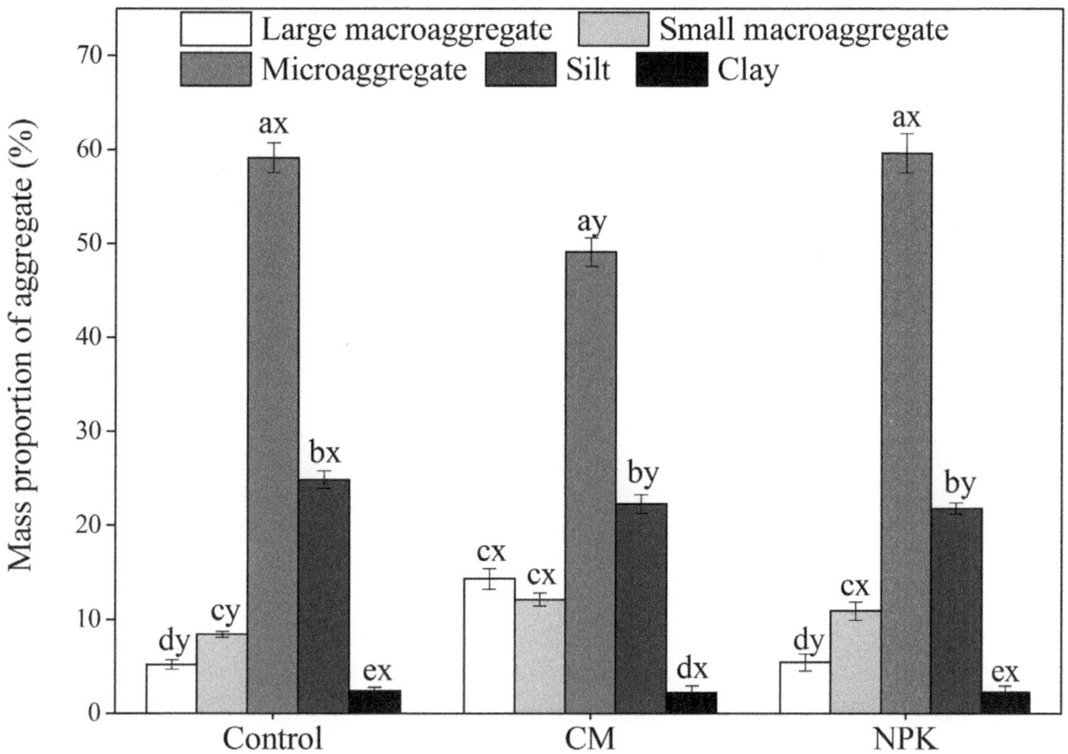

Figure 1. Effect of long-term applications of compost and fertilizer NPK on the mass proportion of aggregates in soil. Vertical bars indicate the standard error of the means ($n = 4$). Different letters denote significant differences between aggregates in the same treatment (a, b, c, d, e) and between treatments with the same aggregate (x, y, z) at $P<0.05$.

Concentrations of Bacterial, Fungal and Actinobacterial PLFAs in Aggregates

The recovery rate of microbial PLFAs in all aggregates compared with soil was 95–100%. The highest concentrations of microbial PLFAs were found in small macroaggregates, ranging from 89.00 to 93.93 nmol g^{-1} aggregate, and the lowest was found in the clay fraction for all treatments (Fig. 3d). The NPK application did not significantly ($P>0.05$) affect the concentration of total microbial PLFAs compared with the control. In contrast, compost significantly ($P<0.05$) increased the concentration of total PLFAs in the microaggregates and clay fraction.

The concentrations of bacterial PLFAs in microaggregates in all treatments varied from 47.97 to 63.07 nmol g^{-1} aggregate, being significantly ($P<0.05$) higher than in other aggregates, while the lowest was in the clay fraction (Fig. 3a). Compost significantly ($P<0.05$) increased the concentrations of bacterial PLFAs in microaggregates, silt and clay fractions, but fertilizer NPK only significantly ($P<0.05$) increased those in the silt fraction compared with the control.

The highest concentrations of fungi, varying from 20.39–21.53 nmol g^{-1} aggregate, were observed in the small macroaggregates, followed by large macroaggregates, while the lowest were measured in the clay fraction, being only 0.81–1.83 nmol g^{-1} aggregate in the three treatments. The concentrations of fungi in large macroaggregates were significantly ($P<0.05$) reduced by NPK and compost application, but were markedly increased in the silt fraction after compost was applied (Fig. 3b).

No actinobacterial PLFAs were found in any of the clay fractions (Fig. 3c). The maximum concentration of actinobacteria was observed in small macroaggregates, and the minimum in the silt fraction. The application of compost significantly ($P<0.05$) reduced their concentration in all aggregates, whereas NPK fertilization principally lowered the concentration in macroaggregates and the silt fraction.

Concentrations of Monounsaturated and Branched PLFAs in Aggregates

The concentration of monounsaturated PLFAs, which were mainly composed of G^- bacterial PLFAs, was highest in microaggregates and lowest in the silt fraction in all treatments (Fig. 4a). Compost significantly ($P<0.05$) reduced the concentrations of monounsaturated PLFAs in all aggregates compared with the control, while fertilizer NPK reduced their levels in small macroaggregates but increased their concentrations in the silt and clay fractions. In contrast, fertilization, regardless of type, significantly ($P<0.05$) elevated the concentrations of branched PLFAs including G^+ bacterial and actinobacterial PLFAs in large and small macroaggregates and clay fractions compared with the control (Fig. 4b). However, compost exerted a larger effect in large and small macroaggregates and weaker influences in the clay fraction than NPK. The ratio of monounsaturated to branched PLFAs (M/B ratio) increased as aggregate size reduced in all treatments, except for small macroaggregates and the silt fraction in the control treatment and small macroaggregates in the NPK treatment (Fig. 4c). Compost application reduced M/B ratios more efficiently in all aggregates than NPK, and the latter significantly ($P<0.05$) increased the M/B ratio in the silt fraction.

Correlation between Microbial PLFAs and the Increase of OC in Aggregates

Compared with the control, the increase of organic C in aggregates was found to be closely related with concentrations of

Table 2. Effect of long-term applications of compost and fertilizer NPK on organic C concentrations (g C kg^{-1} aggregate) in aggregates.

Treatment	Large macroaggregate (>2,000 µm)		Small macroaggregate (250–2,000 µm)		Microaggregate (53–250 µm)		Silt fraction (2–53 µm)		Clay fraction (<2 µm)	
	Concentration (g C kg^{-1})	Increase (%)	Concentration (g C kg^{-1})	Increase (%)	Concentration (g C kg^{-1})	Increase (%)	Concentration (g C kg^{-1})	Increase (%)	Concentration (g C kg^{-1})	Increase (%)
Control	4.32±0.04 cz	-	8.56±0.06 az	-	4.28±0.07 cz	-	4.57±0.22 bz	-	4.04±0.05 dy	-
CM	9.93±0.09 cx	130	20.01±0.29 ax	134	9.67±0.01 dx	126	10.37±0.06 bx	127	9.02±0.07 ex	123
NPK	8.16±0.10 by	89	17.82±0.11 ay	108	6.01±0.16 dy	40	6.72±0.10 cy	47	5.76±0.08 ey	43

Mean ± standard deviation (n = 4).
Increase (%) = (organic C in fertilization treatments – organic C in the control treatment)/organic C in the control treatment × 100.
Different letters denote significant differences between aggregates with the same treatment (a, b, c, d, e) and between treatments with the same aggregate (x, y, z) at $P<0.05$, respectively.

branched PLFAs, fungi, and actinobacteria, as well as the ratio of M/B. The increased OC content of aggregates significantly ($P<0.05$) positively correlated with concentrations of fungal PLFAs for CM and NPK, and that of actinobacteria for CM (Fig. 5). The increase of OC was also found to be marginally ($P<0.1$) positively linked to the concentration of branched PLFAs, but significantly ($P<0.05$) and negatively correlated with the ratio of M/B in CM and NPK treatments.

Discussion

Distribution of Microbial PLFAs in Aggregates

Aggregates in soils provide a spatially heterogeneous micro-habitat for microorganisms characterized by different substrates and oxygen concentrations [38]. In this study, fungi and actinobacteria were mainly located in aggregates >250 µm. In contrast, although the highest abundance of bacteria was found in microaggregates, they were more even distributed across different aggregates than fungi and actinobacteria. However, bacteria in the silt and clay fractions accounted for 79–98% of total microbial PLFAs, but only 46–52% in large and small macroaggregates, indicating that bacteria were abundant across aggregate size classes in soil and dominant in the silt and clay fractions, while fungi and actinobacteria were mainly observed in macroaggregates [39,40]. Using low-temperature scanning electron microscopy, Chenu *et al.* [41] found that actinobacteria and fungi displayed usually on the surface of aggregates greater than 10 µm in size, while bacteria displayed as individual cells or small colonies in clay pores under 2 µm. Thus, it is likely that actinobacteria and fungi were physically prevented from accessing the interior of microaggregates, silt and clay fractions where the habitats' small spaces were limiting for fungi and actinobacteria [42]. In contrast, the protective habitats in microaggregates provided niches for bacteria through exclusion of predators (protozoa) and competition with fungi, because the predation regimen acts as a major structuring force for the bacterial community [20,43].

Sessitsch *et al.* [20] suggested that the availability of nutrients in finer aggregates like silt and clay fractions or in smaller aggregates within macroaggregates could stimulate bacterial growth and thus bacterial diversity. Fungi and actinobacteria have mainly been found in large and small macroaggregates because they have a broader range of extracellular enzymes and prefer the particulate organic matter (POM) or the recalcitrant organic C such as lignin and hemicellulose that is rich in macroaggregates [40,44].

Previous studies have shown that microbial biomass was lowest in the silt and clay fraction and highest in the 1,000–2,000 µm fraction [2,45]. In our study, the concentration of total microbial PLFAs was also highest in small macroaggregates and lowest in the clay fraction, showing a decreasing pattern with reduction of aggregate size from 250–2,000 µm to <2 µm fractions (Fig. 3d). We found that 20-year compost applications significantly ($P<0.05$) increased the concentration of microbial PLFAs in microaggregates and the clay fraction compared with the control, while NPK did not. However, compost-induced stimulation did not drastically alter the distribution pattern of microbial PLFAs in soil aggregates. Previous studies demonstrated that long-term fertilization and conversion from conventional tillage to no-tillage did not affect the distribution pattern of microorganisms in aggregates of a clay loam in Sweden [20] and a subtropical paddy soil in China [22]. These results indicated that the distribution pattern of microorganisms in aggregates was mainly controlled by aggregate size and responded to a lesser extent to management practices like fertilization.

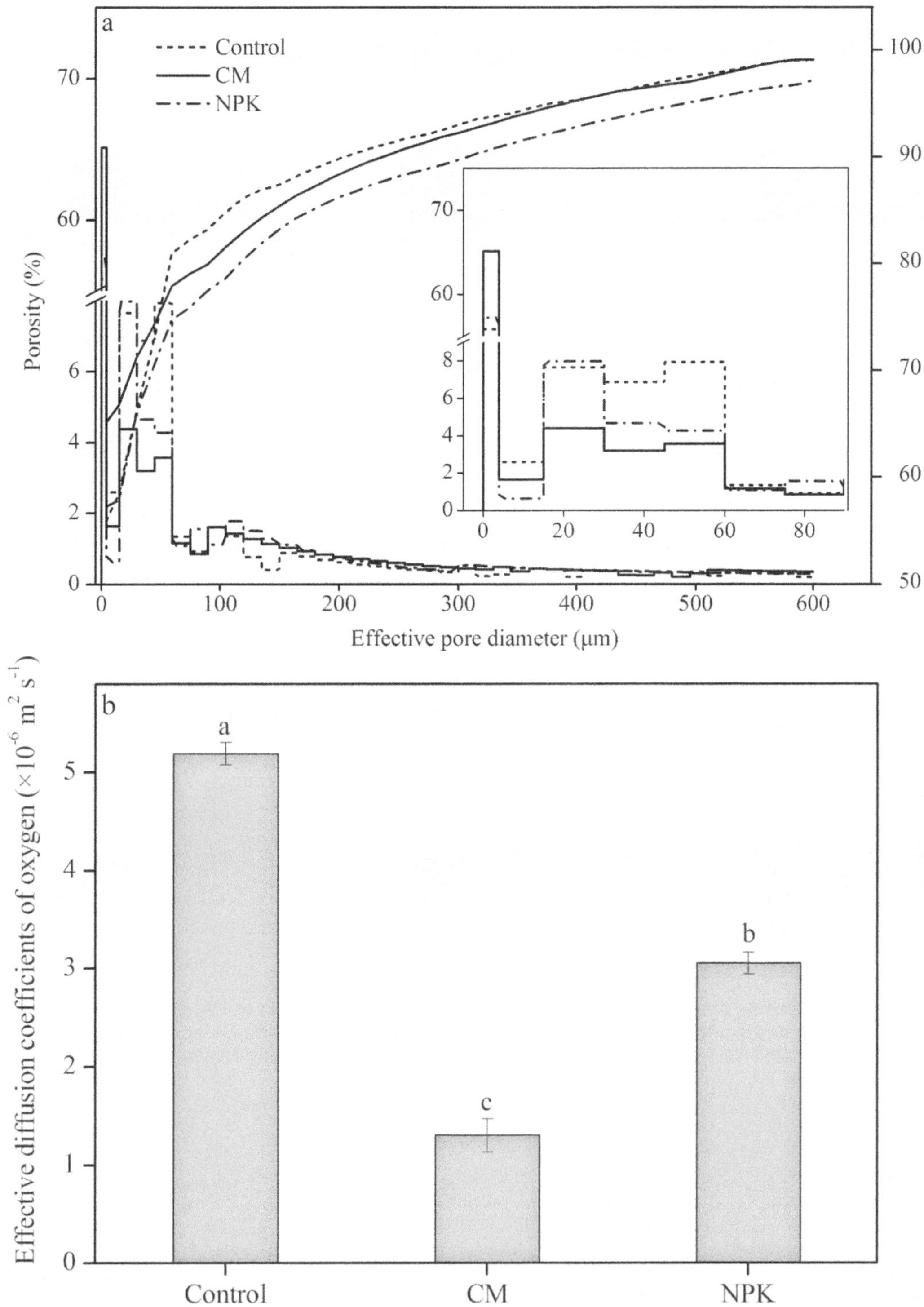

Figure 2. Effect of long-term applications of fertilizer NPK and compost on the proportions of pore volumes of different diameter (a) and effective diffusion coefficient of oxygen (b) in the soil. Vertical bars indicate the standard error of the means ($n=4$). Different lowercase letters denote significant differences between treatments at $P<0.05$.

Relationship between Microbial PLFAs and OC in Aggregates

The 20-year application of compost significantly increased organic C concentrations in all aggregates compared with the control, but fertilizer NPK exerted a much weaker effect (Table 2). The increase of organic C primarily depended on input levels of exogenous organic material and turnover of organic C [7,46]. Our previous study showed that the decomposition rate of per unit organic C in CM-treated soil was lower than in NPK-treated soil [47]. Thus, we argued that more efficient accumulation of organic C in CM-treated soil was not only attributable to higher annual inputs of organic C as compost or crop residues, but also to lower

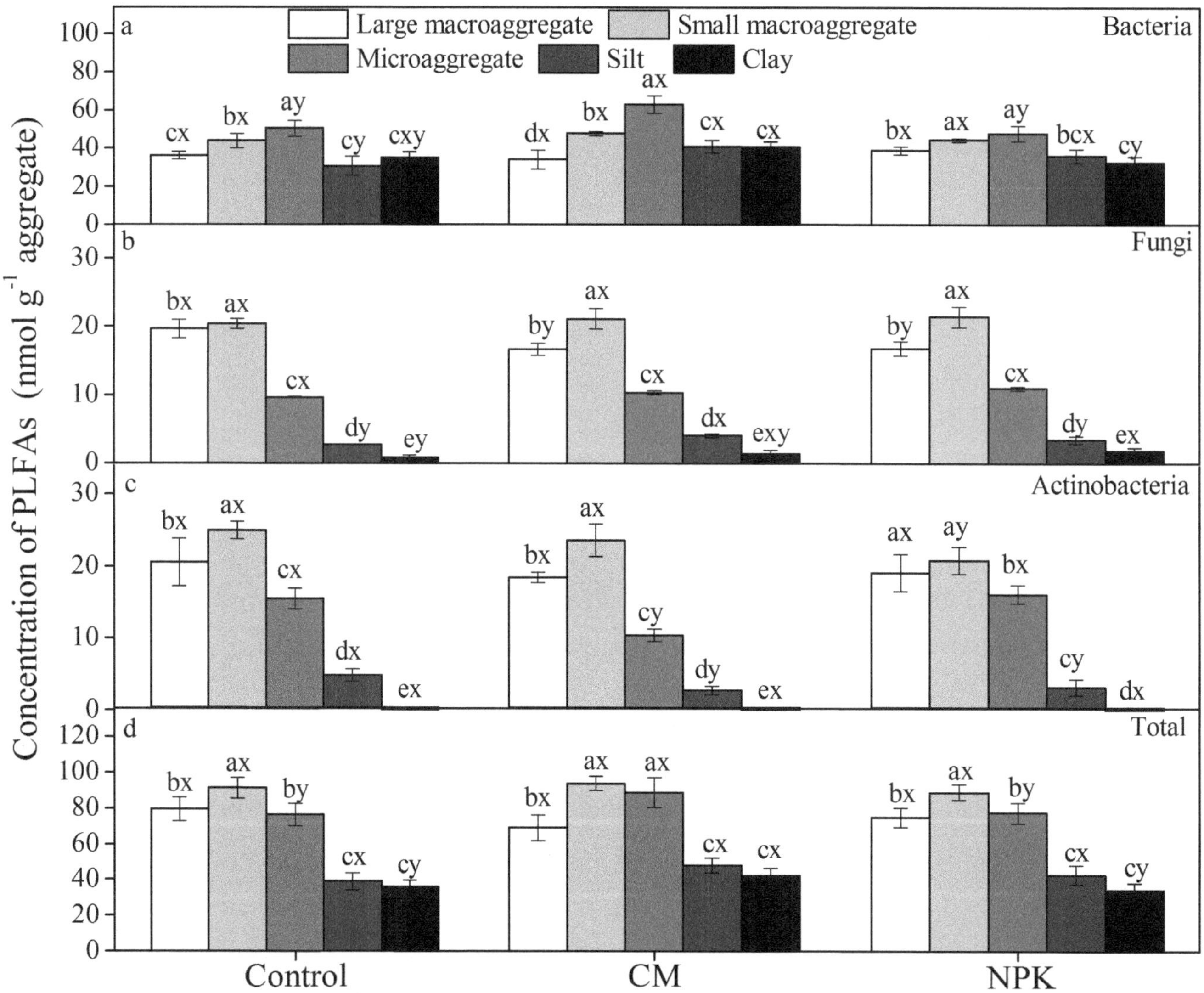

Figure 3. Effect of long-term applications of fertilizer NPK and compost on concentrations (nmol g^{-1} aggregate) of bacterial, fungal, actinobacterial and total PLFA in aggregates. Values are means ($n = 4$) with standard error. Different letters denote significant differences between aggregates with the same treatment (a, b, c, d, e) and between treatments with the same aggregate (x, y, z) at $P<0.05$, respectively.

decomposition rates of per unit organic C. In the present study, fungi and actinobacteria, which prefer particulate organic matter (POM), were reduced more efficiently in compost-treated soil, which might lead to the recalcitrant organic C accumulation in the CM treatment. However, correlation analysis showed that fungi and actinobacteria positively correlated with the increase in organic C in aggregates, possibly because they were mainly distributed in macroaggregates.

Killham *et al.* [48] and Dungait *et al.* [49] pointed out that the turnover of organic C depends on the accessibility of the OC to decomposer organisms or catalytic enzymes rather than its recalcitrance and location in the soil pore network. Organic C turnover was faster in pores with neck diameters greater than 4 μm than in those with smaller neck diameters [50]. Likewise, a previous study found the mineralization ratio of added fructose ^{13}C in pores >291 μm to be 41.4% during a 13-day incubation, being significantly higher than in pores <97 μm [51]. In the present study, organic C in macroaggregates showed a more rapid rate of increase than in other aggregates in spite of the fact that macroaggregates had higher abundances of microorganisms, especially fungi and actinobacteria as discussed above. Although it is possible that more exogenous organic material was accumulated in macroaggregates [14], we found that the M/B ratio declined in parallel with an increase in aggregate size (Fig. 4), and was negatively related to the rise in organic C in aggregates (Fig. 5). Our findings suggest that organic C turnover in aggregates is also dependent on the abundance and composition of the microbial community.

Ruamps *et al.* [51] suggested that the pore-scale distribution patterns of microorganisms did not occur by chance, but were the result of interactions between microorganisms and their habitat. Blagodatsky and Smith [52] found that the aeration of intra-aggregates changed with the formation of aggregates and could be completely anoxic. The increased organic C in microaggregates (free or within macroaggregates) caused an increase in pore-filling organic matter (mainly as POM or amorphous organic materials), and in turn reduced the proportion of large pores [53,54]. As a result, the proportion of pores <4 μm increased significantly in

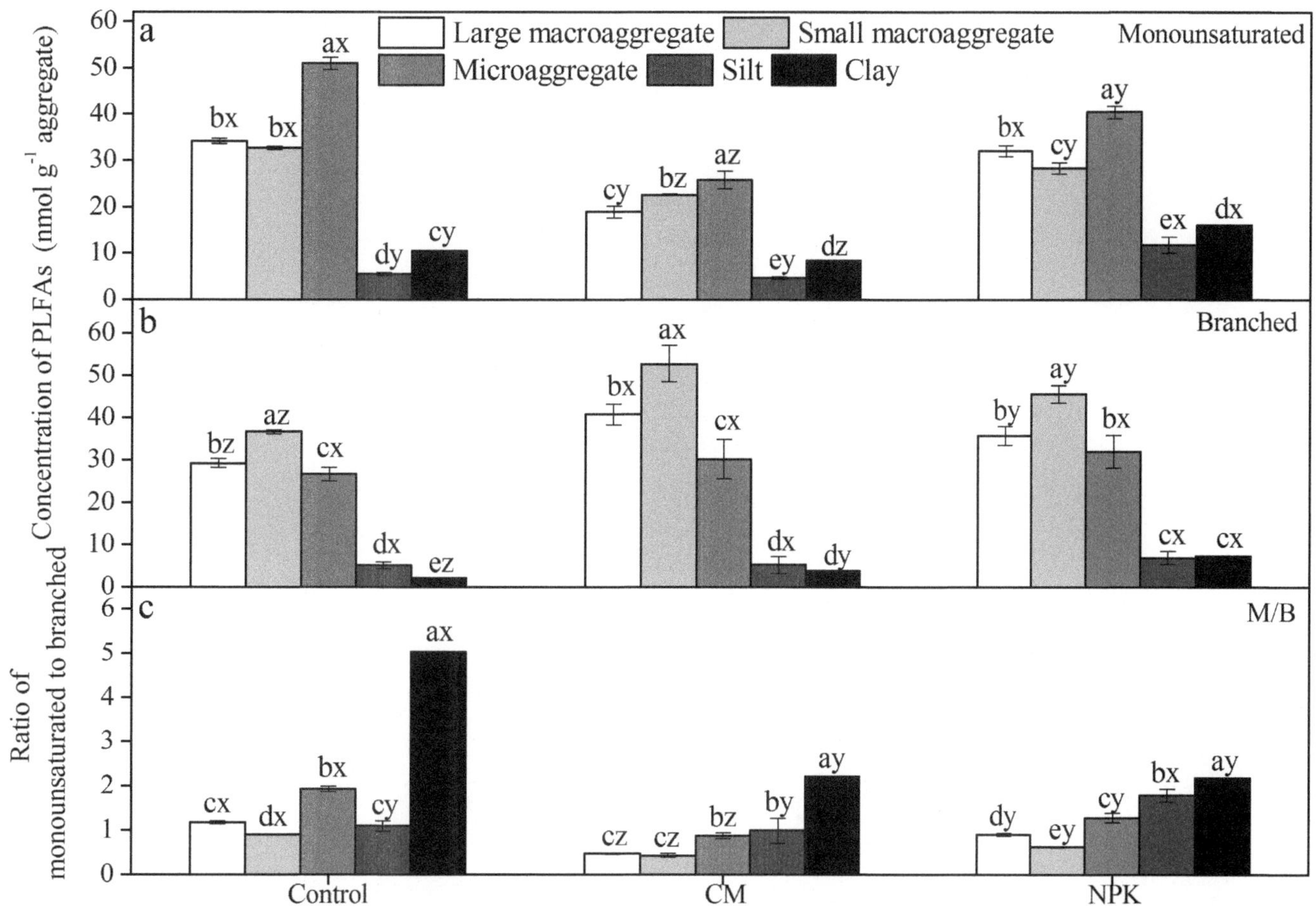

Figure 4. Effect of long-term applications of fertilizer NPK and compost on concentrations (nmol g^{-1} aggregate) of monounsaturated and branched PLFA in aggregates. Values are means ($n=4$) with standard error. Different letters denote significant differences between aggregates with the same treatment (a, b, c, d, e) and between treatments with the same aggregate (x, y, z) at $P<0.05$, respectively.

compost-treated soil while the effective diffusion coefficient of oxygen in both treatments significantly decreased in our study (Fig. 2). Soil aeration and pore networks are known to play a large role in the sequestration and turnover of organic C [34], and when the oxygen concentration in soil air reduced to ≤10%, the accumulation of organic C increased by slowing the oxidation of soluble forms of organic C [55], because aerobes decompose organic C more efficiently than facultative or obligate anaerobes [56]. The M/B ratio has generally been found to decrease with the development of anaerobic conditions or reduction in oxygen availability in soils [36,57]. Monounsaturated PLFAs mainly represent G^- bacteria, which can utilize a variety of organic C sources and promote the decomposition of organic C under well-aerated conditions [58,59], while branched PLFAs being dominantly composed of G^+ bacteria [36], which possess a greater proportion of peptidoglycan, which contains significant quantities of N-acetylglucosamine that is a precursor of relatively decay-resistant organic matter [60]. Thus, we considered that the aeration status would improve as aggregate size reduced [61], which in turn would control the activity and composition of microorganisms and organic C mineralization in aggregates. The more effective accumulation of organic C in macroaggregates than in other aggregates was at least partly due to the higher concentrations of branched PLFAs and low M/B ratios. However, the organic C accumulation was least in microaggregates in NPK treatment, was mainly due to a low input of exogenous organic materials rather than the decomposition of organic C [54].

The increase in organic C in aggregates in NPK-treated soil was significantly ($P<0.05$) lower than in CM-treated soil. Yu *et al.* [54], in an incubation study using the same soils, demonstrated that compost more efficiently improved the stability of organic C, even labile organic C such as carbohydrate, in the silt plus clay fraction compared to NPK. John [62] suggested that the high turnover of organic C in the clay fraction was caused by a relatively high enrichment of organic C from fresh residues and/or microbial biomass. In NPK-treated soil, the newly increased organic C was mainly derived from root residues and exudates, while also from compost in CM-treated soil. Compost had been fermented for 2 months before use. During composting, the proportion of labile, hydrophilic, plant-derived organic compounds gradually reduced. In contrast, that of more stable hydrophobic moieties, including lignin-derived phenols and microbial-derived carbohydrates, increased [63]. When these microbial-derived carbohydrates are incorporated into the soil, they are not usually utilized by microorganisms and can become stabilized by mineral particles. It is likely that low annual input and high decomposability of exogenous organic material in NPK-treated soil resulted in low increases in organic C in the silt and clay fractions compared to CM. Furthermore, the mass proportion of macroaggregates was exponentially related to OC concentra-

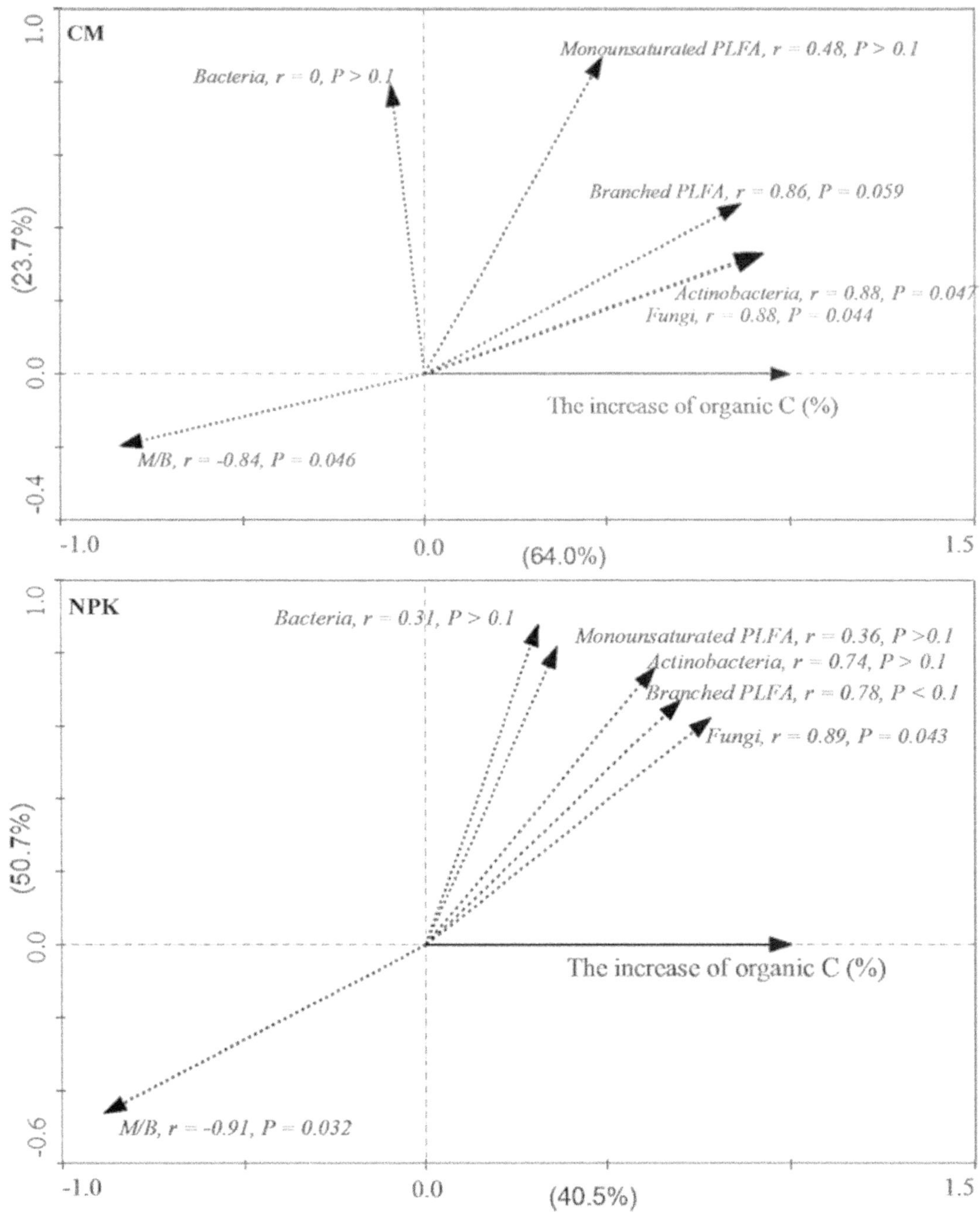

Figure 5. Relationship between increased organic C and the abundance of microbial PLFA (nmol g^{-1} aggregate) in soils. Redundancy analysis (RDA) and regression analyses were used to test relationships between the increased organic C concentration in the CM and NPK treatments compared with the control and the abundance of microorganisms in aggregates in the CM and NPK treatments.

tions in the silt plus clay fractions [54]. Greater input of recalcitrant OC and reduction in monounsaturated PLFAs is likely to have been critical for the observed increase in organic C in the silt and clay fractions. Our findings suggest that organic C accumulation in aggregates and aggregation increased the proportion of pores <4 μm and reduced the effective diffusion coefficient of oxygen, gradually leading to a change in micro–habitats from those favorable for aerobes to those beneficial for facultative or obligate anaerobes in the CM-treated soil [37]; the shift in microbial community composition in turn altered organic C turnover and accumulation in aggregates.

Conclusions

Compost application more effectively increased organic C concentrations in aggregates, and accelerated the formation of macroaggregates than fertilizer NPK. The proportion of pores < 4 μm was significantly ($P<0.05$) increased in CM treatment, while the reduction effective diffusion coefficient of oxygen showed a decrease order: control>NPK>CM ($P<0.05$). The distribution pattern of microorganisms in aggregates was primarily controlled by aggregate size and responded to a lesser extent to management practices like fertilization. The concentration of actinobacteria were more significantly ($P<0.05$) reduced in all aggregates in CM–treated soil than in NPK–treated soil, whereas fungi were only reduced in macroaggregates under compost and NPK application compared with the control, resulting in recalcitrant organic C accumulated effectively in CM treatment. The M/B ratio, which was lowest in CM treatment in all aggregates, increased with the reduction of aggregate size in tested soils. Increased organic C in aggregates in CM– and NPK–treated soils was marginally ($P< 0.10$) positively correlated with the concentration of branched PLFAs but significantly ($P<0.05$) negatively correlated with M/B ratios. More efficient accumulation of organic C in aggregates in CM–treated soil than in NPK–treated soil was probably due to the reduction of monounsaturated PLFAs and increase of branched PLFAs, which resulted from the change of micro–habitats in aggregates from one more favorable for aerobes to that beneficial for facultative or obligate anaerobes with the formation of macroaggregates.

Acknowledgments

The authors thank Dr. Jingyu Huang from Hohai University for his assistance in sample analyses.

Author Contributions

Conceived and designed the experiments: WXD HYY HJZ. Performed the experiments: HJZ HYY. Analyzed the data: HJZ WXD XHH. Contributed reagents/materials/analysis tools: HJZ HYY JLF DYL. Wrote the paper: HJZ WXD XHH.

References

1. Anders MM, Beck PA, Watkins BK, Gunter SA, Lusby KS, et al. (2010) Soil aggregates and their associated carbon and nitrogen content in Winter Annual Pastures. Soil Sci Soc Am J 74: 1339–1347.
2. Helgason BL, Walley FL, Germida JJ (2010) No–till soil management increases microbial biomass and alters community profiles in soil aggregates. Appl Soil Ecol 46: 390–397.
3. Bossuyt H, Six J, Hendrix P (2002) Aggregate–protected carbon in no–tillage and conservation tillage agroecosystems using carbon–14 labeled plant residue. Soil Sci Soc Am J 66: 1965–1973.
4. Wright AL, Hons JM (2004) Soil aggregation and carbon and nitrogen storage under soybean cropping sequences. Soil Sci Soc Am J 68: 507–513.
5. Six J, Paustian K, Elliott ET, Combrink C (2000) Soil structure and organic matter. I. Distribution of aggregate–size classes and aggregate–associated carbon. Soil Sci Soc Am J 64: 681–689.
6. Bach EM, Baer SG, Meyer CK, Six J (2010) Soil texture affects soil microbial and structural recovery during grassland restoration. Soil Biol Biochem 42: 2182–2191.
7. Verchot LV, Dutaur L, Shepherd KD, Albrecht A (2011) Organic matter stabilization in soil aggregates: Understanding the biogeochemical mechanisms that determine the fate of carbon inputs in soils. Geoderma 161: 82–193.
8. Hassink J (1997) The capacity of soils to preserve organic C and N by their association with clay and silt particles. Plant Soil 191: 77–87.
9. Six J, Elliott ET, Paustian K (2000) Soil macroaggregate turnover and microaggregate formation: a mechanism for C sequestration under no–tillage agriculture. Soil Biol Biochem 32: 2099–2103.
10. Mikha MM, Rice CW (2004) Tillage and manure effects on soil aggregate associated carbon and nitrogen. Soil Sci Soc Am J 68: 809–816.
11. Aoyama M, Angers DA, N'Dayegamiye A (1999) Particulate and mineral–associated organic matter in water–stable aggregates as affected by mineral fertilizer and manure application. Can J Soil Sci 79: 295–302.
12. Manna MC, Swarup A, Wanjari RH, Singh YV, Ghosh PK, et al. (2006) Soil organic matter in a west Bengal inceptisol after 30 years of multiple cropping and fertilization. Soil Sci Soc Am J 70: 121–129.
13. Su YZ, Wang F, Suo DR, Zhang ZH, Du MW (2006) Long–term effect of fertilizer and manure application on soil–carbon sequestration and soil fertility under the wheat–wheat–maize cropping system in northwest China. Nutr Cycl Agroecosys 75: 285–295.
14. Kong AYY, Six J, Bryant DC, Denison RF, van Kessel C (2005) The relationship between carbon input, aggregation, and soil organic carbon stabilization in sustainable cropping systems. Soil Sci Soc Am J 69: 1078–1085.
15. Sarkar S, Singh SR, Singh RP (2003) The effect of organic and inorganic fertilizers on soil physical condition and the productivity of a rice–lentil cropping sequence in India. J Agric Sci 140: 419–425.
16. Bhattacharyya R, Prakash V, Kundu S, Srivastva AK, Gupta HS, et al. (2009) Long–term effects of fertilization on carbon and nitrogen sequestration and aggregate associated carbon and nitrogen in the Indian sub–Himalayas. Nutr Cycl Agroecosys 86:1–16.
17. Böhme L, Langer U, Böhme F (2005) Microbial biomass, enzyme activities and microbial community structure in two European long–term field experiments. Agri Ecosyst Environ 109: 141–152.
18. Billings SA, Ziegler SE (2008) Altered patterns of soil carbon substrate usage and heterotrophic respiration in a pine forest with elevated CO_2 and N fertilization. Global Change Biol 14: 1025–1036.
19. Kindler R, Miltner A, Thullner M, Richnow HH, Kästner M (2009) Fate of bacterial biomass derived fatty acids in soil and their contribution to soil organic matter. Org Geochem 40: 29–37.
20. Sessitsch A, Weilharter A, Gerzabek MH, Kirchmann H, Kandeler E (2001) Microbial population structures in soil particle size fractions of a long–term fertilizer field experiment. Appl Environ Microbiol 67: 4215–4224.
21. Jastrow JD, Miller RM, Lussenhop J (1998) Contributions of interacting biological mechanisms to soil aggregate stabilization in restored prairie. Soil Biol Biochem 30: 905–916.
22. Jiang XJ, Wright AL, Wang X, Liang F (2011) Tillage–induced changes in fungal and bacterial biomass associated with soil aggregates: A long–term field study in a subtropical rice soil in China. Appl Soil Ecol 48: 168–173.
23. Alagöz Z, Yilmaz E (2009) Effects of different sources of organic matter on soil aggregate formation and stability: A laboratory study on a Lithic Rhodoxeralf from Turkey. Soil Till Res 103: 419–424.
24. Soil Survey Staff (2010) Keys to soil taxonomy, 11th edn. USDA NRCS, Washington DC, USA 106–111.
25. Carter MR (1993) Soil samples and methods of analysis. Lewis Publishers, Boca Raton., 190–191.
26. van Genuchten MT (1980) A closed–form equation forpredicting the hydraulic conductivity of unsaturated soils. Soil Sci Soc Am J 44: 892–898.
27. Kutílek M, Nielsen DR (1994) Soil Hydrology. Catena Verlag, Cremlingen Destedt.
28. Davis GB, Power TR, Briegel D, Patterson BM (1998) BTEX vapour biodegradation rates in the vadose zone: initial estimates. Groundwater Quality: Remediation and Protection 250: 300–303.
29. Elliot ET (1986) Aggregate structure and carbon, nitrogen and phosphorus in native and cultivated soils. Soil Sci Soc Am J 50: 518–524.
30. Brant JB, Sulzman EW, Myrold DD (2006) Microbial community utilization of added carbon substrates in response to long–term carbon input manipulation. Soil Biol Biochem 38: 2219–2232.
31. Bossio DA, Scow KM, Gunapala N, Graham KJ (1998) Determinants of soil microbial communities: effects of agricultural management, season and soil type on phospholipid fatty acid profiles. Microb Ecol 36: 1–12.
32. Frostegard A, Bååth E (1996) The use of phospholipid fatty acid analysis to estimate bacterial and fungal biomass in soil. Biol Fertil Soils 22: 59–65.
33. Olsson PA (1999) Signature fatty acids provide tools for determination of the distribution and interactions of mycorrhizal fungi in soil. FEMS Microbiol Ecol 29: 303–310.
34. Sundh I, Nilsson M, Borga P (1997) Variation in microbial community structure in two boreal peatlands as determined by analysis of phospholipid fatty acid profiles. Appl Environ Microbiol 63: 1476–1482.
35. Spring S, Schulze R, Overmann J, Schleifer KH (2000) Identification and characterization of ecologically significant prokaryotes in the sediment of freshwater lakes: molecular and cultivation studies. FEMS Microbiol Rev 24: 573–590.
36. Bossio DA, Fleck JA, Scow KM, Fujii R (2006) Alteration of soil microbial communities and water quality in restored wetlands. Soil Biol Biochem 38: 1223–1233.
37. Rajendran N, Matsuda O, Imamura N, Urushigawa Y (1992) Variation in microbial biomass and community structure in sediments of Eutrophic Bays as determined by phospholipid ester–linked fatty acids. Appl Environ Microbiol 58: 562–571.
38. Ladd JN, VanGestel M, Monrozier LJ, Amato M (1996) Distribution of organic C–14 and N–15 in particle–size fractions of soils incubated with C–14, N–15–labelled glucose/NH4, and legume and wheat straw residues. Soil Biol Biochem 28: 893–905.
39. Petersen SO, Debosz K, Schjønning P, Christensen BT, Susanne E (1997) Phospholipid fatty acid profiles and C availability in wet–stable macro–aggregates from conventionally and organically farmed soils. Geoderma 78: 181–196.
40. Kandeler E, Tscherko D, Bruce KD, Stemmer M, Hobbs PJ, et al. (2000) Structure and function of the soil microbial community in microhabitats of a heavy metal polluted soil. Biol Fertil Soils 32: 390–400.
41. Chenu C, Hassink J, Bloem J (2001) Short–term changes in the spatial distribution of microorganisms in soil aggregates as affected by glucose addition. Biol Fertil Soils 34: 349–356.
42. Beguin P, Aubert J (2006) The biological degradation of cellulose. FEMS Microbiol Rev 13: 25–58.

43. Jurgens K, Pernthaler J, Schalla S, Amann R (1999) Morphological and compositional changes in a planktonic bacterial community in response to enhanced protozoan grazing. Appl Environ Microbiol 65: 1241–1250.
44. Kölbl A, Kögel-Knabner I (2004) Content and composition of free and occluded particulate organic matter in a differently textured arable Cambisol as revealed by solid-state ^{13}C NMR spectroscopy. J Plant Nutr Soil Sci 167: 45–53.
45. Singh S, Singh JS (1995) Microbial biomass associated with water-stable aggregates in forest, savanna and cropland soils of a seasonally dry tropical region, India. Soil Biol Biochem 27: 1027–1033.
46. Ding WX, Meng L, Yin YF, Cai ZC Zheng XH (2007) CO_2 emission in an intensively cultivated loam as affected by long-term application of organic manure and nitrogen fertilizer. Soil Biol Biochem 39: 669–679.
47. Yu HY, Ding WX, Luo JF, Geng RL, Ghanni A, et al. (2012) Effects of long-term compost and fertilizer application on stability of aggregates-associated organic carbon in intensively cultivated sandy loam soil. Biol Fertil Soils 48: 325–336.
48. Killham K, Amato M, Ladd J (1993) Effect of substrate location in soil and soil porewater regime on carbon turnover. Soil Biol Biochem 25: 57–62.
49. Dungait JAJ, Hopkins DW, Gregory AS, Whitmore AP (2012) Soil organic matter turnover is governed by accessibility not recalcitrance. Global Change Biol 18: 1781–1796.
50. Strong DT, Wever HD, Merckx R, Recous S (2004) Spatial location of carbon decomposition in the soil pore system. Eur J Soil Sci 55: 739–750.
51. Ruamps LM, Nunan N, Chenu C (2011) Microbial biogeography at the soil pore scale. Soil Biol Biochem 43: 280–286.
52. Blagodatsky S, Smith P (2012) Soil physics meets soil biology: Towards better mechanistic prediction of greenhouse gas emissions from soil. Soil Biol Biochem 47: 78–92.
53. Zhuang J, McCarthy JF, Perfect E, Mayer LM, Jastrow JD (2008) Soil water hysteresis in water-stable microaggregates as affected by organic matter. Soil Sci Soc Am J 72: 212–220.
54. Yu HY, Ding WX, Luo JF, Geng RL, Cai ZC (2012) Long-term application of organic manure and mineral fertilizers on aggregation and aggregate-associated carbon in a sandy loam soil. Soil Till Res 124: 170–177.
55. Zibilske LM, Bradford JM (2007) Oxygen effects on carbon, polyphenols and nitrogen mineralization potential in soil. Soil Sci Soc Am J 71: 133–139.
56. Ding HB, Sun MY (2005) Biochemical degradation of algal fatty acids in oxic and anoxic sediment-seawater interface systems: effects of structural association and relative roles of aerobic and anaerobic bacteria. Marine Chem 93: 1–19.
57. Wixon DL, Balser TC (2013) Toward conceptual clarity: PLFA in warmed soils. Soil Biol Biochem 57: 769–774.
58. Ponder F, Tadros M (2002) Phospholipid fatty acids in forest soil four years after organic matter removal and soil compaction. Appl Soil Ecol 19: 173–182.
59. Feng XJ, Simpson MJ (2009) Temperature and substrate controls on microbial phospholipid fatty acid composition during incubation of grassland soils contrasting in organic matter quality. Soil Biol Biochem 41: 804–812.
60. Simpson AJ, Song G, Smith E, Lam B, Novotny EH, et al. (2007) Unraveling the structural components of soil humin by use of solution-state nuclear magnetic resonance spectroscopy. Environ Sci Technol 41: 876–883.
61. Rappoldt C, Crawford JW (1999) The distribution of anoxic volume in a fractal model of soil. Geoderma 88: 329–347.
62. John B (2003) Carbon turnover in aggregated soils determined by natural ^{13}C abundance. Ph.D. thesis, Georg-August-Universität Göttingen, Germany.
63. Said-Pullicino D, Kaiser K, Guggenberger G, Gigliotti G (2007) Changes in the chemical composition of water-extractable organic matter during composting: distribution between stable and labile organic matter pools. Chemosphere 66: 2166–2176.

11

Impact of Shortened Crop Rotation of Oilseed Rape on Soil and Rhizosphere Microbial Diversity in Relation to Yield Decline

Sally Hilton[1]*, Amanda J. Bennett[1¤a], Gary Keane[1¤b], Gary D. Bending[1], David Chandler[1], Ron Stobart[2], Peter Mills[1¤c]

1 School of Life Sciences, University of Warwick, Wellesbourne, Warwickshire, United Kingdom, **2** The National Institute of Agricultural Botany, Wymondham, Norfolk, United Kingdom

Abstract

Oilseed rape (OSR) grown in monoculture shows a decline in yield relative to virgin OSR of up to 25%, but the mechanisms responsible are unknown. A long term field experiment of OSR grown in a range of rotations with wheat was used to determine whether shifts in fungal and bacterial populations of the rhizosphere and bulk soil were associated with the development of OSR yield decline. The communities of fungi and bacteria in the rhizosphere and bulk soil from the field experiment were profiled using terminal restriction fragment length polymorphism (TRFLP) and sequencing of cloned internal transcribed spacer regions and 16S rRNA genes, respectively. OSR cropping frequency had no effect on rhizosphere bacterial communities. However, the rhizosphere fungal communities from continuously grown OSR were significantly different to those from other rotations. This was due primarily to an increase in abundance of two fungi which showed 100% and 95% DNA identity to the plant pathogens *Olpidium brassicae* and *Pyrenochaeta lycopersici*, respectively. Real-time PCR confirmed that there was significantly more of these fungi in the continuously grown OSR than the other rotations. These two fungi were isolated from the field and used to inoculate OSR and *Brassica oleracea* grown under controlled conditions in a glasshouse to determine their effect on yield. At high doses, *Olpidium brassicae* reduced top growth and root biomass in seedlings and reduced branching and subsequent pod and seed production. *Pyrenochaeta* sp. formed lesions on the roots of seedlings, and at high doses delayed flowering and had a negative impact on seed quantity and quality.

Editor: Leo van Overbeek, Wageningen University and Research Centre, The Netherlands

Funding: Funding was provided by DEFRA (Department for Environment Food and Rural Affairs) project IF028, www.defra.gov.uk/; and HGCA (Home Grown Cereals Authority) project RD-2003-2922, http://www.hgca.com. The funders had no role in study design, data collection and analysis, decision to publish, or preparation of the manuscript.

Competing Interests: Ron Stobart is employed by NIAB TAG. There are no patents, products in development or marketed products to declare. This does not alter the authors' adherence to all the PLOS ONE policies on sharing data and materials.

* E-mail: s.hilton.1@warwick.ac.uk

¤a Current address: *Home-Grown Cereals Authority*, Kenilworth, Warwickshire, United Kingdom
¤b Current address: The National Pollen and Aerobiology Research Unit, University of Worcester, Worcester, Worcestershire, United Kingdom
¤c Current address: Harper Adams University College, Newport, Shropshire, United Kingdom

Introduction

Farmers have traditionally used crop rotation as a method of managing the productivity of the soil and reducing the effects of crop pests and diseases. Often these rotations involved growing four or five different crops in a sequence. However, with the intensification of agricultural production in recent decades, many farmers now specialise in just one or two crop species, using inputs of fertilizers and pesticides to compensate for the lack of rotation [1,2]. This approach is not always successful, and declines in yield have been reported for a number of crops grown associated with continuous cropping or short rotation, including maize, soybean, wheat, potatoes, sugarcane and oilseed rape (OSR) [3–10]. Yield decline in OSR is a particular concern because of rising demand for the crop as a source of biofuel in addition to its cultivation for cooking oil and as an animal feed. OSR is also a very important part of the arable rotation and remains highly profitable when grown as a break crop for cereals. The global demand for biofuel has resulted in a rapid growth in world production of OSR, with 31 million hectares of land under cultivation in 2009, a 79% increase from 1989 [11]. In the UK, the average area used for growing OSR between 2001 and 2011 increased by 56% [12]. This increased demand for OSR has resulted in a higher proportion of OSR in rotations, with standard crop rotations of at least 1 in 5 being shortened by many farmers to 1 in 2. This has resulted in reports of yield decline, which vary from 3–12% in short rotation [8] to 25% in continuous cropping in the field trial used for this study [13]. The causes of yield decline are thought to be complex; various factors have been implicated for a range of crop species, including abiotic factors such as nutrient availability and land management practices, as well as biotic factors such as plant pathogens, autotoxicity of the crop, and changes in the soil or rhizosphere microbial community structure, which can subsequently have a deleterious effect on the crop [2,9,10]. The causal mechanisms for yield decline in OSR are not fully understood [9,10], which is of significant concern given the importance of this crop to the agricultural economy.

In this paper, we investigated the relationship between OSR yield decline and changes in soil/rhizosphere microbial communities. Previous studies have established that plants influence microbial communities through root exudates and residues in the soil and there is evidence of an association between crop type and microbial community composition, including wheat, potato, strawberry and OSR [14,15]. Furthermore, growing a crop continuously or in short rotation has been associated with a decline in the diversity of its rhizosphere microbial community [16–21]. It is possible that crop rotation provides a greater concentration and diversity of organic materials which may lead to a greater diversity of microbial communities. Yet, relatively little is known about the precise nature of these changes and their effects.

Some soil-borne microorganisms can be beneficial to the plant through the suppression of pathogens or by the production of plant growth-promoting substances, potentially resulting in increased vigour and a subsequent increase in yield [2,22]. However, soil microorganisms may also include plant pathogens or deleterious rhizosphere microorganisms (DRMO), which may have a negative impact on plant growth and yield. Major plant pathogens typically cause obvious symptoms of disease, and can usually be controlled through chemical, biological or cultural means. However, the impact of DRMO on yield decline is difficult to quantify as these organisms usually restrict plant growth without any obvious symptoms [22].

This investigation aimed to relate soil/rhizosphere microbial community to yield decline in OSR. Our central hypothesis was that growing OSR continuously or in shortened rotation causes the build-up of populations of specific plant pathogens or DRMO with a concomitant effect on crop yield. The aims of this study were to: (i) determine fungal and bacterial communities associated with OSR grown in rotations of different frequencies of cropping with wheat, where yield decline has been observed; (ii) identify microorganisms that became more prominent with frequent cropping of OSR; (iii) determine if these prominent microorganisms impact on OSR yield using experiments under controlled conditions.

Materials and Methods

Ethics Statement

No specific permits were required for the described field studies. Permission to perform the field studies were granted by the landowner (The Morley Agricultural Foundation).

Table 1. Cropping history of rotations sampled.

	Year of trial			
Rotation	**1**	**2**	**3**	**4**
Continuous OSR	O	O	O	O
'Continuous' wheat	O	W	W	W
OSR 1 in 2	W	O	W	O
OSR 1 in 3	O	W	W	O
Virgin OSR	W	W	W	O

Rhizosphere and bulk soil samples were collected in the 4th year of the trial (O = OSR, W = wheat).

Field Plot Experimental Design and Sampling Strategy

An established field trial based in East Anglia, UK (52° 33′ N and 1° 2′ E), investigating the effect of different frequencies of cropping of OSR (cv. Winner) and winter wheat (cv. Brompton) on OSR yield, was used to provide samples for this project via NIAB TAG and funded by HGCA (Project RD-2003-2922). This field trial showed a yield reduction in the continuous OSR (grown for four consecutive years) of 25% compared with virgin OSR [13] (Fig.S1).

The soil type was a sandy clay loam with a pH of 6.6 and available P, K, Mg and SO_4 of 32.4, 111, 28 and 30.6 mg kg^{-1} respectively. The entire trial area was ploughed and pressed each season ahead of establishment. While specific drilling dates varied according to season, oilseed rape was typically drilled in early September, first winter wheat in the second half of September and subsequent wheat in mid-October [13]. All inputs were consistent with local commercial best practice, although prophylactic approaches have been used when necessary [23]. Oilseed rape treatments included applications of nitrogen (N) and sulphur (S) of 200 kg ha^{-1} N and 30 kg ha^{-1} S [23]. First wheat and subsequent wheat treatments (second wheat and later) received different input programmes but again approaches were typical of local best practice [23].

The field trial was in its fourth year when samples were collected in June 2007; four weeks prior to harvest. Five rotations were used in this study, which are shown in Table 1. Rotations were planted in randomised plots of 24×6 m and each rotation was replicated four times [13]. Each plot was further divided into three equal sub-plots longitudinally. The central sub-plot was used for yield data and the outer two sub-plots were used for destructive sampling.

Bulk soil and rhizosphere samples were collected from the sub-plots of each of the four replicates of the five selected rotation treatments. For each replicate, three plants were excavated from the two sub-plots at approximately 6, 12 and 18 m along the length of the plot (six plants in total per replicate) and pooled. Bulk soil samples were collected at the same intervals, using a 30 cm auger (six samples pooled per replicate). Plants and bulk soil samples were taken back to the laboratory for processing. Roots were shaken free of loose soil and fine roots were cut into approximately 5 mm sections. Fine roots plus closely adhering soil were designated as the rhizosphere and sub-samples (0.5 g) of rhizosphere material were frozen for later molecular analyses. Bulk soil samples were sieved using a 3 mm sieve and sub-samples (0.5 g) were also frozen for molecular analyses.

DNA Extraction and TRFLP Analysis

DNA was extracted from 0.5 g of each bulk soil and rhizosphere sample using the FastDNA® SPIN Kit for Soil (MP Biomedicals LLC, UK), according to the manufacturer's instructions, with the exception that samples were homogenized in a Mini Beadbeater-8 cell disrupter for 3 minutes (Biospec products, Inc., USA) rather than a Fast-Prep machine. DNA samples were amplified with PCR primers universal to the internal transcribed spacer (ITS) region of fungi or the 16S rRNA gene of bacteria. The PCR master mix contained 47 µl Megamix (Microzone Limited, UK) and 10 ng DNA along with 1 µl of a forward primer and 1 µl of a reverse primer. For fungi 25 pmol of PET labelled ITS1f (5′-CTT GGT CAT TTA GAG GAA GTA A-3′) [24] and unlabelled ITS4 (5′-TCC TCC GCT TAT TGA TAT GC-3′) [25] were used. For bacteria 5 pmol of VIC labelled 1087r (5′ –CTC GTT GCG GGA CTT ACC CC 3′) [26] and unlabelled 63f (5′-AGG CCT AAC ACA TGC AAG TC-3′) [27] were used. Thermocycling consisted of an initial denaturation at 95°C for 3 min followed by 30 cycles of 95°C for 30 s, 55°C for 60 s, 72°C for

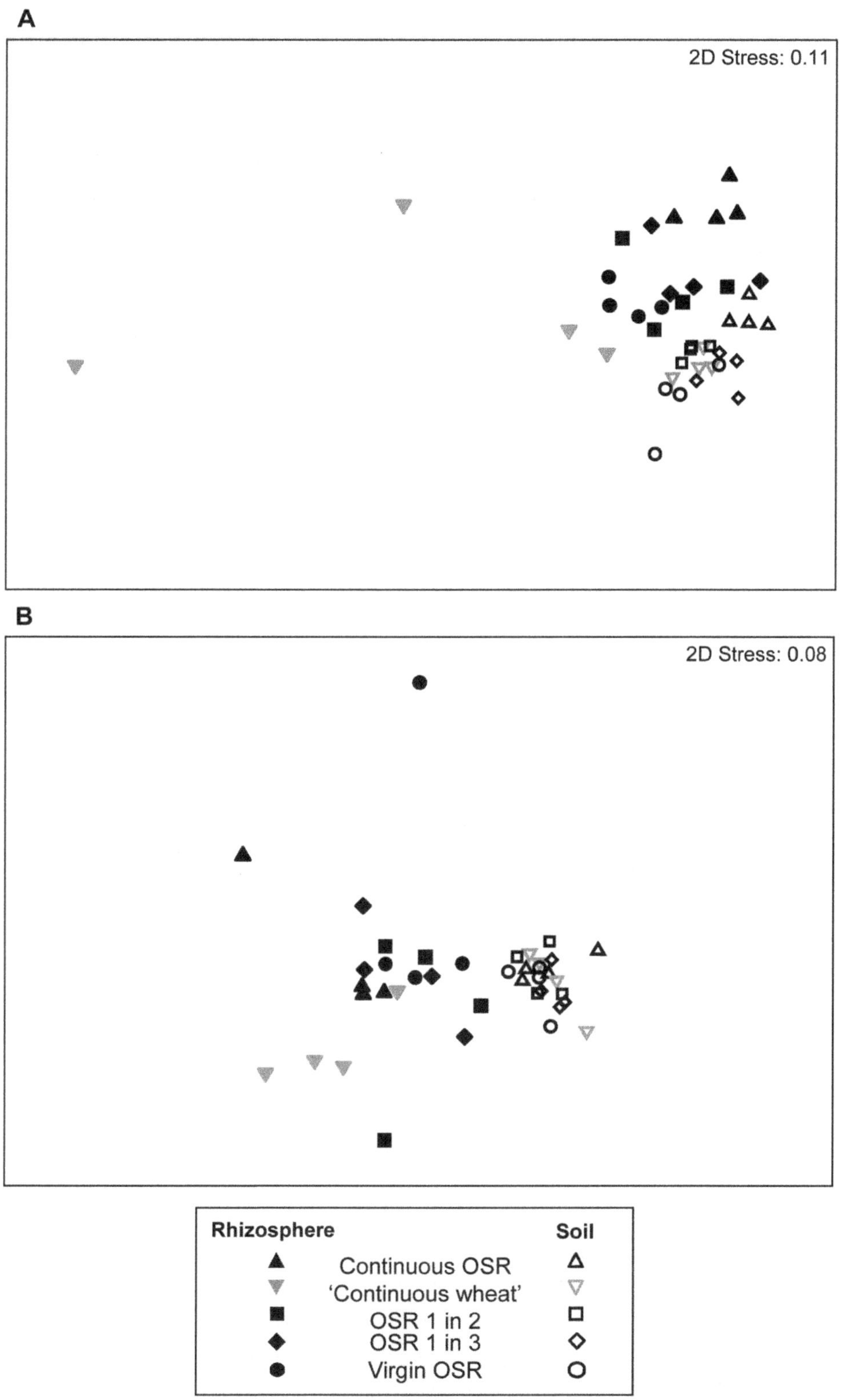

Figure 1. Non-metric multi-dimensional scaling plots represent rhizosphere (solid) and bulk soil (open) DNA profiles generated using primers for: (a) fungi or (b) bacteria, obtained from different rotations of rape (black) and wheat (grey).

60 s. The final extension was at 72°C for 10 min. The PCR products were combined before being purified using a Qiagen PCR purification step so that only one purification, restriction endonuclease digestion and sample for TRFLP analysis was required. Purified DNA (approximately 250 ng) was digested with *Hha*I or *Msp*I for 4 h at 37°C and the reaction terminated by a

Table 2. ANOSIM of fungal and bacterial communities.

Community	Soil type	Treatments compared	P	R
Fungi		Rhizosphere v bulk soil	**0.001**	**0.411**
	Rhizosphere	OSR Rotations	**0.002**	**0.401**
		Crop (OSR v Wheat)	**0.001**	**0.702**
		Continuous OSR v OSR grown after wheat	**0.002**	**0.554**
		Continuous OSR v Virgin OSR	**0.029**	**0.917**
	Bulk soil	OSR Rotations	**0.001**	**0.432**
		Crop (OSR v Wheat)	0.975	−0.284
		Continuous OSR v OSR grown after wheat	**0.008**	**0.454**
		Continuous OSR v Virgin OSR	**0.029**	**0.792**
Bacteria		Rhizosphere v bulk soil	**0.001**	**0.690**
	Rhizosphere	OSR Rotations	0.734	−0.041
		Crop (OSR v Wheat)	0.138	0.204
		Continuous OSR v OSR grown after wheat	0.374	0.047
		Continuous OSR v Virgin OSR	0.057	0.167
	Bulk soil	OSR Rotations	0.845	−0.104
		Crop (OSR v Wheat)	0.362	0.052
		Continuous OSR v OSR grown after wheat	0.178	0.157
		Continuous OSR v Virgin OSR	0.657	−0.073

P-values showing significance of the effects of factors (rotation and crop) on the microbial communities of the rhizosphere and bulk soil for fungi and bacteria. R values close to zero indicate most similarity. Values in bold highlight significant differences ($p = <0.05$).

further incubation at 95°C for 15 min. Aliquots (1 µl) of digested PCR products were mixed with 10 µl of HIDI formamide (Applied Biosystems™, Warrington, UK) and 0.15 µl of internal size standard LIZ 1200 (Applied Biosystems™, Warrington, UK) and then denatured for 5 min at 95°C. TRFLP analysis was carried out on an automated sequencer, ABI PRISM1 3130×l Genetic Analyzer on a 36 cm capillary array (Applied Biosystems™, Warrington, UK). Terminal restriction fragments (TRFs) generated by the sequencer were analysed using GeneMarker 1.60 (SoftGenetics LLC®, USA). To avoid detection of primers and undigested PCR products, TRFs smaller than 70 bp or larger than 500 bp (fungi) or 900 bp (bacteria) were excluded from further analysis. The relative abundance of TRFs was determined by calculating the percentage height of each peak in relation to the total peak height of all peaks within one sample. This relative abundance data was used for statistical analysis.

Cloning and Sequencing

Unlabelled primers were used to amplify fungal or bacterial rhizosphere DNA from pooled samples of four replicate plots of continuous OSR. PCR products were cloned using the QIAGEN PCR cloning plus kit (Qiagen, Crawley, UK). Plasmid DNA from 96 colonies underwent Templiphi™ amplification to provide the sequencing template (GE Healthcare Life Sciences, UK) [28]. Sequencing was carried out using the vector targeted PCR primers M13 forward (−40) and M13 reverse (Qiagen, Crawley, UK) on an automated sequencer (ABI PRISM1 3130×l Genetic Analyzer) using the BigDye® version 3.1 sequencing chemistry. Sequences were assembled and trimmed to the primer sites using the DNAstar, Inc. software suite. *In silico* restriction cut sites were then determined. The sequences were compared with the Genbank database using the BLASTN program [29] and the ribosomal database project (RDP) [30] for phylogenetic comparison. The sequences obtained in this study are available in GenBank under accession numbers GenBank accession JF432891−JF433024.

Identification of TRFs Using the Clone Libraries

The clone libraries were used to identify TRFs which were contributing towards the differences in microbial community structure. *In silico* digests of the clone library data were used to identify TRFs. TRFs may be up to 7 bp different to their *in silico* digest site [31,32], so each DNA clone of interest was digested with the restriction enzyme used for TRFLP analysis (*Hha*I) to confirm TRF sizes. Each TRF was further validated by determining the presence of the predicted size of TRF using a second restriction enzyme (*Msp*I). Identification was only possible for TRFs of high abundance or well-spaced TRFs. Although the presence of another sequence with a restriction site in the same place as an unrelated sequence is always a possibility and can never be completely ruled out, there were no other unrelated sequences in the clone libraries with the same restriction sites for both restriction enzymes.

Statistical Analysis

For statistical analysis, the TRFs produced using the restriction enzyme *Hha*I were used. This restriction enzyme was found to give the largest number of peaks which were also relatively evenly spaced. TRFLP community profiles were expressed in relative abundance and analysed for resemblance using analysis of similarity (ANOSIM) and non-metric multidimensional scaling (non-metric MDS) (PRIMER, version 6, Primer-E) [33]. ANOSIM reports the level of dissimilarity between sample groups (global R) and the associated level of significance (P). R is scaled to be within the range +1 to −1. Positive R values indicate that samples are more dissimilar between groups than within groups. R values close to zero occur if the high and low similarities are

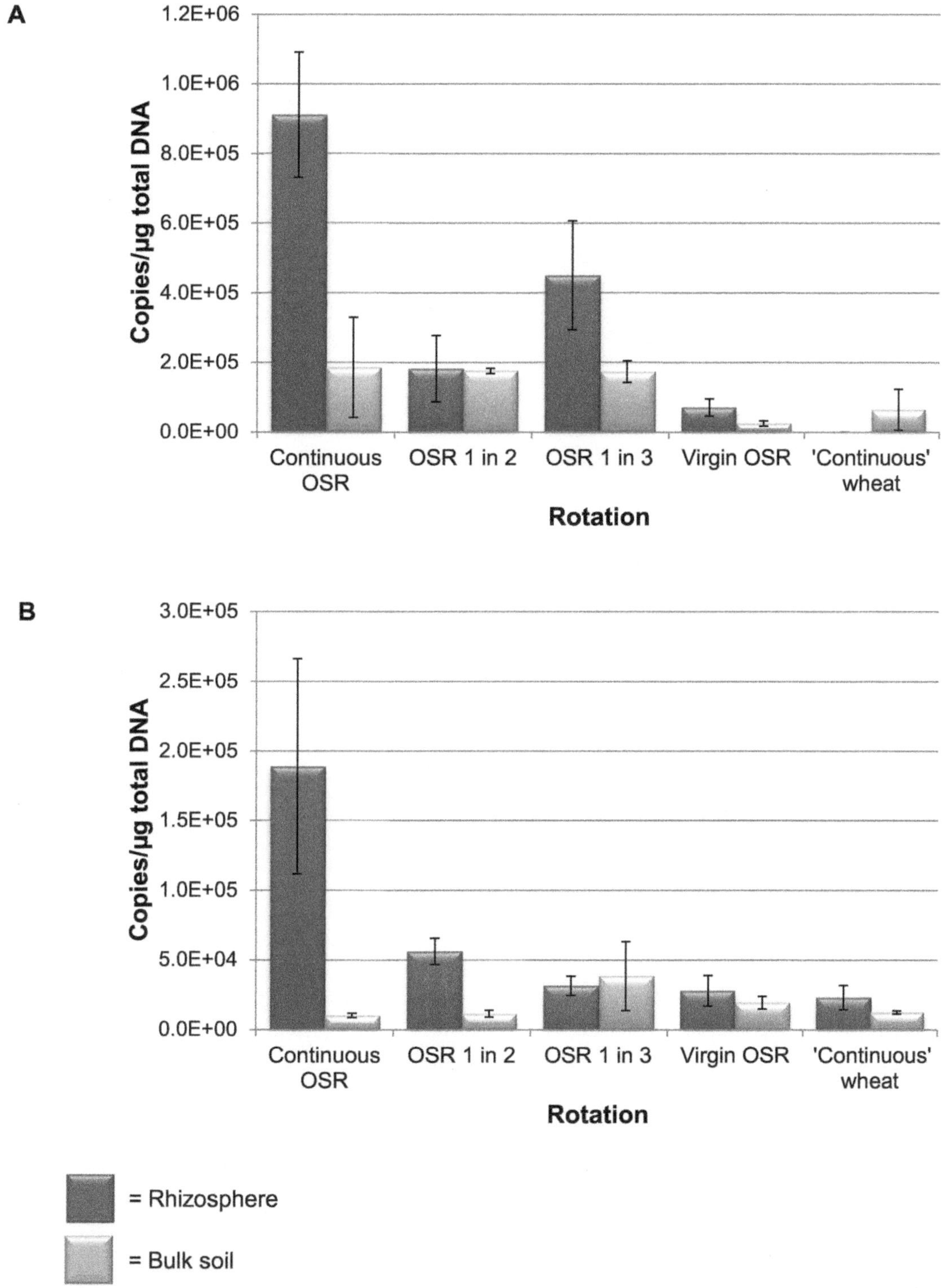

Figure 2. Absolute quantification using specific quantitative PCR primers to (a) *Olpidium brassicae* and (b) *Pyrenochaeta* sp. in different rotations of OSR and wheat (see Table 1 for rotation explanation). Error bars are ± standard errors of the mean for the four replicate plots.

perfectly mixed and bear no relationship to the group. Negative *R* values indicate that dissimilarities within groups are greater than dissimilarities between groups [33]. Significance values were obtained by permutation tests. The relative contribution (%) of each TRF to the similarity matrix structure was assessed using SIMPER (Similarity Percentages - species contributions) [33]. TRFs which had a relative abundance of over 0.5% and real-time PCR copy numbers were tested by ANOVA which was used to compare their abundance across rotations.

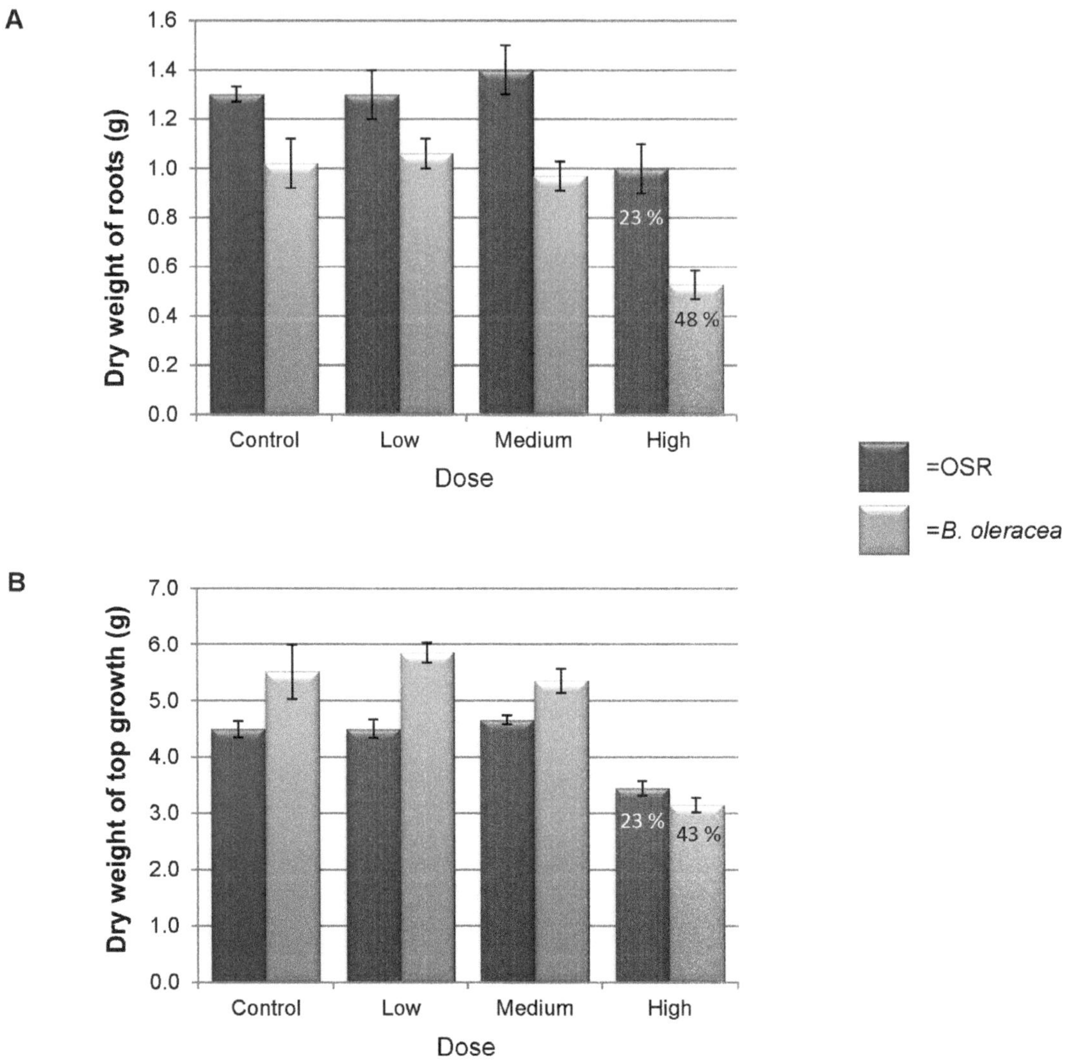

Figure 3. Dry weights of roots and top growth of oilseed rape and *Brassica oleracea* plants grown in a soilless substrate for seven weeks inoculated with zoospores (zs) of *Olpidium brassicae.* (a) roots (b) top growth. Control = 0 zs seedling^{-1}; Low = 2.5×10^2 zs seedling^{-1}; Medium = 5×10^4 zs seedling^{-1}; High = 1×10^7 zs seedling^{-1}. Percentages in the 'High' columns indicate yield reduction compared with the control. Error bars are ± standard errors of the mean for the four replicates.

Real-time PCR

Unique DNA sequences within the ITS region were identified for *Olpidium brassicae* (Ob) and *Pyrenochaeta* sp. (Py) using the fungal clone library and the BLAST database (NCBI). Real-time PCR primers were then designed within these regions using Primer Express 2.0 (Applied Biosystems); ObF (5′-TCT CCT CGT TGG GAA GAC TTG T-3′) and ObR (5′-GAG CTT GAA TTT TTA AGT TCG TCG TT-3′); PyF (5′-CCG CCG GTT GGA CAC TAT AA-3′) and PyR (5′-TCG ATG CCA GAA CCA AGA GAT-3′). To confirm the primer specificity they were used to amplify DNA from rhizosphere samples which had been shown by TRFLP to contain approximately 1% of the total fungal population of either *O. brassicae* or *Pyrenochaeta* sp. The PCR was carried out using the real-time PCR conditions (below) and the products were cloned and sequenced. All of the clones (48) had the correct sequences with no other organisms being amplified.

Cloned *O. brassicae* or *Pyrenochaeta* sp. ITS DNA was quantified using the NanoDrop® ND-1000 spectrophotometer (NanoDrop Technologies). A standard curve of a fivefold dilution of each was made from 32 pg ml^{-1} down to 4.1×10^{-3} pg ml^{-1}. Total rhizosphere or bulk soil DNA (1 ng) was also quantified using the NanoDrop® ND-1000 spectrophotometer (NanoDrop Technologies) and used in the real-time PCR using the SYBR Green PCR master mix (Applied Biosystems). This gave target DNA concentrations within the standard curve. Each reaction was set up in triplicate in a 384-wellplate with the following components: 2×SYBR Green PCR mastermix (10 µl), 1 mM forward primer, 1 mM reverse primer, standard curve DNA or 1 ng total sample DNA, 400 µg ml^{-1} non-acetylated BSA and water added to 20 µl. Real-time PCR was carried out using the ABI Prism 7900 HT sequence detection system (Applied Biosystems) with standard Taqman cycling conditions (40 cycles of 95°C for 15 s followed by 1 min at 60°C). An average of the triplicate results was taken. The quantities of DNA obtained were converted to copy numbers of target DNA/µg total sample DNA.

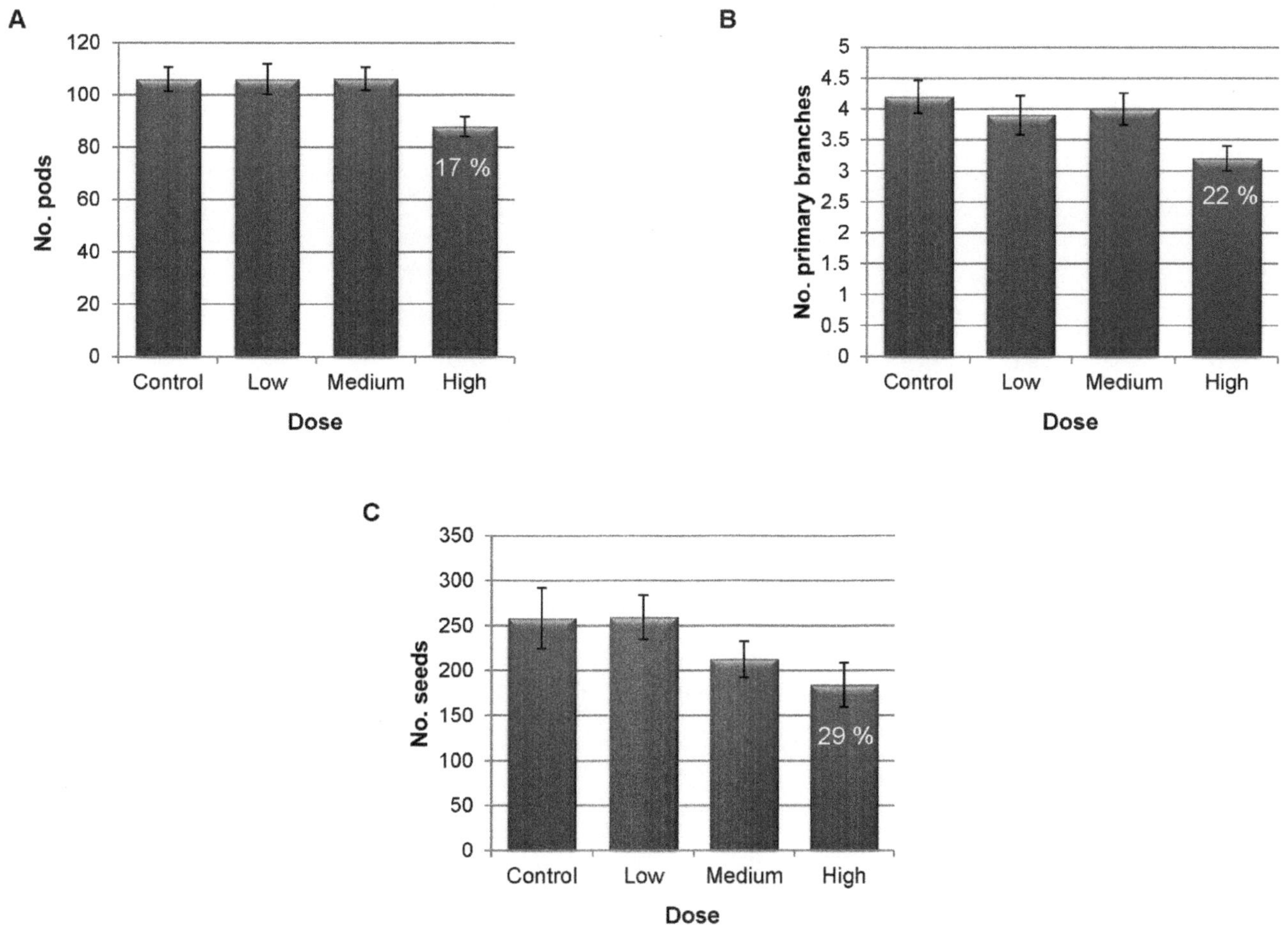

Figure 4. Seed yield components of *Brassica oleracea* plants grown in a soilless substrate for four months after inoculation with zoospores (zs) of *Olpidium brassicae.* (a) pods (b) primary branches (c) seeds. Control = 0 zs seedling^{-1}; Low = 2.5×10^2 zs seedling^{-1}; Medium = 5×10^4 zs seedling^{-1}; High = 1×10^7 zs seedling^{-1}. Error bars are ± standard errors of the mean for the ten replicates.

Isolation of Cultures and Inoculum Preparation

A culture of *O. brassicae* was established through isolation of zoosporangia from infected plant material collected from the field site, and stock cultures were subsequently maintained on oilseed rape seedlings grown in a soilless substrate (sand:terragreen; 50:50 v:v) in the glasshouse. Zoospores were amplified in OSR plants, extracted in GS (0.05 M glycine, 1% sucrose), counted and diluted to provide a range of doses in the glasshouse experiments. Seedlings were grown for 10 days before inoculation with *O. brassicae*, and at this time 1 ml of the required zoospore dilution was pipetted onto the base of each seedling.

Pyrenochaeta sp. was isolated onto tap water agar from a root lesion found on an OSR plant collected from the field site. It was subsequently grown on an agar medium containing OSR leaf extracts. The pure culture of *Pyrenochaeta* sp. would not sporulate in the laboratory and therefore mycelium was used as inoculum for the glasshouse experiments. Mycelium was grown on autoclaved wheat grain for 7 weeks before being macerated in water. Half of the resulting liquid inoculum was autoclaved for use as negative controls. Dilutions of the liquid inoculum were used to provide a range of doses in the glasshouse experiments, and were incorporated into the growing medium at the time of planting.

Glasshouse Experiments

Brassica oleracea was used as a model Brassica for seed yield experiments, as it is a rapid-cycling Brassica suited to glasshouse conditions. The effect of *O. brassicae* and *Pyrenochaeta* sp. on early growth of OSR and *B. oleracea*, and seed yield of *B. oleracea*, was assessed in a soilless substrate using a sand-terragreen mix as a growing medium. This system allowed the candidate microorganisms to be tested in isolation, without the presence of other soil microorganisms.

Early Growth Experiments

OSR (*Brassica napus* cv. Winner) and *Brassica oleracea* (DHSL150) plants were inoculated independently with the selected fungi in a range of doses. The three doses of *Olpidium brassicae* were 'High' = 1×10^7 zoospores seedling^{-1}, 'Medium' = 4×10^4 zoospores seedling^{-1} and 'Low' = 2.5×10^2 zoospores seedling^{-1}. A control was also set up using GS alone. The three doses of *Pyrenochaeta* sp. were 'High' = 4.7×10^4 ($\pm1.0\times10^4$) cfu g^{-1} substrate, 'Medium' = 4.7×10^2 ($\pm1.0\times10^2$) cfu g^{-1} substrate and 'Low' = 4.7 (±1.0) cfu g^{-1} substrate, each of which had a corresponding autoclaved control. Seedlings were grown for 6–7 weeks before top growth and root biomass were recorded, and root material (0.4 g) was taken for DNA extraction and analysis of the inoculated fungi using real-time PCR. For each replicate, two pots per sample were used with 3 plants in each pot to maximise root

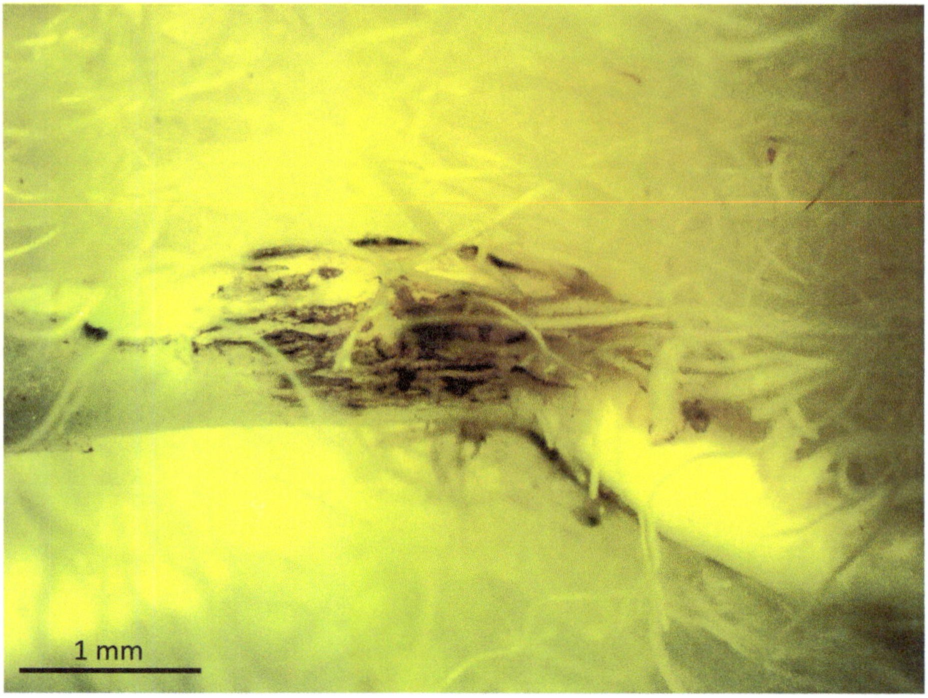

Figure 5. ***B. oleracea*** **root displaying a root lesion after inoculation with** ***Pyrenochaeta*** **sp.**

material for DNA analysis without impacting on overcrowding in the pots. Four complete replicates were set up.

Seed Yield Experiments

B. oleracea plants were again inoculated independently with the selected fungi in a range of doses and were grown to seed set. The three doses of *Olpidium brassicae* were 'High' = 1×10^7 zoospores seedling^{-1}, 'Medium' = 4×10^4 zoospores seedling^{-1} and 'Low' = 2.5×10^2 zoospores seedling^{-1}. A control was also set up using GS alone. The three doses of *Pyrenochaeta* sp. were 'High' = 4.9×10^4 ($\pm0.97\times10^4$) cfu g^{-1} substrate, 'Medium' = 4.9×10^2 ($\pm0.97\times10^2$) cfu g^{-1} substrate and 'Low' = 4.9 (±0.97) cfu g^{-1} substrate, each of which had a corresponding autoclaved control. At harvest, assessments were made on various components of yield, including number of branches, number of pods, number of seed-containing pods, and number and weight of normal (smooth and spherical) or deformed (flattened and/or wrinkled) seeds. Rhizosphere samples (0.4 g) were also taken at harvest for DNA extraction and quantification of inoculated fungi using real-time PCR. For each replicate, one pot per treatment was used with a single plant per pot, and ten complete replicates were set up.

Results

Fungal Community and TRF Identification

Non-metric MDS with ANOSIM analysis of the TRFLP data showed that rotation had a significant effect on the fungal community of the rhizosphere and of the bulk soil (Fig. 1a, Table 2). Within the OSR rotations, the continuous OSR and virgin OSR communities were the most diverse with the other OSR rotations (OSR grown every second or third year) containing communities between the two. The continuous OSR (rhizosphere and bulk soil) had significantly different fungal communities to all other rotations (Fig. 1a, Table 2). Using SIMPER analysis, the TRFs that contributed most towards the differences between the rhizosphere of continuous OSR and OSR grown after wheat were 284 bp and 98 bp which both had a higher relative abundance in the continuous OSR rhizosphere. These TRFs were identified using clone libraries. The TRF at 284 bp was identified as *Olpidium brassicae* as it had 99%–100% identity to *O. brassicae* isolated from cabbage roots (Table S1). Real-time PCR showed that there was significantly more *O. brassicae* in the continuous OSR rhizosphere than all other rotations ($p = 0.001$; Fig. 2a). There was a low amount of *O. brassicae* in the rhizosphere of virgin OSR and intermediate levels in the rotations where OSR was grown every second or third year. *Olpidium brassicae* was not found in the rhizosphere of wheat and low levels of *O. brassicae* were detected in the bulk soil samples.

The TRF at 98 bp was identified as *Pyrenochaeta* sp. as it had 100% identity to a *Pyrenochaeta* sp. isolated from the grass species *Ammophila arenaria* (Table S1) as well as a match at 95% DNA identity to *Pyrenochaeta lycopersici* (causal organism of tomato corky root). Real-time PCR showed that there was significantly more *Pyrenochaeta* sp. in continuous OSR rhizosphere than in all other treatments ($p = 0.012$; Fig. 2b). Very little *Pyrenochaeta* sp. was detected in any of the bulk soil samples and there were no significant differences between rotations.

Using SIMPER analysis, the TRFs that each contributed most towards the differences between the bulk soil of continuous OSR and OSR grown after wheat were 124/125 bp which was identified as *Gibellulopsis nigrescens* (previously named *Verticillium nigrescens* [34]) (Table S1) and 284 bp (*O. brassicae*) which both had a significantly higher relative abundance in the bulk soil of continuous OSR compared to OSR grown after wheat (124/125 bp, $p = 0.001$; 284 bp, $p = 0.042$). A number of other less abundant TRFs were also found to have a significantly higher relative abundance in the continuous OSR bulk soil to other rotations (data not shown).

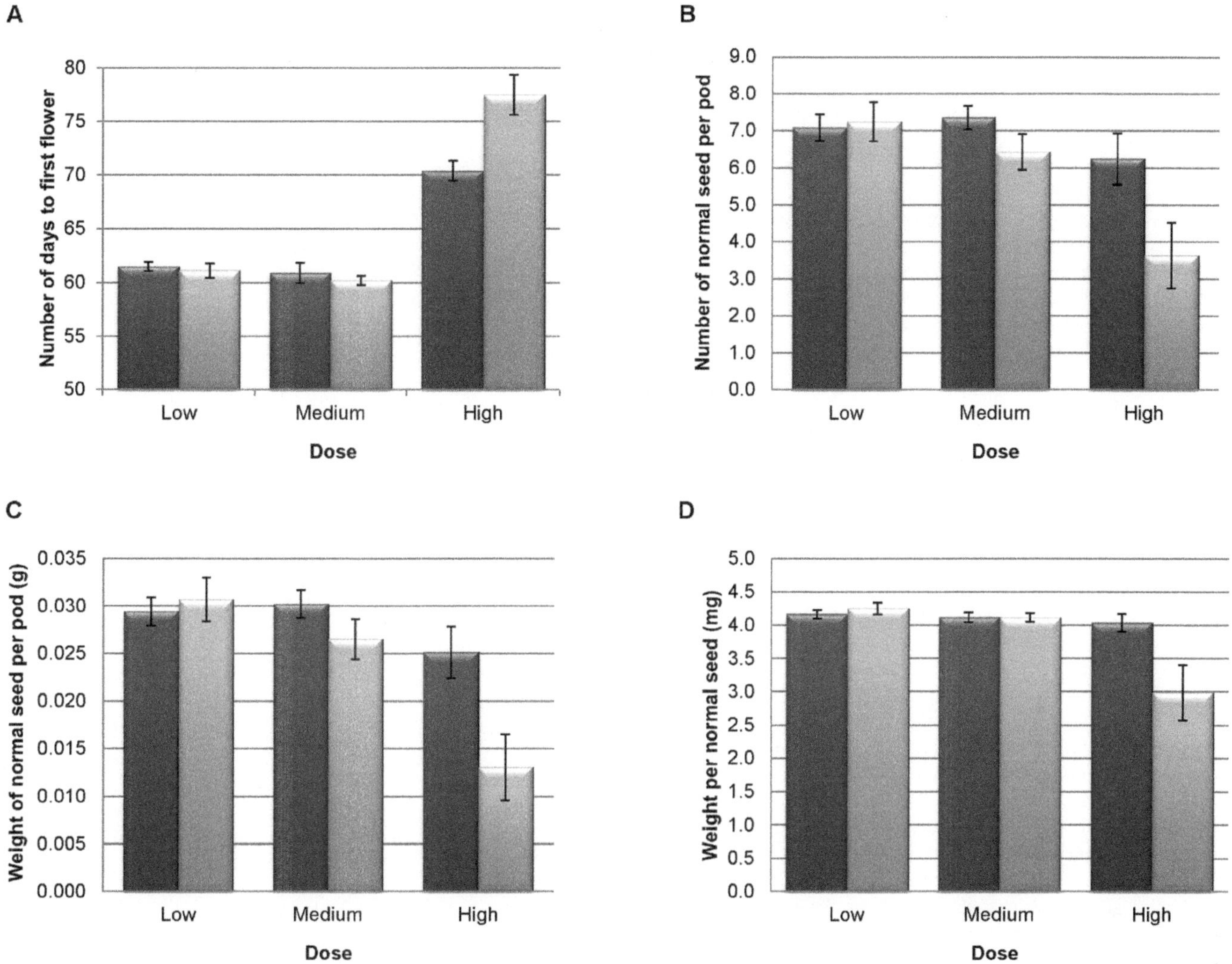

Figure 6. Seed yield components of *Brassica oleracea* plants grown in a soilless substrate for four months containing *Pyrenochaeta* sp. mycelium. (a) days to first flower (b) no. normal seeds/pod (c) weight of normal seeds/pod (d) weight per normal seed. Light grey = live substrate; Low = 4.9 (±0.97) cfu g^{-1} substrate, Medium = 4.9×10^2 ($\pm0.97\times10^2$) cfu g^{-1} substrate, High = 4.9×10^4 ($\pm0.97\times10^4$) cfu g^{-1} substrate. Dark grey = autoclaved substrate control. Error bars are ± standard errors of the mean for the ten replicate plants.

Crop type was shown to have a significant effect on the fungal community in the rhizosphere, but not on the fungal community in the bulk soil (Fig. 1a, Table 2). In the bulk soil there were larger differences within OSR rotations than between OSR rotations and wheat rotations (Fig. 1a, Table 2). This was due to the higher relative abundance of TRFs 124/125 bp (*G. nigrescens*) and 284 bp (*O. brassicae*) in the continuous OSR bulk soil relative to all other rotations. There was also a significant difference in the fungal community of the rhizosphere compared with the bulk soil (Fig. 1a, Table 2). Using SIMPER analysis, the TRFs that contributed most towards the differences between the rhizosphere and bulk soil communities were TRF 284 bp (*O. brassicae*) and TRF 341 bp (identified as *Tetracladium* sp.) (Table S1) which were both found in higher relative abundance in the rhizosphere, and TRF 124/125 bp (*G. nigrescens*) which was found in higher relative abundance in the bulk soil.

Bacterial Community and TRF Identification

Non-metric MDS with ANOSIM analysis of the TRFLP data showed that rotation and crop type did not have a significant effect on the overall bacterial community in the rhizosphere or the bulk soil (Fig. 1b, Table 2).

There was a significant difference in the bacterial community of the rhizosphere compared with the bulk soil (Fig. 1b, Table 2). The TRFs that each contributed most towards the differences between the rhizosphere and bulk soil were TRFs 523 bp (Burkholderiales) and 245 bp (identified as *Pseudomonas fluorescens*), which had a higher relative abundance in the rhizosphere, and TRFs 338 bp and 721 bp (both identified as members of the phylum Acidobacteria Gp6), which had a higher relative abundance in the bulk soil (Table S1).

Glasshouse Experiments Using Model System

Olpidium brassicae: Early growth experiment. After 6–7 weeks growth, a significant reduction in top growth and root biomass occurred in plants inoculated with the High dose, for both OSR and *B. oleracea* ($P\leq0.01$; Fig. 3). Brown discolouration of the roots was also observed in the High dose plants. There was no difference in top growth or root weight between the control plants and the Low or Medium dose treatments.

Olpidium brassicae: Seed yield experiment. At harvest, there was a significant reduction in total number of pods produced in the plants inoculated with the High dose of *O. brassicae* ($p = 0.017$; Fig. 4a). This was found to be predominantly due to a reduced potential for pod production in this treatment as these plants also produced fewer primary branches than the uninoculated control (Fig. 4b). This ultimately resulted in a reduction in the number of seeds produced ($p = 0.015$; Fig. 4c). There was no difference between the control plants and the Low or Medium dose treatments.

Pyrenochaeta sp.: Early growth experiment. After 6–7 weeks growth, there was no effect of *Pyrenochaeta* sp. on the top growth or root biomass at any dose, although the High dose autoclaved inoculum was found to have a phytotoxic effect on *B. oleracea*. Medium and high doses showed brown discolouration of the roots compared to the autoclaved controls. Lesions were observed on the roots of inoculated plants (Fig. 5).

Pyrenochaeta sp.: Seed yield experiment. Plants inoculated with a High dose of *Pyrenochaeta* sp. had a delay in time to first flower (7 days) compared to the equivalent control (Fig. 6a). Subsequent analyses of the seed yield components revealed that the High dose had a negative impact on seed quantity and quality. The High dose resulted in significantly fewer normal seeds per pod compared to its control ($p = 0.044$; Fig. 6b), and a correspondingly greater number of deformed seed in total compared to the autoclaved control ($p = 0.016$). As a result of the reduction in number of normal seeds at the High dose, the seed weight per pod was also significantly less compared to the autoclaved control ($p = 0.019$; Fig. 6c). Not only were fewer normal seeds produced per pod, but those that were produced also weighed less per seed ($p = 0.009$; Fig. 6d). The Medium dose resulted in significantly fewer total number of seeds (including all normal and deformed seeds) compared to its autoclaved control ($p = 0.028$). However, the individual component analysis did not reveal further significant differences in number and weight of seed per pod for the Medium dose. There were no significant effects of the Low dose of *Pyrenochaeta* sp. on any yield component assessed.

Real time PCR. Real-time PCR confirmed the presence of either *O. brassicae* or *Pyrenochaeta* sp. in the roots of the inoculated samples, which was always highest in the 'High' dose inoculated plants. The fungi were not detected in the control samples (data not shown).

Discussion

This study has demonstrated that there is a strong influence of OSR rotation on microbial community structure and results from the field trial support reports that OSR cropping frequency has an impact on yield. Although shifts in microbial populations have previously been identified associated with rotation, consequences for crop growth have not been investigated [16–21]. In this study, we have taken further steps to identify the microbes responsible for the shifts in population, resulting in their isolation and functional studies.

Crop rotation was shown to have a significant effect on the fungal community. This was due predominantly to the increase in abundance of specific fungi in continuously cropped OSR, which suggests that the differences in rotation are due to enrichment of these fungi when OSR is planted. Greatest community divergence was observed between the fungal rhizosphere communities of continuous OSR and virgin OSR, with the other OSR rotations consisting of intermediate communities. Two microorganisms that became more prominent in the rhizosphere with frequent cropping of OSR were identified and selected for further investigation.

The first fungus, which showed the largest increase in abundance in the rhizosphere of continuous OSR compared with virgin OSR, was identified as *Olpidium brassicae* which is a soil-borne fungus within the phylum Chytridiomycota. *Olpidium brassicae* s.l. is a root-infecting obligate plant parasite, which is widespread in temperate regions but is reported to seldom induce obvious symptoms in host plants [35–37]. Two distinct species of *O. brassicae* s.l. have been reported; one that infects brassicas and one that infects non-brassicas, and phylogenetic analysis showed that the species are clearly distinct [35,38]. The non-brassica-infecting *O. brassicae* s.l. has been tentatively renamed *O. virulentus*; this is the species known to transmit destructive plant viruses [35]. The brassica-infecting *O. brassicae* s.l. species has retained the name *O. brassicae*. To date, no viruses have been identified that are associated with *O. brassicae*. Hosts of *O. virulentus* include lettuce, cucumber, tobacco and onion; whereas *O. brassicae* is exclusive to the roots of brassicas, including cabbage, brussel sprouts, broccoli and oilseed rape [38,39]. In this study, it was demonstrated that in a model system, at high doses, *O. brassicae* significantly reduced both top growth and root biomass of OSR and *B. oleracea* seedlings and impacted on *B. oleracea* seed yield by reducing branching and consequently total pod and seed production.

The second fungus identified was *Pyrenochaeta* sp.; an ascomycete which has a 95% DNA identity to *Pyrenochaeta lycopersici;* the causal agent of tomato corky root. Corky root is a severe disease that affects tomato crops worldwide [40]. Visible symptoms of tomato corky root are brown lesions in the roots and swelling of root epidermis with subsequent cracking into a corky texture [40]. The progressive damage to the root system leads to severe losses in fruit yield. Tomato yield reductions of up to 75% have been observed in European greenhouses [41]. In this study, it was demonstrated that in a model system, *Pyrenochaeta* sp. resulted in the development of root lesions, which may correspond to the root lesions found on tomato plants during *P. lycopersici* infection. In high doses, *Pyrenochaeta* sp. also delayed flowering and reduced seed weight, quality and quantity per pod of *B. oleracea*.

Overall, results indicate that the two fungi introduced at high doses, did impact on the seed yield production of the model species *B. oleracea*. Interestingly, the effects on the plants were different; *O. brassicae* reduced the number of branches, thus reducing the potential for seed production, whereas *Pyrenochaeta* sp. affected seed quantity and quality per pod. Although these results are from a model system, they highlight different mechanisms for yield reduction. Whilst the model system (using a soilless substrate under glasshouse conditions) was useful in testing the effects of individual fungal isolates on brassica growth and yield, it is now important to determine the impact of these fungi in soil, where interactions with other microorganisms will occur. The entry points and lesions created by *O. brassicae* and *Pyrenochaeta* sp. respectively, are likely to allow secondary infections which may also contribute to symptoms of yield decline. During this study, other potential pathogens were identified. For example, in the bulk soil of continuously cropped OSR, *Gibellulopsis* (formerly *Verticillium*) *nigrescens* had a greater relative abundance in continuously cropped OSR compared with virgin OSR. The relative importance of other potential pathogens needs to be established as it is likely that the observed yield decline is due to a combination of interacting factors involving a pathogen complex. Furthermore the interactions including any synergistic or antagonistic effects between the pathogens need to be determined.

Yield decline appears to be a nationwide and likely a world-wide phenomenon [2]. It is therefore important to determine whether these potential pathogens are found in field sites other than the site used in this study. A preliminary survey has shown that *Olpidium brassicae* is widespread and has been found to be associated with both winter and spring OSR, at different field sites across the UK, with higher levels found in the majority of intensively cropped OSR field sites (data not shown).

This study has identified two brassica pathogens from the field trial which contribute to yield decline in the model system; *Olpidium brassicae,* which has generally been overlooked as a brassica pathogen, and *Pyrenochaeta* sp., which has not previously been identified as a brassica pathogen. Future research is required to determine the effects and interactions of these fungi under field conditions and to investigate other identified fungi as well as other microorganisms. It is also important to understand how cultural and environmental factors influence these microorganisms in order for sustainable agricultural systems to be developed.

Supporting Information

Figure S1 Yield data from plots within different rotations of OSR (in rotation with wheat), from the fourth year of the field trial (2007). Error bars are ± standard errors.

Table S1 Identification of TRFs using the continuous OSR rhizosphere clone library. NCBI BLAST and Ribosomal Database Project (RDP) (at 80% confidence) were used to classify fungal (a) and bacterial (b) TRFs, respectively. TRF sizes and equivalent cut sites in the clones are shown using *Hha*I and *Msp*I. The accession number of the closest match of the consensus of the clones is shown for the fungal clones. * = An overlapping restriction site occurs resulting in a double peak.

Acknowledgments

We thank Rosie Atwood for carrying out the initial isolations of *O. brassicae* and *Pyrenochaeta* sp. from field samples, and also thank The Arable Group (TAG) and Velcourt for managing the field experimental site (funded by the Home-Grown Cereals Authority), sample collection and helpful discussions.

Author Contributions

Conceived and designed the experiments: PM GDB SH AJB DC RS. Performed the experiments: SH AJB GK. Analyzed the data: SH AJB. Contributed reagents/materials/analysis tools: RS. Wrote the paper: SH AJB GB PM DC.

References

1. Karlen DL, Varvel GE, Bullock DG, Cruse RM (1994) Crop Rotations for the 21st-Century. Advances in Agronomy, Vol 53 53: 1–45.
2. Bennett AJ, Bending GD, Chandler D, Hilton S, Mills P (2012) Meeting the demand for crop production: the challenge of yield decline in crops grown in short rotations. Biological Reviews 87: 52–71.
3. Crookston RK, Kurle JE, Copeland PJ, Ford JH, Lueschen WE (1991) Rotational Cropping Sequence Affects Yield of Corn and Soybean. Agronomy Journal 83: 108–113.
4. Kelley KW, Long JH, Todd TC (2003) Long-term crop rotations affect soybean yield, seed weight, and soil chemical properties. Field Crops Research 83: 41–50.
5. Kirkegaard J, Christen O, Krupinsky J, Layzell D (2008) Break crop benefits in temperate wheat production. Field Crops Research 107: 185–195.
6. Larkin RP, Honeycutt CW (2006) Effects of different 3-year cropping systems on soil microbial communities and rhizoctonia diseases of potato. Phytopathology 96: 68–79.
7. Pankhurst CE, Stirling GR, Magarey RC, Blair BL, Holt JA, et al. (2005) Quantification of the effects of rotation breaks on soil biological properties and their impact on yield decline in sugarcane. Soil Biology & Biochemistry 37: 1121–1130.
8. Christen O, Sieling K (1995) Effect of Different Preceding Crops and Crop Rotations on Yield of Winter Oilseed Rape (Brassica-Napus L). Journal of Agronomy and Crop Science-Zeitschrift Fur Acker Und Pflanzenbau 174: 265–271.
9. Berry PM, Spink JH (2006) A physiological analysis of oilseed rape yields: Past and future. Journal of Agricultural Science 144: 381–392.
10. Weightman R, Gladders P, Berry P (2011) Oilseed Rape. In: Halford NG, Karp A, editors. Energy Crops: Royal Soc Chemistry, Thomas Graham House, Science Park, Cambridge Cb4 4wf, Cambs, Uk. 116–147.
11. FAO (2012) Food and Agriculture Organization of the United Nations.
12. DEFRA (2012) Department for Environment Food and Rural Affairs.
13. Stobart R (2009) Identifying changes in crop performance and microbial populations under frequent cropping with oilseed rape. Aspects of Applied Biology 91: 147–152.
14. Smalla K, Wieland G, Buchner A, Zock A, Parzy J, et al. (2001) Bulk and rhizosphere soil bacterial communities studied by denaturing gradient gel electrophoresis: Plant-dependent enrichment and seasonal shifts revealed. Applied and Environmental Microbiology 67: 4742–4751.
15. Germida JJ, Siciliano SD, Seib AM (1998) Phenotypic plasticity of Pseudomonas aureofaciens (lacZY) introduced into and recovered from field and laboratory microcosm soils. Fems Microbiology Ecology 27: 133–139.
16. Li QH, Wu FZ, Yang Y, Wang XZ (2009) Effects of rotation and interplanting on soil bacterial communities and cucumber yield. Acta Agriculturae Scandinavica Section B-Soil and Plant Science 59: 431–439.
17. Li CG, Li XM, Kong WD, Wu Y, Wang JG (2010) Effect of monoculture soybean on soil microbial community in the Northeast China. Plant and Soil 330: 423–433.
18. Lei JL, Ding J, Wang L, Zhou YH, Yu JQ (2006) Comparison of bacterial diversity in soils of rice-monocropping and vegetable rotation by PCR-RFLP analysis of 16S rRNA gene. Allelopathy Journal 18: 141–152.
19. Larkin RP (2003) Characterization of soil microbial communities under different potato cropping systems by microbial population dynamics, substrate utilization, and fatty acid profiles. Soil Biology & Biochemistry 35: 1451–1466.
20. Lupwayi NZ, Rice WA, Clayton GW (1998) Soil microbial diversity and community structure under wheat as influenced by tillage and crop rotation. Soil Biology & Biochemistry 30: 1733–1741.
21. Alvey S, Yang CH, Buerkert A, Crowley DE (2003) Cereal/legume rotation effects on rhizosphere bacterial community structure in West African soils. Biology and Fertility of Soils 37: 73–82.
22. Schippers B, Bakker AW, Bakker PAHM (1987) Interactions of Deleterious and Beneficial Rhizosphere Microorganisms and the Effect of Cropping Practices. Annual Review of Phytopathology 25: 339–358.
23. Stobart RM, Bingham IJ (2013) Impact of Previous Cropping on Winter Oilseed Rape (including related studies addressing the impact of oilseed rape cropping frequency on components of yield and rooting). Report for HGCA Projects RD-2003–2922, RD-2009–3648 and RD-2009–3649.
24. Gardes M, Bruns TD (1993) Its Primers with Enhanced Specificity for Basidiomycetes - Application to the Identification of Mycorrhizae and Rusts. Molecular Ecology 2: 113–118.
25. White TJ, Bruns TD, Lee S, Taylor J (1990) Amplification and direct sequencing of fungal ribosomal RNA genes for phylogenetics. PCR Protocols: A Guide to Methods and Applications. Innes, M. A.: Academic Press, Inc.
26. Hauben L, Vauterin L, Swings J, Moore ERB (1997) Comparison of 16S ribosomal DNA sequences of all Xanthomonas species. International Journal of Systematic Bacteriology 47: 328–335.
27. Marchesi JR, Sato T, Weightman AJ, Martin TA, Fry JC, et al. (1998) Design and evaluation of useful bacterium-specific PCR primers that amplify genes coding for bacterial 16S rRNA. Applied and Environmental Microbiology 64: 795–799.
28. Bakker AW, Schippers B (1987) Microbial Cyanide Production in the Rhizosphere in Relation to Potato Yield Reduction and Pseudomonas Spp-Mediated Plant Growth-Stimulation. Soil Biology & Biochemistry 19: 451–457.
29. Altschul SF, Gish W, Miller W, Myers EW, Lipman DJ (1990) Basic Local Alignment Search Tool. Journal of Molecular Biology 215: 403–410.
30. Wang Q, Garrity GM, Tiedje JM, Cole JR (2007) Naive Bayesian classifier for rapid assignment of rRNA sequences into the new bacterial taxonomy. Applied and Environmental Microbiology 73: 5261–5267.
31. Kaplan CW, Kitts CL (2003) Variation between observed and true Terminal Restriction Fragment length is dependent on true TRF length and purine content. Journal of Microbiological Methods 54: 121–125.
32. Marsh TL (2005) Culture-independent microbial community analysis with terminal restriction fragment length polymorphism. Environmental Microbiology 397: 308–329.
33. Clarke KR (1993) Nonparametric Multivariate Analyses of Changes in Community Structure. Australian Journal of Ecology 18: 117–143.
34. Zare R, Gams W, Starink-Willemse M, Summerbell RC (2007) Gibellulopsis, a suitable genus for Verticillium nigrescens, and Musicillium, a new genus for V. theobromae. Nova Hedwigia 85: 463–489.

35. Sasaya T, Koganezawa H. (2006) Molecular analysis and virus transmission tests place *Olpidium virulentus*, a vector of Mirafiori lettuce big-vein virus and tobacco stunt virus, as a distinct species rather than a strain of *Olpidium brassicae*. Journal of General Plant Pathology 72: 20–25.
36. Campbell RN (1985) Longevity of Olpidium-Brassicae in Air-Dry Soil and the Persistence of the Lettuce Big-Vein Agent. Canadian Journal of Botany-Revue Canadienne De Botanique 63: 2288–2289.
37. Singh SL, Pavgi MS (1977) Olpidium-Brassicae in Cabbage Roots. Mycopathologia 62: 47–52.
38. Hartwright LM, Hunter PJ, Walsh JA (2010) A comparison of Olpidium isolates from a range of host plants using internal transcribed spacer sequence analysis and host range studies. Fungal Biology 114: 26–33.
39. Herrera-Vasquez JA, Cebrian MD, Alfaro-Fernandez A, Cordoba-Selles MD, Jorda C (2009) Multiplex PCR assay for the simultaneous detection and differentiation of Olpidium bornovanus, O. brassicae, and O. virulentus. Mycological Research 113: 602–610.
40. Varela AR, Ramert B, Martensson A (2009) Potential use of biocontrol agents for control of Pyrenochaeta lycopersici in tomato crops. Acta Agriculturae Scandinavica Section B-Soil and Plant Science 59: 379–384.
41. Hasna MK, Ogren E, Persson P, Martensson A, Ramert B (2009) Management of corky root disease of tomato in participation with organic tomato growers. Crop Protection 28: 155–161.

12

Assessment of the Soil Organic Carbon Sink in a Project for the Conversion of Farmland to Forestland: A Case Study in Zichang County, Shaanxi, China

Lan Mu[1,2], Yinli Liang[1,2]*, Ruilian Han[1,2]

1 Institute of Soil and Water Conservation, Northwest A&F University, Yangling Shaanxi Province, China, 2 Institute of Soil and Water Conservation, Chinese Academy of Sciences and Ministry of Water Resources, Yangling, Shaanxi Province, China

Abstract

The conversion of farmland to forestland not only changes the ecological environment but also enriches the soil with organic matter and affects the global carbon cycle. This paper reviews the influence of land use changes on the soil organic carbon sink to determine whether the Chinese "Grain-for-Green" (conversion of farmland to forestland) project increased the rate of SOC content during its implementation between 1999 and 2010 in the hilly and gully areas of the Loess Plateau in north-central China. The carbon sink was quantified, and the effects of the main species were assessed. The carbon sink increased from 2.26×10^6 kg in 1999 to 8.32×10^6 kg in 2010 with the sustainable growth of the converted areas. The black locust (*Robinia pseudoacacia L.*) and alfalfa (*Medicago sativa L.*) soil increased SOC content in the top soil (0–100 cm) in the initial 7-yr period, while the sequestration occurred later (>7 yr) in the 100–120 cm layer after the "Grain-for-Green" project was implemented. The carbon sink function measured for the afforested land provides evidence that the Grain-for-Green project has successfully excavated the carbon sink potential of the Shaanxi province and served as an important milestone for establishing an effective organic carbon management program.

Editor: Wen-Xiong Lin, Agroecological Institute, China

Funding: This work was financially supported by the National Science and Technology Support Program (2011BAD31B05) and by the National Science and Technology Support Program (2014BAD14B006). The funders had no role in study design, data collection and analysis, decision to publish, or preparation of the manuscript.

Competing Interests: The authors have declared that no competing interests exist.

* E-mail: liangyl@ms.iswc.ac.cn

Introduction

Soil organic carbon (SOC) is an important component of soil that plays a key role in the functions of both natural and agricultural ecosystems. In ecosystem services, SOC is critical for ensuring sustainable food production owing to its nutrient retention function and water-holding capacity [1,2]. The global SOC stock has been estimated to be 1400–1500 Pg C in the upper 100 cm soil layer [3–5], which is approximately twice the amount of C in the atmosphere and three times the amount stored in terrestrial vegetation [6]. Todd-Brown et al [7] reported that the present-day global SOC stocks range from 514 to 3046 Pg C among 11 earth system models (ESMs). Thus, slight reductions in SOC contents due to changes in land-use, soil management, or rates of soil erosion, could significantly raise the CO_2 in the atmosphere. Due to natural drought conditions, intensive human disturbance and severe soil erosion, the hilly-gully area of the Loess Plateau has the lowest soil organic carbon density (SOCD) in China [8]. However, it is possible to increase the organic carbon content and carbon sequestration capacity in the soils of this region through appropriate reforestation of degraded sloping croplands and other ecosystems, whose resilience capacity is intact [1,9].

Changes in land use may alter the land cover patterns in ways that can impact the biomass and soil carbon stocks [10] and alter the rate of C input and output to the soil, ultimately changing the soil C content [11]. Soil C sinks are not permanent and often persist only as long as appropriate management practices (conservation tillage and erosion control) are maintained. When a land-management or land-use change is reversed, the C accumulated as a result of the change is lost, usually more rapidly than it was accumulated [12]. And greater attention to the possibility of encouraging the growth of the forests as a means of removing carbon dioxide (CO_2) from the atmosphere [13]. Several studies have addressed the effects of the conversion of farmland to forestland on the SOC. In a long-term experiment, Guo and Gifford [14] showed that the conversion of forestland or grassland to farmland resulted in significant SOC reduction; the conversion of forestland to both farmland and grassland did not lead to SOC reduction in all cases, but always resulted in significant carbon sequestration from the atmosphere. Ostle et al [15] also found that soil carbon losses occur when grasslands, managed forest lands or native ecosystems are converted to croplands, and soil carbon gains are made when croplands are converted to grasslands, forest lands or native ecosystems. Carbon stored in forest ecosystems represents a substantial portion of the global C stock; worldwide, forests contain ~70% of all plant C and ~20% of all soil C [16]. The conversion of farmland to forestland is therefore an effective method for preventing soil erosion, which also has a large net sink effect on the SOC storage in the hilly-gully areas of China. However, the balance between inputs of organic matter, primarily from vegetation, and losses, as a result of decomposition, leaching and erosion, determines the magnitude of the carbon reservoir of the hilly-gully areas of China. The net CO_2 released via soil and water loss is calculated to be 8.4 g C m^{-2} yr^{-1} [17].

The hilly-gully areas of the Chinese Loess Plateau are known for their agricultural history and severe soil erosion [18]. Since the 1950s, the Chinese government has made substantially efforts to control the soil erosion and restore vegetation to the region. For example, an extensive tree-planting project was undertaken in the 1970s, and integrated soil erosion control was performed on the watershed scale in the 1980s and 1990s. Despite these efforts, the soil erosion remained unchecked, and vegetation had not been substantially restored to the region by the late 1990s [10]. In 1999, an extensive ecological rehabilitation program known as "Grain-for-Green" (conversion of farmland to forestland) was initiated by the Chinese government; this program is currently the largest land retirement project undertaken in the developing world. The Grain-for-Green program was initially intended to reduce soil erosion and desertification and increase the forest cover in China by retiring steep sloping and marginal lands from agricultural production in the hilly-gully area of the Loess Plateau [19]. The program has now been operating for about 10 years and the natural environment in parts of the Loess Plateau is improving as annual crops are replaced by perennial plants. However, limited attention has been paid to the effect of land use conversion on soil carbon sequestration in the Loess Plateau [20,21]. The Grain-for-Green program has potential for C gain through the improved land use conversions [[19] and we hypothesized that the conversion of farmland to forestland is one method that may increase SOC storage in soils in this study region in Loess Plateau.

Therefore, the objective of this paper was to assess the changes in the soil organic carbon content and carbon sink benefit following the implementation of the Grain-for-Green project on the Loess Plateau between 1999 and 2010. The paper also discussed the variation of the soil organic carbon content with the type of vegetation and soil age in the study area and assesses the effect of the land cover change project.

Material and Methods

Research region

Zichang County was located in the northern Shaanxi Province of the Loess Plateau in China within latitudes 36°59′–37°30′N and longitudes 109°11′–110°01′E. The region had a semiarid continental monsoon climate with a mean annual temperature of 9.1°C and mean annual precipitation of 514.7 mm. Over 70% of the precipitation occurred between June and September. The landform had the typical hilly-gully topology of the Loess Plateau, and the valleys and ravines accounted for approximately 94.6% of the total area, in which the gully density was 8.1 km km^{-2}. Soils in the research area originated from parent material of calcareous loess, with a low SOC concentration of less than 10 g kg^{-1}. These belong to the Calcic Cambisol group according to the Food and Agriculture Organization and United Nations Education Scientific and Cultural Organization (FAOUNESCO) soil classification system [10]. The soil type was loessal soil with a pH value of 8.6. The soil bulk density was 1.4 g cm^{-3} and total N content was 0.4 g kg^{-1}. The area of the croplands was 11.08×10^4 ha, accounting for 46.24% of the total land area. In this area, agricultural land occupied 39.5% of the area on slopes exceeding 25° (Table 1); this cultivated steep carried soil load of approximately 229.4×10^4 t yr^{-1} and a water loss of up to 857.8×10^4 m^3 yr^{-1}. The natural vegetation had been largely destroyed by deforestation and cultivation. Slope land reclamation and cultivation along the slope was the primary cause of soil and water loss.

Table 1. The area of the cultivated slope land classification of the ZiChang county.

slope (degree)	area	percentage
	hm^2	%
<5	17927.87	16.9
5~10	3050.33	2.9
10~15	9950.33	9.4
15~25	33247.4	31.3
25~35[a]	32532.93	30.7
>35	9362.93	8.8

[a]The slopes of greater than 25°were cultivated for agriculture occupied 39.5%.

Land degradation, desertification, soil erosion and declining soil fertility had degraded the environment and severely limited the crop yield. In addition, the SOC concentrations in these areas were less than 10 g C kg^{-1}. Consequently, an extensive ecological rehabilitation program known as "Grain-for-Green" was initiated in 1999 in the Loess Plateau region. Zichang was the demonstration county for the larger Shaanxi project. The program in the county had now been implemented for approximately 10 yr (since 1999), and the natural environment was improving as annual crops were replaced by perennial plants and as the vegetation cover increases. The vegetation coverage rate had increased from 6.16% to 32.8%, and the greening rate in the county had reached 62% (Figure 1). And the vegetation reduced the soil and water lost with the amount of approximately 27.5×10^4 t yr^{-1} and 102.9×10^4 m^3 yr^{-1}, respectively (the data come from the Bureau of Soil and Water conservation, Zichang county). These changes had significantly improved the ecological environment, reduced soil and water loss and produced a carbon sink effect in the area [22].

Land cover changes between 1998 and 2010

The main land use change in the Zichang region was the ongoing conversion of cropland to forestland between 1999 and 2010, with the afforestation area occupying up to 6.2×10^4 ha. Figure 2 showed the area over which the land cover was altered during the 10-yr period. 2 sites were selected for the study on the regional scale and a multi-year field study was conducted in the experimental sites at the Zichang experimental station, the institute of soil and water conservation, Chinese Academy of Science. The black locust (*Robinia pseudoacacia L.*) and alfalfa (*Medicago sativa L.*) were the main plantation species during this period.

Site A: alfalfa. Alfalfa is a perennial flowering plant in the pea family Fabaceae cultivated as an important forage crop in many countries around the world. Its root nodules contain bacteria, *Sinorhizobium meliloti* with the ability to fix nitrogen, producing a high-protein feed regardless of available nitrogen in the soil. It has been widely cultivated for restoration in Loess Plateau because of its nitrogen-fixing ability and livestock fodder. This site was situated in a hilly - gully slope which faced north, with a slope of 15°. The planting density was 150 plants /m^2. Three different ages of alfalfa (3, 7 and 12 yr old) were selected for detailed investigation to analyze the effect on the SOC content. The three afforestation chronosequences of alfalfa selected in the

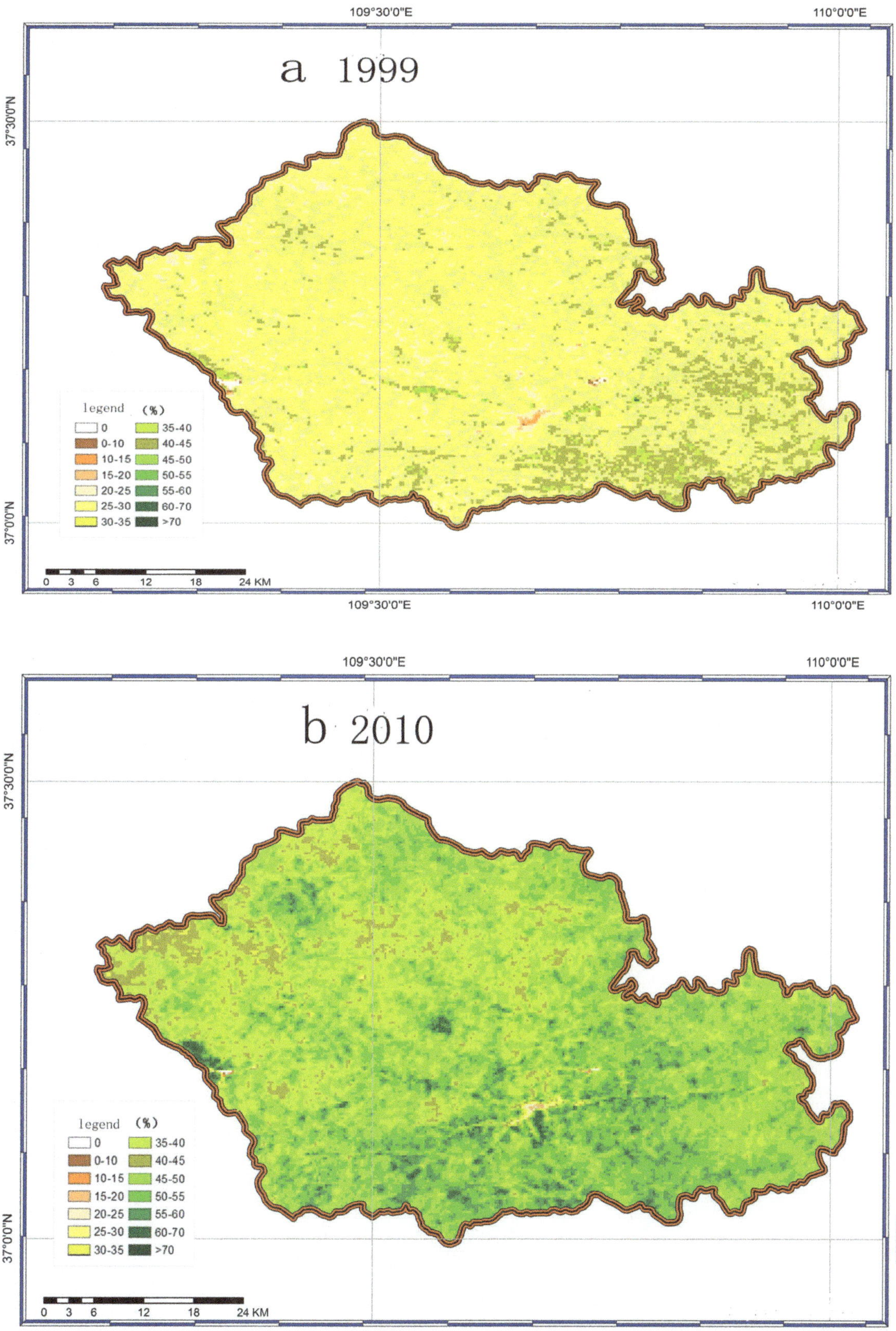

Figure 1. The images of the differences in the remote sensing vegetation coverage between 1999 and 2010 of ZiChang county.

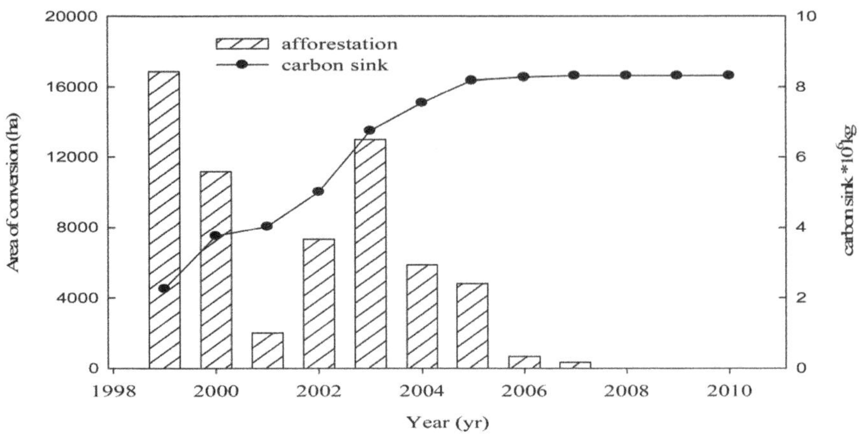

Figure 2. The areas of land cover change (farmland to forestland) and the carbon storage from 1998 to 2010.

same site and the only difference between them is the time when they had been afforested. Each chronosequences of alfalfa had three replications and the total were 3 ages×3 replications = 9 plots.

Site B: black locust. Black locust is a tree of the genus Robinia in the subfamily Faboideae of the pea family Fabaceae which is an exotic nitrogen-fixing tree native to Southeastern North America. It has been widely cultivated for restoration because of its drought resistance, high survival rate, ability to improve the soil nutrient status and remarkable growth rate [23]. At the present, it is the most widely cultivated species in the region. The black locust site was situated in a hilly - gully slope which faced north, with a slope of 19°. And the stand density of black locust was 1667 stands/hm^2. Four afforestation chronosequences of black locust (3, 7, 12 and 15 yr old) were selected for detailed investigation to analyze the effect on the SOC content. The four afforestation chronosequences of black locust selected in the same site and the only difference between them is the time when they had been afforested. Each chronosequences of black locust had three replications and the total were 4 ages×3 replications = 12 plots.

Calculation of forest carbon sink

Plenty of calculation methods for forest carbon sink have been evolved by experts and scholars such as carbon density method, carbon balance model F-CARBON and so on, which are precise but trivial and fall into the category of traditional nature science. In this research, the carbon sink was calculated on the basis of calculation method as forest storage extension suggested by Xi and Li [24] considering of the practicality and maneuverability [25].

Carbon sink was calculated using the following equation:

$$CF = \sum (Sx \times Cy) + \alpha \sum (Sx \times Cy) + \beta \sum (Sx \times Cy),\ Cy = V \times \delta \times \rho \times \gamma$$

Where S_x is the area of forest in the research region; C_y is the carbon density of the forest in the research region; V is the volume per unit area of forest; α is the carbon transfer coefficient of undergrowth plants, which is 0.195 in this research; β is the carbon transfer coefficient of forestland, which is 1.244 in this research; δ is the biomass expanding coefficient, which is 0.5 in this research; ρ is volumetric coefficient, which is 1.90 in this research; Υ is carbon content rate, which is 0.5 in this research. Values of each conversion coefficient in calculations of forest carbon sink potential in this region are taken as default values prescribed by intergovernmental Panel on Climate Change [26].

Soil sample collection and analysis

In November 2010, the soil samples were collected with a 5-cm diameter soil auger and were extracted in 20-cm incremental subsamples, which were subsequently mixed by hand. Soil samples were collected from each soil layer (0–10 cm, 10–20 cm, 20–40 cm, 40–60 cm, 60–80 cm, 80–100 cm and 100–120 cm) at five different locations selected within a 10 m radius surrounding each plot to analyze the soil organic carbon content. The five samples collected for each layer were subsequently mixed by hand, yielding one representative sample for each layer at each site. The total carbon stock for multiple soil layers was calculated by summing the soil carbon stocks of the layers.

All of the soil samples were air-dried, sealed in airtight bags and transported to the Institute for Soil and Water Conservation of the Chinese Academy of Science in Yangling, Shaanxi province to determine the SOC contents. The SOC was determined using the oil bath-K_2CrO_7 titration method after digestion, and the soil bulk density (BD) was determined using the ring tube method suggested by the Chinese Editorial Committee for Soil Analysis [27]. Total N was measured by the Kjeldahl procedure and pH was measured by electronic pH-meter [28].

Statistical analysis

Each SOC contents in composite samples for each plantation species of various depths were averaged at the same ages following afforestation to perform statistical analysis. Analysis of variance was performed using the SPSS 16.0 software. Means were compared by least significant difference (LSD) at $p<0.05$ or $p<0.01$ level. A one-way analysis of variance (ANOVA) was used to analyze the effect of each factor (land use, soil depths, ages and species) on the SOC concentration.

Results

Carbon sink effects following conversion of farmland to forestland

The forest carbon storage increased continuously as the farmland to forestland conversion project was implemented between 1999 and 2010. The carbon sink increased from

2.3×10^6 kg in 1999 to 8.3×10^6 kg in 2010 with the sustainable growth of the converted areas (Figure 2). The total carbon sink increased by 268% compared to that in the first yr of implementation of the project. This result established that the sequestration of carbon (C) is another important environmental effect of afforestation. The land use policy affected the type, distribution, productivity and turnover of vegetation. This policy was therefore a key determinant of whether the land surface was a sink or source of carbon. Forestland had great carbon sequestration potential, especially compared to farmland. The carbon storage induced by the conversion of farmland to forestland in Zichang County was very large.

SOC content for alfalfa and black locust based on the afforestation chronosequence

In the alfalfa soil, there was a significant increase in the SOC content in the 0–20 cm and 40–100 cm layers with increasing age ($p<0.01$; Figure 3a) from 3 yr to 7 yr. Subsequently, the SOC content decreased slightly with increasing age (from 5 to 4.5 g kg^{-1} and from 3.2 to 2.4 g kg^{-1} in the last 5 yr). The SOC did not change appreciably in the 20–40 cm layer (Figure 3a). In contrast with the soil samples from <100 cm depth, the changes in the soil C in the 100–120 cm layer were significantly correlated with plantation age; the SOC decreased continuously with age from 2.7 to 2.0 g kg^{-1}.

The SOC exhibited a different trend with age in the black locust soil compared to the alfalfa soil (Figure 3b). The SOC content decreased in the first 7 yr at depths below 100 cm, from 3.0 to 1.7 g kg^{-1} (for yr 4–7, the concentration was consistently lower than that measured for the first 3 yr). Interestingly, the C content then exhibits a gradually increasing trend between 7 and 15 yr (Figure 3b) reaching 3.8 g kg^{-1} at an age of 15 yr. In contrast, the ecosystem C stores in the 100–120 cm layer of the old plantation were clearly higher than those in the younger afforestation sites, which had increased by 31.3% 15 yr after afforestation. The SOC content in the two afforestation plantations therefore provide evidence for a carbon sink once the farmland was converted to forestland.

A comparison of the SOC contents for plantation species of various depths at the same ages following afforestation

The statistical comparisons in the ANOVA demonstrate that the dependence of the SOC on depth in the areas converted to forestland is significant at a level of $p<0.05$ (Table 2). Between ages of 7 and 12 yr, the SOC content in the alfalfa soil decreases with increasing depth, and the SOC content in the 0–60 cm layer exceeds that at depths above 60 cm. The soil C content is greatest (5.5 and 4.5 g kg¯1) in the 0–10 cm layer. However, in the first 3 yr, the SOC decreases significantly ($p<0.01$, Table 2) with increasing depths of 0–80 cm and then exhibits a smaller increase at depths of 80–100 cm. For black locust, the SOC decreases continuously at depths of 0–100 cm and then increases slightly in the 100–120 cm layer in the 3-, 7-, 12- and 15-yr-old forests. Although the C concentration trend based on depth was different for the two plantations, the SOC content in the topsoil was significantly greater ($p<0.01$, Table 2) than that in the subsoil for both plantations.

To inform organic carbon management strategies, the dependence of the SOC content on the type of vegetation is also quantified (Figure 4). The SOC concentration in the alfalfa soil exceeded that for the black locust soil at a given age in the 0–100 cm and 100–120 cm layers. The soil C for the two plantations shows a highly significant difference between alfalfa and black locust ($p<0.05$ or $p<0.01$, Figure 4). The SOC in the 3-yr-old alfalfa was approximately 36.9% higher than that in the black locust in the 0–100 cm layer. The largest difference between the two species was measured for the 7-yr-old plantation; the SOC contents for this plantation exhibit highly significant differences between the two species at each depth ($p<0.01$, Figure 4). The soil C is similar for both species on the 12-yr-old plantation and exhibits a significant difference ($p<0.05$, Figure 4) compared to the 7-yr-old plantation at each depth.

Relationship between the SOC content, depth and other factors as a function of age following afforestation

A change in land use from farmland to forestland and grassland implies that the annual cycle of cultivating and harvesting crops is replaced by the much longer forest cycle. Therefore, many factors influence the SOC in the alfalfa and black locust soils after the land cover changes. The relationships between the SOC content, soil depth and age are summarized in Figure 5. For the 3-yr-old afforested land, there is a strong correlation ($R^2 = 0.93$, $R^2 = 0.99$, $p<0.01$; Figure 5a, b) between SOC content and depth. As shown in Fig 5, the same trend is obtained for 7- (Figure 5c, d) and 12-yr-old (Figure 5e, f) soil. The SOC is related to the afforestation yr, with the SOC decreasing gradually with increasing age (Figure 5).

Discussion

Effect of the farmland-to-forestland conversion project on the SOC stocks

Land use change can cause a change in land cover and an associated change in the carbon stocks [29]. Each soil has a given carbon carrying capacity, i.e., an equilibrium carbon content depending on the nature of the vegetation, precipitation and temperature [30]. The equilibrium carbon stock arises from a balance between the inflow and outflow to the carbon pool [31]. The equilibrium between the carbon inflow and outflow to the soil is disturbed by any land change until a new equilibrium is reached in the new ecosystem. In our study, we have found that in the farmland-to-forestland conversion project, the soil acts as a carbon sink (from 2.26×10^6 kg in 1999 to 8.32×10^6 kg in 2010). Other studied has got the same results. It has also been estimated that UK forestry and grassland sequester 110±4 kg and 240±200 kg of carbon per hectare yr, respectively, whereas croplands lose on average 140±100 kg of carbon per hectare per yr [32]. Ostle et al [15] indicated that soil carbon accumulates more slowly (decadal) but gains can be made when croplands are converted to grasslands, plantation forest or native woodland. The results of this study partially support the hypothesis that the afforestation of formerly arable land leads to increased C storage in the soil over the short term (approximately 30 yr).

The balance between the inputs (primarily from vegetation) and losses of organic matter (as a result of decomposition, leaching and erosion) determines the magnitude of the area's land carbon reservoir. The net effect of the afforestation of cultivated slope land on the soil C content depends not only on the new C gained, but also on the C lost as a result of the previous management. Although the vegetation carbon stock in the hilly-gully areas are relatively small, plant matter is the single most important source of carbon input to the soil [33]. Therefore, the higher SOC content in the two plantations in our study could be a result of primarily the greater annual litter input and fine root biomass compared to the cropland.

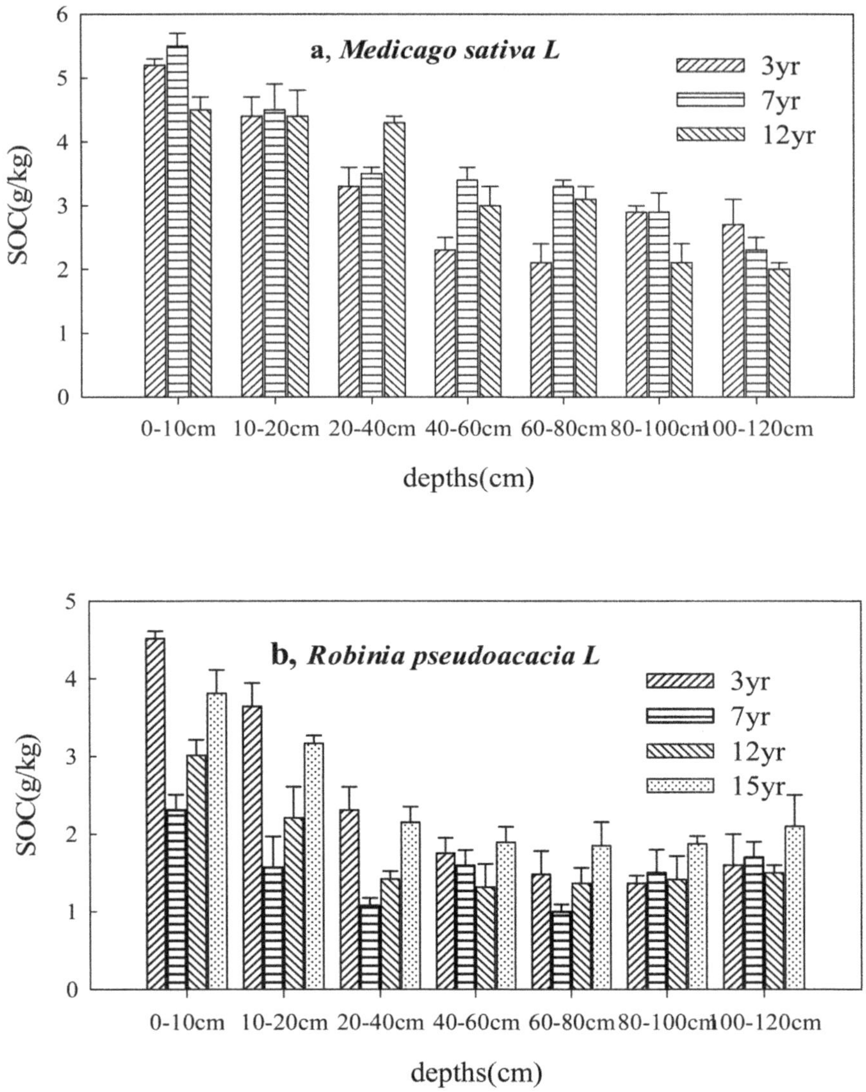

Figure 3. Changes of SOC sequestrations at different ages under alfalfa (*Medicago sativa L*) and black locust (*Robinia pseudoacacia L*) for each depth.

Table 2. SOC[a] at the different depth of the alfalfa (*Medicago sativa L*) and black locust (*Robinia pseudoacacia L*) following the afforestation.

Depth (cm)	alfalfa (*Medicago sativa L*) ($g\ kg^{-1}$)				black locust (*Robinia pseudoacacia L*) ($g\ kg^{-1}$)				
	3 yr	7 yr	12 yr	mean	3 yr	7 yr	12 yr	15 yr	mean
0–10 cm	5.2a	5.5a	4.5a	5.1a	4.5a	2.3a	3.0a	3.8a	3.0a
10–20 cm	4.4b	4.5b	4.4a	4.5a	3.6b	1.6b	2.2ab	3.2b	2.3b
20–40 cm	3.3c	3.5c	4.3a	3.8b	2.3c	1.1c	1.4bc	2.1c	1.5c
40–60 cm	2.3ef	3.4cd	3.0b	2.9c	1.8d	1.6b	1.3c	1.9c	1.6c
60–80 cm	2.01f	3.3cd	3.1b	2.8c	1.5d	1.0c	1.4bc	1.9c	1.4c
80–100 cm	2.9cd	2.9d	2.1c	2.6c	1.4d	1.5b	1.4bc	1.9c	1.6c
100–120 cm	2.7de	2.3e	2.0c	2.3c	1.6d	1.7b	1.5bc	2.1c	1.8cb

[a]Data in the column are mean values (n = 3 for alfalfa and black locust), which are compared among different depths within each ages and are not different at the 5% level of significance if followed by the same letter.

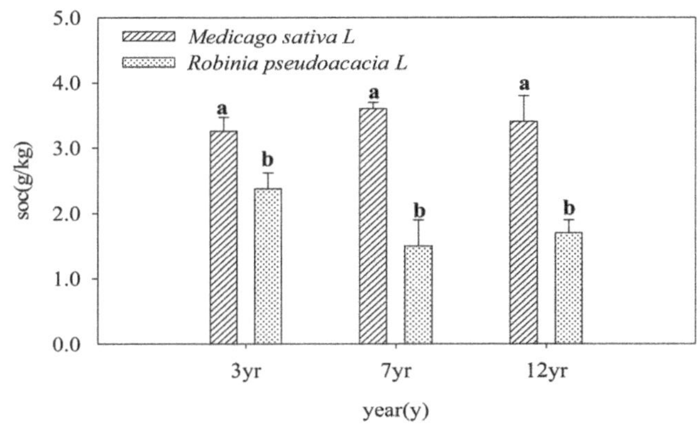

Figure 4. The difference of SOC between alfalfa (*Medicago sativa L*) and black locust (*Robinia pseudoacacia L*) at the same depths under the different ages at the 0–120 cm depths under the 3,7 and 12 ages after the conversion of farmland to forestland.

SOC accumulation for different depths and ages following the afforestation of the cropland

The soil organic carbon content increase in the topsoil (0–40 cm) for both plantations (from 3.8 to 5.1 g kg^{-1} and 1.5 to 3.0 g kg−1) after the conversion of cropland into the forest, is in consistent with other findings for the Loess Plateau [34,10] and other areas of the world [14,35–37] suggesting that the SOC may be highly sensitive to variations in the soil moisture, soil temperature and/or litter input (all of which depend strongly on depth). The increase in the topsoil SOC may be associated with the higher carbon input and lower rates of SOC decomposition and soil erosion associated with alfalfa plantations and black locust forest. It is well established that the top soil is the part of the soil profile that is most susceptible to land use change and disturbance, such as plowing and drainage. The increase in SOC decomposition resulting from tillage disturbance is a key reason for the decrease in SOC following the cultivation of perennial ecosystems [38]. Within limits, soil C increase with increasing soil water and decreasing temperature and the top soil of the increasing water within temperature zones can increase plant production and increase plant litter and root production [39]. In contrast, the cessation of tillage during the establishment of forest on cropland can reduce the SOC decomposition rate, thus increasing the SOC content. Furthermore, the addition of nitrogen derived from nitrogen-fixing species (in this case, alfalfa and black locust) has been well documented and can lower the decomposition of old and new carbon [40], hence favoring SOC sequestration. Moreover, the forest is widely found to be less susceptible to soil erosion compared to the cropland on the Loess Plateau [41], which may contribute to the higher topsoil SOC content following the afforestation of the cropland. And there is also an increase in the carbon input from aboveground litter and fine root biomass in the alfalfa and black locust, which may lead to SOC accumulation during the afforestation.

Understanding the dependence of the SOC content on the plantation age is important for the development of improved biological soil management practices on both local and national scales. This understanding will also help to identify different species that are ecologically compatible and economically useful. The SOC stores in the younger plantation were clearly higher than those in the old afforestation sites, which exhibited decreases of 15.6% and 50% for alfalfa and black locust 12 yr after afforestation. However, for black locust, the SOC continues to increase gradually 15 yr after the conversion of farmland to forestland. A fundamental assumption regarding the effect of land change on the carbon sink is that younger, growing forests sequester carbon more rapidly than older forests. For several yr following a disturbance, the land becomes a carbon source, but as it matures, it becomes a large carbon sink, which slowly declines with age during late succession [42,43]. This trend may owe to the past application of organic fertilizer when the land was used as cropland, producing a higher initial soil C content just before afforestation. When agricultural land is no longer used for cultivation and is allowed to revert to natural vegetation, the balance of C input and output in the agricultural system is broken, as no fertilizer is added to the afforested land.

Factors affecting the SOC sequestration in the "Grain-for-Green" project

The contribution of afforestation to the SOC improvement is controlled by several factors, including the severity of the land degradation, plant species and composition, climate, duration, land use history and management. The effect of the Grain for Green project on SOC sequestration was mainly influenced by ages and depths, which explained 43.2% and 35.6% (Table 3) of the variation of SOC. vegetation species explained 21.2% (Table 3) of the variation in SOC. Xu et al [44] found that the effect of the "Grain-for-Green" project on the SOC sequestration was influenced primarily by the land use type and age, which explained 55.6% and 24.1% of the variation in the SOC, respectively. The SOC content therefore varies across plantations because the C sequestration is affected by the varying land management. Our findings on the carbon sink induced by the project and the relationship between the SOC content and the ages following afforestation supported the conclusion of Xu et al [44]. Topographic factors such as the slope aspect and position were also found to explain 8.5% of the variation in the SOC [44]. Reay ea al [45] indicated that by increasing the productivity, nitrogen enrichment of some natural ecosystems may enhance their capacity to sequester carbon. Clearly, we demonstrate that afforestation has a positive influence on the SOC concentration in our study, depending on the species planted (alfalfa has a higher SOC content than black locust). The species-induced difference in the SOC content and other soil properties may owe to a difference in biomass production and nutrient cycling via litter fall and root turnover [46].

Conclusion

We examined the changes in the carbon sink and SOC content of the topsoil and subsoil following the "Grain-for-Green" (farmland-to-forestland conversion) project. The afforested land (alfalfa and black locust) soil displayed remarkable SOC sequestration compared to the sloping croplands. The two plantations studied exhibited SOC sequestration potential in the top soil (0–100 cm) in the initial 7-yr period, while the sequestration occurred later in the 100–120 cm layer (>7 yr after the "Grain-for-Green" project was implemented). Furthermore, our analyses provide evidence that SOC can be influenced by afforestation in semi-arid region such as the Loess Plateau, China. From a regional perspective, type of land use, age and species all had significant effects on the SOC sequestration in the "Grain-for-Green" project. The implementation of the program has further enhanced the vegetation restoration and ecological conservation and strengthened the management of the Loess Plateau ecosystem.

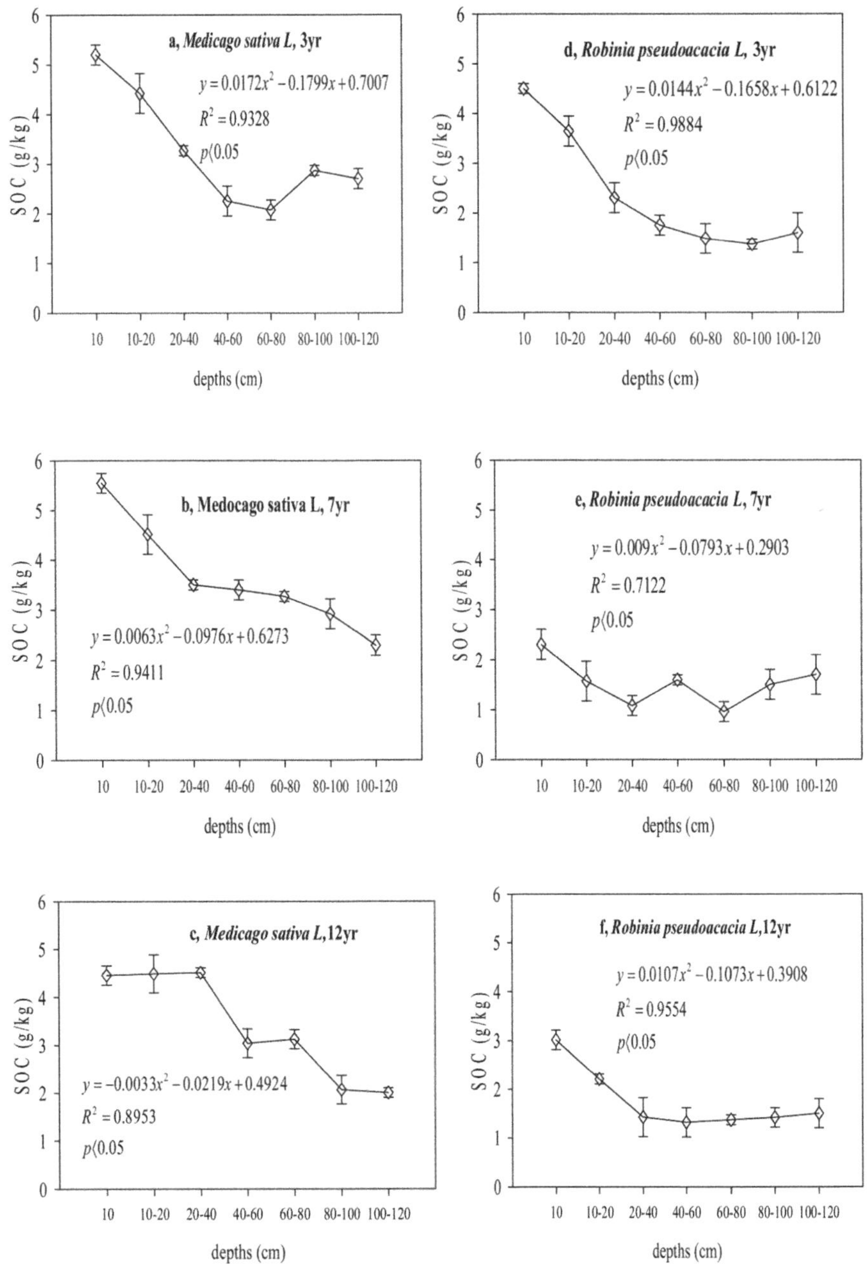

Figure 5. Relationship between the SOC and depths of the alfalfa (*Medicago sativa L*) and the black locust (*Robinia pseduoacacia L*) under 3, 7 and 12 years old following afforestation.

Table 3. Contribution of different factors to the variation of SOC content (variance components, n=45).

	Variance Source		
SOC content	**Afforestation ages**	**Depths**	**Plantation Species**
Variance	16.62	13.69	8.15
Variance percentage to total variance (%)	43.20	35.60	21.20

Author Contributions

Conceived and designed the experiments: YLL LM. Performed the experiments: LM RLH. Analyzed the data: YLL LM. Contributed reagents/materials/analysis tools: YLL LM. Wrote the paper: LM YLL.

References

1. Lal R (2004) Soil carbon sequestration to mitigate climate change. Geoderma 123: 1–22.
2. Lal R (2010) Beyond Copenhagen: mitigating climate change and achieving food security through soil carbon sequestration. Food Secur 4: 1035–1064.
3. Post WM, Emanuel WR, Zinke PJ, Stangenberger AG (1982) Soil carbon pools and world life zones. Nature 298: 156–159.
4. Batjes NH (1996) Total carbon and nitrogen in the soils of the world. Eur.J. Soil Sci 47: 151–163.
5. Tarnocai C, Canadell JG, Schuur EAG, Kuhry P, Mazhitova G, et al. (2009) Soil organic carbon pools in the northern circumpolar permafrost region. Global Biogeochemical Cycles 23: 1–11.
6. Smith P (2004) Soils as carbon sinks- The global context. Soil Use Manage 20: 212–218.
7. Todd-Brown KEO, Randerson JT, Post WM, Hoffman FM, Tarnocai C, et al. (2013) Cause of variation in soil carbon simulations from CMIP5 Earth system models and comparison with observations. Biogeosciences 10: 1717–1736.
8. Yu DS, Shi XZ, Wang HJ, Sun WX, Chen JM, et al. (2007) Regional patterns of soil organic carbon stocks in China. J.Environ.Manage 85: 680–689.
9. Pan GX (2008) Soil organic carbon stock. Dynamics and climate change mitigation of China. Adv.Clim.Change Res 4: 282–289.
10. Chen LD, Gong J, Fu BJ, Huang ZL (2007) Effect of land use conversion on soil organic carbon sequestration in the loess hilly area, loess plateau of china. Ecol Res 22: 641–648.
11. Entry JA, Sojka RE (2002) Management of irrigated agriculture to increase organic carbon storage in soils. Soil Sci Soc Am. J 66: 1957–1964.
12. Smith P, Powlson DS, Glendining MJ (1996) Establishing a European soil organic matter network (SOMNET). In: Powlson DS, Smith P, Smith JU, editors. Evaluation of soil organic matter models using existing, long-term datasets. UK: Harpenden. pp: 429.
13. Lubowski RN, Plantinga AJ, Stavins RN (2005) Land-use change and carbon sinks: Econometric Estimation of the Carbon Sequestration Supply Function. Belfer Centerprograms or projects: Environment and Natural resources, Harvard Project on Climate Agreements 5: 135–152.
14. Guo LB, Gifford RM (2002) Soil carbon stocks and land use change: a meta analysis. Global Change Biol 8: 345–360.
15. Ostle NJ, Levy PE, Evans CD, Smith P (2009) UK land use and soil carbon sequestration. Land use polic 26s: S274–S283.
16. DeGryze S, Six J, Paustian K, Morris SJ, Merckx R, et al. (2004) Soil organic carbon pool changes following land-use conversions. Global Change Biology 10: 1120–1132.
17. Schlesinger WM (1999). Carbon sequestration in soils. Science 284: 2095.
18. Shi H, Shao MA (2000) Soil and Water loss from the Loess Plateau in China. J Arid Environment 45: 9–20.
19. Liu ZP, Shao MA, Wang YQ (2011) Effect of environment factors on regional soil organic carbon stocks across the Loess Plateau region, China. Agricultural, Ecosystemsand Environment 142: 184–194.
20. Xu XL, Zhang KL, Peng WY (2003) Spatial distribution and estimating of soil organic carbon on Loess Plateau. J Soil Water Conservation 17: 13–15.
21. Jia SW, He XB, Chen YM (2004) Effect of land abandonment on soil organic carbon sequestration in loess hilly areas. J Soil WaterConservation 18: 78–80.
22. Chang RY, Fu BJ, Liu GH, Wang SH, Yao XL (2012) The effects of afforestation on soil organic and inorganic carbon: A case study of the Loess Plateau of China. Catena 95:145–152.
23. Jin TT, Fu BJ, Liu GH, Wang Z (2011) Hydrologic feasibility of artificial forestation in the semi-arid Loess Plateau of China. Hydrol.Earth Syst.Sci 15:2519–2530.
24. Xi TT, Li SL (2006) Analysis of forestry carbon mitigation potential in Heilong jiang Province. Problems of Forestry Economics 26: 519–523.
25. Xu YR, Wang DT, Feng ZW (2003) Effect of carbon storage in several artificial vegetations on sea-beach salinity soil in Tianjin, China. Journal of Huazhong Agricultural University 22: 603–607.
26. IPCC (2000) Summary for Policymakers. In: Nakicenovic N, Swart R, editors. Special Report of the Intergovernmental Panel on Climate Change. Cambridge: Cambridge university. pp. 377
27. Chinese Editorial Committee of Soil Analysis (1996) Soil physical and chemical analysis and description of soil profile (in Chinese). China State Standards Press, Beijing.
28. ISS (1978) Physical and chemical analysis methods of soils. Institute of Soil Science. Shanghai: Shanghai Sci Technol Press. pp: 7–59.
29. Bolin B, Sukumar R (2000) Global perspective In: Watson RT, Noble IR, Bolin B, Racindranath NH, Verardo DJ, DokkenDJ, editors. Land use, land use change, and forestry. Cambridge: Cambridge university. pp. 23–51.
30. Gupta RK, Rao DLN (1994) Potential of wastelands for sequestering carbon by reforestation. Current science 66: 378–380.
31. Fearnside PM, Barbosa RI (1998). Soil carbon changes from conversion of forest to pasture in Brazilian Amazonia. Forest ecology and management 108: 147–166.
32. Dawson JJC, Smith P (2007) Carbon losses from soil and its consequences for land-use management. Science of the total environment 382: 165–190.
33. Leake MC, Jennifer HC, George HW, Fan B, Richard M, et al. (2006) Stoichiometry and turnover in single, functioning membrane protein complexes. Nature 443: 355–358.
34. Chang RY, Fu BJ, Liu GH, Liu SG (2011) Soil carbon sequestration potential for "grain for green" project in Loess plateau, China. Environmental management 48: 1158–1172.
35. Laganière J, Angers DA, Par D (2010) Carbon accumulation in agricultural soils after afforestation: a meta analysis. Global change biology 16: 439–453.
36. Paul KI, Polglase PJ, Nyakuengama JG, Khanna PK (2002) Change in soil carbon following afforestation. Forest ecology and management 168: 241–257.
37. Post WM, Kwon KC (2000) Soil carbon sequestration and land use change: processes and potential. Global change biology 6: 317–327.
38. Balesdent J, Chenu C, Balabane M (2000) Relationship of soil organic matter dynamics to physical protection and tillage. Soil and tillage research 53: 215–230.
39. Liski J, Iivesniemi A, Makela CJ (1999) CO_2 emission from soil in response to climate warming are overestimated-the decomposition of old soil organic matter is tolerantof temperature. Ambio 28: 171–174.
40. Binkley D (2005) How nitrogen-fixing trees change soil carbon. In: Binkley D, Menyailo O, editors. Tree species effects on soils: implications for global change. Dordrecht: Kluweracademic publishers. pp. 155–164.
41. Fu BJ, Wang YF, Lu YH, He CS (2009) The effects of land-use combinations on soil erosion: a case study in the loess plateau of China. Progress in physical geography 33: 793–804.
42. Law BE, Sun OJ, Campbell J, Vantuyl S, Thornton PE (2003) Changes in carbon storage and fluxes in a chronosequence of pondersosa pine. Global change biology 9: 510–524.
43. Turner DP, Guzy M (2003) Effects of land use and fine-scale environmental heterogeneity on net ecosystem production over a temperate coniferous forest landscape. Tellus 558: 657–668.
44. Xu MX, Wang Z, Zhang J, Liu GB (2012) Response of soil organic carbon sequestration to the "Grain for Green Project" in the hilly loess plateau regain. Acta Ecologicasinica 32: 5405–5415.
45. Reay DS, Dentener F, Smith P, Grace J, Feely R (2007) Global nitrogen deposition and carbon sinks. Nature Geoscience 1: 430–437.
46. Girmay G, Singh BR, Mitiku H (2008) Carbon stocks in Ethiopian soils in relation to land use and soil management. Land degradation and development 19: 351–367.

Effects of Straw Incorporation on Soil Organic Matter and Soil Water-Stable Aggregates Content in Semiarid Regions of Northwest China

Peng Zhang[1,2]⁹, Ting Wei[1,2]⁹, Zhikuan Jia[1,2]*, Qingfang Han[1,2], Xiaolong Ren[1,2], Yongping Li[3]

1 The Chinese Institute of Water-Saving Agriculture, Northwest A&F University, Yangling, Shaanxi, China, **2** Key Laboratory of Crop Physi-Ecology and Tillage Science in Northwestern Loess Plateau, Ministry of Agriculture, Northwest A&F University, Yangling, Shaanxi, China, **3** Guyuan Institute of Agricultural Sciences, Guyuan, Ningxia, China

Abstract

The soil degradation caused by conventional tillage in rain-fed areas of northwest China is known to reduce the water-use efficiency and crop yield because of reduced soil porosity and the decreased availability of soil water and nutrients. Thus, we investigated the effects of straw incorporation on soil aggregates with different straw incorporation rates in semiarid areas of southern Ningxia for a three-year period (2008–2010). Four treatments were tested: (i) no straw incorporation (CK); (ii) incorporation of maize straw at a low rate of 4 500 kg ha^{-1} (L); (iii) incorporation of maize straw at a medium rate of 9000 kg ha^{-1} (M); (iv) incorporation of maize straw at a high rate of 13 500 kg ha^{-1} (H). The results in the final year of treatments (2010) showed that the mean soil organic carbon storage of the 0–60 cm soil layers were significantly ($P<0.05$) increased with H, M, and L, by 21.40%, 20.38% and 8.21% compared with CK, respectively. Straw incorporation increased >0.25 mm water-stable macroaggregates level, geometric mean diameter, mean weight diameter and the aggregate stability, which were ranked in order of increasing straw incorporation rates: H/M > L > CK. Straw incorporation significantly ($P<0.05$) reduced the fractal dimension in the 0–40 cm soil layers compared with CK. Our results suggest that straw incorporation is an effective practice for improving the soil aggregate structure and stability.

Editor: Raffaella Balestrini, Institute for Plant Protection (IPP), CNR, Italy

Funding: This work was supported by the China Support Program (2012BAD09B03, 2011AA100504, and 2011BAD29B09) for Dryland Farming in the 12th 5-year plan period, the Youth project of National Natural Science Fund (31201156), and the Basic Science Research Fund in Northwest A&F University (QN2013005). The funders had no role in study design, data collection and analysis, decision to publish, or preparation of the manuscript.

Competing Interests: The authors have declared that no competing interests exist.

* E-mail: jiazhk@126.com

⁹ These authors contributed equally to this work.

Introduction

Soil infertility [1], soil erosion, and water deficiency [2] are the major factors that limited crop growth in semiarid areas of northwest China. The rates of crop straw use for fuel and forage have declined significantly since the 1980s and crop straw is increasingly burned after the harvest, which leads to high losses of soil organic substances [3,4], and increased emission of CO_2 that pollute the environment [5]. Furthermore, this practice has led to the degradation of the agricultural ecological environment [6].

The soil organic matter (SOM) content is one of the major factors that affects soil properties and functions including a range of physical characteristics such as the water-holding capacity [7], water infiltration [8], and aggregate stability [9]. SOM is considered to be a major binding agent that stabilizes soil aggregates [10,11]. Soil aggregates are the basic units of the soil structure [12], which are composed of primary particles and binding agents that determine the microbial biomass and mineral nutrient reserves [13–15]. These soil properties are also affected by soil organic matter decomposition processes [16,17].

Many studies have shown that crop straw is rich in organic material and soil nutrients, so it is increasingly considered to be an important natural organic fertilizer [18–20]. Straw can be incorporated to soil either directly or indirectly, which can promote the production of a favorable soil environment. Straw also maintains the physicochemical condition of the soil and improves the overall ecological balance of the crop production system [20,21]. Nelson [22] and Wilhelm *et al.* [23] showed that the incorporation of crop residues into soil significantly prevented soil erosion and enhanced the soil quality. Sonnleitner *et al.* [24] found that straw incorporation also improved the aggregate stability and other soil properties compared with farmyard manure. Mulumba and Lal [25] also reported that the addition of crop residues to cultivated soil had positive effects on the soil porosity, available water content, soil aggregation, and bulk density. Bhagat and Verma [26] showed that the incorporation of crop straws for five years significantly increased the crop yield and improved the soil properties.

The soil improvement effect of straw incorporation has been recognized widely [23–25] but information is still limited on the responses of the SOC and water-stable aggregates under different rates of straw incorporation, particularly in the loessal soil in semiarid areas of northwest China. The theory and technique of straw incorporation in this region have also not been reported. Thus, the present study investigated the effects of different crop straw application rates combined with conventional planting on SOC, the >0.25 mm water-stable macroaggregate rate, and various soil properties in the southern Ningxia region of China.

Materials and Methods

Ethics Statements

The study was carried out on the private land, we rent the farmland from the local farmers, and contracts and deeds are signed. No specific permissions were required in this area to run the experiment as the study sites are farming area without any protection zone, and the farming activities won't hurt the local animals. And we only plant the grain crop in the field, so the field studies did not involve endangered or protected species.

Site description

The experiment was conducted between 2008 and 2010 at the Dryland Agricultural Research Station, Pengyang County, Ningxia, China (106°45'N, 35°79'E and 1800 m a.s.l.). The experimental area was in a hilly and gully region of the Loess Plateau, which was characterized by a semiarid, warm temperate, continental monsoon climate. The average annual rainfall was 435 mm, which fell mainly from June to September. The annual mean evaporation was 1050 mm and the annual temperature average was 8.1°C with a frost-free period of 155 days.

Rainfall during the experimental period was measured using an automatic weather station (WS-STD1, England) at the experimental site. Monthly precipitation distributions during the experimental period are shown in Fig. 1. The total precipitation for 2008, 2009, and 2010 was 390.9, 335.2, and 537 mm, while the precipitation during the maize-growing season was 362, 298.2, and 476.1 mm, respectively.

The soil at the experimental site was a loessal soil with a pH of 8.5. In the 0–40 cm soil layer, the organic matter, total N, P, and K were 8.32 g kg^{-1}, 0.61 g kg^{-1}, 0.58 g kg^{-1}, and 5.4 g kg^{-1}, respectively, while the available N, P, and K were 46.25 mg kg^{-1}, 10.41 mg kg^{-1}, and 104.82 mg kg^{-1}. In 2007, the site was planted with maize prior to the experiment

The experimental field was flat and, according to the FAO/ UNESCO Soil Classification [27], the soil was a Calcic Cambisol (sand 14%, silt 26%, and clay 60%) with low fertility. The key physical properties of the soil layers (0–40 cm depth) are shown in Table 1.

Experimental design and field management

The experiment used a randomized block design with three replicates. Each plot was 3 m wide and 6 m long. The experiment included four straw incorporation rate treatments: (i) no straw incorporation (CK); (ii) incorporation of maize straw at a low rate of 4 500 kg ha^{-1} (L); (iii) incorporation of maize straw at a medium rate of 9000 kg ha^{-1} (M); (iv) incorporation of maize straw at a high rate of 13 500 kg ha^{-1} (H).

The maize straws were mixed manually with the top 25 cm of soil in the field. Before mixing with the soil, the maize straws were chopped into 5 cm pieces and then applied to the soil six months before the crop was planted to facilitate decomposition of the straw. The straw was incorporated into the soil layer on 15 October 2007 and after the crop harvests during 2008–2010.

Ten days before sowing, a basis fertilizer containing 102 kg N ha^{-1} and 90 kg P ha^{-1}, was spread evenly over the each plot and plowed into soil layer. Maize (cv. Shendan 16) was sown at a rate of 5.25 seed m^{-2} on 18 April 2008, 15 April 2009 and 20 April 2010 using a holesowing (3 cm in diameter) machine. An additional 102 kg N ha^{-1} was applied as a top dressing in late June. And on 7 October 2008, 5 October 2008, and 10 October 2010. No irrigation was provided during the experimental years. Manual weeding was performed throughout the experiment.

Sampling and measurement

Rainfall data were recorded using a standard weather station located at the experimental site. After the maize harvest in 2008 and 2010, soil samples were collected for the four incorporation treatments. A soil sample was collected from each plot at depths of 0–20 cm, 20–40 cm and 40–60 cm to determine the soil organic matter. A similar soil sample was collected at depths of 0–10 cm, 10–20 cm, 20–30 cm and 30–40 cm to determine the aggregate stability. The soil samples were collected from four points in each plot replicate and mixed to produce a composite sample. Each soil samples was passed through an 8 mm sieve by gently breaking the soil clods, whereas pebbles and stable clods >8 mm were discarded. Soil samples were air-dried for 24 h in the laboratory before analysis.

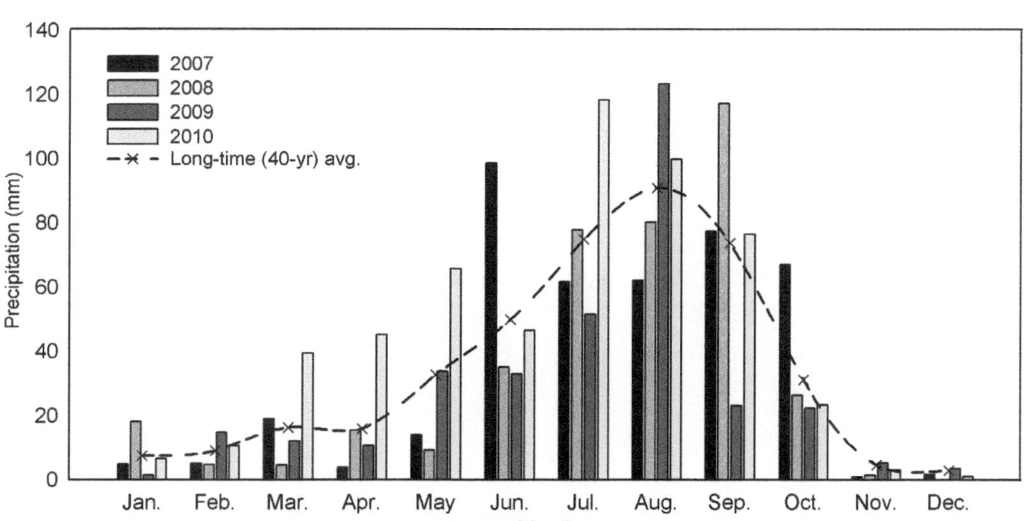

Figure 1. Distribution of mean monthly precipitation at the experimental site during 2007–2010.

Table 1. Physical properties of the tilth soil (0–40 cm depth) in the experimental site.

Depth (cm)	Bulk density (g cm^{-3})	Aggregate size (%)					
		>5 mm	5–2 mm	2–1 mm	1–0.5 mm	0.5–0.25 mm	<0.25 mm
0–10	1.33	0.1	0.27	2.25	4.5	4.47	88.41
10–20	1.33	0.15	0.19	1.38	4.2	4.15	89.93
20–30	1.36	0.1	0.28	1.41	4.02	3.36	90.83
30–40	1.38	0.1	0.22	1.07	3.29	3.8	91.52

Soil organic carbon was determined by the $K_2Cr_2O_7$–H_2SO_4 digestion method, and SOM content was calculated as a portion of SOC which has been described by Wang *et al.* [28].

$$OM_i = O_i \times A \quad (1)$$

$$C_i = \frac{d_i \times \rho_i \times O_i}{10} \quad (2)$$

where OM_i is the soil organic matter content (g kg^{-1}), $O_i$is the soil organic carbon content (g kg^{-1}), C_i is the soil organic carbon storage (Mg C ha^{-1}), $d_i$is the soil depth (cm), ρ_i is the soil bulk density (g cm^{-3}), A is Van Bemmelen coefficient ($A = 2$) [29].

The size distribution of water-stable aggregates was determined by placing a soil sample on a stack of sieves (5,2, 1, 0.5 and 0.25 mm) fitted with a soil aggregate analyzer (Japan, QD24–DIK–2001). The stacked sieves were immersed in water and moved up and down by 3.5 cm at a frequency of 30 cycles $60s^{-1}$ for 15 min. The proportions of aggregates that measured >5, 5–2, 2–0.5, 0.5–0.25 and <0.25 mm were calculated [30].

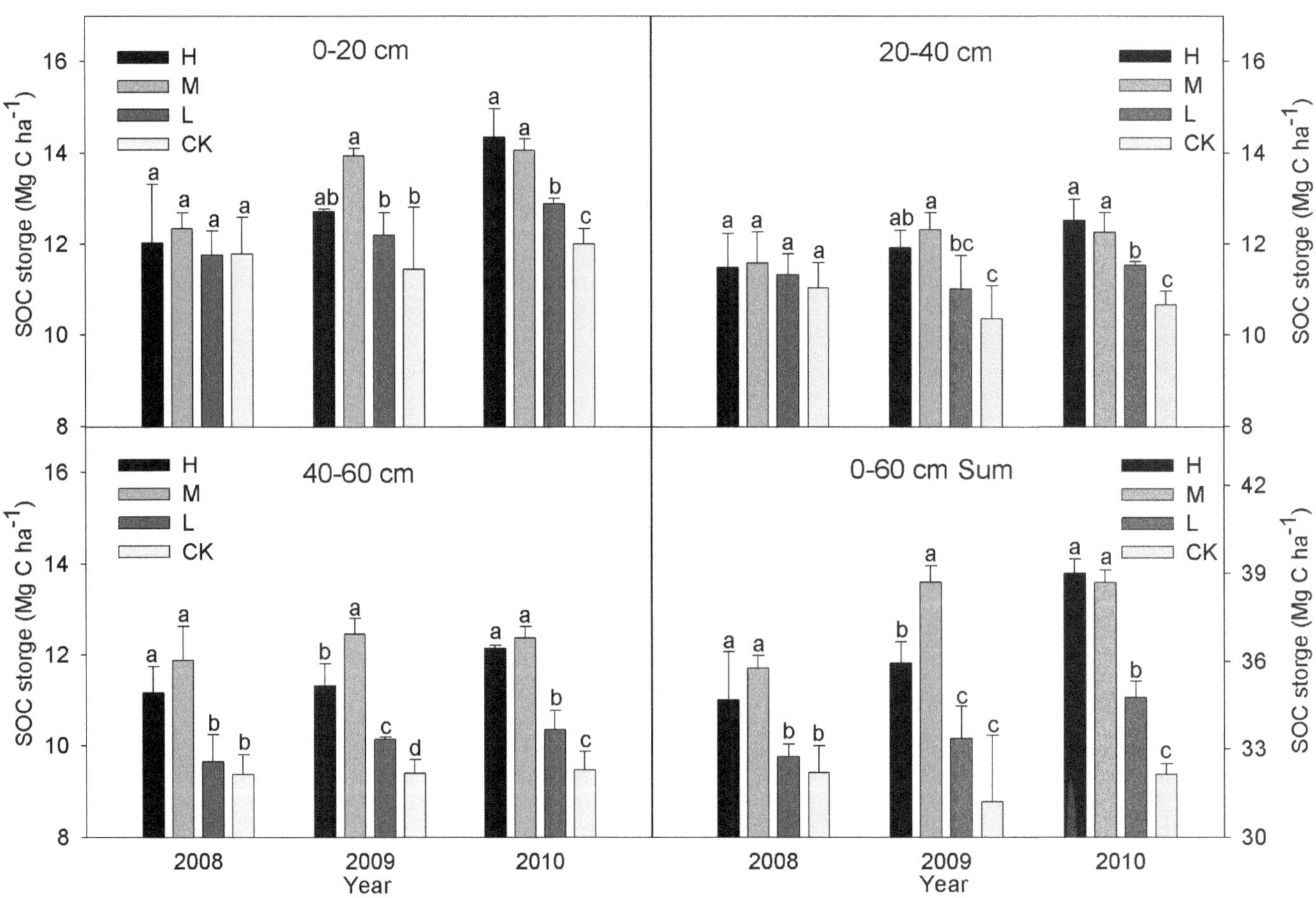

Figure 2. Soil organic carbon storage with different straw incorporation treatments (Mg C ha^{-1}). Note: CK, no straw incorporation; L, incorporation of straw at a low rate of 4 500 kg ha^{-1} maize straw; M, incorporation of straw at a medium rate of 9 000 kg ha^{-1} maize straw; H, incorporation of straw at a high rate of 13 500 kg ha^{-1} maize straw; Sum, the sum value of the 0–60 cm soil layers; Bars with different lower case letters indicate significant differences at $P<0.05$.

Table 2. $WR_{0.25}$ with different straw incorporation treatments (%).

Year	Treatment	Soil depth (cm)				0-40 AVG
		0—10	10—20	20—30	30—40	
2008	H	13.37a[1]±0.33[B2]	10.57a±1.28[3][A]	7.72ab±0.27[B]	7.33a±0.51[A]	9.75a±0.31[B]
	M	11.79b±1.05[A]	9.28a±0.60[B]	7.97a±0.65[A]	6.62ab±0.85[B]	8.91b±0.26[B]
	L	10.10c±0.46[A]	7.39b±0.58[A]	6.53bc±1.08[A]	6.07ab±0.81[A]	7.52c±0.16[A]
	CK	9.12c±0.70[A]	6.93b±0.78[A]	5.89c±0.57[A]	5.68b±0.63[A]	6.91d±0.44[A]
2009	H	14.64a±0.49[AB]	11.46a±0.17[A]	9.51a±0.26[AB]	7.85a±1.00[A]	10.86a±0.19[AB]
	M	14.93a±0.89[A]	11.22a±0.14[AB]	9.34a±0.32[A]	7.06ab±1.46[AB]	10.64a±0.27[AB]
	L	10.39b±0.26[A]	9.48b±1.49[A]	7.02b±1.18[A]	6.48ab±1.10[A]	8.34b±0.32[A]
	CK	9.75b±0.26[A]	7.18c±0.73[A]	6.40b±1.00[A]	5.65b±0.35[A]	7.25c±0.10[A]
2010	H	15.81a±1.20[A]	12.76a±0.44[A]	11.50a±0.20[A]	8.57a±0.05[A]	12.16a±0.59[A]
	M	16.72a±1.36[A]	12.81a±0.05[A]	11.30a±1.10[A]	8.81a±0.44[A]	12.41a±0.72[A]
	L	11.00b±0.78[A]	9.06b±0.81[A]	8.03b±2.14[A]	6.96b±0.81[A]	8.76b±0.66[A]
	CK	10.23b±0.46[A]	7.79b±0.23[A]	6.07b±0.46[A]	5.71c±0.23[A]	7.45c±0.25[A]

Note: CK, no straw incorporation; L, incorporation of straw at a low rate of 4 500 kg ha^{-1} maize straw; M, incorporation of straw at a medium rate of 9 000 kg ha^{-1} maize straw; H, incorporation of straw at a high rate of 13 500 kg ha^{-1} maize straw; AVG, the mean value of the 0–40 cm soil layers.
[1]Values followed by the same lowercase letter in the same line are not significantly different according to Duncan's multiple range test ($P<0.05$) between the four straw incorporation treatments in the same year.
[2]Values followed by the same uppercase letter in the same line are not significantly different according to Duncan's multiple range test ($P<0.05$) between the different years of the same straw incorporation treatment.
[3]Means ± standard deviations.

And the proportions of aggregates were used to calculate the water-stable macroaggregates content with a diameter of >0.25 mm [31], mean weight diameter (MWD) [32,33], geometric mean diameter (GMD) [34], and the soil aggregate stability (WSAR) [35]. These parameters were calculated as follows:

$$macroaggregates(>0.25mm) = \sum_{i=1}^{n} W_i \tag{3}$$

$$MWD = \sum_{i=1}^{n} \overline{X_i} W_i \tag{4}$$

$$GMD = \exp\left(\frac{\sum_{i=1}^{n} W_i \log \overline{X_i}}{\sum_{i=1}^{n} W_i}\right) \tag{5}$$

$$WSAR = WSA/A \times 100\% \tag{6}$$

where $macroaggregates(>0.25mm)$ is the volume of soil particles with a diameter of >0.25 mm, $W_i$is weight of the aggregates in that size range as a fraction of total dry weight of the sample analysed, and n is number of sieves, $\overline{X_i}$ is the mean diameter of aggregates over each sieve size, WSA is the mass of $macroaggregates(>0.25mm)$, and A is the mass of the soil aggregates with a diameter of >0.25 mm.

Fractal dimension D were then obtained to describe the characteristics of soil aggregate size distribution. As suggested by Tyler and Wheatcraft [36] and Zhang *et al.*, [37], the volume of soil with particle diameter$>d_i(d_i>d_{i+1}, i=1,2,\ldots)$ is defined as:

$$V(\delta > d_i) = A\left[1-(d_i/k)^{3-D}\right], \tag{7}$$

where δ is yard measure, k and A are the constants representing size and shape, respectively, and D is the fractal dimension. For a given soil, d_i represents the average value of soil particles between d_i and d_{i+1}.Generally, variations of particle density ρ among different soil particles could be ignored. And hence ρ is a constant. Therefore, another expression (Eq. (8)) is derived from Eq. (7):

$$W(\delta > d_i) = V(\delta > d_i)\rho = \rho A\left[1-(d_i/k)^{3-D}\right], \tag{8}$$

where $W(\delta > d_i)$ is the cumulative mass of particles with sizes $\delta > d_i$, and W_0 is the total mass of any sizes of soil particles. The fractal equation, reflecting the relationship between the mass distributions of soil particles and average particle diameter, can be obtained as follows:

$$\frac{W(\delta > d_i)}{W_0} = 1 - \left(\frac{\overline{d_i}}{d_{max}}\right)^{3-D} \quad or \quad \frac{W(\delta > d_i)}{W_0} = \left(\frac{\overline{d_i}}{d_{max}}\right)^{3-D} \tag{9}$$

Then after regression analysis between $\log(\overline{d_i}/d_{max})$ and $\log((W(\delta<\overline{d_i}))/W_0)$, the fractal dimension D can be calculated.

Statistical analysis

Statistical analysis was carried out by using the SPSS 13.0 (Statistical Package for the Social Sciences) package. The effects of treatments on the measured parameters were evaluated using a one-way ANOVA. Duncan's new multiple range test was used to calculate the least significant difference (LSD) between means

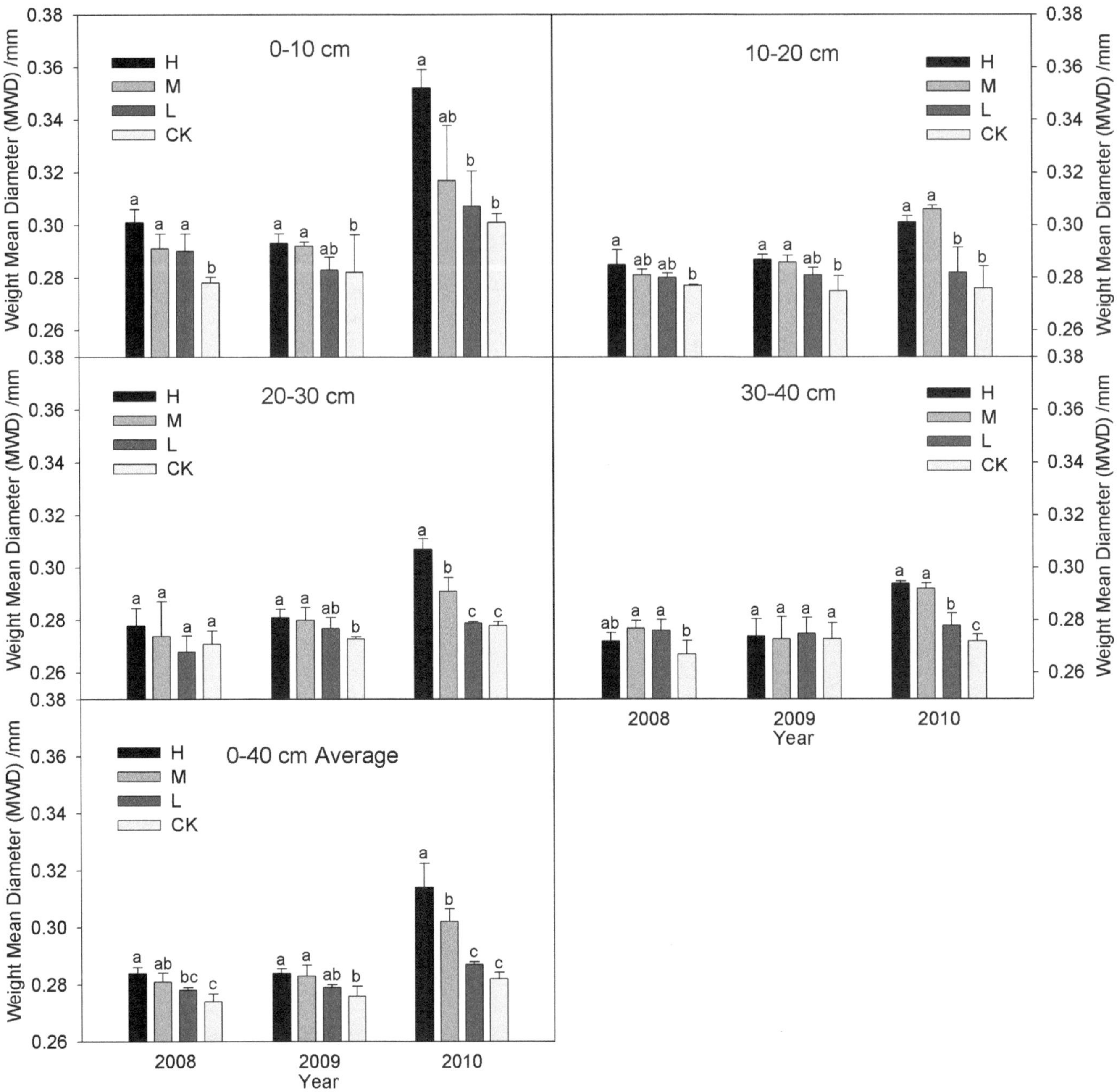

Figure 3. MWD values with wet sieving under the different straw incorporation treatments. Note: MWD, weight mean diameter; CK, no straw incorporation; L, incorporation of straw at a low rate of 4 500 kg ha^{-1} maize straw; M, incorporation of straw at a medium rate of 9 000 kg ha^{-1} maize straw; H, incorporation of straw at a high rate of 13 500 kg ha^{-1} maize straw; Average, the mean value of the 0–40 cm soil layers. Bars with different lower case letters indicate significant differences at $P<0.05$.

when F-values were significant. In all cases, differences were deemed to be significant if $P<0.05$.

Results

Soil organic carbon storage (SOC)

The effect of straw incorporation on SOC storage is shown in Fig. 2, where the soil organic carbon storage increased with the straw incorporation. The sum of SOC storage in 0–60 cm layers with the three incorporation treatments were higher than CK, i.e., 7.71% ($P<0.05$), 11.14% ($P<0.05$) and 1.70% in 2008, 15.15% ($P<0.05$), 24.00% ($P<0.05$) and 6.86% in 2009, and 21.40% ($P<0.05$), 20.38% ($P<0.05$) and 8.21% ($P<0.05$) in 2010, respectively. The SOC storage increased with the number of years of incorporation, i.e., the SOC storage (0–60 cm depth) in 2010 had increased by 6.19–12.48% compared with 2008, and decreased with the soil layer depth, i.e., by 3.75–25.68% in 2008, 11.85–21.70% in 2009 and 13.51–26.64% in 2010. The SOC storage of H and M was slightly higher than CK in 2008, although the difference was significant in 40–60 cm layer only. In 2009, compared with CK, H and M significantly increased the SOC storage by 11.01% and 21.74% ($P<0.05$) in 0–20 cm layer,

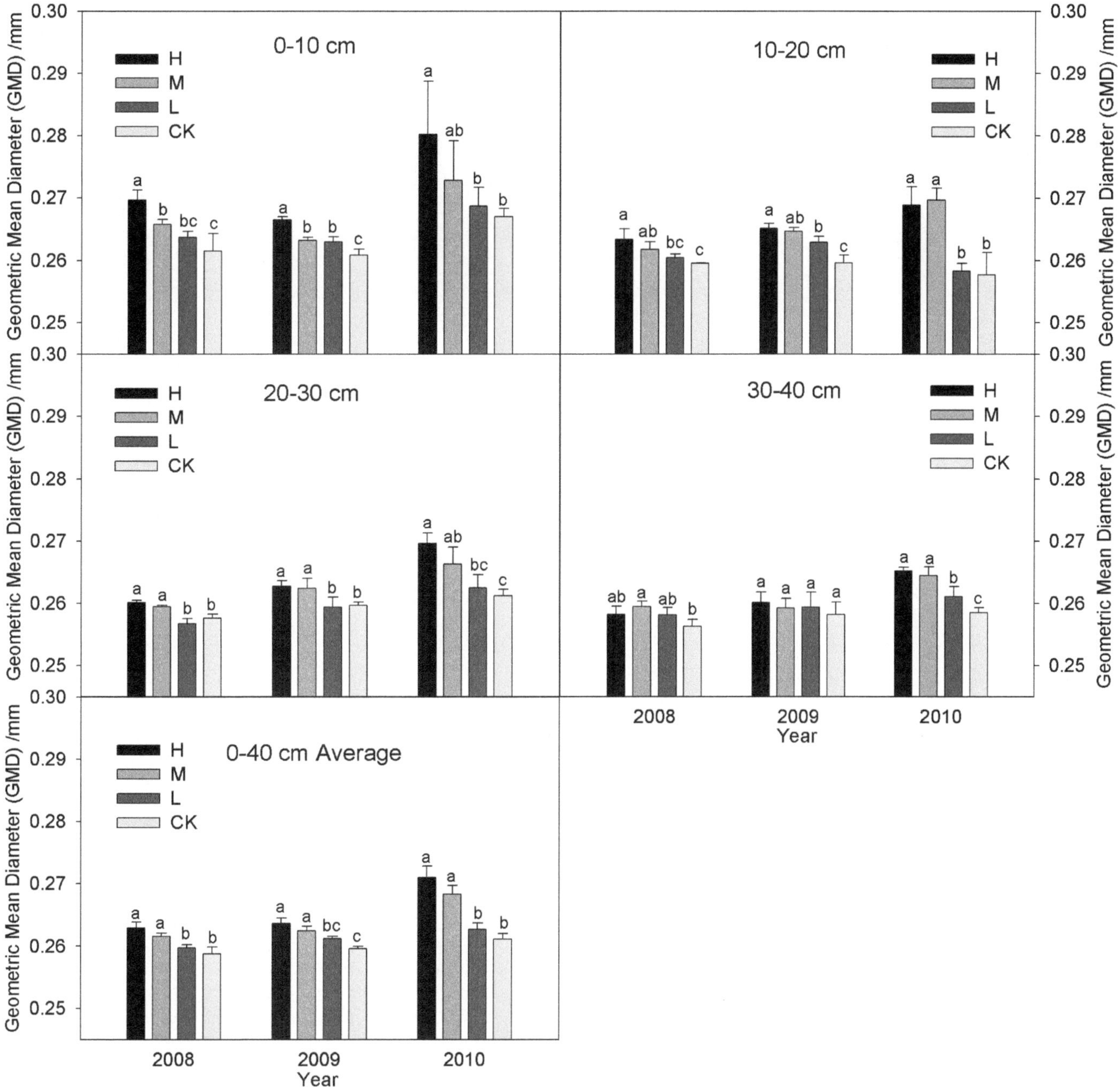

Figure 4. GMD values with wet sieving under the different straw incorporation treatments. Note: GMD, geometric mean diameter; CK, no straw incorporation; L, incorporation of straw at a low rate of 4 500 kg ha^{-1} maize straw; M, incorporation of straw at a medium rate of 9 000 kg ha^{-1} maize straw; H, incorporation of straw at a high rate of 13 500 kg ha^{-1} maize straw; Average, the mean value of the 0–40 cm soil layers. Bars with different lower case letters indicate significant differences at $P<0.05$.

14.94% and 18.81% ($P<0.05$) in 20–40 cm layer, and 20.40% and 32.47% ($P<0.05$) in 40–60 cm layer, respectively, and L was significantly increased by 7.89% ($P<0.05$) in 40–60 cm layer only. There was a significant difference between straw incorporation treatments and CK for each of the three soil layers in 2010, but there were no significant differences between H and M.

Water-stable macroaggregates (>0.25 mm)

The effects of straw incorporation on >0.25 mm macroaggregates are shown in Table 2. The mean >0.25 mm macroaggregates contents of the 0–40 cm layers of incorporation treatments were significantly (P<0.05) higher than CK during 2008–2010, i.e., 8.91–41.14%, 15.13–49.94%, and 17.64–66.58%, respectively. The >0.25 mm macroaggregates content also decreased with the soil layer depth, and increased with the number of years of incorporation,, i.e., the H, M, L and CK in 2010 was 24.72% ($P<0.05$), 39.28% ($P<0.05$), 16.49% and 7.81% higher than that in 2008. Compared with CK, the >0.25 mm macroaggregates content of three straw incorporation treatments in the 0–10 cm layer was increased by 10.75–46.60% in 2008, 6.56–53.13% in 2009, and 7.53–63.44% in 2010, respectively, while the differences between L and CK were not significant ($P>0.05$) during 2008–2010. The

Table 3. Effects on the soil aggregate stability rate with different straw incorporation treatments (%).

Year	Treatments	Soil Depth (cm)				0–40 AVG
		0–10	10–20	20–30	30–40	
2008	H	21.86a[1]±0.89^{A}[2]	16.29a±0.41[3]^{A}	10.83a±0.54^{B}	9.19a±0.65^{A}	14.60a±0.33^{A}
	M	19.62ab±1.09^{A}	15.31b±0.62^{A}	11.16a±0.82^{A}	8.47b±0.21^{A}	13.70b±0.62^{A}
	L	17.83bc±1.56^{A}	12.88c±0.24^{A}	9.70b±0.64^{A}	8.00bc±0.42^{A}	11.83c±0.18^{A}
	CK	16.45c±2.42^{A}	12.52c±0.58^{A}	9.39b±0.80^{A}	7.53c±0.37^{A}	11.79c±0.45^{A}
2009	H	20.32a±0.27^{A}	16.14a±0.54^{A}	13.15a±0.34^{AB}	9.74a±0.99^{A}	14.84a±0.23^{A}
	M	20.02a±0.46^{A}	15.41a±0.50^{A}	12.61a±0.41^{A}	8.66ab±1.28^{A}	14.18b±0.22^{A}
	L	15.76b±0.49^{A}	14.45a±1.78^{A}	10.16b±1.36^{A}	8.28ab±1.01^{A}	12.16c±0.35^{A}
	CK	15.47b±0.30^{A}	11.81b±0.83^{A}	9.59b±1.10^{A}	7.50b±0.41^{A}	11.09d±0.05^{A}
2010	H	20.19a±1.93^{A}	16.48a±1.73^{A}	15.24a±0.12^{A}	10.45a±0.11^{A}	15.59a±0.26^{A}
	M	20.87a±0.68^{A}	15.78a±1.32^{A}	13.99a±1.23^{A}	10.08a±0.43^{A}	15.18a±0.64^{A}
	L	15.72b±0.72^{A}	13.39a±1.81^{A}	12.53b±0.46^{A}	8.66b±0.77^{A}	12.58b±1.00^{A}
	CK	15.11b±0.12^{A}	12.79a±1.70^{A}	12.00b±0.76^{A}	7.52c±0.26^{A}	11.85b±1.32^{A}

Note: CK, no straw incorporation; L, incorporation of straw at a low rate of 4 500 kg ha^{-1} maize straw; M, incorporation of straw at a medium rate of 9 000 kg ha^{-1} maize straw; H, incorporation of straw at a high rate of 13 500 kg ha^{-1} maize straw; AVG, the mean value of the 0-40 cm soil layers.
[1]Values followed by the same lowercase letter in the same line are not significantly different according to Duncan's multiple range test ($P<0.05$) between the four straw incorporation treatments in the same year.
[2]Values followed by the same uppercase letter in the same line are not significantly different according to Duncan's multiple range test ($P<0.05$) between the different years of the same straw incorporation treatment.
[3]Means ± standard deviations.

>0.25 mm macroaggregates content showed the same trends in the 10–20 and 20–30 cm layers, and the H and M treatments had the significant difference ($P<0.05$) with CK. The H, M and L increased by 29.05% ($P<0.05$), 16.55% and 6.87% in 2008, 38.94% ($P<0.05$), 24.96% and 14.69% in 2009, and 50.09% ($P<0.05$), 54.29% ($P<0.05$), 21.89% ($P<0.05$) in 2010, respectively. A linear correlation was found between the >0.25 mm macroaggregates and soil organic carbon ($R^2>0.64$, $P<0.01$).

Mean weight diameter (*MWD*) and Geometric mean diameter (*GMD*)

Fig. 3 & 4 show that the *MWD* and *GMD* values with the three incorporation treatments increased significantly throughout the three-year study. The average *MWD* and *GMD* values under the incorporation treatments were higher in the 0–40 cm layers than CK during 2008–2010, i.e., 1.46–3.65% and 0.39–1.54% in 2008, 1.09–2.90% and 0.62–1.55% in 2009, 1.77–11.35% and 0.77–3.83% in 2010, respectively. There was no significant difference ($P>0.05$) between L and CK during the study period. The *MWD* and *GMD* values decreased with the soil layer depth and increased with the number of years of incorporation, i.e., the H, M, L and CK in 2010 was 10.56% ($P<0.05$) and 3.04% ($P<0.05$), 7.47% ($P<0.05$) and 2.29% ($P<0.05$), 3.24% and 1.15%, and 2.92% and 0.77% higher than that in 2008. The *MWD* and *GMD* values of treatments in the 0–10 and 10–20 cm layers were ranked in the order: H > M > L > CK. The H and M levels were significantly higher than CK while H and M levels were similar during 2008–2010 (Fig. 3–4). The *MWD* and *GMD* exhibited the same trends in 20–30 cm layers and there were no significant ($P>0.05$) differences among incorporation treatments during 2008–2009, the *MWD* values with the H were significantly ($P<0.05$) higher than the other three treatments during 2010, i.e., by 5.50%, 10.04%, and 10.43%, respectively. The *GMD* did not differ significantly among the four treatments, with the exceptions of H and L, and H and CK in 2010. The *MWD* and *GMD* values in the 30–40 cm layers under the incorporation treatments were increased by 1.87%, 3.75% ($P<0.05$), 3.37% ($P<0.05$) and 0.76%, 1.76% ($P<0.05$), 0.78% in 2008, 0.37%, 0.01%, 0.73% and 0.73%, 0.45%, 0.45% in 2009, 8.09% ($P<0.05$), 7.35% ($P<0.05$), 2.21% ($P<0.05$) and 2.32% ($P<0.05$), 1.93% ($P<0.05$), 0.77% ($P<0.05$) in 2010, compared with CK, respectively. There was no significant difference ($P>0.05$) between H and M during the study period.

Soil aggregate stability (*WSAR*)

The straw incorporation significantly affected the *WSAR* after three years (Table 3). The *WSAR* values in the 0–40 cm layers of incorporation treatments was significantly higher than CK during 2008–2010, i.e., 0.34–19.80%, 9.65–33.79%, and 6.16–35.08%, respectively. The *WSAR* also decreased with the soil layer depth, while it increased with the number of years of incorporation. The *WSAR* in the 0–10 cm layers increased significantly with the amount of straw and the *WSAR* was highest with H throughout 2008–2010, i.e., the *WSAR* significantly ($P<0.05$) increased by 28.35%, 28.93% and 25.78% compared with L, respectively, and by 33.62%, 31.35% and 32.89% ($P<0.05$) compared with CK, respectively. Compared with L and CK, the WSAR values with M increased significantly by 20.08–24.67% and 26.87–29.78% during 2008–2010, respectively, but there were no significant ($P>0.05$) differences between H and M, and L and CK. With the four treatments, the *WSAR* values in the 10–20 and 20–30 cm layers were lower than the 0–10 cm layers, which were ranked in the order: H > M > L > CK. There were no significant differences ($P>0.05$) among the four treatments in the 30–40 cm layers throughout the three years, and the *WSAR* maintained at 7.50–10.45%.

Fractal dimension (D)

Fig. 5 shows that the fractal dimensions with the three incorporation treatments decreased significantly ($P<0.05$) after three years of straw incorporation (the R^2 of fitting curve is:

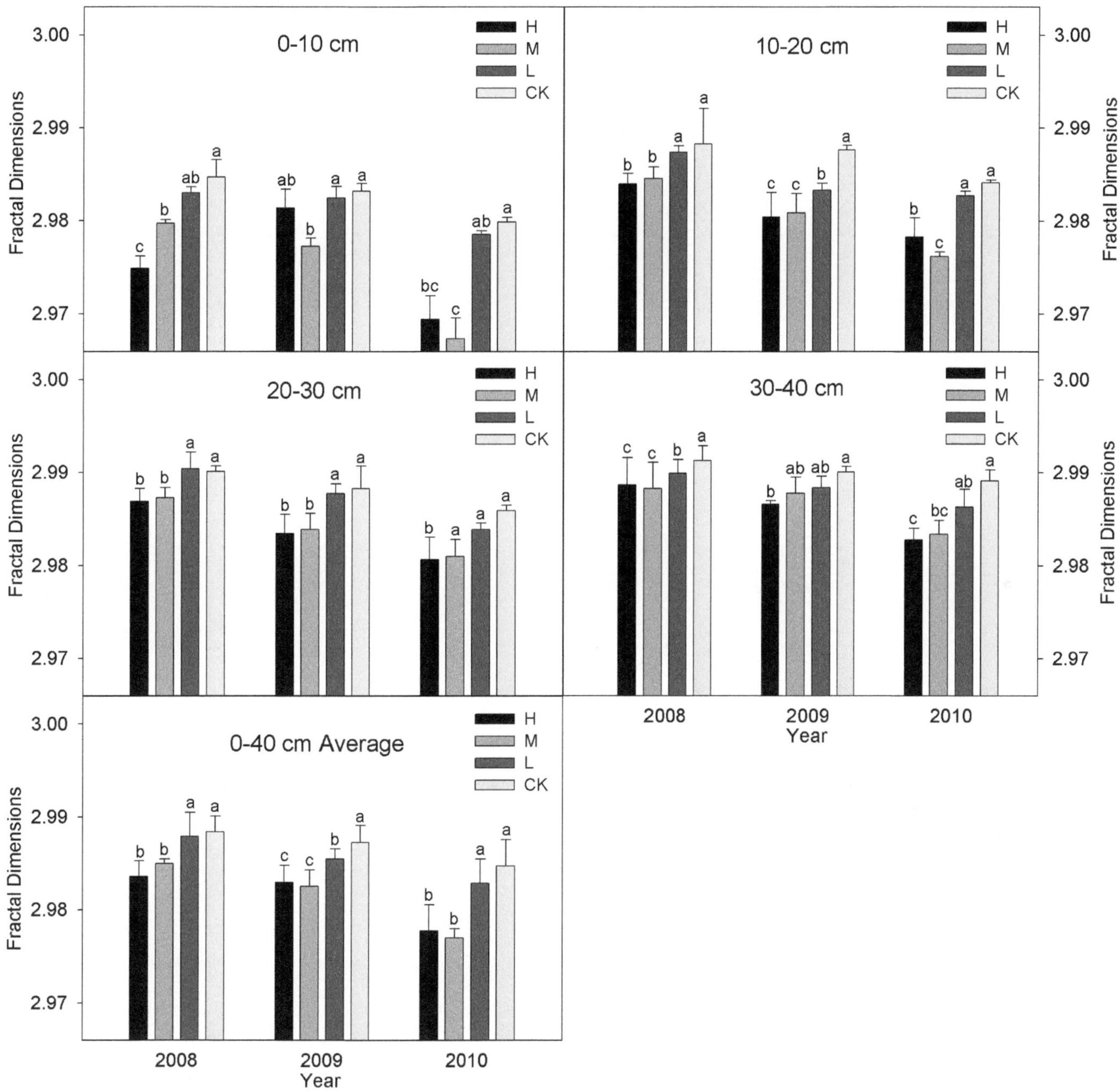

Figure 5. Fractal dimensions (D) of soil aggregates with different straw incorporation treatments. Note: CK, no straw incorporation; L, incorporation of straw at a low rate of 4 500 kg ha^{-1} maize straw; M, incorporation of straw at a medium rate of 9 000 kg ha^{-1} maize straw; H, incorporation of straw at a high rate of 13 500 kg ha^{-1} maize straw; Average, the mean value of the 0–60 cm soil layers; Bars with different lower case letters indicate significant differences at $P<0.05$; $0.92<R^2<0.99$.

0.92~0.99). The fractal dimensions (0–40 cm layer) with the incorporation treatments were lower than those of CK during 2008–2010, i.e., 0.16% ($P<0.05$), 0.11% ($P<0.05$) and 0.02% in 2008, 0.14% ($P<0.05$), 0.16% ($P<0.05$) and 0.06% ($P<0.05$) in 2009, and 0.24% ($P<0.05$), 0.26% ($P<0.05$) and 0.06% in 2010, respectively. The fractal dimensions also increased with the soil layer depth and decreased with the number years of incorporation, i.e., the H, M, L and CK in 2010 was 0.20% ($P<0.05$), 0.27% ($P<0.05$), 0.17% ($P<0.05$) and 0.12% higher than that in 2008. The fractal dimensions in the 0–10 cm layers with the four treatments were ranked in the order: H<M<L<CK in 2008. The fractal dimensions with H and M were significantly ($P<0.05$) lower than CK, i.e., 0.33%, 0.17% and 0.06%, respectively. However, the ranking was M<H<L<CK during 2009–2010, i.e., 0.06%, 0.20% ($P<0.05$), 0.02% in 2009, 0.35% ($P<0.05$), 0.42% ($P<0.05$) and 0.04% in 2010, respectively. While the differences between L and CK were not significant ($P>0.05$) during 2008–2010. Compared with L and CK, the fractal dimensions in the 10–20 cm layers under H was significantly ($P<0.05$) reduced the fractal dimensions by 0.11% and 0.14% in 2008, and 0.10% and 0.24% in 2009,, while M significantly ($P<0.05$) reduced by 0.09% and 0.12% in 2008, and 0.08% and 0.23% in 2009, respectively.

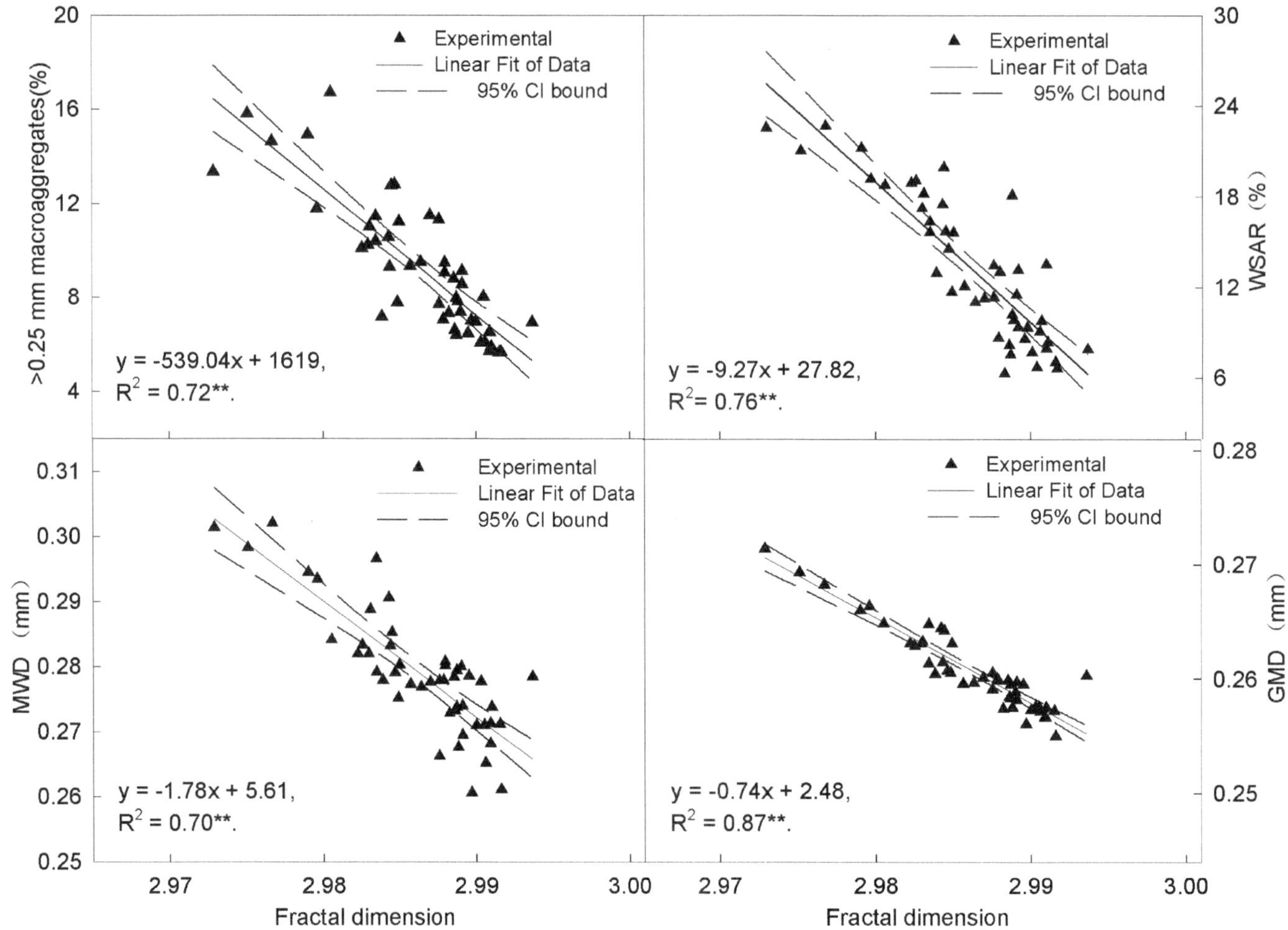

Figure 6. Correlations between D and >0.25 mm macroaggregates, MWD, GMD, WSAR of the soil aggregates. Note: MWD, weight mean diameter of soil aggregates; GMD, geometric mean diameter of soil aggregates; WSAR, soil aggregate stablility rate; D, Fractal dimension of soil aggregates. 95% CI, 95% confidence interval.** $P<0.01$.

The fractal dimension was lowest with M during 2010, i.e., it was decreased by 0.22% ($P<0.05$) compared with L, and by 0.27% ($P<0.05$) compared with CK. The differences among the four treatments decreased gradually in the 20–40 cm layer and the fractal dimension remained at 2.987~2.991.

Data fitting detected a linear correlation between the indexes of soil aggregates and the fractal dimension (Fig. 6).

Discussion

Results of this study demonstrated that application of crop straws had positive effects on the soil physico-chemical properties. On the other hand, there are Loess plateau regions at northwest China with intensive cultivation systems and poor soil management strategies. Therefore, crop residual management is very important for preserving natural ecosystems [38]. Another problem for this area is conventional tillage that reduces the soil water storage and destroys the aggregates and soil structure [39] which prompts deterioration of crop yield.

The results indicated that the SOC storage increased significantly in the 0–60 cm layers after three years of straw incorporation (Fig. 2). Rasmussen and Collins [40] reported that the soil organic matter content was strongly related to the amount of residues added and only weakly related to the type of residue applied. The H and M treatments were significantly different from the L throughout the three years and the soil organic matter level increased as the straw incorporated and decomposed [20], which effectively mitigated the loss of soil organic carbon from in the agroecosystem caused by intensive cropping [41]. The SOC storage decreased gradually in all the treatments with the soil layer depth because the degree of straw incorporation was lower in the deeper layers compared with the surface layers (0–20 cm) of the soil [42]. This was because the amount of straw incorporated in the topsoil was greater than that in the deeper layers [40,43].

Soil aggregates are the basic units of the soil structure and they are composed of primary particles and binding agents [12]. They are also necessary soil conditions for high crop yields [44]. Conventional tillage disturbs the soil and increase the effects of drying–rewetting and freezing–thawing, which increases the susceptibility of the macroaggregate (>0.25 mm) to disruption [45–47]. Pinheiro *et al.* [48] showed that soil exposure with tillage and the lack of residue inputs led to a decline in aggregation and organic carbon, both of which made the soil susceptible to erosion. Our study showed that the straw incorporation of straws determined significantly more and larger soil aggregates than CK, thus indicating an improvement of soil physical quality. This may have been attributable to the significant increase in the SOC storage (an average increase by 8.21–21.40% in 2010), the lower soil bulk density (an average decreased by 1.80–4.13% in 2010, data no shown) [49], and increased soil porosity (an average increased by 1.70–3.90% in 2010, data no shown) [50] after straw incorporation. It also stimulated the activity of soil microorganisms [51] and an abundance of polyose metabolites were produced during the straw decay process [52]. These soil physical and chemical conditions may have accelerated the incorporated SOM decomposing process and increased soil aggregation [24].

Soil aggregate structural stability is widely recognized as a key indicator of soil quality, which is closely related to a number of soil properties, processes, and functions, e.g., the quantity and composition of SOM [48], infiltration capacity [53], soil biotic activity [54] and the resistance to erosion [55,56]. Wei *et al.* [57] showed that the addition of crop residues was the most effective measure for increasing the rhizosphere aggregate stability. Sonnleitner *et al.* [24] and Karami *et al.* [41] also found that straw application improved the aggregate stability and other soil properties. In our study, the soil aggregate stability of the straw incorporation treatments were significantly higher than CK in 2010, and it was decreased with the soil layer depth. These results agreed with studies by Tripathy and Singh [42] and Karami *et al.* [41]. Our results also indicated that straw incorporation was positively related to the physical protection of organic matter [20] and an increased aggregate quantity [58], but it also improved the soil aggregate stability [20] and reduced soil degeneration [22,23].

Many studies have shown that soil is a porous medium with fractal characteristics [59–60]. Thus, fractal theory can be used to describe the complex characteristic of soil structure [61]. Castrignanò and Stelluti [62] reported that a higher fractal dimension indicated the heavier texture of a soil and its inferior permeation properties. This showed that fractal theory is an effective method to describing the soil aggregate distribution [37] and changed with different levels of straw incorporation [31]. The fractal dimensions of the 0–40 cm layers with the four treatments after three years were ranked in the order: H<M<L<CK and the three straw incorporation treatments were significantly different from CK. These results agreed with Zhang *et al.* [37] and Zhang *et al.* [31]. The low values of D indicated a size distribution dominated by a large number of macroaggregates (>0.25 mm) [59,63]. This improvement in the fractal dimension may have been accelerated by the incorporation of straw, which improved the soil structure, increased the SOM content and microbial activity [64], and significantly increased the mount and size of soil aggregates [37]. Our results indicated there were significant improvements in the soil macroaggregates and the aggregate structure after straw incorporation [20].

Conclusion

The incorporation of different amount of straw significant increased the *SOC* storage, >0.25 mm macroaggregates, *MWD* and *GMD* in a semiarid soil. The *SOC* storage, >0.25 mm macroaggregates, *MWD* and *GMD* also increased with higher straw incorporation rates. The fractal dimension decreased with increasing straw incorporation rates. Therefore, the incorporation of straw into the soil in semiarid areas is an effective practice for improving the soil aggregate content and stability.

Acknowledgments

We are grateful to Mr Li Yongping and Liu Shixin for managing the field experiments and professional English editor Jackson who is from UK, and kind help with the language of this manuscript.

Author Contributions

Conceived and designed the experiments: ZJ QH. Performed the experiments: PZ TW. Analyzed the data: PZ. Contributed reagents/materials/analysis tools: XR YL. Wrote the paper: PZ.

References

1. Hu KL, Li BG, Lin QM, Li GT, Chen DL (1999) Spatial variabi lity of soil nutrient in wheat field. Transactions of the Chinese Society of Agricultural Engineering 15:33–38. (in Chinese with English abstract)
2. Mupangwa W, Twomlow S, Walker S (2008) The influence of conservation tillage methods on soil water regimes in semi-arid southern Zimbabwe. Phys Chem Earth, Parts A/B/C 33: 762–767.
3. Biederbeck V, Campbell C, Bowren K, Schnitzer M, McIver R (1980) Effect of burning cereal straw on soil properties and grain yields in Saskatchewan. Soil Sci Soc Am J 44: 103–111.
4. Wuest SB, Caesar-TonThat T, Wright SF, Williams JD (2005) Organic matter addition, N, and residue burning effects on infiltration, biological, and physical properties of an intensively tilled silt-loam soil. Soil Tillage Res 84: 154–167.
5. Duan F, Liu X, Yu T, Cachier H (2004) Identification and estimate of biomass burning contribution to the urban aerosol organic carbon concentrations in Beijing. Atmos Environ 38: 1275–1282.
6. Mandal KG, Misra AK, Hati KM, Bandyopadhyay KK, Ghosh PK, et al. (2004) Rice residue-management options and effects on soil properties and crop productivity. J Food Agric Environ 2: 224–231.
7. Carter MR (2002) Soil quality for sustainable land management. Agron J 94: 38–47.
8. Hillel D (2004) Encyclopedia of Soils in the Environment: London, Elsevier, v. 4, p. 295–303.
9. Six J, Bossuyt H, Degryze S, Denef K (2004) A history of research on the link between (micro) aggregates, soil biota, and soil organic matter dynamics. Soil Tillage Res 79: 7–31.
10. Tisdall J, Oades J (1982) Organic matter and water-stable aggregates in soils. J Soil Sci 33:141–163. doi:10.1111/j.1365-2389.1982.tb01755.x
11. Haynes R, Beare M (1997) Influence of six crop species on aggregate stability and some labile organic matter fractions. Soil Biol Biochem 29: 1647–1653.
12. Scanlon BR, Andraski BJ, Bilskie J (2002) Methods of Soil Analysis: Part 4 Physical Methods: Soil Science Society of America. 643–668 p.
13. Hernández-Hernández R, López-Hernández D (2002) Microbial biomass, mineral nitrogen and carbon content in savanna soil aggregates under conventional and no-tillage. Soil Biol Biochem 34: 1563–1570.
14. Villar M, Petrikova V, Dıaz-Ravina M, Carballas T (2004) Changes in soil microbial biomass and aggregate stability following burning and soil rehabilitation. Geoderma 122: 73–82.
15. Ashagrie Y, Zech W, Guggenberger G (2005) Transformation of a *Podocarpus falcatu* dominated natural forest into a monoculture *Eucalyptus globulus* plantation at Munesa, Ethiopia: soil organic C, N and S dynamics in primary particle and aggregate-size fractions. Agric Ecosyst Environ 106: 89–98.
16. Jastrow J (1996) Soil aggregate formation and the accrual of particulate and mineral-associated organic matter. Soil Biol Biochem 28: 665–676.
17. Chevallier T, Blanchart E, Albrecht A, Feller C (2004) The physical protection of soil organic carbon in aggregates: a mechanism of carbon storage in a Vertisol under pasture and market gardening (Martinique, West Indies). Agric Ecosyst Environ 103: 375–387.
18. Duiker S, Lal R (1999) Crop residue and tillage effects on carbon sequestration in a Luvisol in central Ohio. Soil Tillage Res 52: 73–81.
19. Saroa G, Lal R (2003) Soil restorative effects of mulching on aggregation and carbon sequestration in a Miamian soil in central Ohio. Land Degrad Dev 14: 481–493.
20. Tan DS, Jin JY, H SW, Li ST, He P (2007) Effect of long-term application of K fertilizer and wheat straw to soil on crop yield and soil K under different planting systems. Agric Sci in China 6: 200–207.
21. Li Q, Chen X, Zhang F, Römheld V (2002) Study on balance of phosphorus and potassium in winter wheat and summer maize rotation system. Plant Nutr Fert Sci 8: 152–156. (in Chinese with English abstract)
22. Nelson RG (2002) Resource assessment and removal analysis for corn stover and wheat straw in the Eastern and Midwestern United States—rainfall and wind-induced soil erosion methodology. Biomass Bioenergy 22: 349–363.
23. Wilhelm W, Johnson JMF, Karlen DL, Lightle DT (2007) Corn stover to sustain soil organic carbon further constrains biomass supply. Agron J 99: 1665–1667.
24. Sonnleitner R, Lorbeer E, Schinner F (2003) Effects of straw, vegetable oil and whey on physical and microbiological properties of a chernozem. App Soil Ecol 22: 195–204.
25. Mulumba LN, Lal R (2008) Mulching effects on selected soil physical properties. Soil Tillage Res 98: 106–111.
26. Bhagat R, Verma T (1991) Impact of rice straw management on soil physical properties and wheat yield. Soil Sci 152: 108–115.

27. FAO/UNESCO (1993) 'World soil resources. An explanatory note on the FAO World Soil Resource Map at 1 : 25,000,000 scales.' (FAO: Rome)
28. Wang S, Tian H, Liu J, Pan S (2003) Pattern and change of soil organic carbon storage in China: 1960s–1980s. Tellus B 55: 416–427.
29. Pribyl DW (2010) A critical review of the conventional SOC to SOM conversion factor. Geoderma 156: 75–83.
30. Oades J, Waters A (1991) Aggregate hierarchy in soils. Soil Res 29: 815–828.
31. Zhang P, Jia ZK, Wang W, Lu WT, Gao F, et al. (2012) Effects of straw returning on characteristics of soil aggregates in semi-arid areas in southern Ningxia of China. Sci Agric Sin 45:1513-1520. (in Chinese with English abstract)
32. Van Bavel C (1949) Mean weight diameter of soil aggregates as a statistical index of aggregation. Soil Sci Soc Am Proc. pp. 20–23.
33. Youker R, McGuinness J (1957) A short method of obtaining mean weight-diameter values of aggregate analyses of soils. Soil Sci 83: 291–294.
34. Mazurak AP (1950) Effect of gaseous phase on water-stable synthetic aggregates. Soil Sci 69: 135–148.
35. Hou XQ, Li R, Jia ZK, Han QF, Wang W, et al. (2012) Effects of rotational tillage practices on soil properties, winter wheat yields and water-use efficiency in semi-arid areas of north-west China. Field Crops Res 129: 7–13.
36. Tyler SW, Wheatcraft SW (1992) Fractal scaling of soil particle-size distributions: analysis and limitations. Soil Sci Soc Am J 56: 362–369.
37. Zhang Z, Wei C, Xie D, Gao M, Zeng X (2008) Effects of land use patterns on soil aggregate stability in Sichuan Basin, China. Particuology 6: 157–166.
38. Zhu H, Wu J, Huang D, Zhu Q, Liu S, et al. (2010) Improving fertility and productivity of a highly-weathered upland soil in subtropical China by incorporating rice straw. Plant Soil 331: 427–437.
39. Mele PM, Crowley DE (2008) Application of self-organizing maps for assessing soil biological quality. Agric Ecosyst Environ 126: 139–152.
40. Rasmussen PE, Collins HP (1991) Long-term impacts of tillage, fertilizer, and crop residue on soil organic matter in temperate semiarid regions. Adv Agron 45: 93–134.
41. Karami A, Homaee M, Afzalinia S, Ruhipour H, Basirat S (2012) Organic resource management: Impacts on soil aggregate stability and other soil physico-chemical properties. Agric Ecosyst Environ 148: 22–28.
42. Tripathy R, Singh AK (2004) Effect of water and nitrogen management on aggregate size and carbon enrichment of soil in rice-wheat cropping system. J Plant Nutr Soil Sci 167: 216–228.
43. Prasad R, Power J (1991) Crop residue management. Adv Soil Sci 15: 205–251.
44. Limon-Ortega A, Govaerts B, Sayre KD (2009) Crop rotation, wheat straw management, and chicken manure effects on soil quality. Agron J 101: 600–606.
45. Beare M, Hendrix P, Cabrera M, Coleman D (1994) Aggregate-protected and unprotected organic matter pools in conventional-and no-tillage soils. Soil Sci Soc Am J 58: 787–795.
46. Paustian K, Levine E, Post WM, Ryzhova IM (1997) The use of models to integrate information and understanding of soil C at the regional scale. Geoderma 79: 227–260.
47. Mikha MM, Rice CW (2004) Tillage and manure effects on soil and aggregate-associated carbon and nitrogen. Soil Sci Soc Am J 68: 809–816.
48. Pinheiro E, Pereira M, Anjos L (2004) Aggregate distribution and soil organic matter under different tillage systems for vegetable crops in a Red Latosol from Brazil. Soil Tillage Res 77: 79–84.
49. Lal R (2000) Physical management of soils of the tropics: priorities for the 21st century. Soil Sci 165: 191–207.
50. Pagliai M, Vignozzi N, Pellegrini S (2004) Soil structure and the effect of management practices. Soil Tillage Res 79: 131–143.
51. Kasteel R, Garnier P, Vachier P, Coquet Y (2007) Dye tracer infiltration in the plough layer after straw incorporation. Geoderma 137: 360–369.
52. Pascual JA, García C, Hernandez T (1999) Comparison of fresh and composted organic waste in their efficacy for the improvement of arid soil quality. Bioresour Technol 68: 255–264.
53. Abu-Sharar T, Salameh A (1995) Reductions in hydraulic conductivity and infiltration rate in relation to aggregate stability and irrigation water turbidity. Agric Water Manage 29: 53–62.
54. Roldán A, Salinas-García J, Alguacil M, Caravaca F (2005) Changes in soil enzyme activity, fertility, aggregation and C sequestration mediated by conservation tillage practices and water regime in a maize field. Appl Soil Ecol 30: 11–20.
55. Barthes B, Roose E (2002) Aggregate stability as an indicator of soil susceptibility to runoff and erosion; validation at several levels. Catena 47: 133–149.
56. Ramos M, Nacci S, Pla I (2003) Effect of raindrop impact and its relationship with aggregate stability to different disaggregation forces. Catena 53: 365–376.
57. Wei CF, Gao M, Shao JG, Xie DT, Pan GX (2006) Soil aggregate and its response to land management practices. China Particuology 4: 211–219.
58. Tarafdar JC, Meena SC, Kathju S (2001) Influence of straw size on activity and biomass of soil microorganisms during decomposition. Eur J Soil Biol 37: 157–160.
59. Perfect E, Kay B (1991) Fractal theory applied to soil aggregation. Soil Sci Soc Am J 55: 1552–1558.
60. Rieu M, Sposito G (1991) Fractal fragmentation, soil porosity, and soil water properties: II. Applications. Soil Sci Soc Am J 55: 1239–1244.
61. Bird N, Bartoli F, Dexter A (2005) Water retention models for fractal soil structures. Eur J Soil Sci 47: 1–6.
62. Castrignanò A, Stelluti M (1999) Fractal geometry and geostatistics for describing the field variability of soil aggregation. J Agric Eng Res 73: 13–18.
63. Ding Q, Ding W (2007) Comparing stress wavelets with fragment fractals for soil structure quantification. Soil Tillage Res 93: 316–323.
64. Liu JF, Hong W, Wu CZ (2002) Fractal features of soil clusters under some precious hardwood stands in the central subtropical region, China. Acta Ecol. Sin. 22, 197. (in Chinese with English abstract).

14

Evaluation of Chloropicrin as a Soil Fumigant against *Ralstonia solanacarum* in Ginger (*Zingiber officinale* Rosc.) Production in China

Liangang Mao[1]ⓢ, Qiuxia Wang[1,2]ⓢ, Dongdong Yan[1,2]ⓢ, Taotao Ma[1], Pengfei Liu[1], Jin Shen[1], Yuan Li[1,2], Canbin Ouyang[1,2], Meixia Guo[1,2], Aocheng Cao[1,2]*

1 Department of Pesticides, Institute of Plant Protection, Chinese Academy of Agricultural Sciences, Ministry of Agriculture, Beijing, People's Republic of China, **2** State Key Laboratory for Biology of Plant Disease and Insect Pests, Beijing, People's Republic of China

Abstract

Background: Chloropicrin (Pic) offers a potential alternative to methyl bromide (MB) against *Ralstonia solanacarum* in ginger (*Zingiber officinale* Rosc.) production. MB is scheduled to be withdrawn from routine use by 2015 in developing countries.

Methods: Pic treatments were evaluated in a laboratory study and in three commercial ginger fields.

Results: Laboratory studies showed that the EC_{50} value and EC_{80} value of Pic were 2.7 and 3.7 mg a.i. kg^{-1} soil, respectively. Field trials in highly infested soil revealed that treatments of Pic at the dose of 50 g m^{-2} covered with totally impermeable film (TIF) or polyethylene film (PE) sharply reduced *Ralstonia solanacarum* and maintained high ginger yields. Both of the Pic treatments provided results similar to, or in some cases slightly lower than, MB with respect to *Ralstonia solanacarum* control, plant survival, plant growth and yield. All of the fumigant treatments were significantly better than the non-treated control.

Conclusions: The present study confirms that the Pic is a promising alternative with good efficacy against *Ralstonia solanacarum* for ginger production and could be used in integrated pest management programs in China.

Editor: A. Mark Ibekwe, U. S. Salinity Lab, United States of America

Funding: This research was supported by the Program on Substituted Technology for Methyl Bromide in China (Special Finance of Chinese Ministry of Agriculture, 2110402) and UNDIO Project on Phasing-out of Methyl Bromide in Agriculture Sector in China (TF/CPR-A/08/003). The funders had no role in study design, data collection and analysis, decision to publish, or preparation of the manuscript.

Competing Interests: The authors have declared that no competing interests exist.

* E-mail: caoac@vip.sina.com

ⓢ These authors contributed equally to this work.

Introduction

In 2011 the total ginger production in China, the second largest ginger (*Zingiber officinale* Rosc.) producing country in the world, was 426,032.00 tonnes (t) and the area harvested was 36,007.00 hectare (ha) [1]. Due to successive monocropping of ginger, *Ralstonia solanacarum*, which was originally described by Smith (1896) as the causative agent of bacterial wilt of solanaceous plants [2], has the ability to reduce yields greatly in the cultivation of ginger in China. At present, methyl bromide (MB) is widely used in ginger cultivation in China as a preplant soil fumigant against *R. solanacarum*. MB, however, has to be phased out by 1 January 2015 in developing countries owing to its detrimental effects on the stratospheric ozone layer [3]. The withdrawal of MB from use as an agricultural fumigant has prompted a large amount of research aimed at finding effective and economically acceptable alternatives [4].

The pesticides currently applied against *R. solanacarum* in ginger in China include fumigants and other chemical or biological pesticides, such as chloropicrin (Pic), cupric hydroxide, and *Bacillus cereus*. Pic is becoming one of the most important fumigants for ginger production in China. There is much information in the literature on the ability of Pic to control various soilborne pests effectively in many crops, for example tomatoes [5–9], strawberries [10], peppers [11–13] and others. However, little information has been reported on Pic against *R. solanacarum* in ginger production.

The present work was initiated to test the effects of Pic on *R. solanacarum* in laboratory studies. In addition, three field trials were carried out to evaluate two different Pic treatments as a potential alternative to MB against *R. solanacarum* for ginger production in China.

Materials and Methods

Laboratory studies

The dose-responses of *R. solanacarum* to different Pic concentrations were studied in the laboratory. With the authorization of the institute of Fangshan agricultural science, Beijing, soil samples in the laboratory studies were collected from the top 20 cm of greenhouse soil in Nanhe village, Dashiwo town, Fangshan district, southwest of Beijing. The field studies site does not involve endangered or protected species and the GPS coordinates are 39°33′9.3″N, 115°49′34.1″E. The soil was composed of 61.48% sand, 37.20% silt, and 1.32% clay, with organic matter content 25.38 g kg^{-1} soil and pH 6.49. The soil was sieved through a 2-mm mesh, and then mixed together. The soil moisture was 12.37% (w/w). Particle size analyses were performed using the pipette method [14]. The organic carbon content was determined by wet oxidation using the method of Walkley and Black [15]. The pH was measured in a 1:2.5 soil to H_2O extract. Soil moisture content was measured in the drying oven at 105±5°C until mass constancy was achieved [16]. The soil was sterilized at 121°C for 30 mins using the high-pressure steam sterilization pot. *R. solanacarum*, was grown on triphenyltetrazolium chloride (TTC) selective media and then quantitatively incorporated into the sterilized soil.

300 g *R. solanacarum*-infested soil was placed into each of 0.5 L glass jars [17–19]. The following treatments were applied with three replicates: Pic alone (1.0, 2.0, 4.0, 8.0, 16.0, 32.0, 64.0, 128.0, 256.0 mg a.i. kg^{-1} soil) and non-treated control. Pic was injected into the soil by pipette (Eppendorf, Germany) and then the jars were immediately sealed with covers. The jars were placed in incubators at 28°C for 7 days. The jars were opened to release the residual fumigant for a day and *R. solanacarum* was isolated from the soil quantitatively based on the method described by Kelman [20].

Field trials

In 2012 and 2013, with the authorization of Laiwu agriculture bureauthe, Shandong Province, three field experiments were conducted in three ginger fields located in Xiaoxia village, Zhaili town, Laiwu city, Shandong province, China (trial I: 36°17′31.9″N, 117°28′19.6″E), Hanwangxu village, Zhaili town, Laiwu city, Shandong province, China (trial II: 36°17′52.1″N, 117°26′07.4″E), and Caodaxia village, Zhaili town, Laiwu city, Shandong province, China (trial III: 36°16′30.8″N, 117°30′05.1″E), respectively. The field studies site does not involve endangered or protected species. And there are no specific permissions required for the field sites. All field sites are located in intensive ginger production areas of Laiwu, a region which has a long history of ginger production. However, *R. solanacarum* has become a big problem due to the continuous cropping. The major soil-borne disease is plant death caused by *R. solanacarum*. Details relevant to these trials are given in **Table 1** and **2**.

The following fumigant products were used in the study: (a) Pic 99.5 LD (Dalian Dyestuffs & Chemicals Co., China), a commercial liquid product containing 99.5% Pic; (b) MB 98 TC (Lianyungang Dead Sea Bromine Co. Ltd, Jiangsu Province, China) containing 98% MB and 2% chloropicrin. The soil mulches were 0.04 mm polyethylene film (PE) (Baoding Baoshuo Plastic Co., Ltd., Hebei Province, China) and 0.04 mm totally impermeable film (TIF) (Jiahe Industrial Co., Ltd., Jiangsu Province, China).

Each Pic treated plot area was designed to 50 m^2. Each MB treated plot area and untreated control plot area are both designed to 25 m^2. The treatments were designed as randomized blocks with three replicates (**Table 3**). The tested treatments were MB, used as a reference treatment, Pic, and untreated control. Pic liquid was injected into the soil at 15 cm depth via a manual injection machine at a rate of 50 g m^{-2} (**Figure 1**). The Pic treatments were applied in two different seasons: one was applied in November 2012 (soil temperature at 5 cm depth was about 15 to 21°C during the day) and covered with TIF film; another was applied in March 2013 (soil temperature at 5 cm depth was about 13 to 19°C during the day) and covered with PE film (**Figure 1**). MB was applied by using evaporating iron jars under a PE sheet and allowing it to vaporize in situ (cold method) at a rate of 40 g m^{-2} in March 2013, and a small arc tunnel (40 cm height) was constructed under the mulch with the aim of improving the fumigant distribution [21]. The ginger plants were 40 cm apart, planting distance 25 cm, about 6500 plants per mu. Traditional cultivation techniques were used in all plots.

Soil *R. solanacarum* populations [colony-forming units (cfu) g^{-1} soil] were determined after fumigation from soil at the depth of 0–20 cm. Soil from each plot was sampled from 3 spots along the diagonal lines in a plot. *R. solanacarum* was determined using the same methods as the **Laboratory studies**.

Ginger germination was evaluated at 9 WAT (weeks after transplant). Plant growth (ginger plant height, tillers/plant and mortality) were evaluated at 15 WAT (weeks after transplant) (20 plants per plot). The yield, plant height and root disease severity in ginger were evaluated at the end of the trials. Twenty ginger plants were picked from each plot, and the severity of ginger root disease was assessed separately, based on a disease severity scale of 0–4, where 0 = healthy plant and root, without disease; 1 = black brown rotten roots comprising <25% of the entire root system; 2 = 26–50%; 3 = 51–75%; and 4 = 76–100% black brown rotten roots [22].

Statistical analyses

Laboratory studies. The bacterial control efficacy of treatments can be calculated according to the following equation.

$$Y = \frac{X_1 - X_2}{X_1} \times 100, \quad (1)$$

where Y is the control efficacy on bacteria, X_1 is the bacteria population of the untreated control, X_2 is the bacteria population of fumigant treatments.

The experimental design consisted of a randomized complete block with three replications. A nonlinear dose-response curve, which was analyzed with Origin (Origin Pro 8.0 for Windows),

Table 1. Soil characteristics in the experimental sites.

Site	Soil moisture (%)	pH (1:2.5)	Organic matter (g kg^{-1})	N/NH_4^+ (mg kg^{-1})	N/NO_3^- (mg kg^{-1})	Available K (mg kg^{-1})	Available P (mg kg^{-1})
trial I	14.95	6.78	22.36	38.37	71.09	341.54	392.11
trial II	14.09	6.90	14.17	18.31	43.35	239.56	231.26
trial III	18.20	6.16	17.40	47.42	77.50	479.21	402.20

Table 2. Relevant trial dates and other details.

Site	Fumigant application	Tarp removal	Ginger transplant	Germination investigation	Plant growth evaluation[a]	End of the trial	Preceding crop
trial I	02/11/12; 11/03/13	09/04/13	19/04/13	26/06/13	07/08/13	13/10/13	Ginger
trial II	03/11/12; 12/03/13	10/04/13	22/04/13	27/06/13	08/08/13	13/10/13	Ginger
trial III	03/11/12; 12/03/13	10/04/13	23/04/13	27/06/13	08/08/13	14/10/13	Ginger

[a]Plant growth evaluation included ginger plant height and number of leaves per tiller.

was used to describe the relationship between the *R. solanacarum* mortality (y) and the logarithm of the concentration of fumigant (x):

$$y = A1 + \frac{A2 - A1}{1 + 10^{(LOGx0 - x)p}}, \quad (2)$$

Field trials. The control efficacy on bacteria was calculated using the same equation as the *Laboratory studies.*

The disease scores recorded for each plot were converted into disease indices (% DI) using the formula described by McKinney [22]

$$(\%)DI = \frac{\sum (fv)}{NX}, \quad (3)$$

where f = number of plants in each class, v = class value, N = number of observed plants, and X = highest value in the evaluation scale.

Data were analyzed for ANOVA with SAS (SAS, version 8.0 for Windows). Significant differences among means were determined by Fisher's LSD test at $P = 0.05$ [23,24]. Data for soil bacteria populations were transformed as necessary (square root transformations for small numbers [<100] and log10 for large numbers [> 100] for statistical analyses), but all data are reported as non-transformed values. The data in percentages (mortality and root disease index) were normalized with arcsine square root transformation prior to ANOVA.

Results

Laboratory studies

The nonlinear dose-response curve of *R. solanacarum* to different concentrations of Pic indicated a strong relationship between increases in fumigant concentration and mortality, with R^2 values$\geq$0.994 (**Figure 2**). A nonlinear curve fit model was expressed as follows:

$$y = 38.0 + \frac{60.0}{1 + 10^{(0.4 - x)4.7}}, \quad (4)$$

The nonlinear curve fit (DoseResp) revealed that the concentrations required to control 50% (EC_{50}) and 80% (EC_{80}) of *R. solanacarum* were 2.7 and 3.7 mg a.i. kg^{-1} soil, respectively.

Field trials

Trial I. The untreated controls in trial I were heavily infested by *R. solanacarum* (**Table 4 and Figure 3**). *R. solanacarum* levels were significantly lower in the chemical treatments compared with the untreated control, except for Pic 50 PE treatments applied in March. In trial I the greatest reduction of *R. solanacarum* was provided by MB 40 PE treatments (94.92%), followed by Pic 50

Table 3. Experimental program of the field trials.

Trial	Fumigant and formulation[a]	Rate (g a.i. m^{-2})	Tarp[b]	Application method	Fumigation date	Abbreviation used for the treatment
trial I	Pic 99 LD	50	TIF	Injection	02/11/2012	Pic 50 TIF
	Pic 99 LD	50	PE	Injection	11/03/2013	Pic 50 PE
	MB 98 TC	40	PE	Cold method	11/03/2013	MB 40 PE
	Untreated control	–	–	–	02/11/2012	
trial II	Pic 99 LD	50	TIF	Injection	03/11/2012	Pic 50 TIF
	Pic 99 LD	50	PE	Injection	12/03/2013	Pic 50 PE
	MB 98 TC	40	PE	Cold method	12/03/2013	MB 40 PE
	Untreated control	–	–	–	03/11/2012	
trial III	Pic 99 LD	50	TIF	Injection	03/11/2012	Pic 50 TIF
	Pic 99 LD	50	PE	Injection	12/03/2013	Pic 50 PE
	MB 98 TC	40	PE	Cold method	12/03/2013	MB 40 PE
	Untreated control	–	–	–	03/11/2012	

[a]Abbreviations: Pic = chloropicrin; MB = methyl bromide; TC = Technical.
[b]Abbreviations: PE = polyethylene film; TIF = totally impermeable film.

Figure 1. Application of Pic in field trials: (A) a manual injection machine was used to apply Pic in the soil; (B) the treatments were covered with PE or TIF film.

TIF treatments (85.27%) and Pic 50 PE treatments (75.93%) (**Table 4**).

Ginger germination levels in all treatments were at least 96.97% at 9 WAT, and there was no significant difference between the chemical treatments and the untreated control at this stage (**Table 5**). The fumigation treatments significantly affected plant height, number of leaves per tiller and plant mortality at 15 WAT (**Table 5**). In trial I the ginger grown in MB PE treatment plots had the greatest plant height (91.0 cm), the largest number of leaves per tiller (15) and the lowest plant mortality (0.00%) at 15 WAT. However there was no significant difference between the three fumigant treatments with respect to plant mortality (which ranged from 0.0% to 0.5%), and all fumigant treatments substantially reduced plant mortality compared to the untreated control (60% mortality) (**Table 5**).

Compared to the untreated control, fumigation treatments significantly affected plant height, number of tillers per plant, root disease index and yield at the end of the trials (**Table 6**). Pic 50 TIF treatments produced the greatest plant height (99.8 cm) and the lowest root disease index (0.00%), followed by MB 40 PE and Pic 50 PE. Pic 50 TIF and MB 40 PE treatments showed no significant difference in plant height and root disease index (**Table 6**). There was no significant difference in the number of tillers per plant in all chemical treatments. Ginger grown in the MB 40 PE plots had the lowest plant mortality (0.00%), followed by Pic 50 TIF and Pic 40 PE treatments, but there was no significant difference in plant mortality between MB 40 PE and Pic 50 TIF treatments (**Table 6**). Ginger grown in the untreated plots had the lowest yield (1.50 Kg m^{-2}). The plots treated with MB 40 PE provided the highest yield (6.69 Kg m^{-2}), but this was not statistically different to the two Pic treatments (**Table 6**).

Trial II. The untreated controls in trial II were also heavily infested by *R. solanacarum* (**Table 4**). All chemical treatments reduced *R. solanacarum* at least 75.98%. The lowest level of *R. solanacarum* of was provided by MB 40 PE treatments, followed by Pic 50 TIF and Pic 50 PE treatments. However there was no significant difference in *R. solanacarum* levels between the two different Pic treatments (**Table 4**).

Ginger germination levels in all treatments were at least 97.76% at 9 WAT, and there was no significant difference at this stage (**Table 5**). Fumigation treatments significantly affected plant height, number of leaves per tiller and plant mortality at 15 WAT (**Table 5**). Ginger grown in the untreated controls had the lowest plant height (62.7 cm), the least number of leaves per tiller (5) and the highest plant mortality (86.55%) at 15 WAT. Among the three fumigation treatments there was no statistical difference in plant height, number of leaves per tiller and plant mortality at 15 WAT, and all were significantly different to the control (**Table 5**).

At the end of trial II, the fumigation treatments significantly affected plant height, number of tillers per plant, root disease index and yield (**Table 6**). Plant height was significantly greater in all chemical treatments compared with the untreated control. The greatest plant height was observed in plots treated with MB 40 PE (98.4 cm), followed by Pic 50 TIF and Pic 50 PE. However, there was no significant difference in plant height between the Pic and MB treatments. Similarly, the plots treated with MB 40 PE provided the greatest number of tillers per plant (15), the lowest ginger mortality (0.00%), and the lowest ginger root disease index (0.00%), but these results were not statistically different from the Pic treatments.

Ginger grown in the untreated plots had the lowest yield (0.00 Kg m^{-2}) (**Table 6**). The plots treated with Pic 50 TIF provided the highest yield (7.29 Kg m^{-2}), and this was not statistically different to the yield obtained with MB 40 PE but was

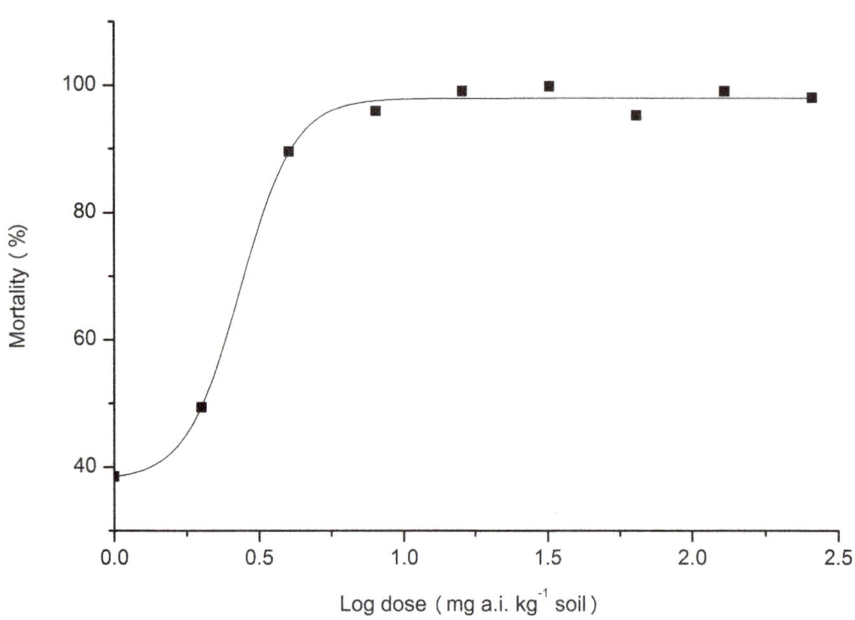

Figure 2. Response of *R. solanacarum* to different Pic concentrations after fumigation of a sandy loam soil for 7 days at 28°C.

Table 4. Effects of soil fumigation on colony-forming units (cfu) of *R. solanacearum* on selective media of 1 g from soil after fumigation.

Site	Treatment[a]	*R. solanacearum*	
		cfu/g[a]	% reduction
trial I	Pic 50 TIF	947b	85.27
	Pic 50 PE	1547ab	75.93
	MB 40 PE	327b	94.92
	Untreated control	6427a	
trial II	Pic 50 TIF	1093b	84.25
	Pic 50 PE	1667b	75.98
	MB 40 PE	293c	95.77
	Untreated control	6940a	
trial III	Pic 50 TIF	340bc	93.82
	Pic 50 PE	583b	89.39
	MB 40 PE	267c	95.15
	Untreated control	5553a	

[a]In each column, data are means of three replication. Means followed by the same letter are not different ($P = 0.05$) according to the LSD test.

significantly higher than the yield obtained with the Pic 50 PE treatment (**Table 6**).

Trial III. Trial III also found that the untreated controls were heavily infested by *R. solanacarum* (**Table 4**). All chemical treatments reduced *R. solanacarum* at least 89.39%. The lowest level of *R. solanacarum* was also provided by MB 40 PE treatments, followed by Pic 50 TIF and Pic 50 PE treatments. However, there was no significant difference in *R. solanacarum* levels between Pic 50 TIF and MB 40 PE treatments (**Table 4**).

At 9 WAT there was no significant difference in ginger germination levels between the treatments except that the levels in MB 40 PE treatments (100%) were significantly higher than the untreated controls (88.10%) (**Table 5**). Fumigation treatments significantly affected plant height, number of leaves per tiller and plant mortality at 15 WAT (**Table 5**). Ginger grown in the untreated controls had the lowest plant height (57.3 cm), the least number of leaves per tiller (3) and the highest plant mortality (70.85%) at 15 WAT. The three fumigation treatments showed no significant difference in plant height, number of leaves per tiller and plant mortality (**Table 5**).

At the end of the trials, fumigation treatments significantly affected plant height, number of tillers per plant, root disease index and yield (**Table 6**). Plant height was significantly greater in all chemical treatments compared with the untreated control. The greatest plant height was observed in plots treated with MB 40 PE (98.0 cm), followed by Pic 50 PE and Pic 50 TIF, however there was no significant difference between the chemical treatments. Similarly, the plots treated with MB 40 PE provided the lowest ginger mortality (0.00%) and the lowest ginger root disease index (0.00%), but this was not statistically different from the Pic treatments. Ginger grown in the untreated plots had the lowest number of tillers per plant (6) and lowest yield (0.00 Kg m^{-2}) (**Table 6**). The plots treated with MB 40 PE provided the greatest number of tillers per plant (17) and the highest yield (7.31 Kg m^{-2}); this was not statistically different from the yield obtained with Pic 50 PE but was significantly higher than the yield obtained with Pic 50 TIF (**Table 6**).

Discussion

After three to five years' successive monocropping of ginger, ginger bacterial wilt caused by soil-borne *R. solanacarum* becomes greater and greater in China and the ginger producers have to plant other vegetables such as Chinese onion (*Agrimonia fistulosum* L.), garlic (*Agrimonia sativum* L.), Chinese cabbage (*Brassica campestris* L. spp.), radish (*Raphanus sativus* L.), maize (*Zea mays* L.) and others. Ginger has a relatively high economic value compared to other vegetables, so the application of a fumigant is regarded as an essential practice to protect ginger plants from soil-borne *R. solanacarum* in China. In our research, laboratory studies and three field trials were conducted to determine effects of several Pic treatments on *R. solanacarum* in ginger production.

Our laboratory studies showed that Pic can provide excellent control of *R. solanacarum* after fumigation of a sandy loam soil for 7 days at 28°C. The specific mechanism by which Pic controls *R. solanacarum* was not examined, however, the current laboratory results indicated that Pic is feasible and effective in controlling *R. solanacarum*.

Figure 3. Ginger plants grown in plots treated with (A) Pic 50 TIF, (B) Pic 50 PE, (C) MB 50 PE and (D) Untreated control at the end of trials, respectively (trial I).

Table 5. Effect of fumigation treatments on germination and plant growth during the ginger crop growth period.

Site	Treatment	Germination (%)	Plant height (cm)	Number of leaves/tiller[b]	Mortality (%)
trial I	Pic 50 TIF	99.56a[a]	80.0ab	11b	0.44b
	Pic 50 PE	99.49a	78.3b	15ab	0.51b
	MB 40 PE	98.10a	91.0a	15a	0.00b
	Untreated control	96.97a	61.7c	5c	60.00a
trial II	Pic 50 TIF	98.00a	87.3a	15a	0.00b
	Pic 50 PE	98.46a	85.7a	15a	0.51b
	MB 40 PE	100.00a	84.7a	16a	0.00b
	Untreated control	97.76a	62.7b	5b	86.55a
trial III	Pic 50 TIF	98.56ab	88.7a	13b	0.44b
	Pic 50 PE	97.44ab	83.7a	17a	1.03b
	MB 40 PE	100.00a	82.7a	14ab	0.00b
	Untreated control	88.10b	57.3b	3c	70.85a

[a]In each column, data are means of three replication. Means followed by the same letter are not different ($P = 0.05$) according to the LSD test.
[b]Number of leaves/tiller were counted from the one who had the most leaves in the all tillers of one ginger plant.

In the present field trials, both Pic treatments (Pic 50 TIF applied in November, and Pic 50 PE applied in March) sharply reduced the colony-forming units (cfu) of *R. solanacarum* on media, and maintained high ginger yields in commercial production. Based on the present field results, the two different Pic treatments provided results similar to MB 40 PE in terms of *R. solanacarum* control after fumigation, plant growth at 15 WAT, plant height, plant mortality, root disease index and ginger yield at the end of the trials. The results confirmed that the two tested Pic treatments both provide an effective alternative to MB. However, based on our present field results, it is too soon to say which one of the Pic treatments is better, and more studies are needed in order to draw a definitive conclusion on this point. Generally speaking, the temperature must be between 4.4 to 26.7°C (40 to 80°F) at the depth of injection when Pic is applied [25]. In our filed trials, when Pic was applied, the soil temperatures at 5 cm depth during the day of two different Pic treatments (in November 2012 and in March 2013) was about 15 to 21°C and 13 to 19°C, respectively (**Figure 1**). In November 2012 treatments, with the winter coming, the soil temperature will become lower and lower, and the lowest soil temperature will be about 0°C. Although from previous laboratory results, it would appear that chloropicrin is effective against *Verticillium* at low temperatures around 50°F±2°F (8.9 to 11.1°C) [26]. In the field application, whether and how low temperatures affect the effect of Pic on *R. solanacarum* is still unclear to us.

The soils in our three field trials were heavily infested by *R. solanacarum*, but at 9 WAT the fumigation treatments did not significantly affect ginger germination levels except for MB 40 PE treatments in trial III. This indicates that ginger bacterial blight did not generally appear in the initial growth stage of ginger, and this could be attributed to the relatively low temperature, low rainfall patterns and low soil moisture during the initial growth stage.

Table 6. Effect of fumigation treatments on ginger plant height, tillers/plant, mortality, root disease, and yield at the end of the trials.

Site	Treatment	Plant height (cm)	Tillers/plant	Mortality (%)	Root disease index (%)	Yield (Kg m^{-2})
trial I	Pic 50 TIF	99.8a[a]	15a	1.25c	0.00c	6.27a
	Pic 50 PE	81.9b	14a	10.00b	26.88b	6.20a
	MB 40 PE	97.9a	16a	0.00c	1.88c	6.69a
	Untreated control	24.9c	2b	88.94a	93.13a	1.50b
trial II	Pic 50 TIF	95.0a	15a	8.50b	12.50b	7.29a
	Pic 50 PE	87.7a	13a	10.55b	12.50b	5.32b
	MB 40 PE	98.4a	15a	0.00b	0.00b	6.24ab
	Untreated control	0.0b	0b	100.00a	100.00a	0.00c
trial III	Pic 50 TIF	90.6a	11b	10.00b	15.63b	4.99b
	Pic 50 PE	93.1a	15ab	5.00b	6.25b	5.49ab
	MB 40 PE	98.0a	17a	0.00b	0.00b	7.31a
	Untreated control	42.2b	6c	90.85a	78.13a	1.87c

[a]In each column, data are means of three replication. Means followed by the same letter are not different ($P = 0.05$) according to the LSD test.

Pic is a strong eye irritant and has a pungent unpleasant smell, which can pose barriers to its adoption. Well-equipped commercial companies generally provide safe soil fumigation services in the United States, Europe, or Japan. However, soil fumigation in China is mostly conducted by individual farmers who generally lack essential application tools and personal protection equipment. Considering the environment emissions and potential human exposure, the gelatin capsule (gel cap) formulation of Pic will offer a promising solution [10,27]. The use of TIF would allow greater retention of Pic than the conventional PE film, trapping the fumigant for a longer period near the soil surface and thereby increasing the dose and prolonging the exposure of soilborne pathogens to the fumigant [28,29]. In the present field trials, Pic 50 TIF treatments applied in November were generally similar or even superior to Pic 50 PE treatments applied in March for controlling *R. solanacarum* and improving plant growth and ginger yield. However, we are not able to attribute this to the use of TIF before removing the effect of the different application seasons. Pic is normally applied in the spring (during March and April) about one month before ginger is transplanted in the field. The soil temperature is low (about 8 to 18°C during the day) during this fumigation period, so it would be desirable to examine in more detail the feasibity of applying Pic in the autumn (during October and November, just after the ginger harvest) when soil temperature is relatively high (about 15 to 25°C during the day).

In summary, our studies determined that the fumigant Pic had similar efficacy to MB in terms of controlling ginger bacterial blight caused by *R. solanacarum*, ginger germination, plant height, plant mortality, number of leaves per tiller and ginger yield. However, more detailed work to identify the most suitable application methods (formulation, rates, and application seasons) and appropriate combinations [30] with other fumigants or biological agents (for example *Trichoderma asperellum*, *Bacillus subtilis* or others) is necessary before Pic treatments can be recommended as an efficient alternative to methyl bromide for ginger production in China.

Acknowledgments

The authors give special thanks to Professor Jie Feng (Institute of Plant Protection, Chinese Academy of Agricultural Sciences) who in our laboratory studies provided *Ralstonia solanacarum* isolated from ginger grown in Shandong, China. We also thank Dr. Melanie Miller for editing the manuscript.

Author Contributions

Conceived and designed the experiments: LM QW DY AC. Performed the experiments: LM DY TM PL JS MG AC. Analyzed the data: LM DY YL CO AC. Contributed reagents/materials/analysis tools: LM QW JS MG. Wrote the paper: LM AC.

References

1. Food and Agriculture Organization of the United Nations (FAO) (2013) FAOSTAT-Agriculture. Available: http://faostat.fao.org/site/339/default.aspx. Accessed 2013 February 4.
2. Yabuuchi E, Kosako Y, Yano I, Hotta H, Nishiuchi Y (1995) Transfer of two *Burkholderia* and an *Alcaligenes* species to *Ralstonia* gen. Nov.: proposal of *Ralstonia pickettii* (Ralston, Palleroni and Doudoroff 1973) comb. Nov., *Ralstonia solanacearum* (Smith 1896) comb. Nov. and *Ralstonia eutropha* (Davis 1969) comb. Nov. Microbiol Immunol 39: 897–904.
3. Bell CH (2000) Fumigation in the 21st century. Crop Prot 19: 563–569.
4. Desaeger JA, Seebold KW, Csinos AS (2008) Effect of application timing and method on efficacy and phytotoxicity of 1,3-D, chloropicrin and metam-sodium combinations in squash plasticulture. Pest Manag Sci 64: 230–238.
5. Gullino ML, Minuto A, Gilardi G, Garibaldi A, Ajwa H, et al. (2002) Efficacy of preplant soil fumigation with chloropicrin for tomato production in Italy. Crop Prot 21: 741–749.
6. Gilreath JP, Santos BM (2008) Managing weeds and nematodes with combinations of methyl bromide alternatives in tomato. Crop Prot 27: 648–652.
7. Gilreath JP, Santos BM (2004) Methyl bromide alternatives for weed and soilborne disease management in tomato (*Lycopersicon esculentum*). Crop Prot 23: 1193–1198.
8. Lombardo S, Longo AMG, Lo Monaco A, Mauromicale G (2012) The effect of soil solarization and fumigation on pests and yields in greenhouse tomatoes. Crop Prot 37: 59–64.
9. Santos BM, Gilreath JP, Motis TN, Noling JW, Jones JP, et al. (2006) Comparing methyl bromide alternatives for soilborne disease, nematode and weed management in fresh market tomato. Crop Prot 25: 690–695.
10. Yan D, Wang Q, Mao L, Xie H, Guo M, et al. (2012) Evaluation of chloropicrin gelatin capsule formulation as a soil fumigant for greenhouse strawberry in China. J Agric Food Chem 60: 5023–5027.
11. Csinos AS, Johnson WC, Johnson AW, Sumner DR, McPherson RM, et al. (1997) Alternative fumigants for methyl bromide in tobacco and pepper transplant production. Crop Prot 16: 585–594.
12. Gilreath JP, Motis TN, Santos BM (2005) *Cyperus* spp. control with reduced methyl bromide plus chloropicrin doses under virtually impermeable films in pepper. Crop Prot 24: 285–287.
13. Gilreath JP, Santos BM, Motis TN, Noling JW, Mirusso JM (2005) Methyl bromide alternatives for nematode and *Cyperus* control in bell pepper (*Capsicum annuum*), Crop Prot 24: 903–908.
14. Schinner F, Öhlinger R, Kandeler E, Margesin R (1995) Methods in soil biology. Verlag Berlin/Heidelberg, Germany: Springer. pp. 386–389.
15. Nelson DW, Sommers LE (1985) Total carbon, organic carbon and organic matter. In Page AL, Miller RH, Keency OR, editors. Methods of soil analysis. Madison, Wisconsin: American Society of Agronomy Publishers. pp. 539–576.
16. Margesin R, Schinner F (2005) Manual for soil analysis - Monitoring and assessing soil bioremediation. Verlag Berlin/Heidelberg Germany: Springer. pp. 47–49.
17. Hutchinson CM, McGiffen Jr ME, Ohr HD, Sims JJ, Becker JO (2000) Efficacy of methyl iodide and synergy with chloropicrin for control of fungi. Pest Manag Sci 56: 413–418.
18. Klose S, Acosta-Martínez V, Ajwa HA (2006) Microbial community composition and enzyme activities in a sandy loam soil after fumigation with methyl bromide or alternative biocides. Soil Biol Biochem 38: 1243–1254.
19. Mao L, Wang Q, Yan D, Xie H, Li Y, et al. (2012) Evaluation of the combination of 1,3-dichloropropene and dazomet as an efficient alternative to methyl bromide for cucumber production in China. Pest Manag Sci 68: 602–609.
20. Kelman A (1954) The relationship of pathogenicity in *Pseudomonas solanacearum* to colony appearance on a tetrazolium medium. Phytopathol 44: 693–695.
21. Cao A, Guo M, Yan D, Mao L, Wang Q, et al. (2013) Evaluation of sulfuryl fluoride as a soil fumigant in China. Pest Manag Sci doi:10.1002/ps.3535
22. McKinney HH (1923) Influence of soil, temperature and moisture on infection of wheat seedling by *Helminthosporium sativum*. J Agric Res 26:195–217.
23. Steel RGD, Torrie JH (1960) Principles and procedures of statistics. New York, America: McGraw-Hill Book Co. Inc. 481 p.
24. Csinos AS, Johnson WC, Johnson AW, Sumner DR, McPherson RM, et al. (1997) Alternative fumigants for methyl bromide in tobacco and pepper transplant production. Crop Prot 16: 585–594.
25. Dow AgroSciences (2014) Telone soil fumigants - A guide to application. Available: http://msdssearch.dow.com/PublishedLiteratureDAS/dh_0062/0901b803800626c7.pdf?filepath=usag/pdfs/noreg/010-31467.&fromPage=GetDoc. Accessed 2014 February 8.
26. Bringhurst RS, Voth V (1960) Chloropicrin diffuses rapidly even at low temperatures. California Strawberry Advisory Board Strawberry News Bulletin, Vol. VI Bulletin 7, February 7, 10, and 17, 1960 (combined).
27. Wang Q, Wang D, Tang J, Yan D, Zhang H, et al. (2010) Gas-phase distribution and emission of chloropicrin applied in gelatin capsules to soil columns. J Environ Qual 39: 917–922.
28. Fennimore SA, Kabir Z, Ajwa HA, Daugovish O, Roth K (2005) Chloropicrin and InLine lethal dose on nutsedge and knotweed under impermeable film. Proceedings Annual International Research Conference on Methyl Bromide Alternatives and Emissions Reductions, Oct 31-Nov 3 2005, San Diego, CA, Abstract 54, pp 1–4.
29. Ou L-T, Thomas JE, Allen LH Jr, Vu JC, Dickson DW (2008) Comparison of surface emissions and subsurface distribution of *cis*- and *trans*-1,3-dichloropropene and chloropicrin in sandy field beds covered with four different plastic films. J Environ Sci Health B 43: 376–381.
30. Stewart ZP, Oxborough RM, Tungu PK, Kirby MJ, Rowland MW, et al. (2013) Indoor Application of Attractive Toxic Sugar Bait (ATSB) in Combination with Mosquito Nets for Control of Pyrethroid-Resistant Mosquitoes. PLOS ONE 8(12): e84168. doi:10.1371/journal.pone.0084168

15

Long-Term Effect of Agricultural Reclamation on Soil Chemical Properties of a Coastal Saline Marsh in Bohai Rim, Northern China

Yidong Wang[1], Zhong-Liang Wang[1,2]*, Xiaoping Feng[1,3], Changcheng Guo[1], Qing Chen[1]

1 Tianjin Key Laboratory of Water Resources and Environment, Tianjin Normal University, Tianjin, China, **2** State Key Laboratory of Environmental Geochemistry, Institute of Geochemistry, Chinese Academy of Sciences, Guiyang, China, **3** College of Urban and Environmental Sciences, Tianjin Normal University, Tianjin, China

Abstract

Over the past six decades, coastal wetlands in China have experienced rapid and extensive agricultural reclamation. In the context of saline conditions, long-term effect of cultivation after reclamation on soil chemical properties has not been well understood. We studied this issue using a case of approximately 60-years cultivation of a coastal saline marsh in Bohai Rim, northern China. The results showed that long-term reclamation significantly decreased soil organic carbon (SOC) (−42.2%) and total nitrogen (TN) (−25.8%) at surface layer (0–30 cm) as well as their stratification ratios (SRs) (0–5 cm:50–70 cm and 5–10 cm:50–70 cm). However, there was no significant change in total phosphorus (TP) as well as its SRs under cultivation. Cultivation markedly reduced ratios of SOC to TN, SOC to TP and TN to TP at surface layer (0–30 cm) and their SRs (0–5 cm:50–70 cm). After cultivation, electrical conductivity and salinity significantly decreased by 60.1% and 55.3% at 0–100 cm layer, respectively, suggesting a great desalinization. In contrast, soil pH at 20–70 cm horizons notably increased as an effect of reclamation. Cultivation also changed compositions of cations at 0–10 cm layer and anions at 5–100 cm layer, mainly decreasing the proportion of Na^+, Cl^- and SO_4^{2-}. Furthermore, cultivation significantly reduced the sodium adsorption ratio and exchangeable sodium percentage in plow-layer (0–20 cm) but not residual sodium carbonate, suggesting a reduction in sodium harm.

Editor: Jin-Song Zhang, Institute of Genetics and Developmental Biology, Chinese Academy of Sciences, China

Funding: This research was supported by the National Science & Technology Pillar Program (2012BAC07B02), the Program for New Century Excellent Talents in University (NCET-10-0954), the National Natural Science Foundation of China (31300381), and the Opening Fund of Tianjin Key Laboratory of Water Resources and Environment (YF11700102). The funders had no role in study design, data collection and analysis, decision to publish, or preparation of the manuscript.

Competing Interests: The authors have declared that no competing interests exist.

* E-mail: wangzhongliang@vip.skleg.cn

Introduction

Coastal wetlands provide essential ecosystem services to people and environment including flood protection, water supply and purification, food productivity, erosion control, wave attenuation, shoreline stabilization, wildlife habitat, biodiversity, climate regulation and amenity [1,2]. Over the past century, natural coastal wetlands all over the world have been rapidly shrunk due to intensive anthropogenic activities [3,4,5,6]. In China, approximately 51% (2.2×10^4 km^2) of coastal natural wetlands were lost or degraded since the 1950s, primarily due to agricultural reclamation [7]. Soil is one of the foundations of ecosystem services of wetland and cropland. Therefore, it is important to study the influence of agricultural cultivation after reclamation on soil properties of the coastal wetlands.

Some studies have evaluated the reclamation-induced changes of soil properties in coastal wetlands. Conversion of coastal wetlands to croplands has been reported to cause radical changes in soil chemical properties because of alterations of hydrology and agricultural activities. First, hydrologic alterations such as ditch drainage and diking led to increase in aeration and decrease in salinity [8,9,10,11]. Aeration accelerated soil organic matter decay [4,12,13] and affected substances' characteristics [5] and mobility through changing redox conditions [8,14]. Changes in salinity further influenced substances' cycle in soils [15,16]. Second, application of fertilizer affected nutrient contents such as carbon [17], nitrogen and phosphorus [14,18]. In addition, agricultural managements including planting, harvesting and tillage also influenced balances of substances inputs and outputs in soils [19,20,21]. Most of these studies were focusing on short-term scales (e.g. [5,8,11,13,14]). In contrast, only few studies were available at long-term scales [4,9,10].

Stratification of soil chemical properties with soil depth is common in many natural ecosystems [4,22]. Different soil horizons of coastal wetlands may have distinct responses to agricultural reclamation. The surface soil is the vital interface that receives intense impact from human disturbance. In support of this concept, chemical characteristics of the surface soil in coastal wetlands were demonstrated to be more sensitive to agricultural reclamation [4,18]. However, quantitative expression of the horizon-induced differences in response of coastal wetland to agricultural reclamation is rare. Stratification ratio (SR) is a quantitative indicator to represent changes of soil profile feature [22]. It is hypothesized that the SR can be used as an indicator of the responses of different soil horizons to human disturbance.

In Bohai Rim of northern China, coastal marsh geogenesises have been realized naturally by land-sea interactions since the middle-late Holocene. Since the founding of New China, natural coastal marshes have been largely reclaimed for agriculture to alleviate the pressure of the increasing population. For instance, approximately 55% of coastal natural wetlands in Tianjin were disappeared since the 1950s. However, it is unclear that how agricultural reclamation affects soil chemical properties and their SRs of the coastal saline marshes in Bohai Rim, northern China.

The objective of this study was to evaluate a long-term (approximately 60-years) impact of agricultural reclamation on soil chemical properties as well as their SR features of a coastal saline marsh in Bohai Rim, northern China.

Materials and Methods

Study site

This study was conducted at a coastal permanent marsh and a cropland cultivated from the marsh in Qilihai (39°18′ N, 117°29′ E, elevation 2 m) of Bohai Rim, northern China. No specific permits were required for the described field studies in the research site. The field studies did not involve endangered or protected species. The study site is flat and low-lying. The region was characterized by a warm temperate semi-humid monsoon climate. Annual mean air temperature was 11.2 °C; total precipitation amount was 500–600 mm $year^{-1}$; evapotranspiration was approximately 1900 mm $year^{-1}$; frost-free period was 180–194 d; and sunshine duration hours were 2600–2800 h. Rainfall occurred synchronously with the heat, with high values in summer (June–August).

The permanent marsh was formed from the land-sea interactions since middle-late Holocene. The permanent marsh was mainly covered by reed (*Phragmites australis*) and generally had a water level of 0–40 cm above surface recently. Croplands were cultivated from the *P. australis* marshes over approximately 60 years. During the first ~20 years (1950s–1960s), paddy was cultivated. Afterwards, the paddy lands converted to dry-lands mainly dominated by corn, cotton and sorghum in rotation. Organic (manure and straw) and chemical fertilizer applications during 1950s–1990s [23] were listed in Table 1. Since the new century, organic and chemical fertilizer applications [24] were presented in Table 2. The current water level of the dry-lands was about 80–100 cm below surface.

Soil sampling and chemical analysis

The *P. australis* marsh and the cropland were adjacent (about 500 m) and had similar altitudes. Soil samples were randomly collected using a soil drill (Eijkelkamp, Netherlands) at three locations in the permanent marsh and cropland in May 2012, respectively. In both marsh and cropland, the distance between any two sampling locations was more than 15 m. In each sampling location, two adjacent soil columns (0–100 cm) were collected and then mixed into one column according to 7 corresponding horizons: 0–10, 10–20, 20–30, 30–50, 50–70, and 70–100 cm.

Soil samples were brought to laboratory and air-dried naturally. The air-dried soil samples were sieved with a 2 mm mesh screen to remove the roots and coarse fraction before chemical analysis. Dilute HCl (0.5 mol L^{-1}) was used to remove soil inorganic carbon. Then deionized water was used to rinse off excess HCl and ensure the neutral pH of approximate 7. The contents of soil organic carbon (SOC) and total nitrogen (TN) were determined by an Elementar (Vario EL III, Elementar, Germany). The concentration of total phosphorus (TP) was determined by sulphuric and perchloric acids digestion and Mo-Sb Anti spectrophotometric method. Soil pH was measured with a pH meter (Star A 420C-01A, Thermo Orion, United States) using a soil:water ratio of 1:2.5 (g g^{-1}). Electrical conductivity (EC) was measured with a EC meter (Star A 420C-01A, Thermo Orion, United States) using a soil:water ratio of 1:5 (g g^{-1}). Soil major soluble cations (Na^+, Ca^{2+}, K^+ and Mg^{2+}) and anions (Cl^-, SO_4^{2-}, HCO_3^- and CO_3^{2-}) were measured using a soil:water ratio of 1:5 (g g^{-1}). Cation concentrations were determined by an Atomic Absorption Spectrophotometer (TAS-990, Beijing Purkinje General Instrument Co., Ltd., China). Concentrations of Cl^- and SO_4^{2-} were determined by an Ion Chromatography (ICS-2100, Dionex Corporation, Sunnyvale, California, United States). Concentrations of CO_3^{2-} and HCO_3^- were measured using neutralization titration method using phenolphthalein and methyl orange.

Data analysis

Sodium adsorption ratio (SAR), an easily measured property that gives information on the comparative concentrations of Na^+, Ca^{2+}, and Mg^{2+} in soil solutions, was calculated using equation (1) [25]. Exchangeable sodium percentage (ESP) was computed using equation (2) [26,27]. Residual sodium carbonate (RSC) was estimated using equation (3) [28]. Comparisons of mean SOC, TN, TP, pH, EC and salinity between the marsh and cropland were analyzed using one-way ANOVA of SPSS 13.0 (SPSS Inc., Chicago, Illinois, United States). Figures were drawn using Origin 8.0 (OriginLabs Corporation, Northampton, Massachusetts, United States) and CorelDraw 9 (Corel Corporation, Canada).

$$SAR = \frac{Na^+}{\sqrt{(Ca^{2+} + Mg^{2+})/2}} \quad (1)$$

$$ESP = 100 \times \frac{(-0.0126 + 0.01475SAR)}{1 + (-0.0126 + 0.01475SAR)} \quad (2)$$

Table 1. Organic and chemical fertilizer applications during 1950s–1990s (source from the reference [23]).

	Organic fertilizer (manure and straw)		Chemical fertilizer	
	Application rate (m^3 ha^{-1})	Proportion (%)	Application rate (kg ha^{-1})	Proportion (%)
1950s–1960s	30	93	53	7
1970s	40	61	600	39
1980s	20	40	690	60
1990s	20	29	1156	71

Table 2. Nitrogen and phosphorus inputs of organic and chemical fertilizer applications in 2007 (source from the reference [24]).

	Organic fertilizer (manure and straw)		Chemical fertilizer	
	Nitrogen (kg ha^{-1})	Phosphorus (kg ha^{-1})	Nitrogen (kg ha^{-1})	Phosphorus (kg ha^{-1})
2007	66	41	269	149

$$RSC = \left(CO_3^{2-} + HCO_3^{-}\right) - \left(Ca^{2+} + Mg^{2+}\right) \quad (3)$$

Results

Soil organic carbon, total nitrogen and phosphorus as well as their stratification ratios

Long-term agricultural reclamation significantly reduced the concentrations of SOC at surface soil (0–30 cm) but not lower soil layers (30–100 cm) (Fig. 1A). The weighted density of SOC at surface soil (0–30 cm) decreased by 10.1 g kg^{-1} (−42.2%). Similarly, the contents of TN of upper soil layers (0–50 cm) but not deeper soil layers (50–100 cm) were also evidently decreased under long-term agricultural cultivation (Fig. 1B). The weighted density of TN at surface soil (0–50 cm) reduced by 0.7 g kg^{-1} (−31.0%). In contrast, there were no significant effects of long-term agricultural reclamation on the contents of TP at the whole soil profile (0–100 cm) (Fig. 1C). The ratios of SOC/TN, SOC/TP and TN/TP were all notably decreased after agricultural reclamation at surface soil (0–30 cm, except for 10–20 cm of SOC/TN) but not lower soil layers (30–100 cm) (Fig. 2).

Stratification ratios of 0–5 cm:50–70 cm and 5–10 cm:50–70 cm of SOC were 12.3 and 3.7 for the marsh and were 3.4 and 3.0 for the cropland (Fig. 3A). Stratification ratios of 0–5 cm:50–70 cm and 5–10 cm:50–70 cm of TN were 4.0 and 1.9 for the marsh and were 1.5 and 1.4 for the cropland (Fig. 3A). Compared to the marsh, SRs of SOC and TN of the cropland were both remarkably dropped, suggesting greater decreases of SOC and TN in the upper soil layers (Fig. 3A and B). In contrast, there was no significant impact of reclamation on the SR of TP (Fig. 3C). The SRs (0–5 cm:50–70 cm) of ratios of SOC/TN, SOC/TP and TN/TP were all significantly decreased after agricultural reclamation (Fig. 4).

Soil EC, salinity and pH

Soil profile EC of the marsh ranged from 1026 to 2365 μs cm^{-1}, with a weighted average of 1488 μs cm^{-1}. Soil profile EC of the cropland varied from 392 to 854 μs cm^{-1}, with a weighted average of 594 μs cm^{-1} (Fig. 5A). Soil profile salinity of the marsh was from 3.2 to 6.9 g kg^{-1} dry weight, with a weighted average of 4.7 g kg^{-1} dry weight. Soil profile salinity of the cropland ranged from 1.3 to 2.5 g kg^{-1} dry weight, with a weighted average of 2.1 g kg^{-1} dry weight (Fig. 5B). Thus, after long-term reclamation, EC and salinity of the whole soil profile (0–100 cm) decreased by 60.1% and 55.3%, respectively. In support of these results, concentrations of water extractable ions showed great declines after reclamation. In contrast, soil pH (20–70 cm) significantly increased after reclamation.

Compositions of cation and anion

Even though the cations of marsh and cropland were both dominated by Na^{+}, the proportion of Na^{+} (0–10 cm) was significantly decreased after reclamation ($p<0.05$) (Fig. 6A). However, the proportion of Ca^{2+} (0–10 cm) and Mg^{2+}+K^{+} (5–20 cm) were significantly increased after reclamation ($p<0.05$) (Fig. 6A). Contrasted to cation, the composition of anion changed greater at 5–100 cm after reclamation, with significant decreases

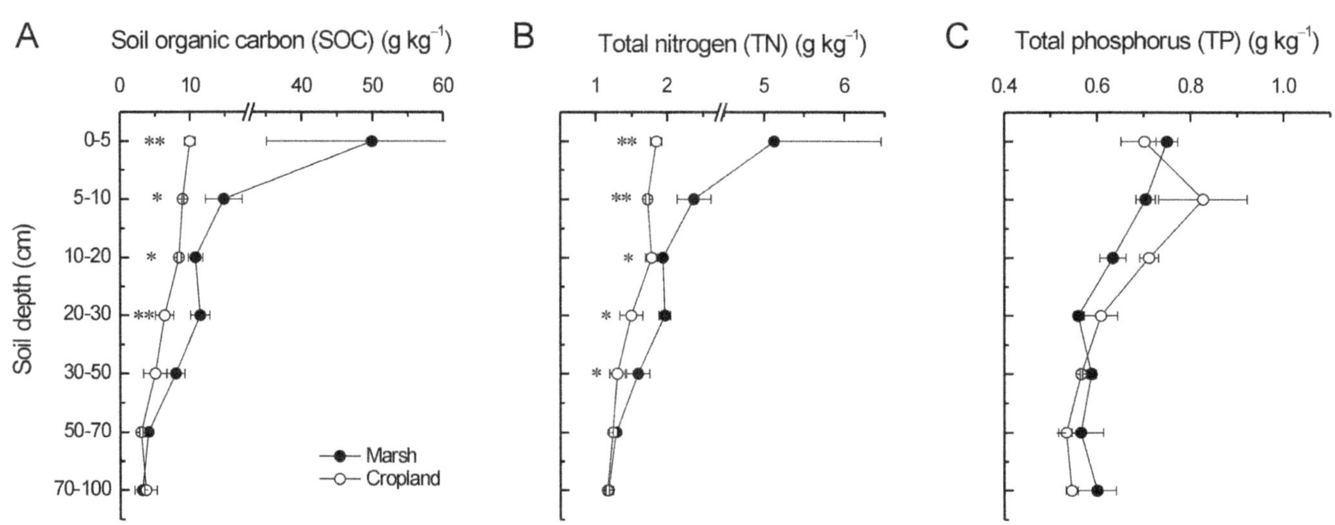

Figure 1. Long-term effects of reclamation on A) SOC, B) TN, and C) TP of the coastal marsh. * and ** indicate significant difference at $p<0.05$ and $p<0.01$ level, respectively.

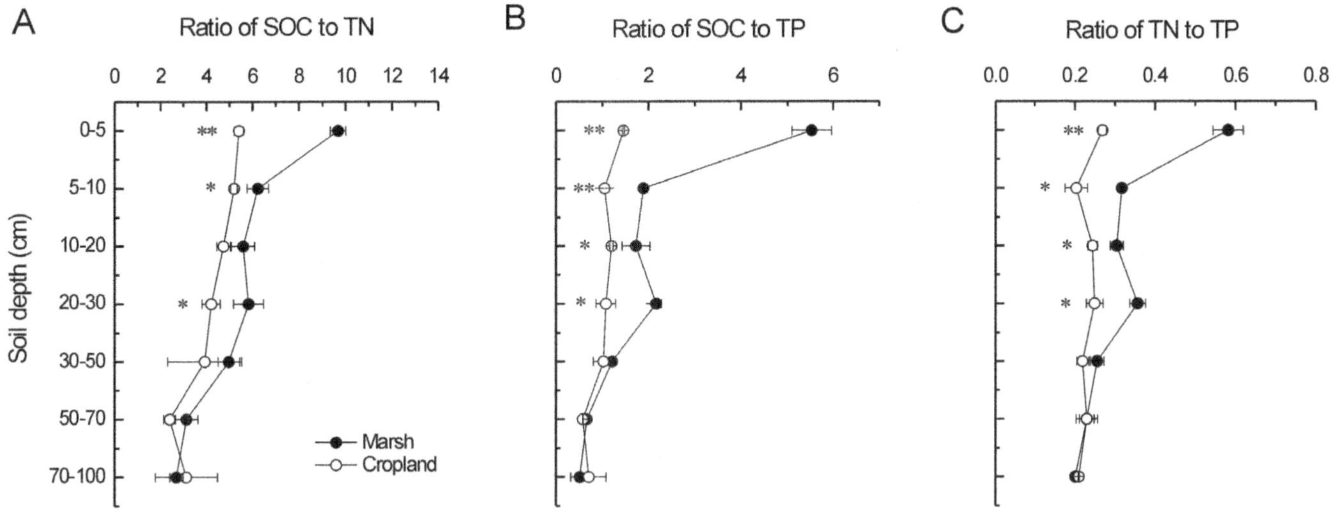

Figure 2. Long-term effects of reclamation on ratios of A) SOC/TN, B) SOC/TP, and C) TN/TP. * and ** represent significant difference at $p<0.05$ and $p<0.01$ level, respectively.

in Cl^- and SO_4^{2-} and increases in $HCO_3^-+CO_3^{2-}$ ($p<0.05$) (Fig. 6B).

Characteristics of alkalization

The SAR and ESP in plow-layer (0–20 cm) significantly reduced after reclamation ($p<0.05$) (Fig. 7A and B). This result suggested a reduction in soil alkalization and sodium harm. However, there was no significant difference in RSC under long-term cultivation (Fig. 7C).

Discussion

SOC, TN and TP as well as their stratification ratios

The soil storages of SOC, TN and TP are counterbalances of inputs (e.g. productivity and fertilizer) and outputs (e.g. loss of mineralization and harvest) during a certain period. The rates of inputs and outputs in soils have been widely recognized to be sensitive to environmental changes (e.g. [29,30,31]) and human disturbance (e.g. [19,20,32]). Agricultural reclamation of wetland substantially altered the soil environmental conditions, thus should change soil stocks of SOC, TN and TP through breaking the balance of inputs and outputs. In our study, even though many organic (manure and straw) and chemical fertilizers were applied for a long-term time (Table 1 and Table 2), SOC and TN decreased strongly in surface soil after reclamation (Fig. 1), suggesting that the outputs were much larger than the inputs. Our result was in agreement with the results in the Yangtze Estuary [10] and the Hangzhou Bay of China [14]. Four reasons may explain the above results. First, the mineralization and accumulation of soil organic matter were both regulated by hydrologic conditions [12,33]. Submerged soils of the marsh were under anaerobic conditions, which had lessened and incomplete decompositions of organic materials and decreased humifications of organic matter. In contrast, cropland soils lacked anaerobiosis and

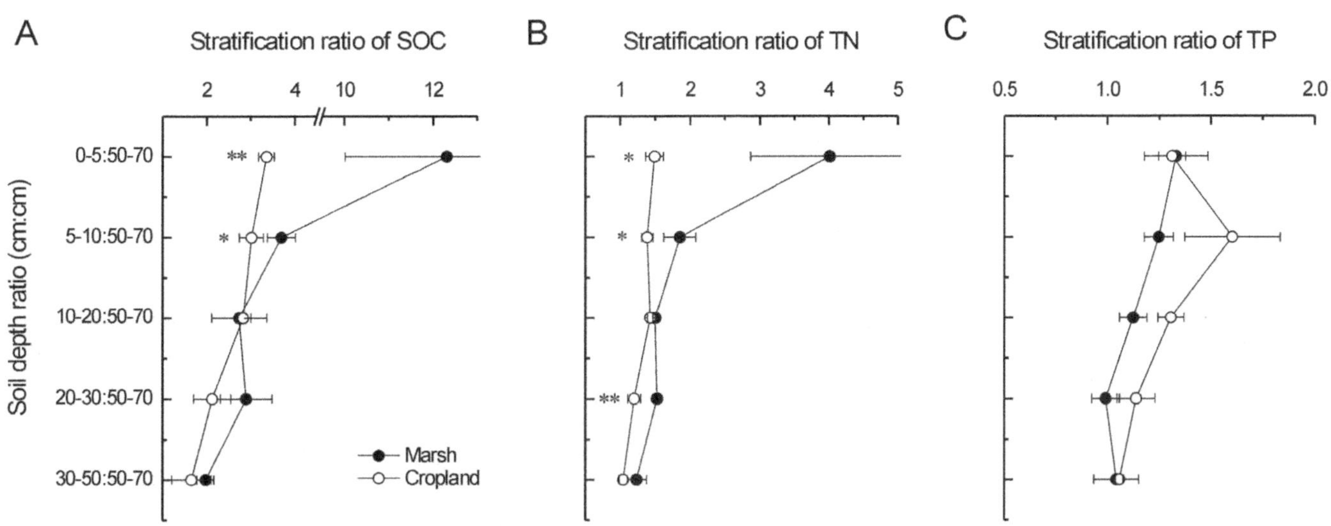

Figure 3. Long-term influences of reclamation on stratification ratios of A) SOC, B) TN, and C) TP. * and ** indicate significant difference at $p<0.05$ and $p<0.01$ level, respectively.

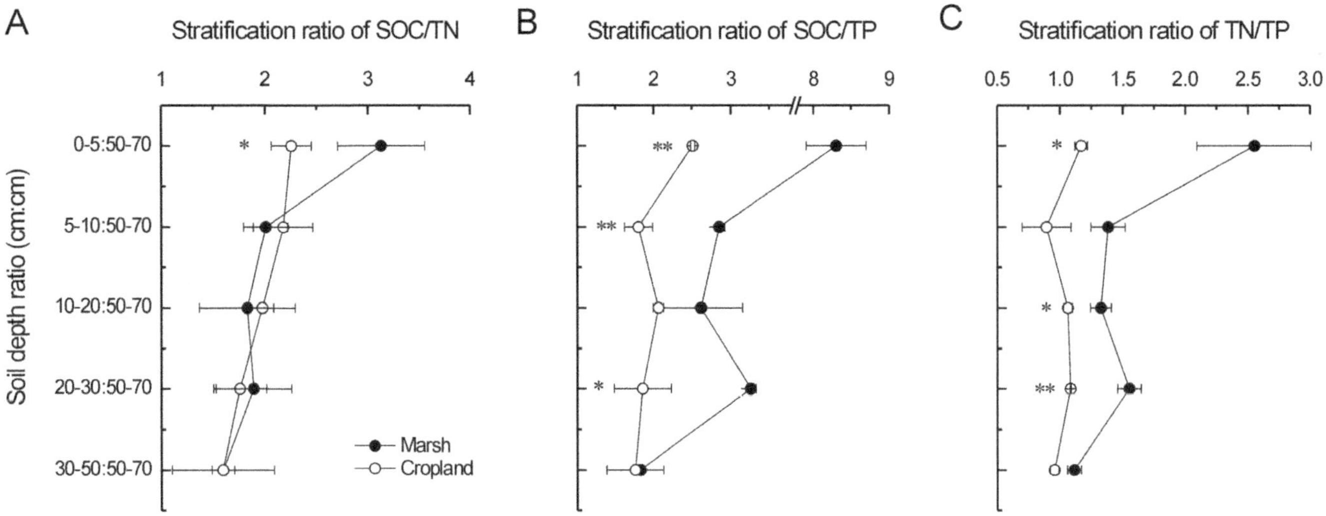

Figure 4. Long-term impacts of reclamation on stratification ratios of A) SOC/TN, B) SOC/TP, and C) TN/TP. * and ** indicate significant difference at $p<0.05$ and $p<0.01$ level, respectively.

had high rates of decomposition due to the aerobic conditions. Compared to cropland soils, there were preferential accumulations of organic matter in submerged soils [12,33]. Second, crop harvesting took lots of organic materials away from farmland ecosystem. Moreover, the net primary productivity of marshes is higher than cropland ecosystems [33]. In addition, decrease in soil salinity caused by agricultural reclamation might accelerate the mineralization of organic matters [16]. In our study, however, there was no significant change of SOC and TN in 50–100 cm soil under long-term reclamation (Fig. 1), indicating limited effects of reclamation on deeper soil horizons. This result was consistent with the results in the Sanjiang Plain [34], the Xingkai Lake [35] and the Hangzhou Bay of China [18]. Compared to SOC and TN, there was no significant variation in TP in whole soil profile after long-term reclamation (Fig. 1). However, Huang et al. [13] found significant decreases in TP caused by conversions of marshes to drylands. In marsh and cropland ecosystems, biogeochemical cycle of phosphorus is different from carbon and nitrogen. Phosphorus has poor mobility and is easy to be fixed in the soil. The input and output of phosphorus in the soil is mainly as fertilizer application and crop harvesting, respectively [36]. Thus, reclamation-induced impacts on TP may be explained by the balance of fertilizer application and crop harvesting.

Stratification ratio is a quantitative indicator to represent the responses of soil profile feature to anthropogenic disturbance [22], which was mainly applied in farmlands (e.g. [19,20]). However, the application of SR is not common in costal wetlands. Our results showed that SRs (0–5 cm:50–70 cm and 5–10 cm:50–70 cm) of SOC and TN decreased remarkably after long-term agricultural reclamation. In support of this result, chemical properties of the surface soil were reported to be more sensitive to reclamation than that of the deeper layers in two coastal wetlands [4,18]. These results are reasonable because the surface

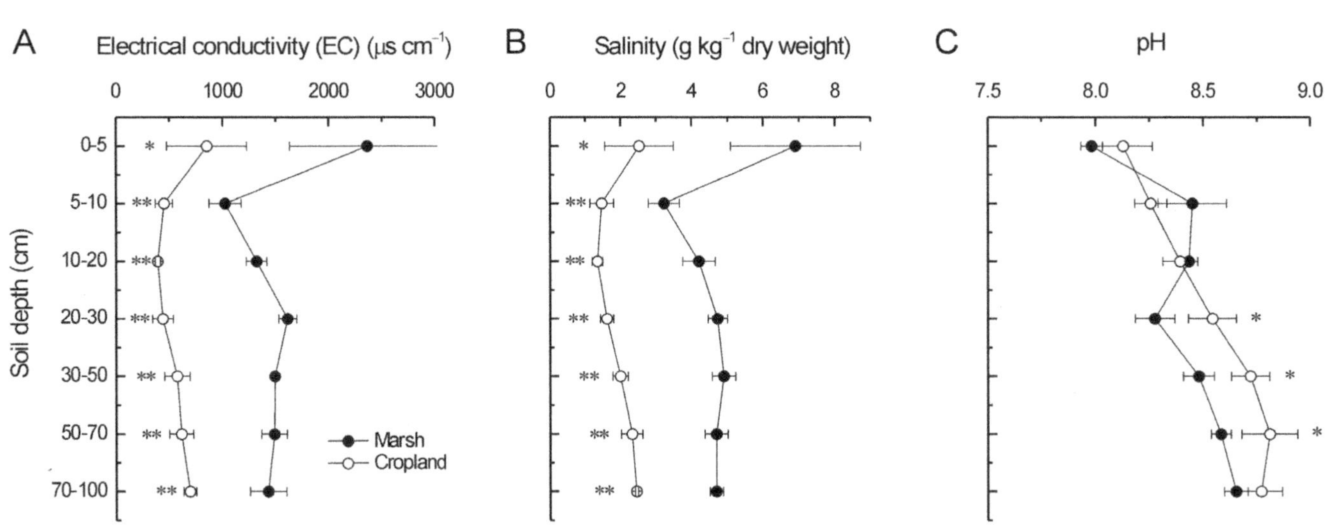

Figure 5. Long-term effects of reclamation on A) EC, B) salinity, and C) pH of coastal soils. * and ** indicate significant difference at $p<0.05$ and $p<0.01$ level, respectively.

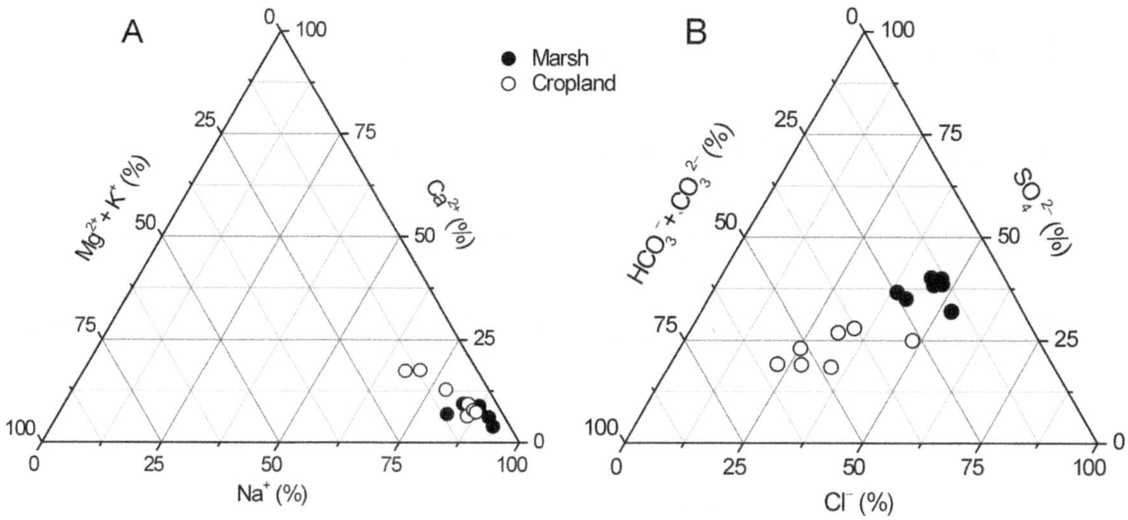

Figure 6. Long-term effects of agricultural cultivation on the compositions of A) cations and B) anions.

soils are the critical zones that receive intensive influences from anthropogenic activities.

Salinity, ion composition and alkalization

Our results showed that soil salinity and EC across the whole profile (0–100 cm) dropped significantly after agricultural reclamation (Fig. 5 A and B). This result was in agreement with the results in the Yellow River Delta [13], the Jiangsu province [11], the Yangtze Estuary [10], the Hangzhou Bay of China [14] and the Biscay Bay of Spain [9]. A reliable interpretation for this result is the effective leaching of the soil exchangeable ions induced by dewatering after reclamation. However, each of the major ions (anions and cations) in soil solutions has some unique properties which affect its production and mobility [37]. For example, the production of bicarbonate (HCO_3^-), one of the most common anions, is regulated by soil CO_2 pressure and pH. The sulfate (SO_4^{2-}) is involved in both biological and inorganic chemical reactions, whereas the chloride (Cl^-) is relatively uninvolved in these reactions and thus has high mobility [37]. The cations are held by the negatively charged clay and organic matter particles in the soil through electrostatic forces. Therefore, ion compositions were proposed to be sensitive to disturbance due to the specific mobility of each ion. In support of this concept, we found that the proportion of Na^+ (0–10 cm), Cl^- and SO_4^{2-} (5–100 cm) decreased significantly after reclamation. However, the proportion of Ca^{2+} (0–10 cm), Mg^{2+}+K^+ (5–20 cm) and HCO_3^-+CO_3^{2-} (5–100 cm) increased significantly after reclamation (Fig. 6). Knowing these reclamation-induced properties of the major ions, it is possible to protect and manage coastal saline soils of the wetlands and croplands as well as their ecosystem services. For example, reclamation-induced decreases in water table and soil salinity were beneficial to crop cultivation. High mobility of Na^+ further reduced the harm of salt ions after reclamation. Even though overlying water or saturated soil are basic properties of coastal wetlands, reclamation-induced hydrologic alterations such as ditch

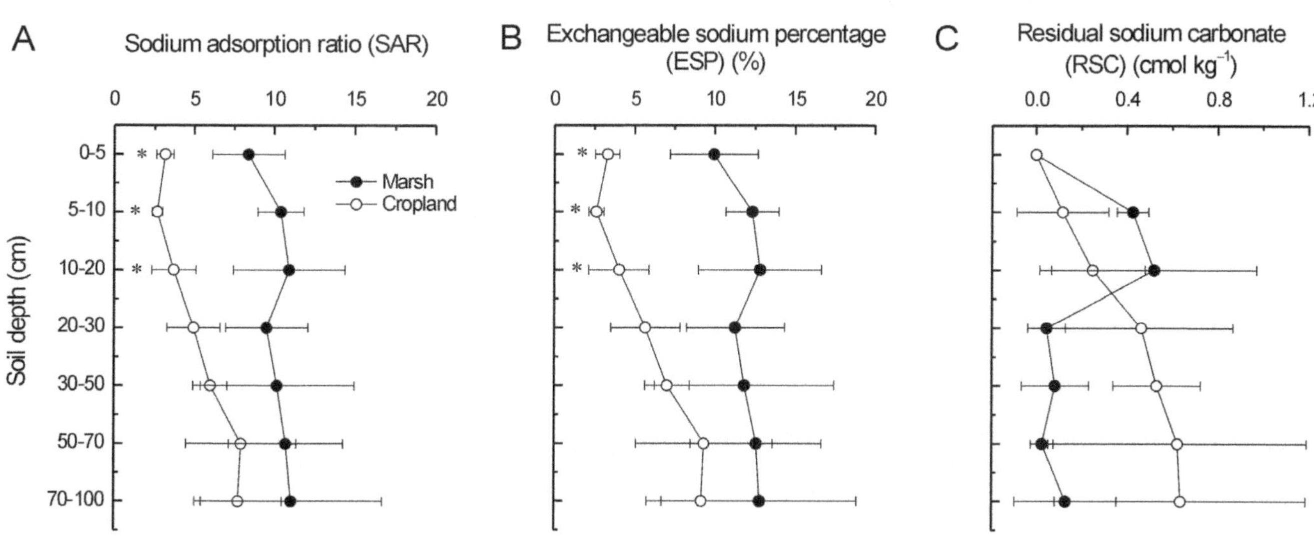

Figure 7. Long-term effects of reclamation on A) SAR, B) ESP, and C) RSC of the coastal marsh. * indicates significant difference at $p<0.05$.

drainage and diking result in decreases in salinity, which promotes plant activities and crop yields.

The SAR, ESP and RSC are three indicators of the degree of soil alkalization. The SAR of a soil extract takes into consideration that the adverse effect of Na^+ is moderated by the presence of Ca^{2+} and Mg^{2+}. When the SAR rises above 12 to 15, serious physical soil problems arise and plants have difficulty absorbing water [38]. In our study, all SARs of the marsh and cropland were below 12. Our results showed that agricultural reclamation significantly decreased the SAR and ESP but not RSC in the plow-layer (0–20 cm), suggesting a reduction in Na^+ harm (Fig. 7). These results were consistent with those in the Cape Cod, Massachusetts of USA [8] and the Hangzhou Bay of China [14]. Moreover, significant decreases of SAR and ESP in the upper soil horizons further supported the rapid desalinization.

Conclusions

In coastal marshes of northern China, long-term agricultural reclamation significantly reduced SOC and TN at surface soil (0–30 cm) as well as their SRs (0–5 cm:50–70 cm and 5–10 cm:50–70 cm). However, there was no significant change in TP as well as its SRs. Reclamation notably decreased ratios of SOC/TN, SOC/TP and TN/TP at surface soil (0–30 cm) as well as their SRs (0–5 cm:50–70 cm). A significant desalinization of whole soil profile occurred during cultivation. Soil pH (20–70 cm) markedly increased as an impact of reclamation. Compositions of cations (0–10 cm) and anions (5–100 cm) changed mainly as decreases in proportion of Na^+, Cl^- and SO_4^{2-}. Furthermore, reclamation significantly reduced the sodium adsorption ratio and exchangeable sodium percentage in plow-layer (0–20 cm) but not residual sodium carbonate, suggesting a reduction in sodium harm.

Acknowledgments

The authors thank Mr. Yuntao Shang for his help in laboratory analysis. The authors also appreciate Dr. Jin-Song Zhang and an anonymous reviewer for their helpful comments that improved this paper.

Author Contributions

Conceived and designed the experiments: ZW YW. Performed the experiments: YW XF CG QC. Analyzed the data: YW XF. Contributed reagents/materials/analysis tools: ZW YW. Wrote the paper: YW.

References

1. Costanza R, d'Arge R, de Groot R, Farber S, Grasso M, et al. (1997) The value of the world's ecosystem services and natural capital. Nature 387: 253–260.
2. Shepard CC, Crain CM, Beck MW (2011) The Protective Role of Coastal Marshes: A Systematic Review and Meta-analysis. PLoS ONE 6(11): e27374. doi:10.1371/journal.pone.0027374.
3. Cashin GE, Dorney JR, Richardson CJ (1992) Wetland alteration trends on the North Carolina coastal plain. Wetlands 12: 63–71.
4. Ellis S, Atherton JK (2003) Properties and development of soils on reclaimed alluvial sediments of the Humber estuary, eastern England. Catena 52: 129–147.
5. Santín C, de la Rosa JM, Knicker H, Otero XL, Álvarez MÁ, et al. (2009) Effects of reclamation and regeneration processes on organic matter from estuarine soils and sediments. Org Geochem 40: 931–941.
6. An S, Tian Z, Cai Y, Wen T, Xu D, et al. (2013) Wetlands of Northeast Asia and High Asia: an overview. Aquat Sci 75: 63–71.
7. An S, Li H, Guan B, Zhou C, Wang Z, et al. (2007) China's natural wetlands: past problems, current status, and future challenges. Ambio 36: 335–342.
8. Portnoy JW (1999) Salt marsh diking and restoration: biogeochemical implications of altered wetland hydrology. Environ Manage 24: 111–120.
9. Fernández S, Santín C, Marquínez J, Álvarez MA (2010) Saltmarsh soil evolution after land reclamation in Atlantic estuaries (Bay of Biscay, North coast of Spain). Geomorphology 114: 497–507.
10. Cui J, Liu C, Li Z, Wang L, Chen X, et al. (2012) Long-term changes in topsoil chemical properties under centuries of cultivation after reclamation of coastal wetlands in the Yangtze Estuary, China. Soil Till Res 123: 50–60.
11. Jin X, Huang J, Zhou Y (2012) Impact of coastal wetland cultivation on microbial biomass, ammonia-oxidizing bacteria, gross N transformation and N_2O and NO potential production. Biol Fertil Soils 48: 363–369.
12. Bridgham SD, Updegraff K, Pastor J (1998) Carbon, nitrogen, and phosphorus mineralization in northern wetlands. Ecology 79: 1545–1561.
13. Huang L, Bai J, Chen B, Zhang K, Huang C, et al. (2012) Two-decade wetland cultivation and its effects on soil properties in salt marshes in the Yellow River Delta, China. Ecol Inform 10: 49–55.
14. Iost S, Landgraf D, Makeschin F (2007) Chemical soil properties of reclaimed marsh soil from Zhejiang Province P.R. China. Geoderma 142: 245–250.
15. Weston NB, Dixon RE, Joye SB (2006) Ramifications of increased salinity in tidal freshwater sediments: geochemistry and microbial pathways of organic matter mineralization. J Geophys Res-Biogeoscience 111, G01009. doi:10.1029/2005JG000071.
16. Wong VNL, Greene RSB, Dalal RC, Murphy BW (2010) Soil carbon dynamics in saline and sodic soils: a review. Soil Use Manage 26: 2–11.
17. Lou Y, Wang J, Liang W (2011) Impacts of 22-year organic and inorganic N managements on soil organic C fractions in a maize field, northeast China. Catena 87: 386–390.
18. Wu M, Shao X, Hu F, Jiang K (2008) Effects of reclamation on soil nutrients distribution of coastal wetland in south Hangzhou Bay. Soils 40: 760–764. (in Chinese with English abstract)
19. Du Z, Ren T, Hu C (2010) Tillage and residue removal effects on soil carbon and nitrogen storage in the North China Plain. Soil Sci Soc Am J 74: 196–202.
20. Lou Y, Xu M, Chen X, He X, Zhao K (2012) Stratification of soil organic C, N and C:N ratio as affected by conservation tillage in two maize fields of China. Catena 95: 124–130.
21. Han G, Xing Q, Yu J, Luo Y, Li D, et al. (2013) Agricultural reclamation effects on ecosystem CO_2 exchange of a coastal wetland in the Yellow River Delta. Agr Ecosyst Environ doi:org/10.1016/j.agee.2013.09.012.
22. Franzluebbers AJ (2002) Soil organic matter stratification ratio as an indicator of soil quality. Soil Till Res 66: 95–106.
23. Liu H, Mou S (1999) Farmland fertilization conditions and countermeasures in Tianjin. Sci Technol Tianjin Agr For 2: 43–45. (in Chinese)
24. Zhu M (2011) Study on agricultural NPS loads of Haihe Basin and assessment on its environmental impact. Beijing: Doctoral Dissertation, Institute of Agricultural Economics and Development, Chinese Academy of Agricultural Sciences. (in Chinese with English abstract)
25. Robbins CW (1984) Sodium adsorption ratio-exchangeable sodium percentage relationships in a high potassium saline-sodic soil. Irrigation Sci 5: 173–179.
26. Oster JD, Sposito G (1980) The gapon coefficient and the exchangeable sodium percentage-sodium adsorption ratio relation. Soil Sci Soc Am J 44: 258–260.
27. Shi Y, Xin D (1983) Water-salt movement and comprehensive management of droughts, floods and saline-alkali in the Huang-Huai-Hai Plain. Hebei People's Publishing House, Shijiazhuang. (in Chinese)
28. Wan H, Yang J, Yu R (1983) Investigation on calculating method of soil alkalinity in the Huang-Huai-Hai Plain. Soils 23: 319–325. (in Chinese)
29. Wang Y, Wang Z-L, Wang H, Guo C, Bao W (2012) Rainfall pulse primarily drives litterfall respiration and its contribution to soil respiration in a young exotic pine plantation in subtropical China. Can J Forest Res 42: 657–666.
30. Wang Y, Wang H, Ma Z, Dai X, Wen X, et al. (2013) The litter layer acts as a moisture-induced bidirectional buffer for atmospheric methane uptake by soil of a subtropical pine plantation. Soil Biol Biochem 66: 45–50.
31. Wang Y, Wang H, Wang Z-L, Ma Z, Dai X, et al. (2014) Effect of litter layer on soil-atmosphere N_2O flux of a subtropical pine plantation in China. Atmos Environ 82: 106–112.
32. Macreadie PI, Hughes AR, Kimbro DL (2013) Loss of 'Blue Carbon' from Coastal Salt Marshes Following Habitat Disturbance. PLoS ONE 8(7): e69244. doi:10.1371/journal.pone.0069244.
33. Sahrawat KL (2003) Organic matter accumulation in submerged soils. Adv. Agron. 81: 169–201.
34. Wang L, Song C, Ge R, Song Y, Liu D (2009) Soil organic carbon storage under different land-use types in Sanjiang Plain. China Environ Sci 29: 656–660. (in Chinese with English abstract)
35. Huo L, Zou Y, Guo J, Lü X (2013) Effect of reclamation on the vertical distribution of SOC and retention of DOC. Environ Sci 34: 283–287. (in Chinese with English abstract)
36. Shen J, Yuan L, Zhang J, Li H, Bai Z, et al. (2011) Phosphorus dynamics: from soil to plant. Plant Physiol 156: 997–1005.
37. Johnson DW, Cole DW (1980) Anion mobility in soils: Relevance to nutrient transport from forest ecosystems. Environ Int 3: 79–90.
38. Munshower FF (1994) Practical handbook of disturbed land revegetation. Florida: Lewis Publishers.

Tungsten Distribution in Soil and Rice in the Vicinity of the World's Largest and Longest-Operating Tungsten Mine in China

Chunye Lin[1]*, Ruiping Li[1], Hongguang Cheng[1], Jing Wang[2], Xiao Shao[1]

1 State Key Joint Laboratory of Environmental Simulation and Pollution Control, School of Environment, Beijing Normal University, Beijing, China, **2** Key Laboratory of Land Use, China Land Surveying and Planning Institute, Ministry of land and Resources, Beijing, China

Abstract

The objective of this study is to investigate tungsten (W) contamination in soil and its enrichment in rice in the area of the world's largest and longest-operating W mines in China. Root zone soil and rice plants were sampled at 15 sites in the agricultural field adjacent to W mines and analyzed for Al, Fe, Mn, Sc, and W contents and W chemical forms in the soil samples and W contents in the rice root, stem, leaf, and grain samples. Results showed that W content in the soil ranged from 3.99 to 43.7 mg kg^{-1}, with more than 90% of W in the residual fraction, showing its low mobility and bioavailability. Average W contents in the rice root, stem, leaf, and grain were 7.06, 2.34, 4.76, 0.17 mg kg^{-1}, respectively. In addition, they were linearly independent of W content and chemical forms in the soil. Average enrichment factor values were 0.39, 0.13, 0.28, and 0.01 for the root, stem, leaf, and grain, respectively. In can be concluded that W mining activity in the Dayu county contaminated the nearby agricultural soil and led to W bioaccumulation in the rice. This may pose a health risk to residents via food and soil ingestion, which should be a focus of scrutiny.

Editor: Jin-Song Zhang, Institute of Genetics and Developmental Biology, Chinese Academy of Sciences, China

Funding: This study was supported by the National Natural Science Foundation of China (41371441, http://www.nsfc.gov.cn/), and the Ministry of Environmental Protection Funded Project (201309044, 201109064; http://www.zhb.gov.cn/). The funders had no role in study design, data collection and analysis, decision to publish, or preparation of the manuscript.

Competing Interests: The authors have declared that no competing interests exist.

* E-mail: c.lin@bnu.edu.cn

Introduction

In recent years, the biogeochemistry of tungsten (W) has become a matter of increasing concern due to the scrutiny of a children leukemia cluster in Nevada, its toxicity to organisms, and ubiquitous presence of this element in the environment as a result of geogenic and anthropogenic processes [1–5]. Whereas accumulation of some heavy metals in soils and organisms has been well studied [6–11], the researches on W accumulation in soils and plants are extremely limited.

In general, the content of W in non-polluted soils is low, ranging from 0.4 to 5.0 mg kg^{-1} [6]. However, a few previous studies showed that the concentration of W in the soils around the W mining and/or smelting sites was elevated due to the emission of W during the mining and/or smelting processes [12–18]. For instance, high average W concentration (56 mg kg^{-1}) was reported for the topsoils in the vicinity of mining/smelting sites in North Queensland, Australia [14]. An investigation detected similar levels (10–67 mg kg^{-1}) of W content in top soils collected from four random locations in Fallon, Nevada within a few miles from a W refining plant [3]. Up to 150 mg kg^{-1} of W in the soil at an industrial production site for W trioxide in Switzerland was detected compared to geogenic background concentrations of 1 to 2.5 mg kg^{-1} [19].

Mobility and bioavailability of inorganic contaminants in soils depend on their chemical forms. However, studies on the chemical forms and bioavailability of W in soils are extremely limited. Wilson and Pyatt [15,17] investigated chemical forms and bioavailability in the soils adjacent to W mines, showing that the bioavailability and mobility of W was much lower in the acidic soils than in alkaline soils. Soils surrounding a W ore-processing plant contained total W in the range of 100 to 200 mg kg^{-1}, of which 30% is water soluble, 15% bound to Fe oxides, and 5% fixed by organic matter [6].

W uptake by plants, especially by agricultural crops, is of concern because of the potential for W to enter the food supply [20]. Land plants growing in the uncontaminated soils by W generally contain low W, being less than 0.1 mg kg^{-1} [16,19,21–23]. High concentration of W in the contaminated soils generally elevated the concentration of W in the wild land plants such as trees, shrubs, and grasses [12–18]. For example, *Calluna Vulgaris* growing in the soil of Carrock Fell Mine, UK containing 1169 mg kg^{-1} of W had 655, 48.9, and 124 mg kg^{-1} of W in its root, woody part, and leaves, respectively [15]. In addition, the concentration of W in the plants growing in the tungstate or metallic W added soils increased with the concentration of W in the soils [2,20,24–30]. For example, bioaccumulation of W into *Brassica Oleracae* leaves was dose-dependent with soil W concentration, ranging from 278 to 1420 mg kg^{-1} of W [30].

The literature findings discussed above are confusing as to W chemical forms in the soils. No one determined W concentration, distribution, and enrichment in rice, one of major food around the world.

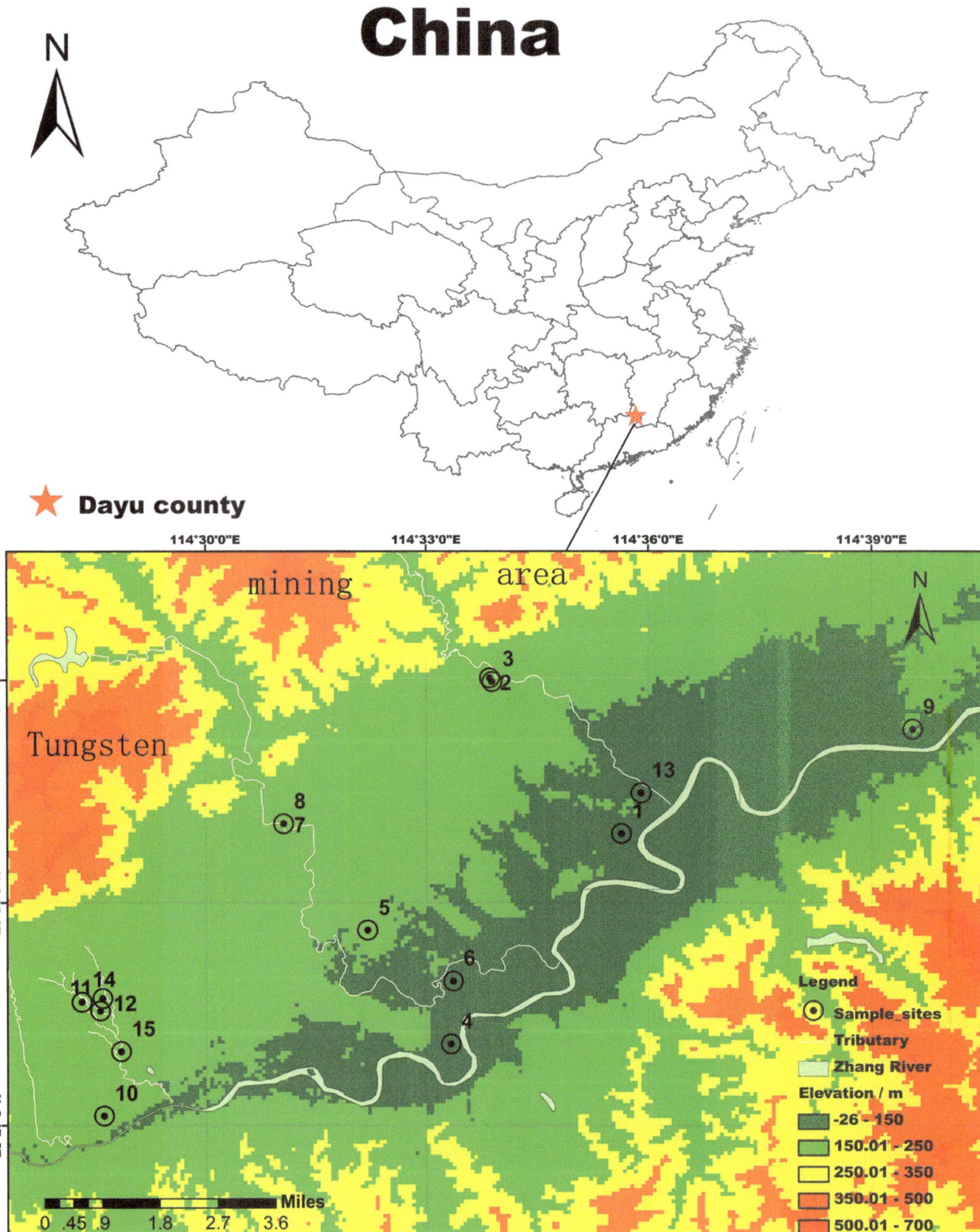

Figure 1. Schematic graph of W mine locations and soil sampling areas.

Table 1. Contents of mineral matrix elements and W in the soil.

Sample	pH	Mn	Sc	Al	Fe	W
		mg/kg	mg/kg	mg/kg	mg/kg	mg/kg
1	5.60	718.60	14.21	7.69	5.04	8.21
2	5.57	181.70	8.51	4.17	1.30	43.65
3	5.26	161.80	6.42	2.72	1.73	16.22
4	5.64	435.10	9.96	5.52	2.26	28.00
5	5.07	168.60	5.96	2.78	1.77	12.59
6	5.42	316.50	10.19	4.84	2.69	17.04
7	5.45	145.40	5.91	2.47	0.93	15.20
8	5.38	158.50	6.49	2.90	1.02	15.48
9	5.36	405.30	10.11	5.79	2.73	37.95
10	5.92	179.40	8.74	4.19	1.68	10.41
11	5.37	68.76	5.87	2.60	1.31	5.27
12	5.51	77.79	6.96	3.32	1.21	3.99
13	4.95	180.40	8.42	4.29	2.40	4.26
14	5.00	82.84	6.06	2.76	1.33	4.98
16	5.24	170.00	8.92	3.96	1.88	25.36
Mean	5.38	230.05	8.18	4.00	1.95	16.57
Median	5.38	170.00	8.42	3.96	1.73	15.20
Max	5.92	718.60	14.21	7.69	5.04	43.65
Min	4.95	68.76	5.87	2.47	0.93	3.99
Jiangxi[1]		328.00	10.31	8.60	2.88	5.28
China[1]		583.00	11.10	6.62	2.94	2.48
World[2]		1000.00	7.00	7.10	4.00	1.50

1: Data are cited from reference [34].
2: Data are cited from reference [35].

Table 2. Proportion of W in each chemical forms in the soil.

Sample	EXC	CARB	OX	OM	RES
	%	%	%	%	%
1	0.02	0.02	0.08	2.87	97.01
2	0.13	0.03	0.43	7.62	91.80
3	0.02	0.01	0.23	6.16	93.58
4	0.02	0.01	0.38	6.27	93.33
5	0.02	0.01	0.20	6.31	93.46
6	0.03	0.01	0.21	5.14	94.60
7	0.11	0.02	0.36	7.62	91.88
8	0.10	0.02	0.37	10.25	89.26
9	0.03	0.01	0.35	5.51	94.09
10	0.08	0.01	0.24	7.26	92.41
11	0.05	0.01	0.30	7.25	92.39
12	0.11	0.03	0.45	6.14	93.27
13	0.01	0.00	0.12	5.21	94.66
14	0.06	0.02	0.46	8.27	91.19
15	0.02	0.01	0.31	3.98	95.69
Mean	0.05	0.01	0.30	6.39	93.24
Median	0.03	0.01	0.31	6.27	93.33
Max	0.13	0.03	0.46	10.25	97.01
Min	0.01	0.00	0.08	2.87	89.26

China is the world's largest W producer and consumer. World tungsten supply is dominated by Chinese production and exports. In 2010, world W mine production was 61,000 t, of which 52,000 t is from China. Ganzhou in the south of Jiangxi province, being the birth place of Chinese W industry, is extremely rich in W sources. W output in Ganzhou has been at the top in China since W was discovered in Xihuashan mine in 1907 in the Dayu county, Ganzhou.

The objectives of this study were to investigate W content and distribution in the agricultural soils and rice in the vicinity of the largest W mine in China, with the longest W mining history. In addition, relationships between W contents in rice and soils were developed. This information is important to quantify W exposure to habitants via food chain.

Materials and Methods

Study site

The study site is situated in Dayu county, the southern Jiangxi province of southern China. It is characterized by subtropical monsoon climate, with average annual precipitation and temperature of 1591.5 mm and 18.5°C, respectively. Major crops are paddy rice and maize. There are three major tungsten mines in Dayu county: Xihuashan, Dangping, and Piaotang, among which Xihuashan is the first tungsten mine operated in China (Fig. 1.). Tungsten occurs in wolframite (Fe,MnWO_4) in these mines. Local farmers usually irrigate cultivated field with river water originated from the W mining area. No specific permissions were required for the studied area (25°27′08″−25°33′02″N, 114°28′19″−114°39′33″E). This field study did not involve endangered or protected species.

The soil at Dayu county generally contains soil organic matter of 2.76% to 6.26% and is characterized with acidic pH in the range of 5.09 to 6.28 and sandy loam texture [31].

Soil and plant sampling and analysis

Root zone soil and the rice plants were sampled at 15 sites in the agricultural fields adjacent to W mines. In order to investigate the transfer of W from soils to rice plant, we think it better to collect root zone soil samples. When rice is ready for harvest, rice plants were pulled out together with their roots and root-attached soils. The plants were vigorously shaken by hand and then washed with a little deionized water to remove the root-attached soil. The removed soil and rice plants were put into the plastic bags separately and brought to lab. The soil samples were air-dried in lab, crushed, passed through 0.149 mm sieve (100 mesh), and stored in glass bottles. The rice plants were washed in deionized water and oven-dried at 70°C. Then, root, stem, leaf, and grain (without husk) of the rice plants were separated, ground, and stored in glass bottles.

The pH value of each soil sample was analyzed in a 1:10 solid/liquid ratio suspension (left for ~0.5 h) using a combination pH electrode (Orion, USA). Portions of the soil samples were digested with HNO_3–HF–$HClO_4$ [32]. The Al, Fe, and Mn in the extracts were measured with ICP-AES (IRIS Intrepid II, Thermo Electron), while W and Sc were measured with ICP-MS (X Series II, Thermo Electron). The soil mineral matrix elements (Al, Fe,

Table 3. W contents in rice root, stem, leaf, and grain and its enrichment factor in them.

Sample	W content				W enrichment factor			
	Root	Stem	Leaf	Grain	Root	Stem	Leaf	Grain
	mg/kg	mg/kg	mg/kg	mg/kg				
1	2.34	0.70	0.85	0.12	0.285	0.086	0.104	0.015
2	25.76	7.50	16.66	0.57	0.590	0.172	0.382	0.013
3	7.09	1.64	9.64	0.33	0.437	0.101	0.594	0.020
4	10.27	4.34	5.33	0.36	0.367	0.155	0.190	0.013
5	4.64	1.42	3.71	0.08	0.369	0.113	0.295	0.006
6	5.83	1.49	1.83	0.08	0.342	0.087	0.107	0.005
7	4.26	2.34	4.13	0.09	0.280	0.154	0.272	0.006
8	5.44	2.21	3.78	0.09	0.351	0.143	0.244	0.006
9	21.75	6.30	14.41	0.40	0.573	0.166	0.380	0.010
10	2.30	1.45	2.58	0.07	0.221	0.139	0.248	0.006
11	1.57	1.12	2.57	0.05	0.297	0.212	0.488	0.010
12	1.90	0.72	1.45	0.04	0.478	0.181	0.364	0.010
13	2.23	0.23	0.59	0.02	0.524	0.055	0.138	0.005
14	2.24	0.59	1.75	0.07	0.449	0.118	0.351	0.014
15	8.31	3.10	2.19	0.18	0.328	0.122	0.086	0.007
Mean	**7.06**	**2.34**	**4.76**	**0.17**	**0.393**	**0.134**	**0.283**	**0.010**
Median	**4.64**	**1.49**	**2.58**	**0.09**	**0.367**	**0.139**	**0.272**	**0.010**
Max	**25.76**	**7.50**	**16.66**	**0.57**	**0.590**	**0.212**	**0.594**	**0.020**
Min	**1.57**	**0.23**	**0.59**	**0.02**	**0.221**	**0.055**	**0.086**	**0.005**

Mn, and Sc) were determined in order to show the basic geochemical compositions of the soils. Together with digestion and measurement of our soil samples, one reference soil (GSD7a) (triplicate), provided by Institute of Geophysical and Geochemical Exploration, Chinese Academy of Geological Sciences, was digested and analyzed to check the analytical quality. Average relative errors were 3.6 (1.9′5.8)%, 4.0(3.5~4.5)%, 3.9(3.2~4.7)%, 6.5(4.1~8.2)%, and −1.1(−6.2~2.0)% for Al, Fe, Mn, Sc, and W, respectively.

Portions of the root, stem, leaf, and grain samples were digested with concentrated HNO_3 and H_2O_2, and then concentrations of W in the extracts were measured with ICP-MS (X Series II, Thermo Electron).

In order to evaluate mobility and potential bioavailability of W, chemical forms of W in the soil samples were analyzed using a selectively sequential dissolution (SSD) method developed by Tessier et al. [33]. The chemical forms were operationally fractioned into the five phases: (1) soluble and exchangeable fraction (EXC), (2) bound to carbonate minerals (CARB), (3) bound to oxides (OX) (4) bound to organic matter (OM), and (5) residual fraction (RES). The selective sequential extractions were conducted in centrifuge tubes (polypropylene, 50 mL). After each successive extraction, separation was performed by centrifuging at 10,000 rpm for 10 min using a Xiang Yi centrifuge at room temperature. The supernatants were then removed with a pipette and filtered through a 0.45-µm membrane. The W concentration in each solution was determined using ICP-MS (X Series II, Thermo Electron). In this study, only the first four forms were measured and the contents of W in the residual fraction were calculated by total content minus sum of the first four forms.

Results and Discussion

Contents of mineral matrix elements and tungsten in the soil

Table 1 summarizes levels of pH, some mineral matrix elements and W in the soils collected in the agricultural fields near the W mines. Soil pH value ranged from 4.95 to 5.92, showing its acidic property. The mineral matrix elements Mn, Sc, Al, and Fe contents in the soil ranged from 68.8 to 718.8 mg kg^{-1}, 5.87 to 14.21 mg kg^{-1}, 2.47 to 7.69%, and 0.93 to 5.04%, respectively. The mean and median contents of Mn, Sc, Al, and Fe were generally lower than their background contents in the soils of the Jiangxi province, China, and world (except Sc) [34,35]. However, mean and median contents of W in the soil were 16.57 and 15.20 mg kg^{-1}, respectively; much higher than its background contents in the soils of the Jiangxi province, China, and world [34,35]. Therefore, the agricultural soil adjacent to W mines in Dayu county is contaminated by W.

In general, soils in the areas of W mining and/or smelting contained high W, e.g., 24.7–78.4 mg kg^{-1} for Mt Carbine Mine (Queensland, Australia) [14], 10 to 67 mg kg^{-1} for Fallon (Nevada, USA) [3], 1593 mg kg^{-1} for Carrock Fell Mine (Cumbria, UK), 920 mg kg^{-1} for Trumbull Old Mine (Connecticut, USA), 116 mg kg^{-1} for Devon Great Consols (Tmar, UK) [16], and 150 mg kg^{-1} for an industrial production site for W trioxide in Switzerland [19]. In the fallowland soil down the hill from the Höllberg W mine in Germany, W content ranged 15.9 to 101.4 mg kg^{-1} [18]. W content ranged from 3.99 to 43.65 mg kg^{-1} in the agricultural top soils adjacent to W mines in Dayu county, similar to the above mentioned contents.

Table 4. Relationship between the contents of W in plants (Y, mg W kg^{-1} dry weight) and in soils (X, mg kg^{-1}) for W mining field investigations and pot experiments with W-spiked soils.

Plant	Soil W conc. mg/kg	Duration	Equation	R^2	n	Ref.
Field investigation						
Rice	3.99–43.65					This study
Root			Y = 0.5692X − 2.3728	0.9141	15	
Stem			Y = 0.1703X − 0.4797	0.9496	15	
Leaf			Y = 0.3404X − 0.8781	0.7178	15	
Grain			Y = 0.0121X − 0.0306	0.8161	15	
Taraxacum officinale	15.9–101.4					[18]
Root			Y = 0.1757X − 1.229	0.8898	7	
Leaf			Y = 0.0132X+0.1979	0.8259	7	
Trifolium pratense	18.2–97.7					[18]
Root			Y = 0.1059X+2.5793	0.6139	7	
Leaf			Y = 0.0121X+0.1885	0.7048	7	
Pot experiment with W-spiked soils						
Nothofagus menziesii	0–125	8 weeks				[26]
Root			Y* = 32.205X+182.28	0.9778	12	
Stem (woody)			Y* = 0.7051X+19.313	0.7509	12	
Leaf			Y* = 0.9234X+9.8788	0.9009	12	
Sunflower	0–3900	28 days				[20]
Root			Y = 1.7098X+157.46	0.9863	5	
Leaf			Y = 0.0384X+16.234	0.9633	5	
Ryegrass	0–1000	9 months				[2]
Shoot			Y = 0.1916X+11.267	0.9935	5	
Wheat	0–120	up to fruiting				[28]
Root			Y = 0.88X − 2.66	0.9716	6	
Shoot			Y = 17.50X − 204.44	0.9288	6	
Cowpea	0–25	up to fruiting				[29]
Root			Y = 0.73X+6.3152	0.9661	6	
Shoot			Y = 0.23X+2.7685	0.9609	6	

*mg W kg^{-1} ash weight, R^2: correlation coefficient, n: number of samples, Ref.: reference.

Chemical forms of W in the soils

Generally, W in the soils was mostly bound in the RES fraction (89 to 97% of total W), while 2.9 to 10.3% of total W was associated with soil organic matter (Table 2.). Tungsten associated with EXC, CARB, and OX fraction was generally less than 0.5% of total W.

Heavy metals in soils are present in different chemical forms with a wide variety of solubility or bioavailability. Partitioning of trace metals among various soil fractions in a given soil depends on the soil fractions, affinity to the metals, soil composition and fraction concentrations in the soils, metal concentration in soil solution, and soil pH etc. [36]. Researches on W chemical forms are extremely limited. Wilson and Pyatt [15] investigated W chemical forms in the acidic spoils in the vicinity of an abandoned W mine, showing about 99% of W in the RES fraction. However, in the calcareous soils in the vicinity of an ancient metalliferous mine in the Corbières Area (Southwestern France), W content ranged from 20.5 to 77.9 mg kg^{-1}, of which about 41.2%, 23.0%, 19.1%, and 16.8% were bound in the exchangeable and carbonates, Fe/Mn oxides, soil organic matter, and residual fractions, respectively [17]. Therefore, W is more readily mobilized under alkaline conditions [17,37]. In general, mobility and bioavailability of W might be relatively low in the acidic soils of Dayu county.

Contents of W in the rice plants

The contents of W in the root, stem, leaf, and grain ranged from 1.57 to 25.76, 0.23 to 7.50, 0.59 to 16.66, and 0.02 to 0.57 mg kg^{-1}, respectively (Table 3). The W content and its enrichment factor in rice follow the order: root > leaf > stem > grain. This order is similar to that for wild *Calluna Vulgaris*, *Taraxacum officinale*, and *Trifolium pretense* reported by Wilson and Pyatt [15] and Jiang et al. [18], i.e., W content in root > W content in leaf > W content in stem. In addition, content and enrichment in the root of *Sunflowers*, *Ryegrass(Lolium perenne)*, *Nothofagus menziesii*, *wheat*, and *cowpea* growing in W-spiked soils were also higher than those in their leaves [2,12,20,28,29]. The common range of W in terrestrial plants was <0.001 to 0.15 mg kg^{-1} [6]. For example, the winter wheat grain, the spring barley grain, the wild berry grown in Sweden contained, in average, 0.006, 0.005, <0.01 mg kg^{-1} of

Table 5. Correlation matrix among mineral matrix element, W content and chemical forms in soil, and W contents in rice root, stem, leaf, and grain.

	pH	Mn	Sc	Al	Fe	TW	EXC-W	CARB-W	OX-W	OM-W	RES-W	R-W	S-W	L-W	G-W
pH	1.000	0.341	0.406	0.384	0.130	0.242	0.302	0.330	0.247	0.277	0.238	0.156	0.283	0.149	0.238
Mn		1.000	0.914*	0.933*	0.915*	0.237	−0.095	0.031	0.118	0.077	0.248	0.181	0.174	0.058	0.256
Sc			1.000	0.982*	0.903*	0.255	−0.024	0.118	0.143	0.063	0.269	0.211	0.185	0.014	0.234
Al				1.000	0.897*	0.284	−0.023	0.115	0.188	0.101	0.297	0.263	0.240	0.078	0.280
Fe					1.000	−0.002	−0.291	−0.161	−0.142	−0.205	0.014	−0.010	−0.083	−0.127	0.042
TW						1.000	0.708*	0.768*	0.969*	0.933*	1.000*	0.956*	0.975*	0.847*	0.903*
EXC-W							1.000	0.969*	0.807*	0.861*	0.692*	0.768*	0.773*	0.736*	0.680*
CARB-W								1.000	0.849*	0.869*	0.755*	0.811*	0.811*	0.742*	0.776*
OX-W									1.000	0.964*	0.964*	0.958*	0.990*	0.856*	0.897*
OM-W										1.000	0.922*	0.922*	0.951*	0.870*	0.871*
RES-W											1.000	0.953*	0.971*	0.841*	0.901*
R-W												1.000	0.968*	0.927*	0.916*
S-W													1.000	0.879*	0.899*
L-W														1.000	0.911*
G-W															1.000

Note: R-W: root W content, S-W: stem W content, L-W: leaf W content, G-W: grain W content. * Significant at $p = 0.01$ level.

W, respectively [23,38]. The content of W in the Danish onion ranged from 0.006 to 0.039 mg kg^{-1}, with an average of 0.017 mg kg^{-1} [22]. Twenty-two crop materials from Iowa contained 0 to 0.35 mg kg^{-1} of W [39]. However, plants grown in the W-contaminate soils usually contain high content of W. The maximum W levels of 50.7 mg kg^{-1} were found in cabbage, 41.2 mg kg^{-1} in onions and 12.3 mg kg^{-1} in potatoes grown in the W-contaminated soil adjacent to the W trioxide production plant [19].

Rice is the main cereals food for the local residents in Dayu county. Assuming that local residents consume 0.4 kg rice per day per person, maximal dietary exposure to W for consumers of contaminated rice was estimated to be up to 0.23 mg W d^{-1} per person corresponding to the highest W content in rice grain. On the other hand, the exposure to W for the residents via soil ingestion was estimated to be 0.002 W mg d^{-1} per person corresponding to the highest W content in the soil, assuming that soil ingestion rate is 50 mg d^{-1} per person [40]. At an industrial production site for W trioxide in Switzerland, dietary exposure to W for consumers of contaminated vegetables (cabbage, onions, and potatoes) growing in garden was estimated to be up to 1.5 mg W d^{-1} per person compared to background dietary exposure of approximately 0.01 mg W d^{-1} per person [19]. Therefore, the major way of W exposure to the residents at W contaminated sites is the dietary ingestion. Although permissible exposure limit for W in workplace atmospheric environments in USA (5 mg m^{-3}) and drinking water in the former USSR (0.05 mg L^{-1}) were established, the threshold value for W content in any plant has not been established up to now from the viewpoint of food safety [2]. Tungsten health risk caused by the rice ingestion can not be quantitatively estimated at current stage.

The enrichment factors, defined as W contents in root, stem, leaf, and grain divided by its content in the soil, were 0.221 to 0.590 for root, 0.055 to 0.212 for stem, 0.086 to 0.594 for leaf, and 0.005 to 0.020 for grain. Previous studies showed the enrichment factor of W in the wild shrub shoots growing in the W mining areas (20.5–77.9 mg W kg^{-1} soil) was generally 0.01 to 0.8 [17]. The enrichment factors of W in the root and the leaf were 0.05–0.28 and 0.01–0.03, respectively, for *Taraxacum officinale* and *Trifolium pretense* growing in the fallowland soil (15.9 mg W kg^{-1} soil) down the hill from the Höllberg W mine in Germany [18]. Therefore, the W enrichment factor values in the rice root and leaf in this studied area were higher than those for the herb *Taraxacum officinale* and *Trifolium pretense.*

The contents of W in the root, stem, leaf, and grain of rice linearly increased with the content of W in the soil (Table 4), with the determination coefficients of 0.7178 to 0.9496. Several previous studies observed the dependence of plant W content on the soil W contents [3,18,28–30]. Previous experiments in field showed W contents in the roots and leaves of maize were significantly correlated with the W content in the soil adjacent to an old tungsten mine (Pechtelsgrün) in Germany [18]. In detail, W contents in the roots and leaves of maize increased from about 1 to 16 mg kg^{-1} and 0.4 to 1.5 mg kg^{-1}, respectively, with the increase of soil W content from about 20 to 100 mg kg^{-1} [18]. In addition, W concentrations of roots and leaves both of wild *Taraxacum officinale and Trifolium pratense* were significantly correlated with soil W concentration. And there was a significant correlation between root and leaf tungsten concentrations for both species [18]. Pot experiment in greenhouse also demonstrated that W contents in root and shoot of sunflowers, ryegrass, wheat, and cowpea showed a very strong linear dependence on the W content in the W-spiked soils [3,20,28,29] (Table 4).

Correlation among elemental contents, W chemical forms, and W contents in rice

The correlation matrix in Table 5 shows the relationships between elemental contents in soil, W chemical forms in soil, and W contents in rice. pH value was positively correlated to the contents of Mn, Sc, Al, Fe, W, W chemical forms, and W in rice, whereas these correlations are not significant at $p = 0.01$ level. In general, the mineral matrix elements Mn, Sc, Al and Fe in the soil were correlated to one another, but they were not correlated to W content and chemical forms in the soil and the W contents in the rice. This might further inferred that the Mn, Sc, Al, and Fe contents in the soil mainly originated from geogenic sources, while W contents in the soil and rice were impacted by W mining activities. Tungsten contents in the soil and rice and W chemical forms in the soil were correlated to one another. These correlations showed that W mining activities in the Dayu county not only increased the total W content in soil, but also increased the W contents in each chemical form and in rice.

Plants are known to take up (possibly in anionic form, WO_4^{2-}) and accumulate W in substantial amounts [2,25]. Whereas 89% to 97% of W in the studied soil was associated to the residual fraction, non-residual content of W was 0.23 to 3.58 mg kg^{-1}, similar to W content (1.50 mg kg^{-1}) in worldwide soils [35]. In addition, W content in the non-residual fractions was positively correlated to the total content of W in the studied soil. Therefore, the contents of W in rice were positively correlated to the total content of W in the soil. Previous studies also showed that the extent of W accumulation in plants appears to be directly related to the W content of the soils whereas chemical forms of W in the soils were not measured [2,3,18,20,26,28,29].

Conclusions

About 100 years' mining W activity significantly elevated W contents in the soil adjacent to the W mines of the Dayu county. The rice grown in the soil accumulated W, with the linear dependence of W contents in the rice root, stem, leaf, and grain on the soil W content and chemical forms. The enrichment factor of W decreased in the order: root, leaf, stem, and grain. Most of W (about 93% in average) the soil was associated to the residual fraction. Tungsten contents and chemical forms in the soil and W contents in the rice root, stem, leaf, and grain were correlated to one another, but not correlated to the mineral matrix elemental (Mn, Sc, Al, and Fe) contents, due to the soil W contamination by W mining activities. The W exposure to local habitants via dietary and soil should be assessed further.

Author Contributions

Conceived and designed the experiments: CL HC JW. Performed the experiments: RL XS. Analyzed the data: CL RL HC JW. Contributed reagents/materials/analysis tools: CL RL XS. Wrote the paper: CL RL HC JW. Sample collection: CL HC XS.

References

1. inmaus C, Lu M, Todd RL, Smith AH (2004) Probability estimates for the unique childhood leukemia cluster in Fallon, Nevada, and risks near other U.S. military aviation facilities. Environ Health Perspect 112:766–771.
2. Strigul N, Koutsospyros A, Arienti P, Christodoulatos C, Dermatas D, et al. (2005) Effects of tungsten on environmental systems. Chemosphere 61:248–258.
3. Koutsospyros A, Braida W, Christodoulatos C, Dermatas D, Strigul N (2006) A review of tungsten: From environmental obscurity to scrutiny. J Hazard Mater 136:1–19.
4. Strigul N (2010) Does speciation matter for tungsten ecotoxicology?. Ecotox Environ Safe 73:1099–1113.
5. Adamakis IDS, Panteris E, Eleftheriou EP (2012) Tungsten toxicity in plants. Plants 1:82–99.
6. Kabata-Pendias A, Mukherjee AB (2007) Trace Elements from Soil to Human. Springer, Berlin.
7. Fritsch C, Coeurdassier M, Giraudoux P, Raoul F, Douay F, et al. (2011) Spatially expicit analysis to biota: Influence of soil contamination and landscape. PLoS ONE 6(5): e20682. doi:10.1371/journal.pone.0020682.
8. Zheng SA, Zheng X, Chen C (2012) Leaching behavior of heavy metals and transformation of their speciation in polluted soil receiving simulated acid rain. PLoS ONE 7(11): e49664. doi:10.1371/journal.pone.0049664.
9. Zhu H, Guo J, Chen M, Feng G, Yao Q (2012) Burkholderia dabaoshanensis sp. nov., a heavy-metal-tolerant bacteria isolated from Dabaoshan mining area soil in China. PLoS ONE 7(12): e50225. doi:10.1371/journal.pone.0050225.
10. Jiang M, Zeng G, Zhang C, Ma X, Chen M, et al. (2013) Assessment of heavy metal contamination in the surrounding soils and surface sediments in Xiawangang River, Qingshuitang District. PLoS ONE 8(8): e71176. doi:10.1371/journal.pone.0071176.
11. Huang Z, Pan XD, Wu PG, Han JL, Chen Q (2013) Health risk assessment of heavy metals in rice to the population in Zhejiang, China. PLoS ONE 8(9): e75007. doi:10.1371/journal.pone.0075007.
12. Quin BF, Brooks RR (1972) The tungsten contents of some plants from a mineralized area in New Zealand. NZ J Sci 15:308–312.
13. Quin BF, Brooks RR (1974) Tungsten concentrations in plants and soils as a means of detecting scheelite-bearing ore-badies in New Zealand. Plant Soil 41:177–188.
14. Pyatt FB, Pyatt AJ (2004) The bioaccumulation of tungsten and copper by organisms inhabiting metalliferous areas in North Queensland, Australia: an evaluation of potential health implications. J Environ Health Res 3:13–18.
15. Wilson B, Pyatt FB (2006) Bioavailability of tungsten in the vicinity of an abandoned mine in the English Lake District and some potential health implications. Sci Total Environ 370:401–408.
16. Wilson B, Pyatt FB (2009) Persistence and bioaccumulation of tungsten and associated heavy metals under different climatic conditions. Land Contam Reclam 17:93–100.
17. Wilson B, Pyatt FB (2009) Bioavailability of tungsten and associated metals in calcareous soils in the vicinity of an ancient metalliferous mine in the Corbieres area, Southwestern France. J Toxicol Env Health A 72:807–816.
18. Jiang F, Heilmeier H, Hartung W (2007) Abscisic acid relations of plants grown on tungsten enriched substrates. Plant Soil 301:37–49.
19. Brueschweiler B, Waber U, Gupta S (2009) Tungsten, a new vegetable contaminant needs further elaborated evaluation. Toxicol Letters 189:S219.
20. Johnson DR, Inouye LS, Bednar AJ, Clarke JU, Winfield LE, et al. (2009) Tungsten bioavailability and toxicity in sunflowers (Helianthus Annuus L.). Land Contam Reclam 17:141–151.
21. Bowen HJM (1960) The determination of tungsten in biological material by activation analysis. Biochem J 77:79–82.
22. Bibak A, Behrens A, Stefan Stürup S, Knudsen L, Gundersen V (1998) Concentrations of 63 major and trace elements in Danish agricultural crops measured by inductively coupled plasma mass spectrometry. 1. Onion (Allium cepa Hysam). J Agric Food Chem 46: 3139–3145.
23. Rodushkina I, Ödmana F, Holmströmb H (1999) Multi-element analysis of wild berries from northern Sweden by ICP techniques. Sci Total Environ 231:53–65.
24. Davies EB, Stockdill SMJ (1956) A pasture response to sodium tungstate on a New Zealand soil. Nature 78:866.
25. Wilson DO, Cline JF (1966) Removal of plutonium-239, tungsten-185 and lead-210 from soils. Nature 209:941–942.
26. Quin BF, Brooks RR, Reay PF (1972) The accumulation of tungsten by Nothofagus menziessi. Plant Soil 36:699–703.
27. Quin BF, Hoglund JH (1976) The effect of tungstate and nitrogen source on the dry weight and nitrogen yields, and molybdenum and tungsten content, of white clover (Trifolium Repens). Plant Soil 45:201–212.
28. Kumar A, Aery NC (2011) Effect of tungsten on growth, biochemical constituents, molybdenum and tungsten contents in wheat. Plant Soil Environ 57:519–525.
29. Kumar A, Aery NC (2012) Effect of tungsten on growth, dry-matter production, and biochemical constituents of cowpea. Commun Soil Plant Anal 43:1098–1107.
30. Kennedy AJ, Johnson DR, Seiter JM, Lindsay JH, Boyd RE, et al. (2012) Tungsten toxicity, bioaccumulation, and compartmentalization into organisms representing two trophic levels. Environ Sci Technol 46:9646–9652.
31. Cheng HG, Duan XL, Liu SQ, Lin CY, Shao X (2013) Thallium, arsenic, and mercury contamination of soil near the world's largest and longest-operating tungsten mine. Polish J Environ Studies 22:301–305.
32. Lin C, He MC, Zhou YX, Guo W, Yang ZF (2008) Distribution and contamination assessment of heavy metals in sediment of the Second Songhua River, China. Environ Monit Assess 137:329–342.
33. Tessier A, Campbell PGC, Bisson M (1979) Sequential extraction procedure for the speciation of particulate trace metals. Anal Chem 51:844–851.
34. Wei FS, Chen JS, Wu YY, Zheng CJ, Jiang DZ (1990) Background contents of elements in China soils. Publishing House of Chinese Environmental Sciences, Beijing.
35. Bowen HJM (1979) Environmental chemistry of the elements. Academic Press, London.
36. Lin CY, Negev I, Eshel G, Banin A (2008) In Situ Accumulation of Copper, Chromium, Nickel, and Zinc in Soils Used for Long-term Waste Water Reclamation. J Environ Qual 37:1477–1487.
37. Bednar AJ, Boyd RE, Jones WT, McGrath CJ, Johnson DR, et al. (2009) Investigations of tungsten mobility in soil using column tests. Chemosphere 75:1049–1056.

38. Eriksson JE (2001) Concentrations of 61 trace elements in sewage sludge, farmyard manure, mineral fertilizers, precipitation, and in soil and crops. Swedish EPA Rep 5159.
39. Fu MH, Tabatabai MA (1988) Tungsten contents of soils, plants and sewage sludges in Iowa. J Environ Qual 17:146–148.
40. USEPA (2011) Exposure Factors Handbook: 2011 Edition. EPA/600/R-090/052F, www.epa.gov.

Impacts of Agricultural Management and Climate Change on Future Soil Organic Carbon Dynamics in North China Plain

Guocheng Wang, Tingting Li*, Wen Zhang, Yongqiang Yu

State Key Laboratory of Atmospheric Boundary Layer Physics and Atmospheric Chemistry, Institute of Atmospheric Physics, Chinese Academy of Sciences, Beijing, China

Abstract

Dynamics of cropland soil organic carbon (SOC) in response to different management practices and environmental conditions across North China Plain (NCP) were studied using a modeling approach. We identified the key variables driving SOC changes at a high spatial resolution (10 km×10 km) and long time scale (90 years). The model used future climatic data from the FGOALS model based on four future greenhouse gas (GHG) concentration scenarios. Agricultural practices included different rates of nitrogen (N) fertilization, manure application, and stubble retention. We found that SOC change was significantly influenced by the management practices of stubble retention (linearly positive), manure application (linearly positive) and nitrogen fertilization (nonlinearly positive) – and the edaphic variable of initial SOC content (linearly negative). Temperature had weakly positive effects, while precipitation had negligible impacts on SOC dynamics under current irrigation management. The effects of increased N fertilization on SOC changes were most significant between the rates of 0 and 300 kg ha^{-1} yr^{-1}. With a moderate rate of manure application (i.e., 2000 kg ha^{-1} yr^{-1}), stubble retention (i.e., 50%), and an optimal rate of nitrogen fertilization (i.e., 300 kg ha^{-1} yr^{-1}), more than 60% of the study area showed an increase in SOC, and the average SOC density across NCP was relatively steady during the study period. If the rates of manure application and stubble retention doubled (i.e., manure application rate of 4000 kg ha^{-1} yr^{-1} and stubble retention rate of 100%), soils across more than 90% of the study area would act as a net C sink, and the average SOC density kept increasing from 40 Mg ha^{-1} during 2010s to the current worldwide average of ~55 Mg ha^{-1} during 2060s. The results can help target agricultural management practices for effectively mitigating climate change through soil C sequestration.

Editor: Ben Bond-Lamberty, DOE Pacific Northwest National Laboratory, United States of America

Funding: This work was jointly supported by the Ministry of Science and Technology of China (Grant No. 2010CB950603), the CAS Strategic Priority Research Program (Grant No. XDA05050507), the National Natural Science Foundation of China (Grant No. 41021004), and the State Key Program of National Natural Science of China (Grant No. 41130104). The funders had no role in study design, data collection and analysis, decision to publish, or preparation of the manuscript.

Competing Interests: The authors have declared that no competing interests exist.

* E-mail: litingting@mail.iap.ac.cn

Introduction

Cultivation generally leads to dramatic changes in soil organic carbon (SOC) by influencing the processes involved with soil C production and decomposition [1,2]. Changes in agricultural SOC are characterized by dynamic exchange processes which is strongly linked to environmental conditions such as air temperature, precipitation, soil pH and texture; and agronomic management such as crop rotation, stubble retention, tillage regimes, and application of chemical fertilizers and animal manure [3,4]. The complex interactions between management practices and environmental conditions, as well as the lack of spatiotemporal continuity in SOC monitoring data over meaningfully large areas, hamper our ability to accurately predict the regional cropland SOC change and identify the factors that control the SOC dynamics.

A modeling approach has advantages in estimating spatiotemporal changes in SOC under various management practices and environmental conditions [5]. A number of process-based carbon models such as DNDC [6], RothC [7], and CENTURY [8] have been used to simulate SOC dynamics in agricultural systems at both national and continental scales [9–11]. Recently, Huang et al. [12] developed a biogeophysical model, Agro-C, which has been validated and further used to assess the long-term agricultural SOC changes on a national scale in China [5] and Australia [2].

The North China Plain (NCP) is one of the most important agricultural production areas in China, covering a total cropland area of ~18 Mha, with a typical continuous winter wheat-summer maize cropping system (Figure 1). During the past several decades, significant changes in both climate and agricultural management practices have taken place in NCP. For example, Ding et al. [13] reported that both temperature and precipitation had shown an increasing trend across the northern part of China during the past 50 years. Improved management practices such as crop residue retention and application of mineral fertilizers and/or organic manure have been promoted as strategies to both increase crop productivity and optimize soil fertility [14]. Although an overall net increase in SOC across the croplands of NCP during the past 30 years has been reported [5], there remains large uncertainties in future cropland SOC dynamics due to the unclear future changes in both climate and management practices. For instance, it is generally recognized that the current global warming will increase the rates of organic matter decomposition, thereby accelerating climate change through C cycle feedbacks [15]. However, in areas with relatively low mean annual temperature and without water and nutrient deficiencies (e.g., the North China Plain, typically with irrigation and N fertilization to support crop production), increasing air temperature could potentially increase crop productivity and result in more input C into soils, thereby offsetting the loss in SOC through soil respiration. More recently, Yu et al. [16] simulated the Chinese cropland SOC changes during the next 40 years, by simply designing three levels of carbon input based on the historic changing trend in annual amounts of crop residues and manure. However, the quantity of input C is strongly linked to management such as stubble retention and N fertilization, which needs to be considered during a comprehensive assessment of long-term SOC dynamics, and determined to provide strategies in effectively managing the farming systems. Moreover, since it takes several decades or even a century to reach a new equilibrium of SOC following a change in management practice [17,18], the role of soils to sequester C (source or sink) as affected by environmental variables and management practices needs to be identified on a long time scale, e.g., around 100 years.

In this study, we quantified the influence of both climatic variables and management practices on SOC across croplands of North China Plain. The SOC was simulated under 900 combinations of climate, N fertilization, stubble retention, and manure application scenarios at a daily time step, from 2011 to 2100, and at a high spatial resolution, using the pre-calibrated and validated Agro-C model. The impacts of management practices and climatic variables on SOC dynamics were then statistically analyzed. This would not only provide a better understanding of the factors and processes regulating C cycling and balance in the agro-ecosystems, but would also provide insight into the effectiveness of methods for enhancing soil quality and mitigating climate change.

Materials and Methods

1.1 Agro-C

Agro-C [12] is a biogeophysical model, and it consists of two submodels, Crop-C and Soil-C. The Crop-C submodel simulates processes involved with crop photosynthesis, autotrophic respiration, and net primary production (NPP); and the Soil-C submodel simulates soil heterotrophic respiration via the decomposition of both input C (e.g., crop residues, roots and manure) and SOC. Changes in SOC are then determined by balancing the loss of SOC with the gain of input C. The input C is split into two components of labile-C and resistant-C, and the SOC pool is divided into two sub-pools named light-C and heavy-C. The light-C sub-pool is hypothesized to be more biologically reactive, while the heavy-C sub-pool is much more resistant to decomposition. Decomposition of each component and sub-pool is described by first-order kinetics at predefined potential rates (d^{-1}) of 2.6×10^{-2}, 8.4×10^{-4}, 2.5×10^{-4} and 1.8×10^{-5} for labile-C, resistant-C, light-C and heavy-C, respectively. The actual decomposition rate of each component and sub-pool is further modified by soil parameters such as temperature, moisture, texture and pH [12]. The C flow between different pools was determined based on the following assumptions: (i) the decomposition of both the labile-C and the resistant-C converts a fraction of the C into the light-C pool, (ii) a fraction of the decomposed light-C is transferred into

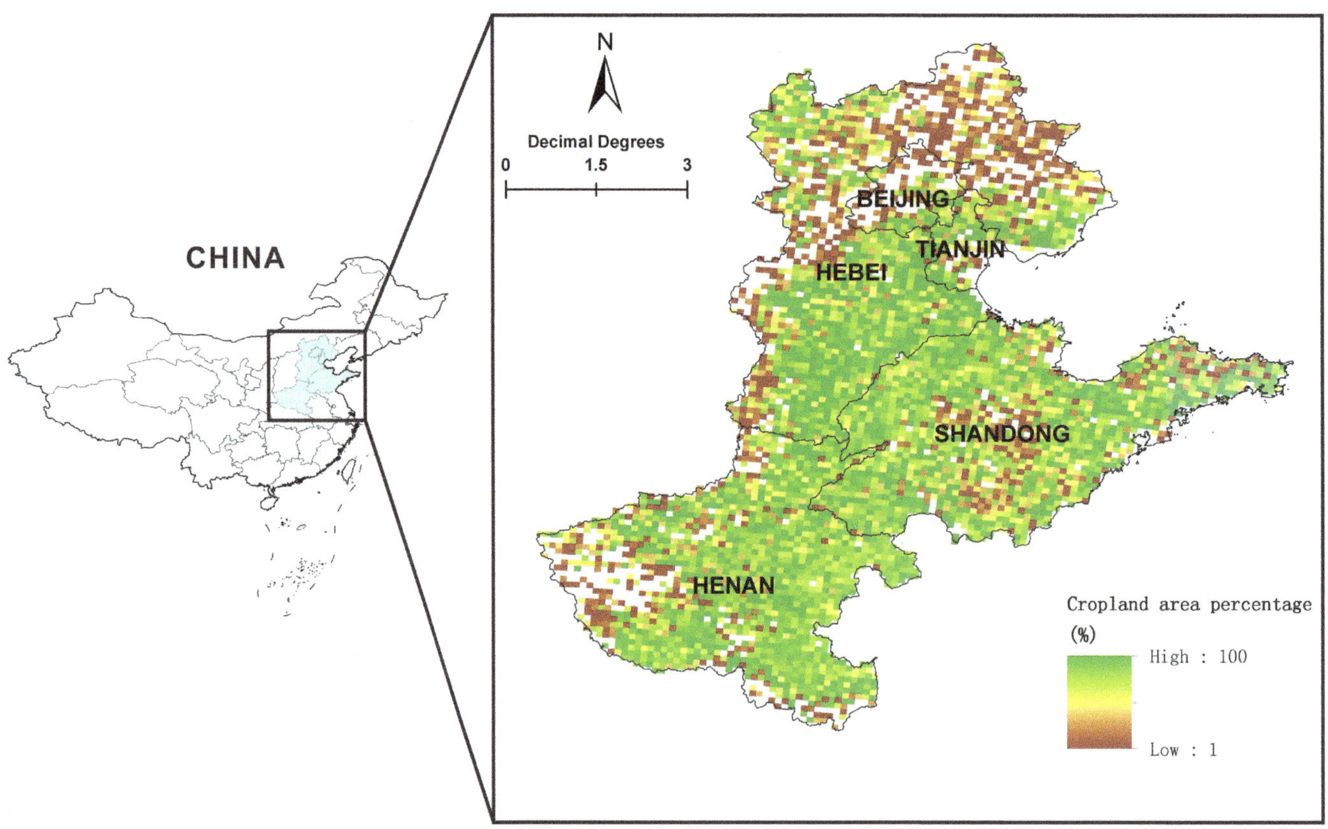

Figure 1. Spatial distribution of croplands in the North China Plain.

the heavy-C pool, and (iii) the decomposition of the heavy-C pool only produces CO_2. More detailed description of the Agro-C model was shown in Yu et al. [5] and Huang et al. [12].

In our previous studies [5,12], Agro-C has already been parameterized using observed data of several field measurements, and validated against independent datasets across a vast area representing different cropping systems, soil and climate conditions in China. The results indicate that the calibrated Agro-C model can reasonably capture observed aboveground biomass and changes in SOC under various agricultural management practices at different sites across china [5,12]. In the present study, we used the previously calibrated Agro-C model to simulate regional SOC dynamics of agro-ecosystems in North China Plain (NCP).

1.2 Spatial data

The changes in SOC under different climatic and management scenarios were simulated with a daily step from 2011 to 2100. Up-scaling of the Agro-C model was accomplished by first rasterising the model inputs (e.g., climate data, initial soil properties and agricultural management information) with 10 km×10 km gridded datasets across the North China Plain and then running the model grid by grid across the study area.

Input soil properties in Agro-C include the concentrations of organic carbon and total nitrogen, bulk density, clay and sand fraction, and pH in the topsoil to 30 cm depth. These soil properties, with a 10 km×10 km spatial resolution, were firstly obtained from China Soil Scientific Database [19], representing more than 7000 soil profile measurements across China made in the late 1970s/early 1980s. Using Agro-C, we then simulated these soil properties grid by grid on a daily step from 1981 to 2010, based on the historical climate data and recorded agricultural management practices [5]. The model outputs of soil properties, including the amount of C in different pools, in the end of 2010 were further compiled as the initial model input of soil properties of the present study.

Gridded daily climate data such as maximum and minimum temperature, precipitation, and relative humidity from 2010 to 2100 were derived from the projections of the climate model FGOALS (Document S1). Due to the coarser spatial resolution (2.8° longitude×2.8° latitude) of outputs from the FGOALS model, the projections for the future GHG concentration scenarios [i.e., a set of four representative concentration pathways (RCPs) named RCP2.6, RCP4.5, RCP6.0 and RCP8.5 respectively; Figure S1 and Figure S2] of the FGOALS were therefore statistically downscaled [20,21] to a resolution of 10 km×10 km (Document S1). Additionally, as a driving variable in Agro-C, atmospheric carbon dioxide (CO_2) concentration affects the process of crop photosynthesis [12], which under each future climate scenario was obtained from Clarke et al. [22] and used as model inputs in this study.

The spatial distribution of croplands was computed based on the National Land Cover Data Sets (NLCD) of China that was developed from Landsat TM digital images around 2000 [23]. Due to a lack of yearly cropland data with sufficient spatial distribution, we assumed that the area of croplands in North China Plain did not significantly change over the study period.

1.3 Model initialization

In Agro-C model, the half-life residence time for labile-C, resistant-C, light-C and heavy-C were 0.1 y, 2.3 y, 7.6 y, and 105.4 y [5], respectively. To obtain initial fractions for the different C pools in 1981, a 'spin-up' procedure was performed by first setting the initial fractions of light-C and heavy-C as 0.25 and 0.75 [5], respectively, and then running the model until a steady state of SOC pool was achieved. In this, an equilibrium run was not stopped until the difference of SOC pools between two successive years was less than 0.1% of the pools themselves. The length of spin-up procedure differed across different places, and generally lasted about 100 years. The average time duration of the spin-up seemed relatively short but is reasonable. Most of Chinese croplands, particularly in North China Plain, have generally experienced a very long history of cultivation. When the spin-up procedure was performed, the average amount of crop stubble retention and manure amendment over the period between 1980 and 1989 was used as organic C input, due to the unavailability of organic C input before 1980. The average of 1960–1990 climate were used as driving data, because of the unavailability of climate data before 1960. The other driving datasets such as nitrogen fertilizer and irrigation were set to 1980–1989 average values. We then calculated the initial values of different C pools based on the total SOC information obtained from the above mentioned soil database and the spin-upped fractions of different sub-C pools. The dynamics of different C pools in each grid from 1981 to 2010 on a daily time step were then modeled based on the historical climate data and recorded agricultural management practices [5]. The simulated values of above-mentioned four C pools in the end of 2010 were further compiled as the model inputs of the present study.

1.4 Agricultural management scenarios

Nitrogen (N) fertilization, application of manure, and crop stubble retention through tillage are the major management practices that affect SOC dynamics in croplands of North China Plain (NCP). A database containing the county-level yearly cultivation acreage and total amount of synthetic N application (Chinese Academy of Agricultural Sciences [24]), shows that N fertilization rates in 2010 across croplands of NCP varied from 0 to 800 kg ha^{-1} (Figure S3A). The environmental pressures that are related to excessive mineral N fertilization in Chinese croplands decrease the possibility of increasing its use in the future [25]. Another province-level database [26] shows that farm manure application rates during the past 30 years varied from 1000 to 4000 kg ha^{-1} across the studied area (Figure S3B). In the croplands of NCP, the average stubble retention rate over the period of 2000–2010 varied from ~40% to ~60% across space [27]. In general, an increasing fraction of crop residues were being left in the fields instead of being taken away for household fuel and animal feed in China over the past several decades [25], which was mainly attributed to the financial motivations of the government policies. To quantify the effects of both climate change and agricultural management practices on SOC, we simulated SOC changes within the winter wheat-summer maize systems under 225 combinations of N fertilization, manure application and stubble retention rates, under each of the above-mentioned four climate scenarios.

In this study, management practices included 9 nitrogen application rates (0–800 kg N ha^{-1} yr^{-1} in 100 kg N ha^{-1} yr^{-1} increments, i.e., N:0, N:100, N:200, N:300, N:400, N:500, N:600, N:700, and N:800). Apart from N:0, we specified that 55% of the N fertilizers were applied during summer-maize growing seasons, with another 45% applied during winter-wheat growing seasons. For each growing season, 40% of the N fertilizers were applied at crop emergence, another 40% applied at tillering, and the rest applied at heading for wheat and silking for maize. We also specified five manure application rates (0–4000 kg ha^{-1} yr^{-1} in 1000 kg ha^{-1} yr^{-1} increments, i.e., M:0, M:1000, M:2000, M:3000, and M:4000), and five stubble retention rates (0–100% in 25% increments, i.e., R:0, R:25, R:50, R:75, and R:100). Apart from M:0, 72% of the manure was applied before sowing of summer maize, with the rest applied before sowing of winter wheat. Stubble retention rates denoted the percentage of aboveground straw and leaf biomass incorporated into the soil by tillage after harvesting, with the rest directly removed from the system. Combinations of management practices are abbreviated in this study as M:x, N:x, R:x. The M:0, N:0, R:0 denotes the management practice with no N fertilization, no manure application, and no stubble retention.

Because of a lack of detailed crop cultivar and phenology data on the regional scale, we simply assumed that the same wheat and maize cultivars were used throughout the studied area, and crop parameters were set as default values of the pre-calibrated Agro-C model. Sowing dates of maize and wheat were based on local crop calendars. From the south to north in NCP, the sowing date ranged from 6 June to 15 June for maize, and ranged from 5 October to 11 November for wheat [28,29]. Crop growth stops at physiological maturity as determined by the accumulated temperature. In addition, according to the local agronomic management, we assumed that the soil moisture would arrive at field capacity using irrigation module when the modeled soil moisture in the 30 cm depth drops to 70% of the field capacity during the growing seasons [5].

In total, we ran 4,086,000 (225 management scenarios ×4 climate scenarios ×4540 grids) Agro-C simulations. Each simulation quantified SOC content in the top 30 cm of soil from 2011 to 2100 on a daily step.

1.5 Identifying controls on SOC dynamics

In the present study, we assessed the impacts of management practices, climatic and soil variables on SOC change using Spearman's rank correlation coefficient (*rho*). Selected climatic variables included mean annual temperature (hereafter simply denoted as *temperature*) and mean annual precipitation (hereafter simply denoted as *precipitation*) since these two variables have been reported uncorrelated and significantly represent the spatial variation in a range of climate patterns [30]. For correlation analysis, the long-term daily climate variables were summarized to mean annual values under each climate scenario for each grid. Selected soil parameters included the main model edaphic inputs, e.g., pH, initial SOC, clay and total nitrogen content. Change in SOC of the top 30 cm (ΔSOC) is calculated as the difference in SOC between 2100 and 2011. Spearman's rank correlation coefficient was then calculated between ΔSOC and the management practices and environmental variables across the full set of Agro-C simulations. The sign of *rho*, positive or negative, indicates the direction of association between the independent and dependent variables. The absolute magnitude of *rho*, between 0 and 1, suggests the strength of correlation between the two variables.

We also investigated the effects of the initial SOC content and each of the three management practices on ΔSOC by producing boxplots characterizing the influence of each variable on ΔSOC including the variance calculated across agricultural management practices, grids, and the four climate scenarios. All analyses were performed using statistical and graphical software R 3.0.1 [31].

Results

2.1 Correlations between environmental variables, management practices and ΔSOC

Initial SOC was strongly but negatively correlated (median *rho* = −0.87) with ΔSOC (Figure 2). Initial soil clay content showed a weak negative correlation (median *rho* = −0.21) with ΔSOC (Figure 2). The climatic variables, e.g., temperature and precipitation, displayed a weak positive (median *rho* = 0.21) and negligible correlation with ΔSOC, respectively (Figure 2). Agricultural management such as stubble retention (median *rho* = 0.45) and manure application (median *rho* = 0.81) both showed a strong and positive correlation, whereas N fertilization showed a negligible correlation with ΔSOC (Figure 2).

Figure 3 presented the impacts of initial SOC content, manure application, N fertilization and stubble retention, respectively, on ΔSOC. On average, initial SOC content seemed to linearly and negatively correlate with ΔSOC (Figure 3A), whereas manure application and stubble retention had a relatively linear positive effect (Figure 3B and D) on ΔSOC. N fertilization increased ΔSOC most significantly ($P<0.05$) at rates between 0 and 300 kg ha^{-1} yr^{-1}, with a limited influence at higher fertilization rates. Across the sets involved with different climate scenarios, the median ΔSOC increased slightly, although not significantly ($P>0.05$), from RCP2.6 to RCP8.5 (Figure 4).

2.2 Spatiotemporal effects of agricultural practices on SOC

On average of the four climate scenarios, the response of ΔSOC to N fertilization, manure application and stubble retention varied over the study area, and cropland soils in Hebei and northern part of Shandong Province would accumulate more carbon than the rest of the study area if agricultural practices were optimized (Figure 5 and 6). Increasing the manure application rate led to soil across more areas turning into a net C sink, regardless of the other two management practices (Figure 5 and 6). N fertilization could not significantly halt the declining SOC trend unless more crop residues were incorporated into soils. For example, when stubble retention rate increased from 0 to 100% under M:2000, SOC increased in 11.4%, 27.7%, 40.7%, 53.9%, and 63.9% of the study area under N:0 (Figure 5), whereas increased in 13.5%, 44.5%, 62.5%, 71.3%, and 76.6% of the study area under N:300 (Figure 6). If the amount of manure application rate reached 4000 kg ha^{-1} yr^{-1} and all stubble incorporated into soils (i.e., M:4000 and R:100), SOC increased in

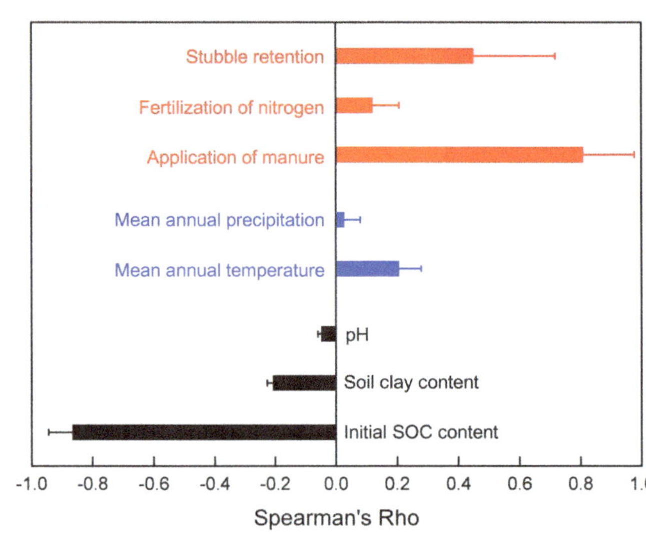

Figure 2. Spearman's rank correlation coefficients between ΔSOC (2011–2100, Mg C ha^{-1}) and climate (blue), soil (black), and management (red) variables. All the tests were significant ($P<0.001$).

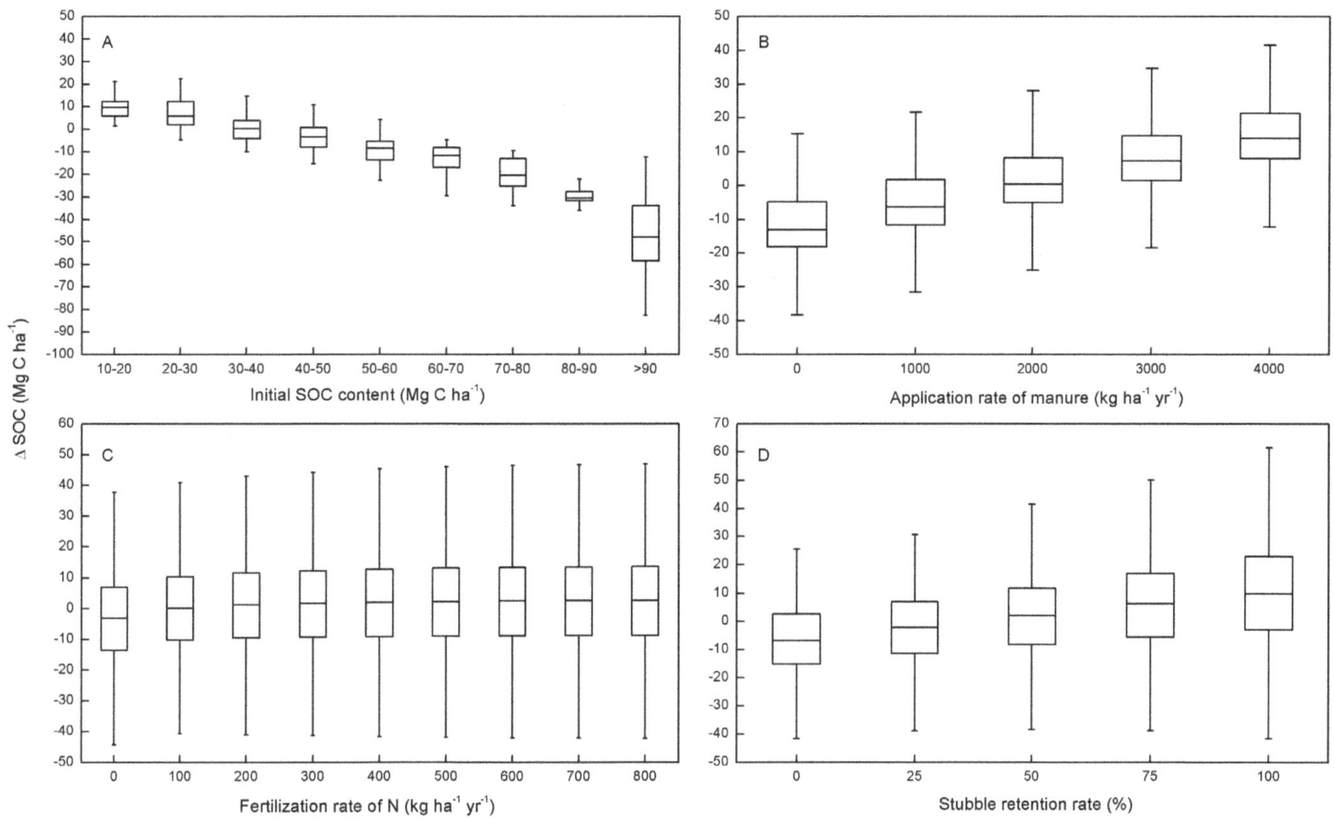

Figure 3. Response of ΔSOC (2011–2100, Mg C ha^{-1}) to initial SOC content (A), manure application (B), N fertilization (C), and stubble retention (D). Boxplots show the median and interquartile range, with whiskers extending to the most extreme data point within 1.5×(75–25%) data range.

more than 90% of the study area under both N:0 and N:300 (Figure 5 and 6).

Under M:0, N:0, R:0, the average SOC content across croplands of NCP kept decreasing from 35.7 Mg ha^{-1} during the 2010s to 16.1 Mg ha^{-1} during the 2090s, on average of the four climate scenarios (Figure 7). However, adopting a moderate management practice (e.g., M:2000, N:100, R:50) could maintain the SOC level during the next 90 years (Figure 7). Under M:4000, N:300, R:100, the average SOC density kept increasing from 39.5 Mg ha^{-1} during the 2010s to 54.6 Mg ha^{-1} during the 2060s, with a relatively lower increasing rate afterward (Figure 7).

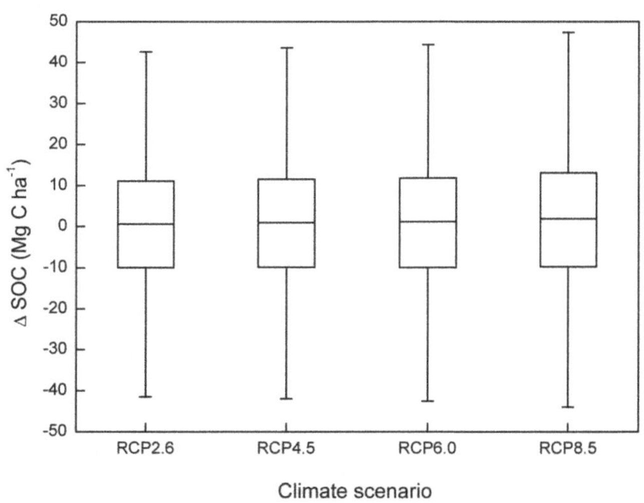

Figure 4. Response of ΔSOC (2011–2100, Mg C ha^{-1}) to different climate scenarios. Boxplots show the median and interquartile range, with whiskers extending to the most extreme data point within 1.5×(75–25%) data range.

Discussion

3.1 Interpretation and implication of the results

Initial SOC content was identified as strongly and negatively influencing SOC change (Figure 2). This is because, in model initialization, organic C input were taken to be the average of crop stubble and manure from 1980 to 1989. However, in the following model simulations with various kinds of management scenarios, great land use changes took place and the equilibrium of soil C pools got changed. In this, under otherwise similar environmental and managed conditions, soils with higher initial SOC content would experience greater SOC loss, and vice versa [32]. This negative correlation between ΔSOC and initial SOC content has also been documented in other studies [33–35]. Although higher soil clay content has been reported to benefit the stabilization of soil organic matter (SOM) in many studies [36–38], we identified a weak negative correlation between ΔSOC and soil clay content in the present study (Figure 2). This is because, other than the benefits of SOM stabilization, higher soil clay content could also reduce the amount of C inputs to soil [12], thereby showing a negative impact on total SOC variations in the present study.

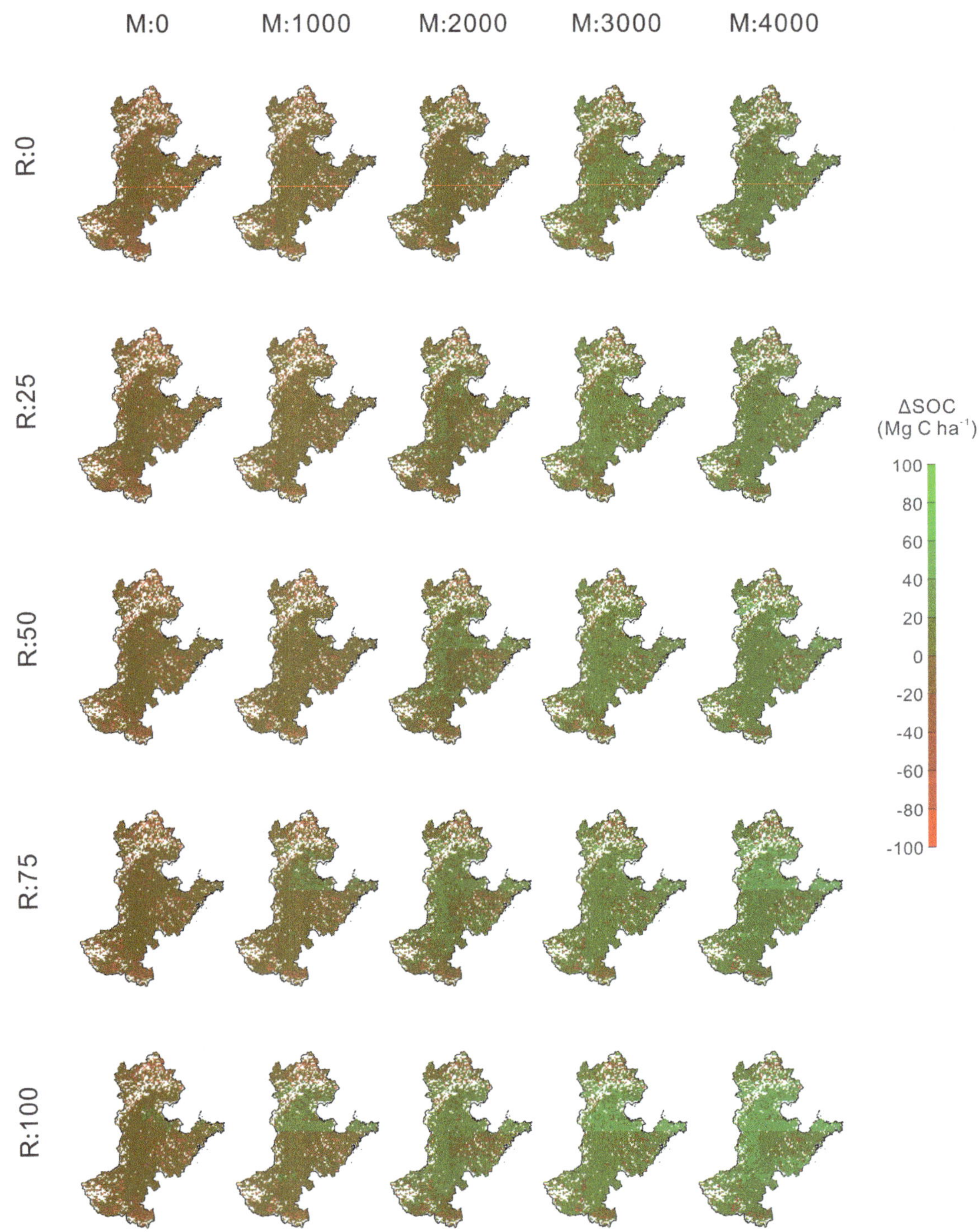

Figure 5. Spatial pattern of the effects of manure application and stubble retention on ΔSOC (2011–2100, Mg C ha^{-1}) under N:0. For a certain combination of management practices, ΔSOC in each grid represents the mean value of the results under four climate scenarios.

In Agro-C, high soil temperature and soil water content increase the SOC decomposition rate which underpins the negative relationship of SOC accumulation with both temperature and rainfall [12]. However, in NCP, we found that the mean annual temperature and precipitation were not strongly correlated with ΔSOC (Figure 2), and higher temperature seemed to benefit SOC accumulation (Figure 4 and Figure S1). This is because the influences of temperature and precipitation on SOC dynamics depend on other factors such as soil and management practices [39]. For example, artificial irrigation, which was adopted across all sets of scenarios in this study, would certainly overshade the possible impacts of precipitation on SOC change. And the elevated atmospheric CO_2 concentration accompanied with higher temperature increased the crop productivity, thereby compensating the elevated amount of decomposed SOC induced by climate warming [40]. Furthermore, Zhang and Huang [41] reported a positive correlation between air temperature and the production of cereal crops in China during the past 30 years. This

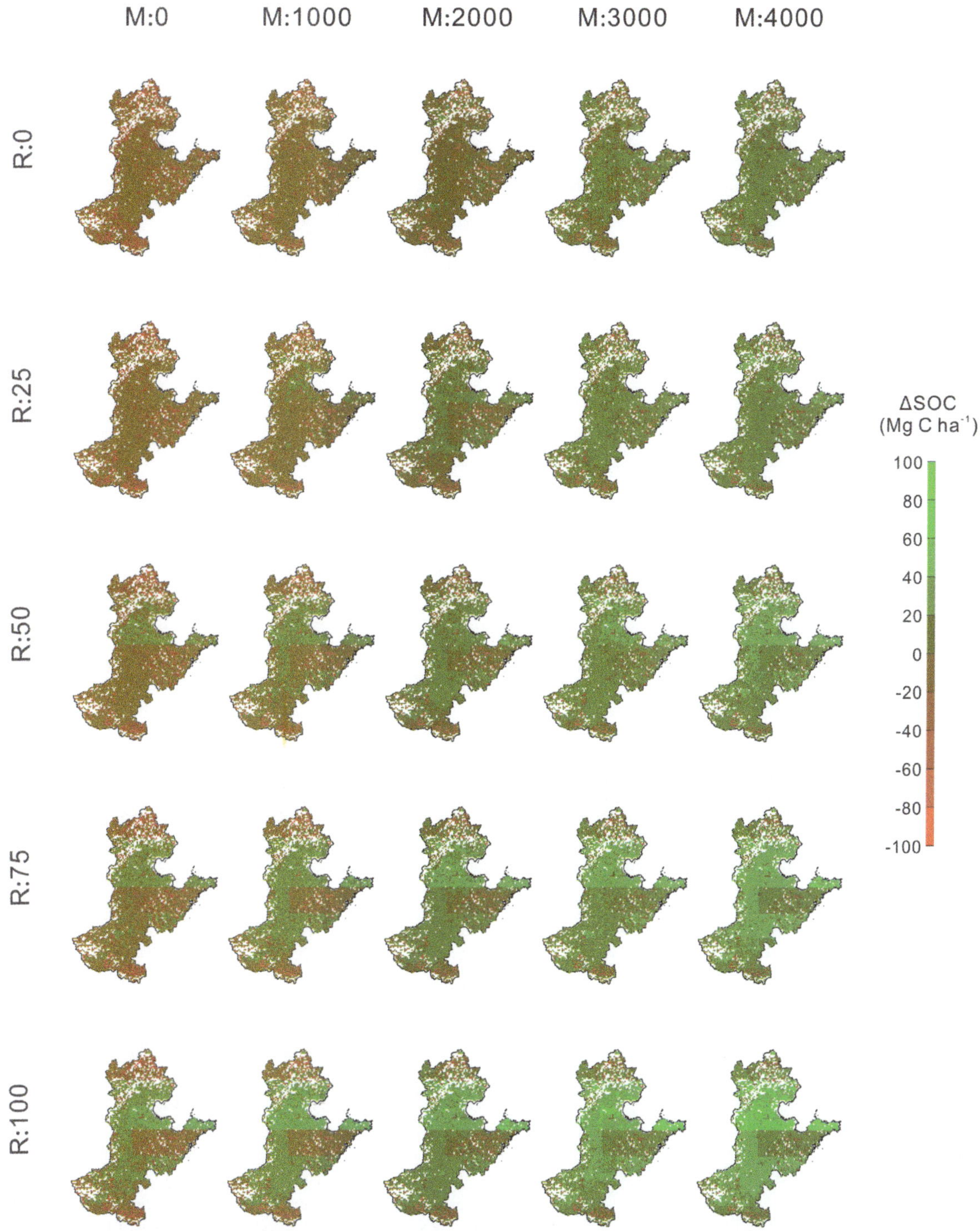

Figure 6. Spatial pattern of the effects of manure application and stubble retention on ΔSOC (2011–2100, Mg C ha^{-1}) under N:300. For a certain combination of management practices, ΔSOC in each grid represents the mean value of the results under four climate scenarios.

indicated that climate warming could potentially cause a higher amount of C input from crop residues to soils, with which our findings was consistent.

Manure application and stubble retention directly add C into soil, and were identified as the most two predominant agricultural management practices driving SOC (Figure 2 and 3). Although it has been reported that the manure application levels has been decreasing during the past 50 years, and the changing trends might continue in the future across Chinese croplands [25]. The increasing amount of incorporated crop residues, caused by both enhanced stubble retention rate [27] and elevated amount of crop NPP [42], could still potentially override the weakening manure application rates and increase SOC.

N fertilization enhances crop residue and root biomass production (Figure 8), thus increases the amount of C incorporated into soil. However, the impact of N fertilization depends on whether nutrient is limiting. In this study, we found that N fertilization showed a negligible correlation with changes in SOC

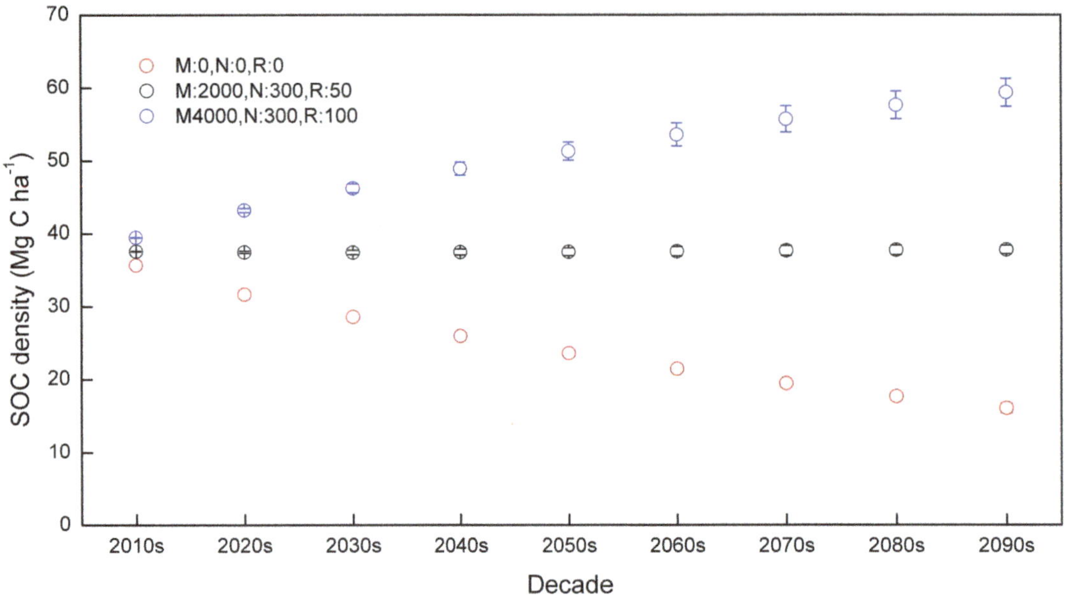

Figure 7. Temporal change of SOC (Mg C ha^{-1}) under three combinations of manure application, nitrogen fertilization and stubble retention. Open circles represent mean values of SOC under four climate scenarios, with standard deviations represented by error bars on the circles.

(Figure 2). This is mainly because changes in SOC is dominated by manure application and stubble retention in most of the scenarios. However, under scenarios with high stubble retention rates and low manure application rates, effects of N fertilization could be more obvious (Figure 5 and 6). Additionally, our results also showed that although N fertilization had a positive effect on SOC accumulation, the effect did not further increase if N fertilization rate exceeds 300 kg ha^{-1} yr^{-1} (Figure 3C). This is because with a N fertilization rate higher than 300 kg ha^{-1} yr^{-1}, nutrient was not a limiting factor of crop growth, and enhancing N inputs could not significantly elevate the amount of soil C inputs from crop residues and roots (Figure 8). Additionally, excessive use of N fertilizer is currently very common in NCP [43], which has aroused concerns about its negative impacts on ecosystems and environments [44]. We suggest that a N fertilization rate of about 300 kg ha^{-1} yr^{-1} should be adopted to both maintain SOC accumulation and reduce the environmental risks caused by excessive use of N fertilizers. As the benefit of N fertilization on SOC sequestration could be offset by N_2O emissions from the soil and greenhouse gas (GHG) emissions from fossil fuel combustion during the processes involving fertilizer production and transport [45], there remains a need to further investigate the proper rates and timing of N fertilization under different climate and soil conditions.

With a high (M:4000, N:300, R:100) C input management, the temporal change in SOC generally experienced a fast to low increase, with the most gain of soil C happened in the first 50 years of simulations (Figure 7). This is consistent with a number of other studies [1,46], which suggest that soil C would reach a new steady state after 20–50 years of cultivation if management practices were significantly changed. Moreover, our results revealed that under M:4000, N:300, R:100, the average SOC density of NCP croplands could achieve 55 Mg ha^{-1} during 2060s, which is similar to the current worldwide average agricultural SOC density [47]. However, the average SOC density would keep decreasing under M:0, N:0, R:0, and the "new" steady state of SOC was not attained even after 90 years of simulation (Figure 7). This could be attributed to the limited C input to balance the SOC loss by decomposition. With a much lower decomposition rate, the

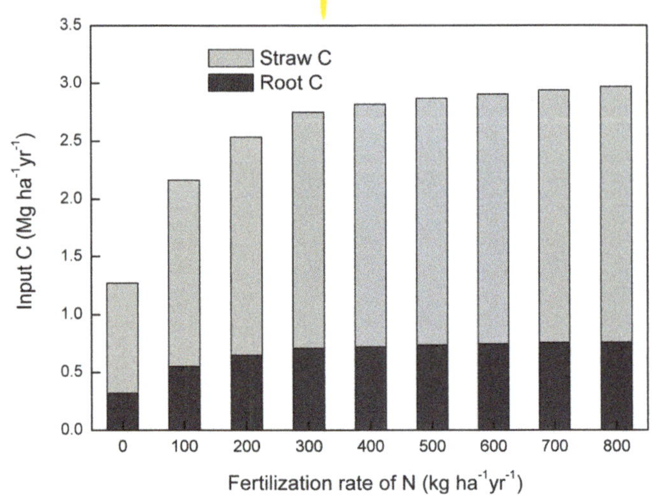

Figure 8. Response of mean annual C inputs from crop residues and roots (Mg C ha^{-1}) to different N fertilization rates.

proportion and absolute amount of heavy-C would have significant impact on long-term SOC balance.

3.2 Uncertainties and limitations

Several uncertainties and limitations should be noticed in interpreting the simulation results in this study. First, several management scenarios simulated in this study represent the most extreme conditions and may not happen in the real world. For example, zero inputs of N and C to soil, because it is improbable to extract all roots and residues from the fields, or prevent atmospheric N deposition. Second, the possible improvement in future crop varieties, which could potentially affect the quality and quantity of soil C inputs, was not considered in this study. For example, if the crop variety improvement increased the production of either crop residues or roots, this could lead to an underestimation of our results in the future SOC accumulation. Third, the effects of conservation tillage on SOC were not simulated in this study, due to a lack of the detailed information on annual conservation tillage practice. However, it could still be hypothesized that our results could represent most parts of the studied areas, because the average fraction of the area with conservation tillage practices in Chinese croplands is only about 9% in 2009 [5]. Moreover, a meta-analysis of global data on the effects of tillage on SOC change [48] indicates that conversion from conventional tillage to conservation tillage only redistributed more C into the top layer and less C into the deeper layer, and the total SOC stock in the upper 40 cm soil profile was not significantly changed. Fourth, the rainfed condition was not simulated because artificial irrigation has been traditionally adopted to support crop production in NCP. However, the current irrigation-supported agricultural production might be affected by future resource and environment related pressures [49]. In this, other environmental factors such as precipitation and temperature may play a more important role in regulating both crop productivity and SOC dynamics [50]. Last but not least, certain issues related to state-of-the-art model limitations should be mentioned. For instance, in a first-order decay model (e.g., Agro-C), there is a general linear relation between C input and SOC variation, which usually contradicts the fact that increase of C inputs by crop residues may have variable effects on SOC change in the real world [51]. Future efforts are needed from model developers to implement currently uncertain mechanisms, such as SOC saturation, priming, and different sensitivity of SOC pools to temperature, which will enable a better understanding the influence of environment on agricultural SOC dynamics.

Supporting Information

Figure S1 Climate change in North China Plain (NCP) from 2011 to 2100 under the RCP2.6 (A), RCP4.5 (B), RCP6.0 (C) and RCP8.5 (D) scenarios. The solid circles represent the annual temperature, and the open triangles represent annual precipitation.

Figure S2 Spatial distribution of mean annual temperature and precipitation increments in North China Plain (NCP) between 2011 and 2100 under four climate scenarios.

Figure S3 The range and density of N fertilization (A) and manure application (B) rates in North China Plain (NCP). The N fertilization data represent the year of 2010 on a county scale, and the manure application data represent that from 1981 to 2010 on a provincial scale. The datasets were derived from online database and published literatures. *n* shows the sample size.

Document S1 Description of FGOALS and the RCP scenarios for AR5.

Acknowledgments

Many thanks to Dr. Yongqiang Yu from the State Key Laboratory of Numerical Modeling for Atmospheric Sciences and Geophysical Fluid Dynamics (LASG), Institute of Atmospheric Physics, Chinese Academy of Sciences, for providing the access to the FGOALS modeling result datasets.

Author Contributions

Conceived and designed the experiments: GW TL. Performed the experiments: GW TL. Analyzed the data: GW TL. Contributed reagents/materials/analysis tools: TL WZ YY. Wrote the paper: GW.

References

1. Lal R (2004) Soil carbon sequestration impacts on global climate change and food security. Science 304: 1623–1627.
2. Wang G, Huang Y, Wang E, Yu Y, Zhang W (2013) Modeling Soil Organic Carbon Change across Australian Wheat Growing Areas, 1960–2010. PloS one 8: e63324.
3. Brady NC, Weil RR (2008) The nature and properties of soils (14th ed). Upper Saddle River: Pearson Education Publication.
4. Van Wesemael B, Paustian K, Meersmans J, Goidts E, Barancikova G, et al. (2010) Agricultural management explains historic changes in regional soil carbon stocks. Proceedings of the National Academy of Sciences 107: 14926–14930.
5. Yu Y, Huang Y, Zhang W (2012) Modelling soil organic carbon change in croplands of China, 1980–2009. Global and Planetary Change 82–83: 115–128.
6. Li C, Frolking S, Harriss R (1994) Modeling carbon biogeochemistry in agricultural soils. Global biogeochemical cycles 8: 237–254.
7. Jenkinson D, Andrew S, Lynch J, Goss M, Tinker P (1990) The Turnover of Organic Carbon and Nitrogen in Soil. Philosophical Transactions: Biological Sciences 329: 361–368.
8. Parton W, Scurlock J, Ojima D, Gilmanov T, Scholes R, et al. (1993) Observations and modeling of biomass and soil organic matter dynamics for the grassland biome worldwide. Global biogeochemical cycles 7: 785–809.
9. Li CS, Zhuang Y, Frolking S, Galloway J, Harriss R, et al. (2003) Modeling soil organic carbon change in croplands of China. Ecological Applications 13: 327–336.
10. Falloon P, Smith P, Bradley R, Milne R, Tomlinson R, et al. (2006) RothCUK–a dynamic modelling system for estimating changes in soil C from mineral soils at 1-km resolution in the UK. Soil Use and Management 22: 274–288.
11. Ogle SM, Breidt F, Easter M, Williams S, Killian K, et al. (2010) Scale and uncertainty in modeled soil organic carbon stock changes for US croplands using a process-based model. Global Change Biology 16: 810–822.
12. Huang Y, Yu Y, Zhang W, Sun W, Liu S, et al. (2009) Agro-C: A biogeophysical model for simulating the carbon budget of agroecosystems. Agricultural and Forest Meteorology 149: 106–129.
13. Ding Y, Ren G, Shi G, Gong P, Zheng X, et al. (2006) National assessment report of climate change (I): climate change in China and its future trend. Advances in Climate Change Research 2: 3–8 (in Chinese).
14. Huang Y, Sun W (2006) Changes in topsoil organic carbon of croplands in mainland China over the last two decades. Chinese Science Bulletin 51: 1785–1803.
15. Bond-Lamberty B, Thomson A (2010) Temperature-associated increases in the global soil respiration record. Nature 464: 579–582.
16. Yu Y, Huang Y, Zhang W (2013) Projected changes in soil organic carbon stocks of China's croplands under different agricultural managements, 2011–2050. Agriculture, Ecosystems & Environment 178: 109–120.
17. West TO, Post WM (2002) Soil organic carbon sequestration rates by tillage and crop rotation: A global data analysis. Soil Science Society of America Journal 66: 1930–1946.
18. Luo Z, Wang E, Sun OJ, Smith CJ, Probert ME (2011) Modeling long-term soil carbon dynamics and sequestration potential in semi-arid agro-ecosystems. Agricultural and Forest Meteorology 151: 1529–1544.
19. Yu D, Shi X, Wang H, Sun W, Chen J, et al. (2007) Regional patterns of soil organic carbon stocks in China. Journal of Environmental Management 85: 680–689.
20. Bertacchi Uvo C, Olsson J, Morita O, Jinno K, Kawamura A, et al. (2001) Statistical atmospheric downscaling for rainfall estimation in Kyushu Island, Japan. Hydrology and Earth System Sciences 5: 259–271.
21. Kidson JW, Thompson CS (1998) A comparison of statistical and model-based downscaling techniques for estimating local climate variations. Journal of Climate 11: 735–753.
22. Clarke L, Edmonds J, Krey V, Richels R, Rose S, et al. (2009) International climate policy architectures: Overview of the EMF 22 International Scenarios. Energy Economics 31: S64–S81.
23. Liu J, Liu M, Zhuang D, Zhang Z, Deng X (2003) Study on spatial pattern of land-use change in China during 1995–2000. Science in China Series D: Earth Sciences 46: 373–384.

24. Huang Y, Tang Y (2010) An estimate of greenhouse gas (N_2O and CO_2) mitigation potential under various scenarios of nitrogen use efficiency in Chinese croplands. Global Change Biology 16: 2958–2970.
25. Zhang W, Yu Y, Huang Y, Li T, Wang P (2011) Modeling methane emissions from irrigated rice cultivation in China from 1960 to 2050. Global Change Biology 17: 3511–3523.
26. National Development and Reform Commission (2007) The People's Republic of China, Initial National Communication on Climate Change (In Chinese). China Planning Press, Beijing, China.
27. Gao L, Ma L, Zhang W, Wang F, Ma W, et al. (2009) Analysis on the Quantities and Utilization of Crop Straw and its Nutrient in Huang-Huai-Hai Region. Chinese Agricultural Science Bulletin 25: 186–193 (in Chinese with English summery).
28. Wu D, Yu Q, Lu C, Hengsdijk H (2006) Quantifying production potentials of winter wheat in the North China Plain. European Journal of Agronomy 24: 226–235.
29. Wu D, Yu Q, Wang E, Hengsdijk H (2008) Impact of spatial-temporal variations of climatic variables on summer maize yield in North China Plain. International Journal of Plant Production 2: 71–88.
30. Bryan BA (2003) Physical environmental modeling, visualization and query for supporting landscape planning decisions. Landscape and urban planning 65: 237–259.
31. R Development Core Team (2013) R: A language and environment for statistical computing. Vienna, Austria: R Foundation for Statistical Computing.
32. Sanderman J, Baldock JA (2010) Accounting for soil carbon sequestration in national inventories: a soil scientist's perspective. Environmental Research Letters 5: 034003.
33. Saby N, Arrouays D, Antoni V, Lemercier B, Follain S, et al. (2008) Changes in soil organic carbon in a mountainous French region, 1990–2004. Soil Use and Management 24: 254–262.
34. Goidts E, Wesemael Bv, Van Oost K (2009) Driving forces of soil organic carbon evolution at the landscape and regional scale using data from a stratified soil monitoring. Global Change Biology 15: 2981–3000.
35. Zhao G, Bryan BA, King D, Luo Z, Wang E, et al. (2013) Impact of agricultural management practices on soil organic carbon: simulation of Australian wheat systems. Global change biology 19: 1585–1597.
36. Oades J (1988) The retention of organic matter in soils. Biogeochemistry 5: 35–70.
37. Amato M, Ladd J (1992) Decomposition of ^{14}C-labelled glucose and legume material in soils: Properties influencing the accumulation of organic residue C and microbial biomass C. Soil Biology and Biochemistry 24: 455–464.
38. Burke I, Yonker C, Parton W, Cole C, Schimel D, et al. (1989) Texture, climate, and cultivation effects on soil organic matter content in US grassland soils. Soil Science Society of America Journal 53: 800–805.
39. Giardina CP, Ryan MG (2000) Evidence that decomposition rates of organic carbon in mineral soil do not vary with temperature. Nature 404: 858–861.
40. Smith J, Smith P, Wattenbach M, Zaehle S, Hiederer R, et al. (2005) Projected changes in mineral soil carbon of European croplands and grasslands, 1990–2080. Global Change Biology 11: 2141–2152.
41. Zhang T, Huang Y (2012) Impacts of climate change and inter-annual variability on cereal crops in China from 1980 to 2008. Journal of the Science of Food and Agriculture 92: 1643–1652.
42. Huang Y, Zhang W, Sun W, Zheng X (2007) Net primary production of Chinese croplands from 1950 to 1999. Ecological Applications 17: 692–701.
43. Liu X, Ju X, Zhang F, Pan J, Christie P (2003) Nitrogen dynamics and budgets in a winter wheat–maize cropping system in the North China Plain. Field Crops Research 83: 111–124.
44. Richter A, Burrows JP, Nüβ H, Granier C, Niemeier U (2005) Increase in tropospheric nitrogen dioxide over China observed from space. Nature 437: 129–132.
45. Knorr W, Prentice I, House J, Holland E (2005) Long-term sensitivity of soil carbon turnover to warming. Nature 433: 298–301.
46. West TO, Six J (2007) Considering the influence of sequestration duration and carbon saturation on estimates of soil carbon capacity. Climatic Change 80: 25–41.
47. Batjes NH (1999) Management options for reducing CO_2-concentrations in the atmosphere by increasing carbon sequestration in the soil: National Institute of Public Health and the Environment.
48. Luo Z, Wang E, Sun OJ (2010) Can no-tillage stimulate carbon sequestration in agricultural soils? A meta-analysis of paired experiments. Agriculture, Ecosystems & Environment: 224–231.
49. Wang E, Yu Q, Wu D, Xia J (2008) Climate, agricultural production and hydrological balance in the North China Plain. International Journal of Climatology 28: 1959–1970.
50. Piao S, Ciais P, Huang Y, Shen Z, Peng S, et al. (2010) The impacts of climate change on water resources and agriculture in China. Nature 467: 43–51.
51. Powlson DS, Glendining MJ, Coleman K, Whitmore AP (2011) Implications for soil properties of removing cereal straw: results from long-term studies. Agronomy Journal 103: 279–287.

The Impact of Selective-Logging and Forest Clearance for Oil Palm on Fungal Communities in Borneo

Dorsaf Kerfahi[1,2], Binu M. Tripathi[1], Junghoon Lee[2], David P. Edwards[3]*, Jonathan M. Adams[1]*

1 Department of Biological Sciences, Seoul National University, Seoul, Republic of Korea, 2 School of Chemical and Biological Engineering, Interdisciplinary Program of Bioengineering, Seoul National University, Seoul, Republic of Korea, 3 Department of Animal and Plant Sciences, University of Sheffield, Sheffield, United Kingdom

Abstract

Tropical forests are being rapidly altered by logging, and cleared for agriculture. Understanding the effects of these land use changes on soil fungi, which play vital roles in the soil ecosystem functioning and services, is a major conservation frontier. Using 454-pyrosequencing of the ITS1 region of extracted soil DNA, we compared communities of soil fungi between unlogged, once-logged, and twice-logged rainforest, and areas cleared for oil palm, in Sabah, Malaysia. Overall fungal community composition differed significantly between forest and oil palm plantation. The OTU richness and Chao 1 were higher in forest, compared to oil palm plantation. As a proportion of total reads, Basidiomycota were more abundant in forest soil, compared to oil palm plantation soil. The turnover of fungal OTUs across space, true β-diversity, was also higher in forest than oil palm plantation. Ectomycorrhizal (EcM) fungal abundance was significantly different between land uses, with highest relative abundance (out of total fungal reads) observed in unlogged forest soil, lower abundance in logged forest, and lowest in oil palm. In their entirety, these results indicate a pervasive effect of conversion to oil palm on fungal community structure. Such wholesale changes in fungal communities might impact the long-term sustainability of oil palm agriculture. Logging also has more subtle long term effects, on relative abundance of EcM fungi, which might affect tree recruitment and nutrient cycling. However, in general the logged forest retains most of the diversity and community composition of unlogged forest.

Editor: Andrew Hector, University of Oxford, United Kingdom

Funding: This project was funded by a National Research Foundation grant (NRF-2013-031400), Ministry of Education, Science and Technology, South Korea. The funders had no role in study design, data collection and analysis, decision to publish, or preparation of the manuscript.

Competing Interests: Dr. Jonathan M. Adams is a PLOS ONE Editorial Board member. This does not alter the authors' adherence to PLOS ONE Editorial policies and criteria.

* Email: david.edwards@sheffield.ac.uk (DPE); foundinkualalumpur@yahoo.com (JMA)

Introduction

Tropical forests are one of the world's most important reservoirs for biodiversity [1]. They contain an exceptional concentration of the world's species, but are being reduced in area faster than any other ecosystem [2]. Roughly half of the world's natural extent of tropical forest has been logged or converted to different land uses [3]. Some 403 million hectares of tropical rainforests have been included in timber estates and slated for selective logging [4], and between 2000 and 2010, approximately 13 million hectares of forest within the tropics were cleared for agricultural activities [5,6], including oil palm plantations [7].

Anthropogenic disturbances in tropical forests are causing a dramatic decline in global biodiversity, and in associated biological processes that maintain the productivity and sustainability of ecosystems [8]. Several studies have shown that selective logging does not drastically impact the overall species richness and diversity of tropical forest [9–12], however, it has been shown that the impact of selective logging could be anything between fairly mild and severe depending on the intensity of logging [13] with changes in the composition of species, as forest-interior specialists decline and edge-tolerant, gap specialists increase in abundance [14,15]. In contrast, the conversion of both primary and logged forest to agricultural land uses has been to shown to have a far greater negative impact on biodiversity than does logging. Conversion to agriculture results in a major reduction in biodiversity, again across a host of animal and plant taxa [16]. The conversion of primary and logged forest to agricultural plantations also results in a substantial decrease in the functional diversity of tropical ecosystems, with implications for the provision of ecosystem functions, whereas logging has lesser impacts on these metrics [17,18]. As well as affecting plants and larger animals, land use change also affects the soil biota. Land use change affects soil pH, carbon and nutrient content [19,20], causing shifts in soil microbial communities [21–23].

Fungi constitute one of the most diverse and dominant groups of organisms in soil, and they play important ecological roles in the ecosystem as decomposers, pathogens and plant mutualists [24,25]. Understanding the structure and diversity of soil fungal communities is fundamental to the understanding of their function in the ecosystem and their impact on plant communities [26]. However, while minimal work has been done in the tropics to assess the effect of land use changes on soil bacterial communities [23,27], until recently relatively little was known about the impacts of tropical land use change on soil fungal communities. Various studies have suggested that forest clearance to tree plantations or agricultural crops shifts soil fungal communities, linked to strong changes in soil properties [22,28]. However, previous studies of forest clearance to other forms of agriculture on forest fungal communities have mostly been limited to techniques that give relatively low taxonomic resolution (i.e. T-RFLP and PCR-DGGE). Nevertheless, a recent study by McGuire et al. [29], which used high throughput sequencing to analyze soil fungal communities in Southeast Asian tropical forests in west Malaysia, showed that conversion of primary forest to oil palm plantations alters fungal community composition and function, whereas primary and logged forests were more similar in composition and nutrient cycling potential. However, there is a need for further studies to understand the impacts of logging cycles on soil fungal communities and of the conversion of logged forest to agriculture, since logged forests now dominate the tropics [3] and are much more likely to be converted to agriculture than primary forests [2,6].

In this study, we also focused on the rainforests of the Sundaland region, of Southeast Asia, but some 1,800 km away in east Malaysia (Borneo). Across the Sundaland region, the primary forest has been subject to differing degrees of logging intensity. Much of the region's forest (about 50%) has never been logged, while many areas have been subject to one or two logging cycles [30]. Also, oil palm is one of the most rapidly expanding crops in this region. This provides an opportunity to study the effect of different intensities of logging on the soil fungal community and also to evaluate if conversion of forest to oil palm plantation has a stronger impact on soil fungi than logging, as is the case for numerous macroscopic taxa. Our objective here was to understand whether land use change has an impact on the structure and diversity of fungal communities in the Yayasan Sabah (YS) logging concession in Malaysian, Borneo. We compared the fungal communities in forests with different logging histories (unlogged, once-logged and twice-logged), and oil palm plantations. We examined whether there are differences in α and â-diversity, as well as community composition. These results may provide important information for soil management policies, and estimation of ecological impact of land use change in this region.

Materials and Methods

Study area

The study area is located within the Yayasan Sabah (YS) logging concession and contiguous oil palm plantation areas, in Sabah, Malaysian Borneo (4°58′ N, 117°48′ E). The forests in this area are naturally dominated by valuable timber tree species belonging to the family Dipterocarpaceae [31]. Due to logging for the wood industry and clearance for palm oil plantations, the area of forest in Borneo - as elsewhere in the tropics - has been dramatically reduced in recent decades [32].

Fieldwork was conducted in the Ulu Segama-Malua Forest Reserve (US-MFR) and adjacent oil palm estates in Sabah, Borneo. Some areas of forest were logged between 1970 and 1990, and some of these were then re-logged between 2000 and 2007. During the first logging rotation, approximately 113 m^3 per hectare (range 73 m^3 to 166 m^3) of commercially valuable trees > 0.6 m diameter were extracted. During the second logging rotation, an additional 31 m^3 per hectare (range 15 m^3 to 72 m^3) of timber were removed [10,31]. Selectively logged forest in the US-MFR is adjacent to the 45,200 ha Danum Valley Conservation Area (DVCA) and Palum Tambun Watershed Reserve, containing large areas of unlogged forest [10,16]. Oil palm plantations are situated to the north and south of the US-MFR, with mature palms of 20 to 30 years old, planted at a density of 100 trees per hectare.

Soil sampling and DNA extraction

From September to October 2012, twenty-four transects each of 200 m in length were located across four different land uses: unlogged (primary) forests, once-logged and twice-logged forests, and oil palm plantations, with six transects per habitat. Within each habitat, distances between transects ranged from 500 m to 65 km, whereas across habitats, distances between transects ranged from 1 km to 67 km. From each transect, at 50 m intervals, approximately 50 g from the top 5 cm of soil (excluding the leaf litter layer) was taken in a sterile plastic bag using a trowel, giving five samples of soil per transect. The trowels were thoroughly cleaned with ethanol between successive transect sampling. All soil samples were then sieved (2 mm) in laboratory to homogenize the sample and stored at −20°C until DNA extraction [10]. Twenty-four soil DNA extractions (one for each transect) were performed using 0.3 g of soil, with the Power Soil DNA extraction kit (MO BIO Laboratories, Carlsbad, CA, USA) following the directions described by the manufacturer.

PCR amplification and pyrosequencing

Fungal DNA was amplified using ITS primers targeting the internal transcribed spacer (ITS) region 1 and 2. Forward primers comprised the 454 Fusion Primer A-adaptor, a specific multiplex identifier (MID) barcode, and the ITS1F primer (5′-CTTGGTCATTTAGAGGAAGTAA-3′) [33], while the reverse primer was composed of the B-adapter and ITS4 primer (5′-TCCTCCGCTTATTGATATGC-3′) [34]. Polymerase chain reactions (PCR) were performed in 50 µl reactions using the following temperature program: 95°C for 10 min s; 30 cycles of 95°C for 30 s, 55°C for 30 s, 72°C for 30 s; and 72°C for 7 min. The PCR products were purified using the QIAquick PCR purification kit (Qiagen) and quantified using PicoGreen (Invitrogen) spectrofluorometrically (TBS 380, Turner Biosystems, Inc. Sunnyvale, CA, USA). 50 ng of purified PCR product for each sample were combined in a single tube and sent to Macrogen Inc. (Seoul, Korea) for sequencing using 454/Roche GS FLX Titanium Instrument (Roche, NJ, USA).

Sequence processing

Initial quality filtering and denoising were performed following the 454 SOP in the mothur pipeline [35]. The ITS1 region was verified and extracted using the ITS1 extractor for fungal ITS sequences [36]. Putative chimeric sequences were detected and removed via the Chimera Uchime algorithm contained within mother [37]. Operational taxonomic units (OTUs) were assigned using the QIIME implementation of UCLUST [38], with a threshold of 97% pairwise identity. OTUs were classified taxonomically using the classify command in mothur at 80% Naïve Bayesian bootstrap cutoff with 1000 iterations against the UNITE database [39]. Ectomycorrhizal (EcM) fungi were determined by matching taxonomy assignments with established EcM lineages as determined by recent phylogenetic and stable isotope data [40]. The 454 sequence run has been deposited in the NCBI Sequence Read Archive under accession number SRP041467.

Statistical analysis

To correct for differences in number of reads, which can bias diversity estimates, all samples were rarified to 3,347 reads per sample. To test for effects of land use types on the OTU richness

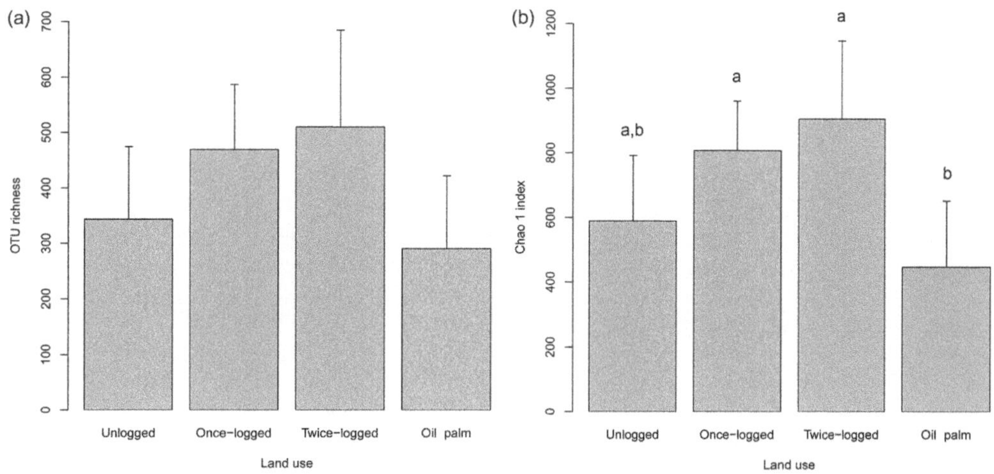

Figure 1. Diversity indices of the fungal community across different land uses in Sabah, Malaysian Borneo. (a) OTU richness and (b) Chao1 index. Pairwise comparisons are shown; different letters denote significant differences between groups at P<0.05.

and diversity indices, we used a linear model (LM) for normal data or generalized linear model (GLM) for non-normal data, considering land use as the major factor. We used the same procedure to test whether relative abundance of the most abundant phyla differed among different land use types. We also assessed the effect of land use on the relative abundance at the order and genus levels within those phyla that showed significant differences due to land use. Post-hoc Tukey tests were used for pairwise comparisons. When neither a linear nor a generalized linear model fitted the data, we used a Kruskall-Wallis test to assess the effect of land use on the relative abundance of fungal taxa, with the Bonferroni correction to assess pairwise comparisons.

To test whether species composition results may have been influenced by pseudoreplication within study sites, we used a Mantel test (Mantel Nonparametric Test Calculator 2.0) [41] to compare transect matrices of fungal compositional to geographic

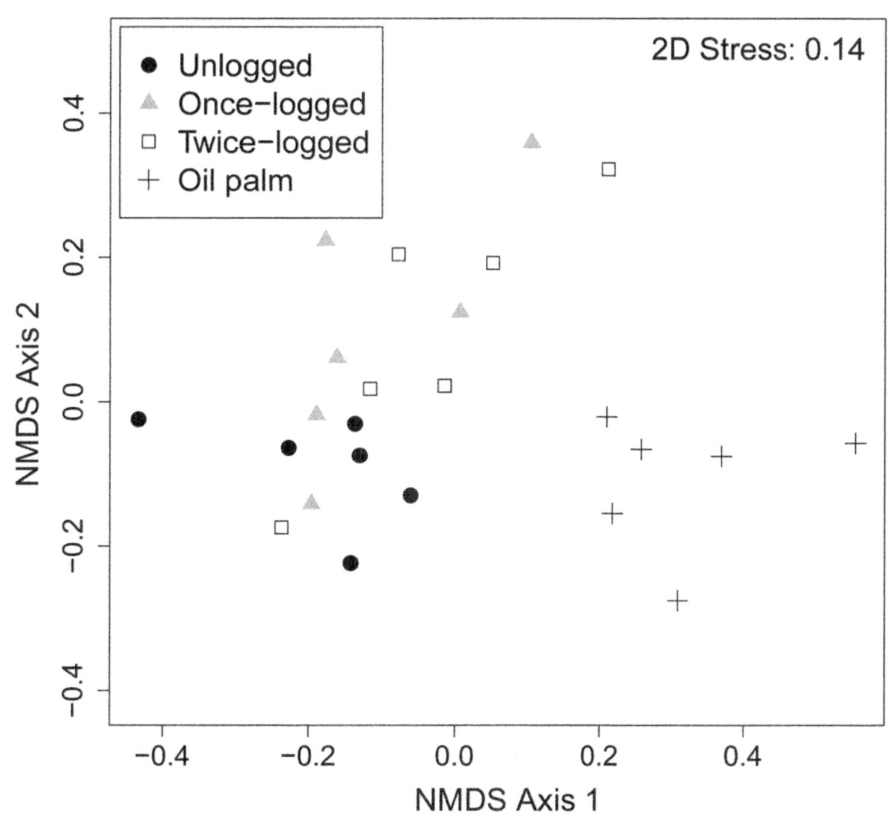

Figure 2. Non-metric multidimensional scaling (NMDS) ordination showing clustering of fungal communities among different land uses in Sabah, Malaysian Borneo.

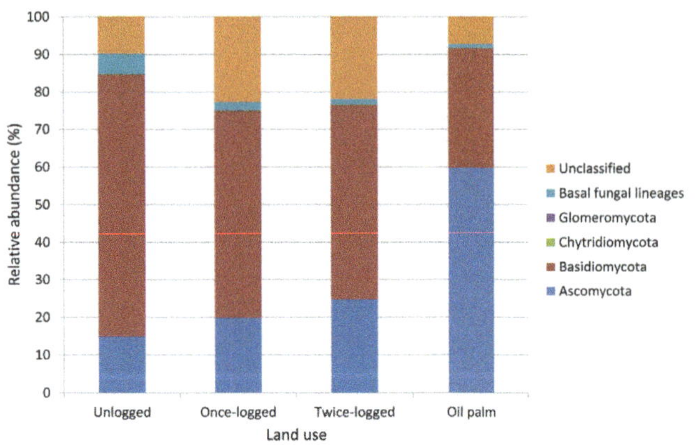

Figure 3. Relative abundance of dominant fungal phyla among different land uses in Sabah, Malaysian Borneo.

distance between pairs of transects within a site and between pairs of transects across the entire dataset [42,43].

The OTU-based community similarity was calculated by using the Bray-Curtis index [44]. Non-metric multidimensional scaling (NMDS), to visualize the change in species composition across the land use types, was conducted in Primer-E software (Version 6, Plymouth, UK). Then, we tested the difference among different land use types using an analysis of similarity (ANOSIM).

We measured β-diversity amongst land use types following Anderson et al. [45], which is defined as the variation in community structure without defining a particular gradient or direction. Therefore, we estimated true β-diversity following Whittaker [46] in [47] for every land use type. In addition, we calculated β-diversity as the average distance from each site to the group centroid [45]. The betadisper function in R was used to test if β-diversity shows any difference between land use types. True β-diversity (i.e. $S/\bar{\alpha}$) for each pair of samples within each of the four land uses was estimated by the following equation:

$$\frac{S}{\bar{\alpha}}=\frac{a+b+c}{(2a+b+c)/2}$$

Where S is the total number of OTUs in two samples, $\bar{\alpha}$ is the average number of OTUs for both samples, a is shared OTUs between both samples, b is OTUs found only in sample 1 and c are OTUs found only in sample 2. To compare true β-diversity among land uses, we used a linear model using land use as factor, and sample as random factor to control for pseudoreplication, as every sample is used in more than one comparison within each land use. Post-hoc Tukey tests were used for pairwise comparisons among different land uses.

Results

A total of 114,744 quality sequences were obtained from the 24 soil samples, with coverage ranging from 3,347–6,456 sequences. After rarifying to 3,347 reads per sample, we obtained a total of 80,328 sequences, and of these around 84% sequences were classified up to phylum level with a total of 5,327 OTUs (defined at ≥97% sequence similarity level). Fungal OUT richness (i.e. number of OTUs) was marginally significantly different across land use types ($F_{3,24} = 2.98$, $P = 0.05$; Fig. 1a), with lowest levels of OTU richness observed in oil palm plantations compared to primary and logged forests. Predicted OTU richness calculated using the Chao1 estimator was significantly higher in logged forests than oil palm plantations (F3,24 = 4.74, P = 0.01; Fig. 1b), whereas Shannon index did not show any variation among land uses (F3,24 = 1.93, P = 0.15).

The NMDS plots of pairwise Bray–Curtis dissimilarities showed that fungal communities were clustered significantly across land use types (ANOSIM: R = 0.51, P<0.001; Fig. 2). Mantel tests showed no effect of distance on the composition of fungal communities across different land use types in Borneo (all P> 0.07). The majority of fungal sequences recovered in our study belonged to the Basidiomycota and Ascomycota, with relative abundances of 52% and 29%, respectively (Fig. 3). The basal

Table 1. Comparison of relative abundance of the dominant fungal orders within the phyla Ascomycota and Basidiomycota among land uses[a].

Taxa	F or χ^{2b}	df	P	Pairwise comparisons[c]
Ascomycota				
Helotiales	3.67	3, 24	0.02	Once-logged > oil palm
Hypocreales	5.54	3, 24	0.006	Once-logged/twice-logged/unlogged < oil palm
Pleosporales	6.42	3, 24	0.003	Once-logged/unlogged < oil palm
Basidiomycota				
Agaricales	10.9	3, 24	0.0001	Once-logged/twice-logged/unlogged > oil palm
Russulales	11.3	3, 24	0.0001	Unlogged > Once-logged/twice-logged/oil palm
Sebacinales	11.1*	3	0.01	Unlogged/twice-logged > oil palm
Sporidiobolales	15.1*	3	0.001	Once-logged/twice-logged/unlogged < oil palm
Thelephorales	10.0*	3	0.01	Unlogged/twice-logged > oil palm
Trichosporonales	13.3	3, 24	0.0001	Once-logged/twice-logged/unlogged > oil palm

[a]Only orders for which significant differences were found are shown.
[b]Effect of land use on relative abundance evaluated by linear or generalized linear model or by the Kruskal-Wallis test (*).
[c]Pairwise comparisons by *post hoc* Tukey test for linear/generalized linear models or *P* values Bonferroni-corrected for Kruskal-Wallis. Differences were considered significant at a P value of <0.05.

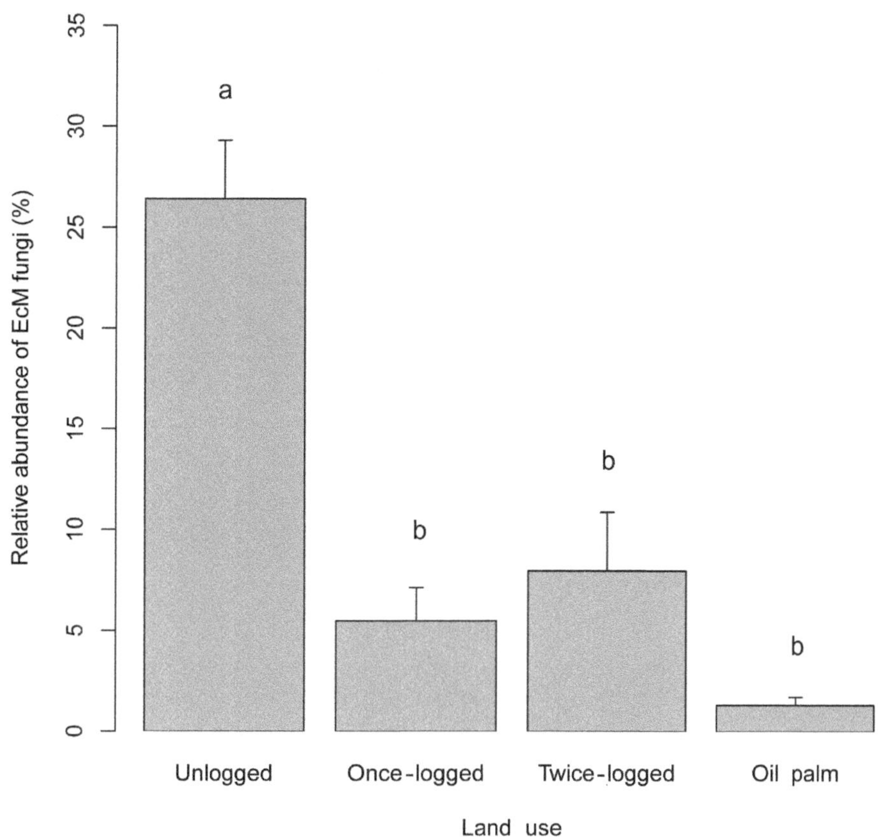

Figure 4. Relative abundance (means ± SD) of ectomycorrhizal (EcM) fungal sequences among different land uses in Sabah, Malaysian Borneo. Pairwise comparisons are shown; different letters denote significant differences between groups at P<0.05.

fungal lineages represented 2%, followed by Glomeromycota and Chytridiomycota with less than 1%, and 15% of the detected sequences were unclassified (Fig. 3). We found a significant change in the relative abundance of the two most dominant phyla: Basidiomycota ($F_{3,24} = 4.23$, $P = 0.01$) and Ascomycota ($F_{3,24} = 4.81$, $P = 0.01$) along different land uses. The abundance of Basidiomycota was greater in the unlogged forest, intermediate in the logged forest, and least in the oil palm. Conversely, the oil palm plantations had greater abundance of Ascomycota compared to forest soils (Fig. 3). At the order level, there were significant differences in the relative abundances of the most dominant orders (Table 1). The results revealed that the relative abundance of *Agaricales*, *Russulales*, *Thelephorales*, *Trichosporonales*, *Sebacinales* and *Helotiales* were significantly higher in the forest than oil palm plantations (P<0.05; Table 1). However, *Hypocreales*, *Sporidiobolales* and *Pleosporales*, were more abundant in oil palm plantations than unlogged and logged forests (P<0.05; Table 1).

A total of 11,421 sequences belonged to known groups of ectomycorrhizal (EcM) fungi, with the EcM fungi representing around 10% (11,421 sequences) of the total detected fungal sequences, with 180 OTUs. The relative abundance of EcM sequences was significantly different across land use types, with highest and lowest relative abundances observed in primary forest (mean relative abundance = 26%) and oil palm plantations (mean relative abundance = 0.5%), respectively ($\chi^2 = 18.04$, $P < 0.001$; Fig. 4). From our soil samples, we identified 14 genera belonging to EcM fungi with *Russula* as the most dominant genus (65% of total EcM sequences), followed by *Tomentella*, *Sebacina* and *Lactarius*. The relative abundance of these four dominant EcM genera combined were significantly higher in forest soils compared to oil palm plantations (P<0.05). For *Russula* alone, we also found a difference in abundance between unlogged (greatest abundance), once-logged (less abundant), and twice-logged forest (lowest abundance of *Russula*) (P<0.05 in each case).

The β-diversity, measured as the average distance of all samples to the centroid in each land use type, did not show significant difference among land uses ($F_{3,24} = 2.77$, $P = 0.06$). However, there was a significant effect of forest conversion to oil palm plantations on fungal true β-diversity (i.e. S/α; $F_{3,24} = 3.85$, $P = 0.01$), with oil palm having lowest true β-diversity compared to unlogged and logged forests (Fig. 5). Logging did not produce any significant change in true β-diversity.

Discussion

Our results showed that fungal OTU richness and Chao1 index differed among land uses. Oil palm plantations had lower OTU richness, which can be attributed to the effects of anthropogenic intervention and forest conversion. This contrasts with the findings on bacterial communities of Lee-Cruz et al. [27] in the same study site, where they found that OTU richness and diversity indices did not differ among land uses, and that α-diversity was similar in forests and oil palm plantations.

We found that the structure of fungal communities differed most fundamentally between forests and oil palm plantations. This finding mirrors that of McGuire et al. [29], who found that the community of soil fungi collected across three different land uses in Malaysia differed between oil palm plantations and forests. Changes in fungal community composition in logged forest and

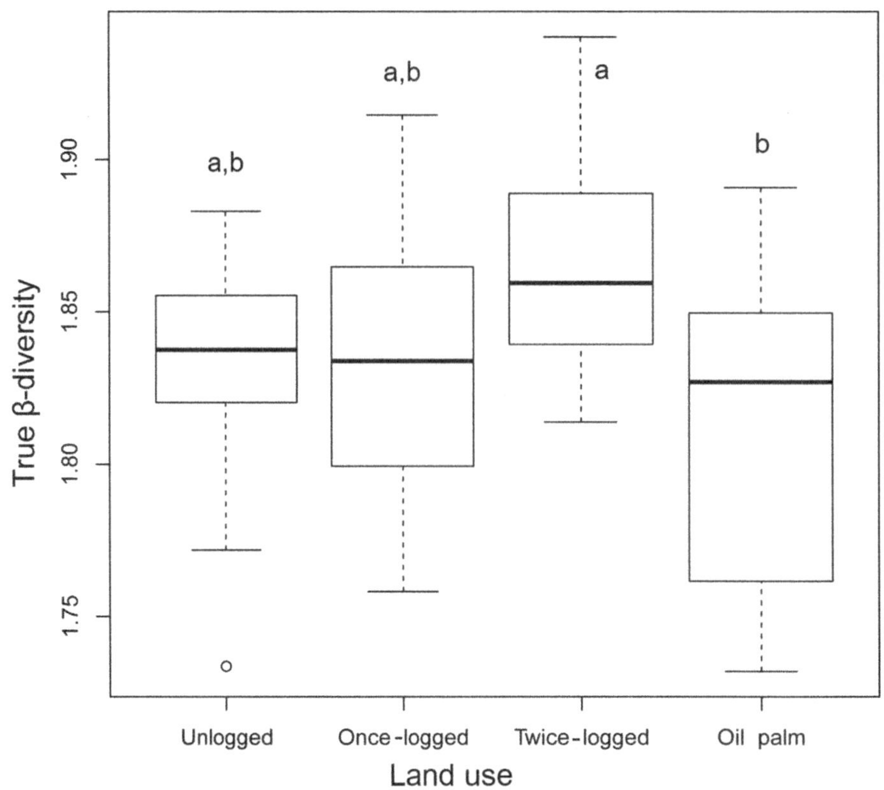

Figure 5. Fungal community true β-diversity (i.e. $S/\bar{\alpha}$) among the four land uses in Sabah, Malaysian Borneo. Boxes show the lower quartile, the median and the upper quartile. Pairwise comparisons are shown; different letters denote significant differences between groups at $P < 0.05$.

former forest areas could directly impact the functioning of soil communities and their ability to provide key ecosystem services, such as decomposition and nutrient recycling [22]. Given that fertilizer prices are predicted to rise dramatically in the coming decades [48], these results suggest that improvements in agricultural methods by establishing diversified farms could be necessary for sustaining vital soil biodiversity and ecosystem-service values.

The abundance pattern of fungal taxa detected here is similar to the soils of the Neotropics and elsewhere, where Basidiomycota and Ascomycota are also the most prevalent groups [49–53]. The most abundant orders of Basidomycota in the forest areas we sampled were *Russulales*, *Agaricales*, *Sporidiobolales*, *Telephorales*, *Trichosporonales* and *Sebacinales*. They all became less abundant in oil palm plantations compared to primary forest. These orders are generally reported to be the most varied and abundant groups of fungi in forests around the world [40,54,55], they have also been characterized as being lignicolous, saprobic or mycorrhizal, associated with litter decomposition, and they are known to degrade plant-derived cellulose [56]. In our forest plots, the ectomycorrhizal orders *Russulales* and *Telephorales* were the most common. This pattern has also been found in forests from the Neotropics and African tropics [57–59].

The lower abundance of Basidiomycota in oil palm soils may be due to the lack of large quantities of coarse woody debris, often derived from roots or branches of forest trees, in oil palm plantations. However, it may be worth investigating whether the change in abundance of Basidiomycota - and change in their overall community structure - with conversion of forest to oil palm may have implications for long term nutrient processing in the oil palm soils. Since Basidiomycota as a group are often able to break down relatively recalcitrant substrates and changes in their abundance and community composition might impede nutrient recycling in oil palm soils. The converse increase in Ascomycota, may be seen in terms of the relatively lignin-poor and nutrient-rich character of most of the organic matter reaching the soil from roots and leaves of oil palm, and from the herbaceous weedy layer that grows under the palms.

We found that both a history of logging, and forest conversion to oil palm plantation, resulted in shifts in EcM fungal communities. The most drastic effects were with forest conversion to oil palm. However, logging history also had detectable effects EcM fungi relative abundance. Ectomycorrhizas are one of the most important widespread types of mycorrhiza in forests of the cool temperate and boreal latitudes [60], and they also form an important group and often the dominant ecologically and economically important minority of Dipterocarpaceae family trees in tropical Asia [61–65].

The lower abundance of EcM fungi in logged forests might be due to a thinner, more incomplete root mat following past logging disturbance. The much lower abundance of EcM fungi in oil palm plantation soils could be partly due to the much lower abundance of potential host roots (e.g. Dipterocarpaceae which are absent from the palm plantations). Such fungi have generally been found to recover slowly from disturbance even when potential host plants are present [66]. Among the detected genera in our study, we found that *Russula*, *Sebacina*, *Lactarius* and *Tomentella* showed significant impact of land use change. *Russula* was relatively the most abundant EcM genus in our samples: this genus has been found to be common on roots of dipterocarp forests, tropical and southern hemisphere angiosperm forests [62,67,68].

This study also investigated the impacts of land use change on the turnover of fungal communities across space (β-diversity), and results suggest that there is a spatial homogenization of fungal communities in oil palm agriculture compared to forest: fungal β-diversity in oil palm plantations was lower than forest. Habitat conversion to agriculture also reduces β-diversity of soil bacteria in Amazonia [69]. In contrast, across the same land use system and sites in Borneo as we studied here, there was actually an increase in β-diversity of bacteria with both logging and conversion to oil palm [27].

There are some caveats that accompany our findings. We worked in only one biogeographic region and on only one form of agriculture. It is thus important to replicate this work in other logging systems and in other key expanding crops, including soya, sugar cane, and cacao. We also did not explicitly demonstrate the impacts of changing fungal composition on soil ecosystem functions and services: this is a major knowledge gap, with critical importance to the development of sustainable logging and, in particular, agricultural systems in the tropics.

In conclusion, the conversion of both primary and logged forest to oil palm drives a change in the overall fungal community, including EcM fungi abundance, and an associated decrease in total fungal community beta-diversity. This finding invites further studies that investigate the long-term implications of such changes for agricultural sustainability. There was a more subtle long-term impact of logging on fungal communities. Most measurable features of the unlogged forest fungal community remained unchanged after logging. However, there were significant changes in EcM fungal abundance due to logging, which could have a pervasive impact since EcM fungi are thought to play a key role in tree growth and community structure. Despite this, the lack of drastic changes in the overall forest fungal community structure following logging strengthens the view that logged forest is not necessarily an irretrievably damaged and drastically altered system, and that protecting it from conversion to oil palm may still have considerable conservation benefits.

Acknowledgments

We thank Yayasan Sabah, Danum Valley Management Committee, Sabah Parks, the state Secretary, Sabah Chief Minister's Departments, and the Sabah Biodiversity Council for permission to conduct research. We also, thank the Royal Society's Southeast Asian Rainforest Research Program (SEARRP) and the Borneo Rainforest Lodge for logistical support and site access.

Author Contributions

Conceived and designed the experiments: DPE JMA. Performed the experiments: DK BMT DPE. Analyzed the data: DK BMT DPE JL. Contributed reagents/materials/analysis tools: DK BMT DPE JMA JL. Contributed to the writing of the manuscript: DK BMT DPE JMA.

References

1. Whitmore TC (1990) An introduction to tropical rain forests. Oxford: Clarendon Press. 226 pp.
2. Asner GP, Rudel TK, Aide TM, Defries R, Emerson R (2009) A contemporary assessment of change in humid tropical forests. Conserv Biol 23: 1386–1395.
3. Wright SJ (2005) Tropical forests in a changing environment. Trends Ecol Evol 20: 553–560.
4. Blaser J (2011) Status of tropical forest management 2011: International Tropical Timber Organization.
5. Food and Agriculture Organization of the United Nations (2010) Global forest resources assessment 2010 : main report. Rome: FAO. 340 p.
6. Hansen MC, Stehman SV, Potapov PV, Loveland TR, Townshend JR, et al. (2008) Humid tropical forest clearing from 2000 to 2005 quantified by using multitemporal and multiresolution remotely sensed data. Proc Natl Acad Sci U S A 105: 9439–9444.
7. Morris RJ (2010) Anthropogenic impacts on tropical forest biodiversity: a network structure and ecosystem functioning perspective. Philos Trans R Soc Lond B Biol Sci 365: 3709–3718.
8. Achard F, Eva HD, Stibig H-J, Mayaux P, Gallego J, et al. (2002) Determination of Deforestation Rates of the World's Humid Tropical Forests. Science 297: 999–1002.
9. Gibson L, Lee TM, Koh LP, Brook BW, Gardner TA, et al. (2011) Primary forests are irreplaceable for sustaining tropical biodiversity. Nature 478: 378–381.
10. Edwards DP, Larsen TH, Docherty TD, Ansell FA, Hsu WW, et al. (2011) Degraded lands worth protecting: the biological importance of Southeast Asia's repeatedly logged forests. Proc Biol Sci 278: 82–90.
11. Berry NJ, Phillips OL, Lewis SL, Hill JK, Edwards DP, et al. (2010) The high value of logged tropical forests: lessons from northern Borneo. Biodiver Conserv 19: 985–997.
12. Woodcock P, Edwards DP, Newton RJ, Vun Khen C, Bottrell SH, et al. (2013) Impacts of intensive logging on the trophic organisation of ant communities in a biodiversity hotspot. PLoS One 8: e60756.
13. Burivalova Z, Şekercioğlu ÇH, Koh LP (2014) Thresholds of Logging Intensity to Maintain Tropical Forest Biodiversity. Curr Biol 24: 1–6.
14. Hamer K, Hill J, Benedick S, Mustaffa N, Sherratt T, et al. (2003) Ecology of butterflies in natural and selectively logged forests of northern Borneo: the importance of habitat heterogeneity. J Appl Ecol 40: 150–162.
15. Cleary DF, Boyle TJ, Setyawati T, Anggraeni CD, Loon EEV, et al. (2007) Bird species and traits associated with logged and unlogged forest in Borneo. Ecol Appl 17: 1184–1197.
16. Edwards DP, Hodgson JA, Hamer KC, Mitchell SL, Ahmad AH, et al. (2010) Wildlife-friendly oil palm plantations fail to protect biodiversity effectively. Conserv Lett 3: 236–242.
17. Edwards F, Edwards D, Larsen T, Hsu W, Benedick S, et al. (2013) Does logging and forest conversion to oil palm agriculture alter functional diversity in a biodiversity hotspot? Anim Conserv 17: 163–173.
18. Baraloto C, Hérault B, Paine C, Massot H, Blanc L, et al. (2012) Contrasting taxonomic and functional responses of a tropical tree community to selective logging. J Appl Ecol 49: 861–870.
19. McGrath DA, Smith CK, Gholz HL, Oliveira FdA (2001) Effects of Land-Use Change on Soil Nutrient Dynamics in Amazônia. Ecosystems 4: 625–645.
20. Murty D, Kirschbaum MUF, McMurtrie RE, McGilvray H (2002) Does conversion of forest to agricultural land change soil carbon and nitrogen? a review of the literature. Global Change Biol 8: 105–123.
21. Cornejo FH, Varela A, Wright SJ (1994) Tropical Forest Litter Decomposition under Seasonal Drought: Nutrient Release, Fungi and Bacteria. Oikos 70: 183–190.
22. Lauber CL, Strickland MS, Bradford MA, Fierer N (2008) The influence of soil properties on the structure of bacterial and fungal communities across land-use types. Soil Biol Biochem 40: 2407–2415.
23. Tripathi B, Kim M, Singh D, Lee-Cruz L, Lai-Hoe A, et al. (2012) Tropical Soil Bacterial Communities in Malaysia: pH Dominates in the Equatorial Tropics Too. Microb Ecol 64: 474–484.
24. Orgiazzi A, Lumini E, Nilsson RH, Girlanda M, Vizzini A, et al. (2012) Unravelling soil fungal communities from different Mediterranean land-use backgrounds. PloS one 7: e34847.
25. Wu TH, Chellemi DO, Martin KJ, Graham JH, Rosskop EN (2007) Discriminating the effects of agricultural land management practices on soil fungal communities. Soil Biol Biochem 39: 1139–1155.
26. Martin F, Cullen D, Hibbett D, Pisabarro A, Spatafora JW, et al. (2011) Sequencing the fungal tree of life. New Phytol 190: 818–821.
27. Lee-Cruz L, Edwards DP, Tripathi BM, Adams JM (2013) Impact of Logging and Forest Conversion to Oil Palm Plantations on Soil Bacterial Communities in Borneo. Appl Environ Microbiol 79: 7290–7297.
28. Lupatini M, Jacques RJS, Antoniolli ZI, Suleiman AKA, Fulthorpe RR, et al. (2013) Land-use change and soil type are drivers of fungal and archaeal communities in the Pampa biome. World J Microbiol Biotechnol 29: 223–233.
29. McGuire KL, D'Angelo H, Brearley FQ, Gedallovich SM, Babar N, et al. (2014) Responses of soil fungi to logging and oil palm agriculture in Southeast Asian tropical forests. Microbial Ecol.
30. Wilcove DS, Giam X, Edwards DP, Fisher B, Koh LP (2013) Navjot's nightmare revisited: logging, agriculture, and biodiversity in Southeast Asia. Trends Ecol Evol 28: 531–540.
31. Fisher B, Edwards DP, Giam XL, Wilcove DS (2011) The high costs of conserving Southeast Asia's lowland rainforests. Front Ecol Environ 9: 329–334.
32. Gibbs HK, Ruesch AS, Achard F, Clayton MK, Holmgren P, et al. (2010) Tropical forests were the primary sources of new agricultural land in the 1980s and 1990s. Proc Natl Acad Sci U S A 107: 16732–16737.
33. Gardes M, Bruns TD (1993) ITS primers with enhanced specificity for basidiomycetes–application to the identification of mycorrhizae and rusts. Mol Ecol 2: 113–118.

34. White TJ, Bruns T, Lee S, Taylor J (1990) Amplification and direct sequencing of fungal ribosomal RNA genes for phylogenetics. PCR protocols: a guide to methods and applications 18: 315–322.
35. Schloss PD, Gevers D, Westcott SL (2011) Reducing the effects of PCR amplification and sequencing artifacts on 16S rRNA-based studies. PLoS One 6: e27310.
36. Nilsson RH, Veldre V, Hartmann M, Unterseher M, Amend A, et al. (2010) An open source software package for automated extraction of ITS1 and ITS2 from fungal ITS sequences for use in high-throughput community assays and molecular ecology. Fungal Ecol 3: 284–287.
37. Edgar RC, Haas BJ, Clemente JC, Quince C, Knight R (2011) UCHIME improves sensitivity and speed of chimera detection. Bioinformatics 27: 2194–2200.
38. Edgar RC (2010) Search and clustering orders of magnitude faster than BLAST. Bioinformatics 26: 2460–2461.
39. Abarenkov K, Nilsson RH, Larsson KH, Alexander IJ, Eberhardt U, et al. (2010) The UNITE database for molecular identification of fungi - recent updates and future perspectives. New Phytol 186: 281–285.
40. Tedersoo L, May T, Smith M (2010) Ectomycorrhizal lifestyle in fungi: global diversity, distribution, and evolution of phylogenetic lineages. Mycorrhiza 20: 217–263.
41. Liedloff A (1999) Mantel nonparametric test calculator. Version 2.0. School of Natural Resource Sciences, Queensland University of Technology, Australia.
42. Ghazoul J (2002) Impact of logging on the richness and diversity of forest butterflies in a tropical dry forest in Thailand. Biodiver Conserv 11: 521–541.
43. Ramage BS, Sheil D, Salim HM, Fletcher C, MUSTAFA NZA, et al. (2013) Pseudoreplication in tropical forests and the resulting effects on biodiversity conservation. Conserv Biol 27: 364–372.
44. Magurran AE (2004) Measuring biological diversity. Maldan, MA: Blackwell Pub. viii, 256 p.
45. Anderson MJ, Crist TO, Chase JM, Vellend M, Inouye BD, et al. (2011) Navigating the multiple meanings of beta diversity: a roadmap for the practicing ecologist. Ecol Lett 14: 19–28.
46. Whittaker RH (1960) Vegetation of the Siskiyou mountains, Oregon and California. Ecol Monogr 30: 279–338.
47. Koleff P, Gaston KJ, Lennon JJ (2003) Measuring beta diversity for presence-absence data. J Anim Ecol 72: 367–382.
48. Piesse J, Thirtle C (2009) Three bubbles and a panic: An explanatory review of recent food commodity price events. Food Policy 34: 119–129.
49. O'Brien HE, Parrent JL, Jackson JA, Moncalvo JM, Vilgalys R (2005) Fungal community analysis by large-scale sequencing of environmental samples. Appl Environ Microbiol 71: 5544–5550.
50. Wubet T, Christ S, Schöning I, Boch S, Gawlich M, et al. (2012) Differences in soil fungal communities between European beech (Fagus sylvatica L.) dominated forests are related to soil and understory vegetation. PloS one 7: e47500.
51. Gomes NC, Fagbola O, Costa R, Rumjanek NG, Buchner A, et al. (2003) Dynamics of fungal communities in bulk and maize rhizosphere soil in the tropics. Appl Environ Microbiol 69: 3758–3766.
52. Oros-Sichler M, Gomes NC, Neuber G, Smalla K (2006) A new semi-nested PCR protocol to amplify large 18S rRNA gene fragments for PCR-DGGE analysis of soil fungal communities. J Microbiol Methods 65: 63–75.
53. Bridge PD, Newsham KK (2009) Soil fungal community composition at Mars Oasis, a southern maritime Antarctic site, assessed by PCR amplification and cloning. Fungal Ecol 2: 66–74.
54. Geml J, Laursen GA, Timling I, McFarland JM, Booth MG, et al. (2009) Molecular phylogenetic biodiversity assessment of arctic and boreal ectomycorrhizal Lactarius Pers. (Russulales; Basidiomycota) in Alaska, based on soil and sporocarp DNA. Mol Ecol 18: 2213–2227.
55. Geml J, Laursen GA, Herriott IC, McFarland JM, Booth MG, et al. (2010) Phylogenetic and ecological analyses of soil and sporocarp DNA sequences reveal high diversity and strong habitat partitioning in the boreal ectomycorrhizal genus Russula (Russulales; Basidiomycota). New Phytol 187: 494–507.
56. Kuramae EE, Hillekens RHE, de Hollander M, van der Heijden MGA, van den Berg M, et al. (2013) Structural and functional variation in soil fungal communities associated with litter bags containing maize leaf. Fems Microbiol Ecol 84: 519–531.
57. Ba AM, Duponnois R, Moyersoen B, Diedhiou AG (2012) Ectomycorrhizal symbiosis of tropical African trees. Mycorrhiza 22: 1–29.
58. Smith ME, Henkel TW, Aime MC, Fremier AK, Vilgalys R (2011) Ectomycorrhizal fungal diversity and community structure on three co-occurring leguminous canopy tree species in a Neotropical rainforest. New Phytol 192: 699–712.
59. Tedersoo L, Sadam A, Zambrano M, Valencia R, Bahram M (2010) Low diversity and high host preference of ectomycorrhizal fungi in Western Amazonia, a neotropical biodiversity hotspot. Isme Journal 4: 465–471.
60. Smith SE, Read DJ (2010) Mycorrhizal symbiosis: Academic press.
61. Brearley FQ (2012) Ectomycorrhizal Associations of the Dipterocarpaceae. Biotropica 44: 637–648.
62. Peay KG, Kennedy PG, Davies SJ, Tan S, Bruns TD (2010) Potential link between plant and fungal distributions in a dipterocarp rainforest: community and phylogenetic structure of tropical ectomycorrhizal fungi across a plant and soil ecotone. New Phytol 185: 529–542.
63. Moyersoen B, Becker P, Alexander I (2001) Are ectomycorrhizas more abundant than arbuscular mycorrhizas in tropical heath forests? New Phytol 150: 591–599.
64. Haug I, Weiß M, Homeier J, Oberwinkler F, Kottke I (2005) Russulaceae and Thelephoraceae form ectomycorrhizas with members of the Nyctaginaceae (Caryophyllales) in the tropical mountain rain forest of southern Ecuador. New Phytol 165: 923–936.
65. Natarajan K, Senthilarasu G, Kumaresan V, Riviere T (2005) Diversity in ectomycorrhizal fungi of a dipterocarp forest in Western Ghats. Curr Sci 88: 1893–1895.
66. Peay KG, Schubert MG, Nguyen NH, Bruns TD (2012) Measuring ectomycorrhizal fungal dispersal: macroecological patterns driven by microscopic propagules. Mol Ecol 21: 4122–4136.
67. Tedersoo L, Jairus T, Horton BM, Abarenkov K, Suvi T, et al. (2008) Strong host preference of ectomycorrhizal fungi in a Tasmanian wet sclerophyll forest as revealed by DNA barcoding and taxon-specific primers. New Phytol 180: 479–490.
68. Riviere T, Diedhiou AG, Diabate M, Senthilarasu G, Natarajan K, et al. (2007) Genetic diversity of ectomycorrhizal Basidiomycetes from African and Indian tropical rain forests. Mycorrhiza 17: 415–428.
69. Rodrigues JL, Pellizari VH, Mueller R, Baek K, Jesus Eda C, et al. (2013) Conversion of the Amazon rainforest to agriculture results in biotic homogenization of soil bacterial communities. Proc Natl Acad Sci U S A 110: 988–993.

19

Evaluation of the CENTURY Model Using Long-Term Fertilization Trials under Corn-Wheat Cropping Systems in the Typical Croplands of China

Rihuan Cong[1,2], Xiujun Wang[3,4], Minggang Xu[1]*, Stephen M. Ogle[5], William J. Parton[5]

1 Ministry of Agriculture Key Laboratory of Crop Nutrition and Fertilization, Institute of Agricultural Resources and Regional Planning, Chinese Academy of Agricultural Sciences, Beijing, China, **2** College of Resources and Environment, Huazhong Agricultural University, Wuhan, China, **3** State Key Laboratory of Desert and Oasis Ecology, Xinjiang Institute of Ecology and Geography, Chinese Academy of Sciences, Urumqi, China, **4** Earth System Science Interdisciplinary Center, University of Maryland, College Park, Maryland, United States of America, **5** Natural Resource Ecology Laboratory, Colorado State University, Fort Collins, Colorado, United States of America

Abstract

Soil organic matter models are widely used to study soil organic carbon (SOC) dynamics. Here, we used the CENTURY model to simulate SOC in wheat-corn cropping systems at three long-term fertilization trials. Our study indicates that CENTURY can simulate fertilization effects on SOC dynamics under different climate and soil conditions. The normalized root mean square error is less than 15% for all the treatments. Soil carbon presents various changes under different fertilization management. Treatment with straw return would enhance SOC to a relatively stable level whereas chemical fertilization affects SOC differently across the three sites. After running CENTURY over the period of 1990–2050, the SOC levels are predicted to increase from 31.8 to 52.1 Mg ha^{-1} across the three sites. We estimate that the carbon sequestration potential between 1990 and 2050 would be 9.4–35.7 Mg ha^{-1} under the current high manure application at the three sites. Analysis of SOC in each carbon pool indicates that long-term fertilization enhances the slow pool proportion but decreases the passive pool proportion. Model results suggest that change in the slow carbon pool is the major driver of the overall trends in SOC stocks under long-term fertilization.

Editor: Julio Vera, University of Erlangen-Nuremberg, Germany

Funding: Financial supports are from the National Science Foundation of China (41171239) and the National Basic Research Program (2011CB100501). The funders had no role in study design, data collection and analysis, decision to publish, or preparation of the manuscript.

Competing Interests: The authors have declared that no competing interests exist.

* E-mail: mgxu@caas.ac.cn

Introduction

Soil organic carbon (SOC) is one of the most important terrestrial pools for C storage. It is estimated that the total soil carbon pool is around 1400–1500 Pg C, which is approximately three times greater than the atmospheric pool (750 Pg C) [1,2]. The SOC pool represents a dynamic equilibrium resulting from changes in gains and losses. Even small changes in SOC at a site may potentially add up to significant changes in large-scale carbon cycling across a region [3]. Furthermore, SOC is relatively dynamic and can be greatly influenced by agricultural practices. Increases in SOC storage in cropland soils would benefit soil productivity and environmental health [4,5], and so alternative farming management practices have been evaluated to identify their potentials for increasing SOC in the agroecosystems [4–7].

Long-term experiments are crucial for determining fundamental crop, soil and ecological processes and their impacts on the environment [6–9]. Data from long-term experiments provide a unique resource to investigate long-term influences of climate, crop rotation and crop residue management on soil fertility [6–12]. However, SOC change is affected by complex interactions that vary across space and time depending on the environmental conditions and agricultural management practices. A weakness of long-term experiments is that they are typically restricted to small subset of the entire set of environmental conditions and management practices that exists [13].

Process-based models are an effective way to evaluate SOC changes across a broader set of environmental conditions and management practices [14]. In recent decades, the development and evaluation of soil organic matter models has improved the understanding of factors controlling SOC dynamics, and thus increased our ability to predict future SOC trends. A number of SOC models have been developed, but applying these models requires adequate evaluation with measured SOC trends from experimental for different environmental conditions and management practices [15]. For example, the CENTURY model [16] has been widely used to simulate SOC changes under different management conditions in long-term experiments (e.g., [17], [18] and [19]). With the development of CENTURY, the model has been successfully employed in long-term fertilizer, irrigation, pest management, and site-specific farming applications [20,21]. In China, CENTURY model has been used in grassland [22], forest [23], and regional farmland [24]. However, CENTURY modeling research was still limited in farmland especially under the double cropping rotations and in the acidic soil.

Here, we evaluate the CENTURY with data from three long-term experiments with wheat-corn cropping rotations and different fertilization practices. Specifically, our objectives were

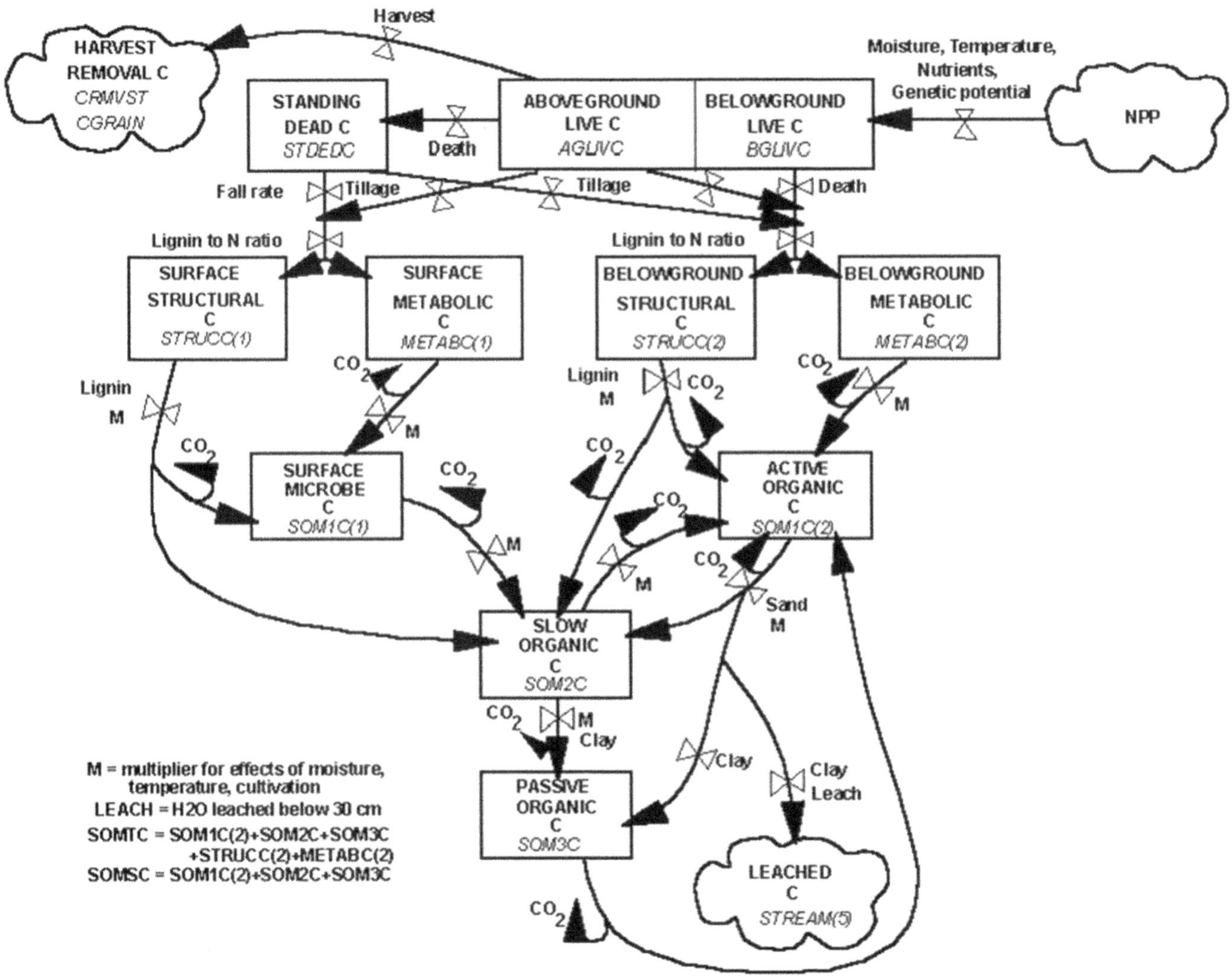

Figure 1. The pools and flows of carbon in the CENTURY model. The diagram showed the major factors which control the flows [29].

(i) to evaluate the performance of CENTURY with evaluation of modeled SOC stocks for different fertilizations and under acidic soil; (ii) to study the effect of fertilization practices on different SOC pools in the modeling framework; and (iii) to predict soil carbon potential under long-term fertilization.

Materials and Methods

Long-term Experiment

Three long-term experiments were utilized for this study, which were located at Changping (40°13′N, 116°15′E), Yangling (34°17′N, 108°00′E) and Qiyang (26°45′N, 111°52′E) in China. Climate conditions varied from semi-humid (Changping site) to

Table 1. Land management practices for the long-term experimental sites during the different blocks and periods used in the CENTURY model simulations.

No.	Periods	Management practices	Repeating sequence
1	up to 1900 (4000 yrs)	Native grassland, low-intensity grazing	1 year
2	1901–1960	Corn, low crop yield cultivar, ploughing, remove straw, applying manure	1 year
3	1961–1988	Wheat-corn rotation, low crop yield cultivar, ploughing, remove straw, applying chemical fertilizers and manure	2 years
4	1989–1989	Uniform tillage to set up long-term experiments	1 year
5	1990–2050	Long-term fertilization experiments	2 years

doi:10.1371/journal.pone.0095142.t001

Table 2. Carbon input (kg $ha^{-1}yr^{-1}$) from manure and straw residue in each period used in the CENTURY model at Changping, Yangling, and Qiyang sites.

Sites	1901–1960	1961–1988	1989–1989	1990–2050 (long-term experiment)		
				NPKM	hNPKM	NPKS
Changping	500	500	–	3150	4725	1000
Yangling	500	500	–	3327	4991	1998
Qiyang	500	500	–	5838	8757	1052

humid warm-temperate (Yangling sites) to humid subtropical climate (Qiyang site). Annual mean temperature was 13.1°C at the Changping site, 14.9°C at the Yangling site, and 18.1°C at the Qiyang site. Annual precipitation was generally low at Changping (515 mm) and Yangling (525 mm) sites but 1445 mm at Qiyang. However, annual evaporation was much higher, varying from 993 mm to 1470 mm [25].

The experimental sites had double cropping systems, i.e., winter wheat and summer corn. Winter wheat was seeded in early November and harvested in early May at the Qiyang site. For the other two sites, winter wheat was seeded around October 20th, and harvested around June 1st. The wheat seeding rates ranged from 165 to 225 kg ha^{-1}. Summer corn was sown around June 10th, with the rate of 63 000–75 000 seeds per hectare, and harvested around October 1st at most sites. For the Qiyang site, summer corn was sown in holes between the wheat strips in early April, and harvested in the middle of July. Thus, there was a short period of inter-cropping for wheat and corn at the Qiyang site. However, in the CENTURY modeling, wheat was harvested in April and corn was sown in May at the Qiyang site since inter-cropping could not be fitted in the model. We believed that there would be little impact on the belowground biomass of wheat and corn, and thus to the soil organic carbon turnover.

Seven treatments were selected in this study: (i) control (no fertilizer); (ii) mineral nitrogen (N); (iii) mineral nitrogen and phosphorus combination (NP); (iv) mineral nitrogen, phosphorus and potassium combination (NPK); (v) NPK combinations with livestock or farmyard (i.e., livestock manure mixed with soil and/ or crop residue) manure (NPKM); (vi) 1.5 times' application rate of NPKM (hNPKM); and (vii) mineral NPK combined with crop straw (NPKS). The total nitrogen applied (i.e., mineral plus organic) was equal (i.e., nitrogen balanced) in each of the fertilizer treatments (i.e., N, NP, NPK, NPKM and NPKS treatments). The NPKM and hNPKM treatments had 30% of total N applied from mineral fertilizer and the remaining 70% from organic manure. The treatment plots were initially randomized and isolated by 100-cm-cement baffle plates. There was no replicate for the treatments at these sites due to field availability. However, the plot size was relative large (196 m^2–400 m^2) at each study site. The durations of experiments were from 1990 to 2005 at the Changping site, from 1990 to 2008 at the Qiyang sites, and from 1990 to 2009 at the Yangling site.

Urea, calcium superphosphate, and potassium chloride were used as sources of mineral N, P, and K, respectively. All P and K fertilizers and nearly half of the mineral N fertilizer were applied as basal fertilizers prior to seeding. The remaining mineral N fertilizer was applied as top dressing during the growing season. Manure was generally applied once each year before wheat sowing. However, at the Qiyang site, 30% of manure was applied before wheat seeding and remaining 70% was applied before corn seeding. The sources of organic manure were farmyard manure mixed with crop residue at Changping site, cattle manure at Yangling site, and pig manure at Qiyang site [25]. For the NPKS treatment, half amount (2000 kg ha^{-1} yr^{-1}) of both wheat and

Table 3. Soil classification and initial physical and chemical properties (0–20 cm) in 1989.

Properties		Changping	Yangling	Qiyang
Soil classification (FAO)		Haplic Luvisol	Calcaric Regosol	Eutric Cambisol
Parent material		Diluvial Alluvium	Loess	Quaternary Red Clay
Clay mineral type		Hydromica, Montmorillonite	Hydromica, Montmorillonite	Kaolinite
Bulk density	(g cm^{-3})	1.58	1.30	1.19
Total porosity	(%)	40.4	49.6	51.7
Field capacity	(%)	24.8	21.2	23.7
Texture		silt loam	silt loam	clay
Clay (<2 μm)	(%)	10.2	16.8	40.9
Silt (2–50 μm)	(%)	72.0	51.6	27.7
Sand (50–2000 μm)	(%)	16.2	31.6	31.4
Soil pH		8.7	8.6	5.7
SOC	(g kg^{-1})	7.1	6.3	8.6
TN	(g kg^{-1})	0.80	0.83	1.07
C/N ratio		8.9	7.6	8.0

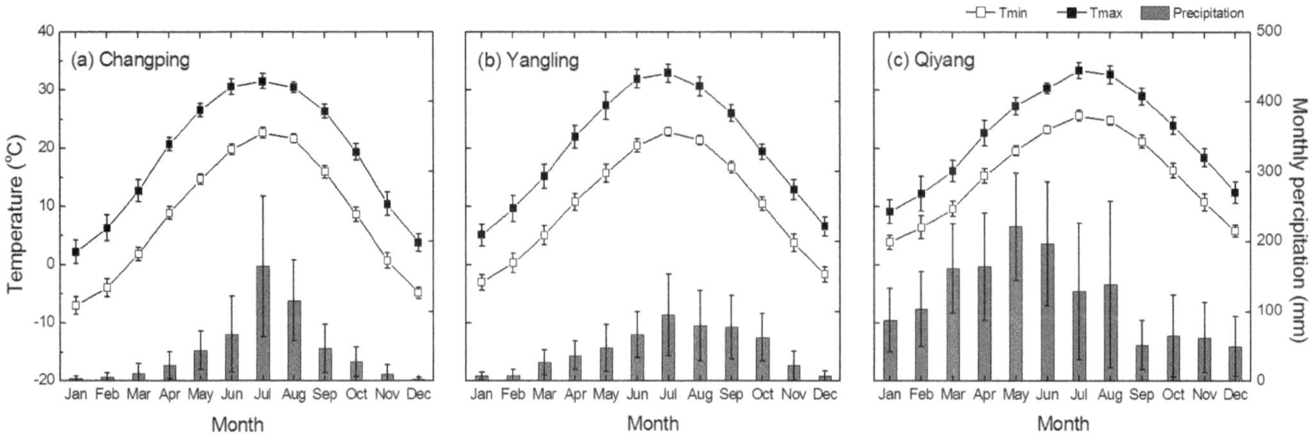

Figure 2. Average monthly precipitation, maximum (Tmax) and minimum (Tmin) temperatures from 1990 to 2010 at (A) Changping, (B) Yangling, and (C) Qiyang. Error bars are standard deviations for the mean values from 1990–2010.

corn straw was incorporated at the Qiyang site, whereas 2250 kg ha^{-1} yr^{-1} and 4500 kg ha^{-1} yr^{-1} of corn straw were incorporated at the Changping and Yangling sites, respectively [26]. Carbon and nutrient contents of manure are measured in 2009 and 2010. Nutrient contents of straw are measured annually over the period of experiment.

There was no irrigation at the Qiyang site since sufficient precipitation occurs during the growing season (i.e., from March to August). However, at the other two sites, irrigations were applied 2–3 times during wheat growing season: 5 mm before seeding, 4 mm for the over wintering stage if needed, and 5 mm at jointing stage. Irrigation was applied once during corn growing season, with 5 mm before seeding.

Plots were ploughed to a depth of 20 cm twice a year, usually in early June at most sites after wheat harvest and early October after corn harvest. In contrast, at the Qiyang site, the whole plot area was ploughed (20 cm depth) in late October before wheat seeding. Then, area between the wheat belts was ploughed (20 cm depth) in early April before the corn seeding.

Aboveground biomass was removed with negligible stubble left in the field. Thus, organic carbon input was mainly from manure application (i.e., NPKM and hNPKM treatments), residue

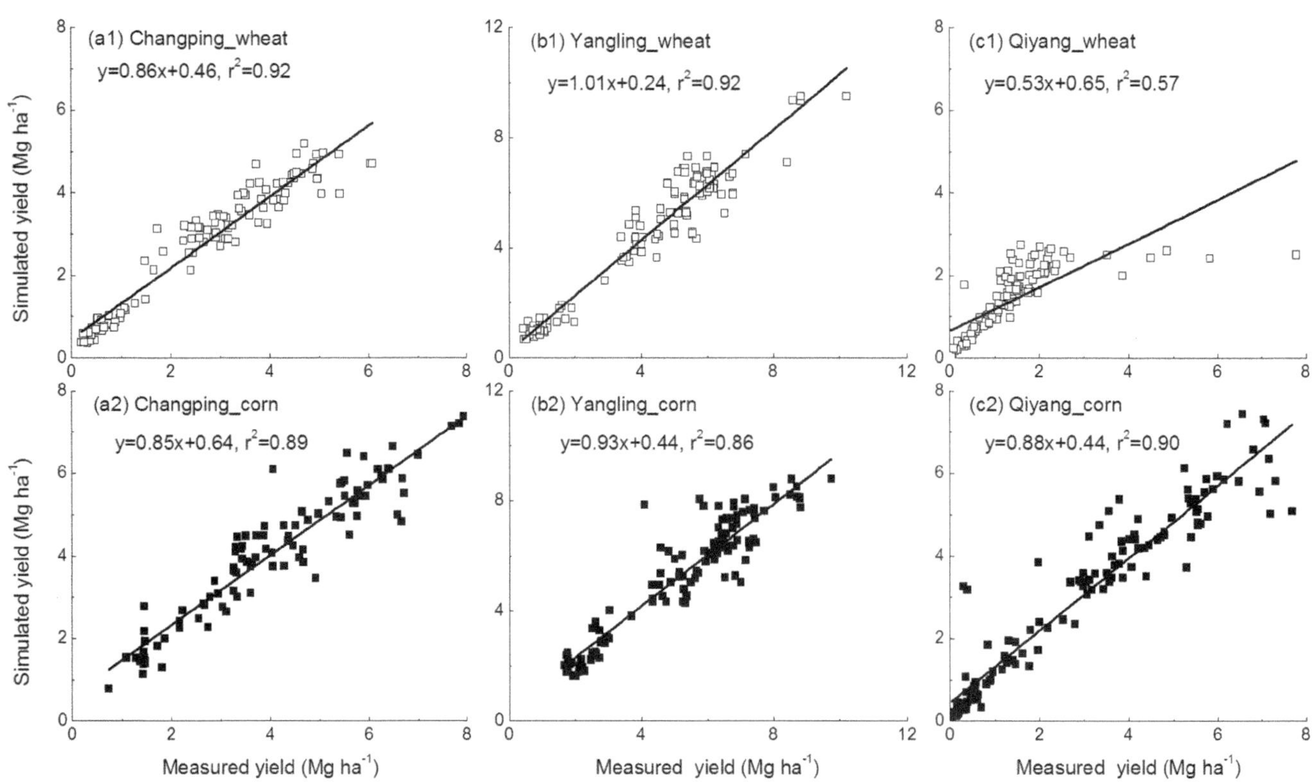

Figure 3. Correlationship between simulated and observed crop (i.e., □wheat and ■corn) grain yield data under the control, N, NP, NPK, NPKM, hNPKM and NPKS treatments at (A) Changping, (B) Yangling, and (C) Qiyang sites.

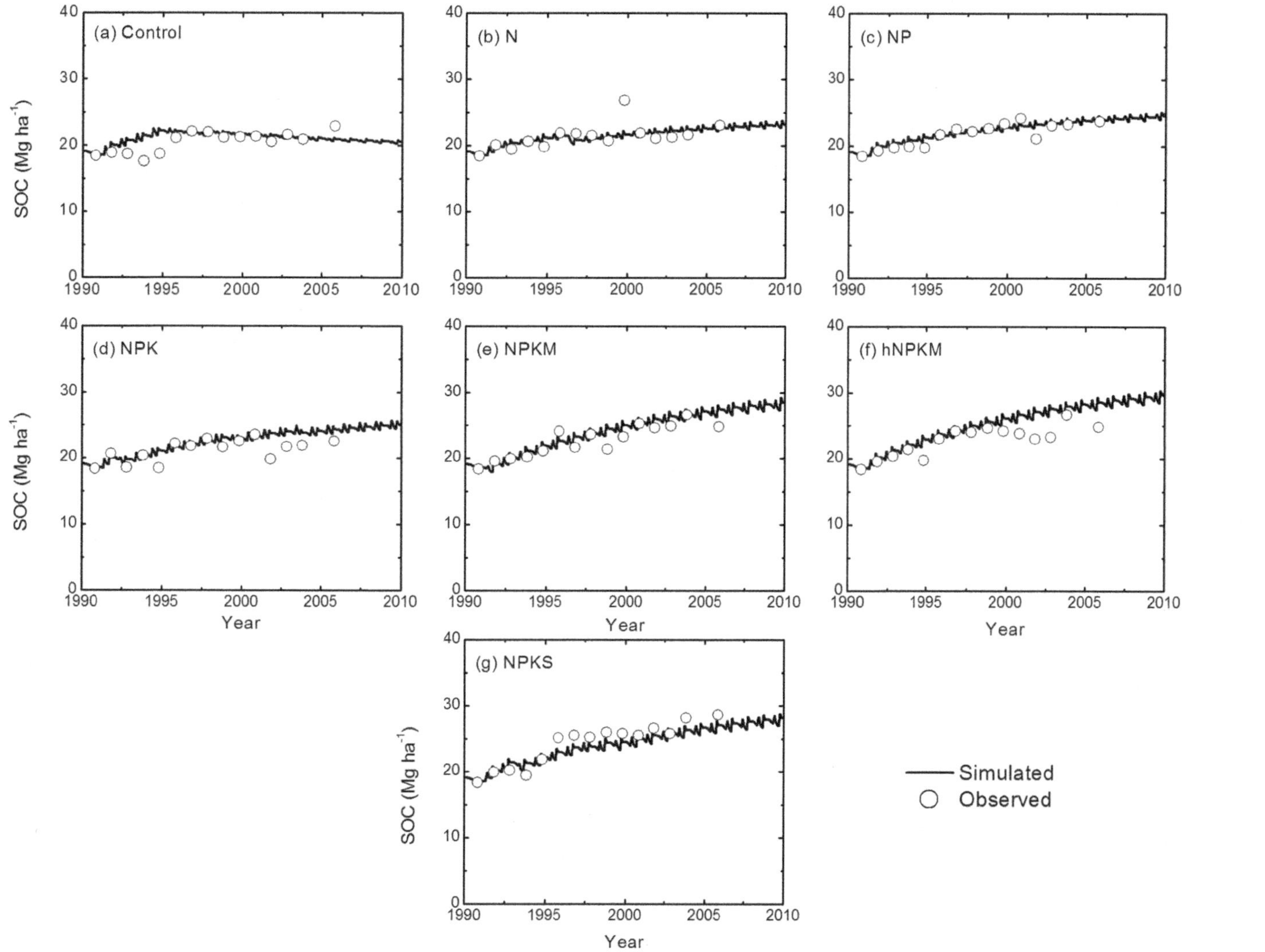

Figure 4. Simulated and measured soil organic carbon (SOC) stocks (0–20 cm) under the control, N, NP, NPK, NPKM, hNPKM and NPKS treatments at Changping.

addition (i.e., NPKS treatment) and from root system during crop growing season. All harvested biomass was removed from the plots for determining crop grain and residue yields. The grain and residue samples were air-dried, threshed, oven-dried at 70°C to a uniform moisture level, and then weighed separately.

During the experimental periods, surface soil (0–20 cm) was collected in each plot annually approximately 15 days after corn harvest. Five cores of soil from each plot were randomly taken, and soils were mixed thoroughly, and air dried for seven days after removing discrete plant residues and visible soil organism. Air-dried soil was sieved through 2 mm screen to determine pH (1:1w/v water). Representative subsamples were crushed to 0.25 mm for measurements of SOC [27]. Soil bulk density was measured in situ once every five or ten years. Surface soil SOC stock expressed as Mg ha^{-1} was calculated by multiplying the SOC content (g kg^{-1}) by averaged soil bulk density (g cm^{-3}) and depth (20 cm).

Model Description and Application

The simulation of SOC dynamics was performed with the CENTURY model (version 4.5) which was described by Parton and Rasmussen [28]. As shown in Fig. 1 [29], the arrows showing the CO_2 evolved in the transformations was indicative of the microbial growth efficiencies. The first-order decay rates for each of the pools corresponded to turnover times of roughly 3 and 0.5 years for the structural and metabolic components; 1.5 years for the active fraction, 25 years for the slow pool, and 1000 years for the passive pool [28]. The actual turnover times of each carbon pool was a function of the maximum turnover time of specific carbon pool and DEFAC. The value of DEFAC was calculated by multiplying the soil moisture factor (function of precipitation and stored soil water) and the soil temperature factor (function of the average monthly soil surface temperature). The turnover rate of active carbon pool was also a function of soil texture (higher for sandy soils), while the stabilization of active carbon into slow carbon was a function of the silt plus clay content [28].

Model simulations were set-up based on the historical crop rotations and farm practices' investigation from the local farmers at these sites. Five distinct periods were modeled (Table 1), including (1) an initialization period, i.e., 4000 years of native vegetation to reach an equilibrium; (2) the first cultivation period, i.e., plowing of the native grassland in 1901, and planting of corn

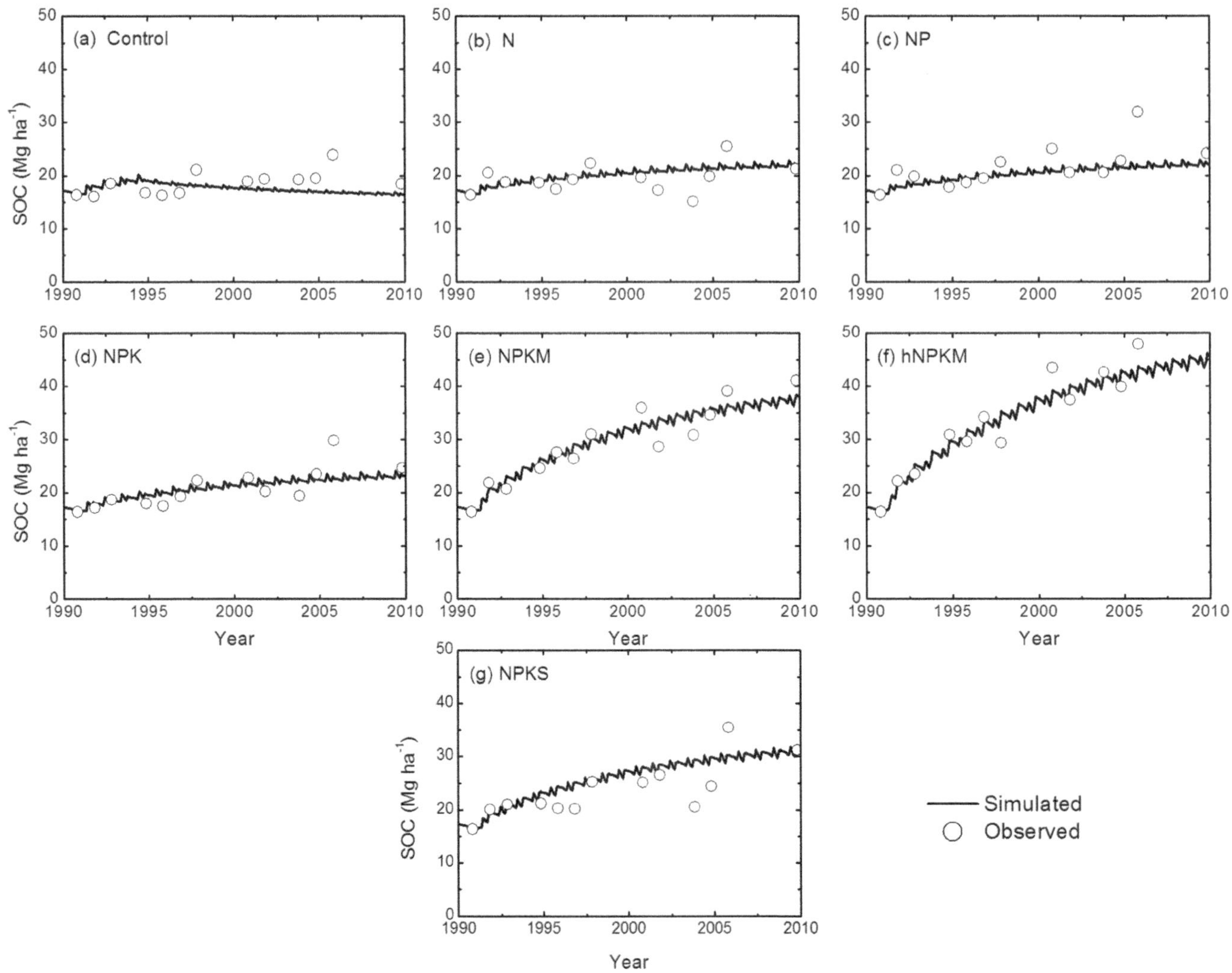

Figure 5. Simulated and measured soil organic carbon (SOC) stocks (0–20 cm) under the control, N, NP, NPK, NPKM, hNPKM and NPKS treatments at Yangling.

with removal of straw until 1960; (3) a second cultivation period, i.e., corn-wheat cropping with removal of straw from 1961 to 1988; (4) pre-experimental treatment period, i.e., uniform tillage with no crop in 1989 prior to establishing the long-term experiment; and (5) experiment period, i.e., the long-term corn-wheat fertility experiment from 1990 to 2050. Management during the experiment (1990–2010) was repeated for remainder of the experimental period until the end time of simulation to evaluate future trends. As shown in Table 2, carbon inputs from manure application were around 500 kg C ha^{-1} before the long-term experiment. Then carbon input varied based on measured data at different sites.

Weather data from 1990–2010 were obtained from China meteorological sharing service system (http://cdc.cma.gov.cn/). For the future simulations (i.e., 2011–2050), we used monthly mean climate data from the 1990–2010 period. Soil properties parameters were soil texture (sand, silt and clay content), soil pH, and bulk density (Table 3). Soils from the Changping and Yangling sites had high soil pH (~8.5) and silt loam texture with similar clay minerals (e.g., Hydromica and Montmorillonite). On the contrary, soil at the Qiyang site, developed from Quaternary red clay, had a lower soil pH (5.7) and heavy texture with the main clay mineral of Kaolinite.

As shown in Fig. 2, Changping and Yangling sites had similar climate condition whereas temperature and precipitation were much higher at the Qiyang site. The humid and warm climate at Qiyang site would accelerate decomposition of soil organic matter [28]. After setting up all the information (e.g., climate condition, soil texture and etc.) and modeling crop growth successfully, we modeled soil organic carbon pool by calibrating the dec4 value (i.e., maximum decomposition rate of soil organic matter with slow turnover) in the fix.100 file. By raising the parameter 10% higher for every test compared with the measured SOC, we finally got the sound fixed parameter as 0.0045 at the Qiyang site but 0.0023 (i.e., default value in the CENTURY model) at the other two sites [28].

For the plant production submodel, there were pools for live shoots and roots, and standing dead plant material. The potential production was a function of a genetic maximum (PRDX(1)) defined for each crop and 0–1 scalars depending on soil temperature, moisture status, shading by dead vegetation, and seedling growth. The maximum potential production of a crop, unlimited by temperature, moisture or nutrient stresses, was

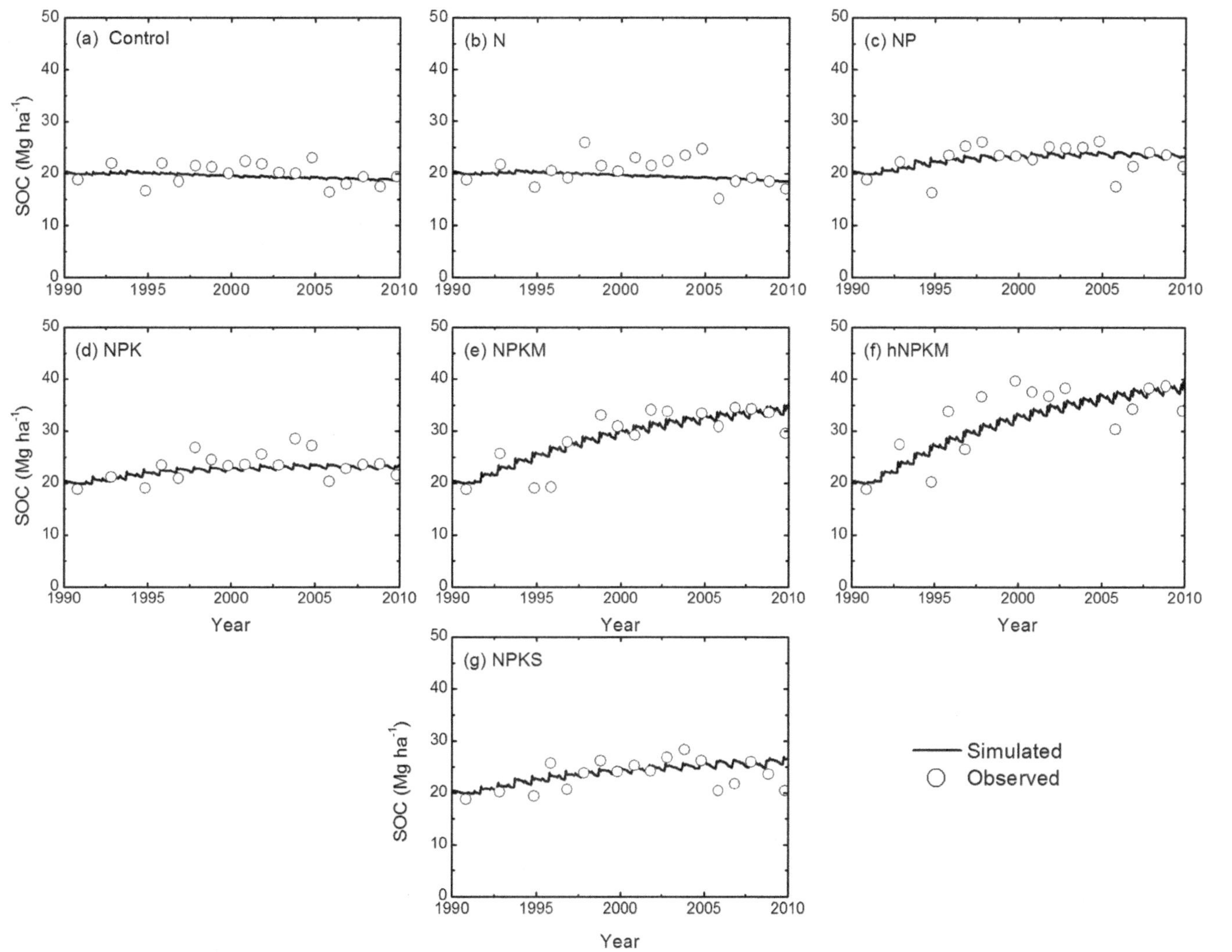

Figure 6. Simulated and measured soil organic carbon (SOC) stocks (0–20 cm) under the control, N, NP, NPK, NPKM, hNPKM and NPKS treatments at Qiyang.

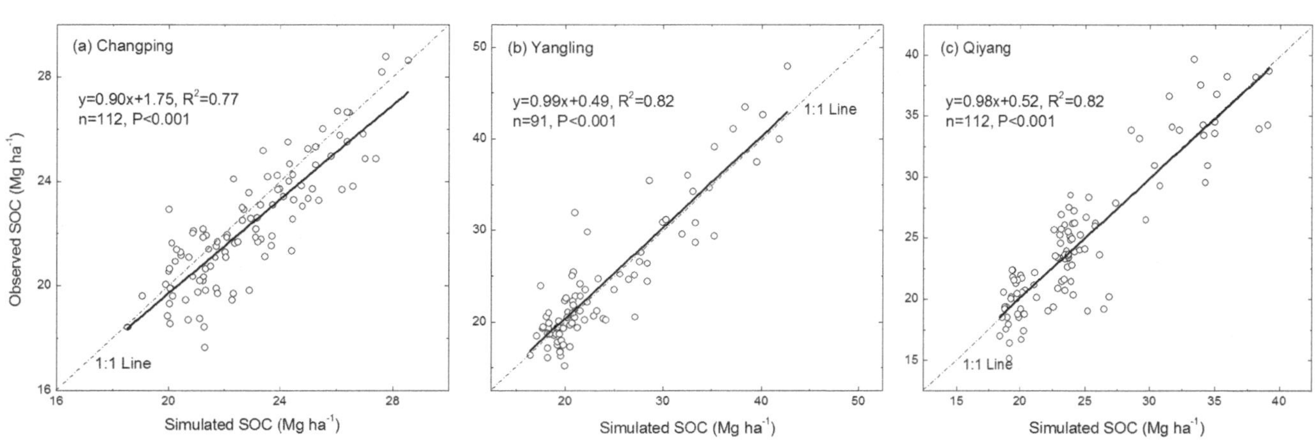

Figure 7. Correlation between simulated SOC and measured SOC under the control, N, NP, NPK, NPKM, hNPKM and NPKS treatments at (A) Changping, (B) Yangling, and (C) Qiyang sites.

doi:10.1371/journal.pone.0095142.g007

Table 4. Statistics comparing simulated and measured SOC stocks at the three long-term experimental sites.

Site	Parameter[a]	Treatment						
		Control	N	NP	NPK	NPKM	hNPKM	NPKS
Changping	*n-RMSE*	6%	5%	5%	6%	7%	4%	6%
(n = 16×7)	*d*	0.57	0.81	0.9	0.81	0.91	0.97	0.91
	R^2	0.01	0.76	0.83	0.46	0.80	0.90	0.76
Yangling	*n-RMSE*	9%	12%	9%	12%	9%	9%	14%
(n = 13×7)	*d*	0.5	1.00	0.83	0.77	0.96	0.97	0.85
	R^2	0.39	0.19	0.43	0.59	0.89	0.88	0.61
Qiyang	*n-RMSE*	8%	7%	6%	9%	10%	11%	7%
(n = 16×7)	*d*	0.37	0.51	0.75	0.61	0.89	0.88	0.82
	R^2	0.07	0.16	0.47	0.34	0.69	0.64	0.61

[a]Note: *n-RMSE*, the normalized-root mean square err; *d*, the index of agreement; R^2, the sample correlation coefficient.

primarily determined by the level of photosynthetically active radiation, the maximum net assimilation rate of photosynthesis, the efficiency of conversion of carbohydrate into plant constituents, and the maintenance respiration rate [30]. Thus, the parameter for maximum potential production (PRDX(1)) had both genetic and environmental components. However, in CENTURY, the seasonal distribution of production was primarily controlled by the temperature response function rather than the seasonal variation in photosynthetically active radiation, so the maximum potential production parameter would reflect aboveground crop production in optimal summer conditions. This parameter would frequently be used to calibrate the predicted crop production for different environments, species, and varieties [31].

The CENTURY model had no function for pH effects on aboveground growth. Based on the 20 years of field measurement at the Qiyang site, soil pH had no impact on crop growth as long as the value remains above pH 5.5, and would completely stop growth in wheat and corn if the value declined to less than pH 4.0 [32]. We added the following soil pH factor (f_{pH}) in the crop growth sub-model to account for this effect:

$$\begin{cases} f_{pH} = 0 & \text{if pH} \leq 4.0 \\ f_{pH} = \dfrac{(pH-\delta)}{(\delta-\beta)} + 1 & \text{if } 4.0 \leq \text{pH} \leq 5.5 \\ f_{pH} = 1 & \text{if pH} \geq 5.5 \end{cases} \qquad (1)$$

where δ and β were the maximum (i.e., $\delta = 5.5$) and minimum (i.e., $\beta = 4.0$) pH value that effected plant growth, respectively.

Model Evaluation

We compared the simulated SOC (0–20 cm) with field measured values. Both visual examination of graphic output and several statistical tests were used to evaluate the CENTURY model performance. Four statistical parameters were selected [33–36] for the evaluation, including (*i*) the sample coefficients of determination (R^2), (*ii*) the normalized-root mean square error (*n-RMSE*, equation 2), and (*iii*) the index of agreement (*d*, equation 3), as a descriptive measure of the average relative error [37].

$$n-RMSE = \frac{1}{\overline{M}}\sqrt{\sum_{i=1}^{n}(S_i - M_i)^2/n} \qquad (2)$$

where M_i were the measured values, S_i were the simulated values, $\overline{M}$ was the mean of the measured data and n was the number of the paired values. The index of agreement is computed with the following equation:

$$d = 1 - \frac{\sum_{i=1}^{n}(M_i - S_i)^2}{\sum_{i=1}^{n}(|S_i'| + |M_i'|)^2} \qquad (3)$$

where $S_i' = S_i - \overline{M}$ and $M_i' = M_i - \overline{M}$.

The classical R^2 statistic ($0 < R^2 < 1$) provided the percentage of data variance that was accounted for by the model. The *RMSE* evaluated the difference between observed and modeled values in the original units of the data, while the normalized-*RMSE* (*n-RMSE*, equation 2) removed the influence of the units, and placed the results on a percentage scale for comparison of model performance among variables with different units. For an ideal fit,

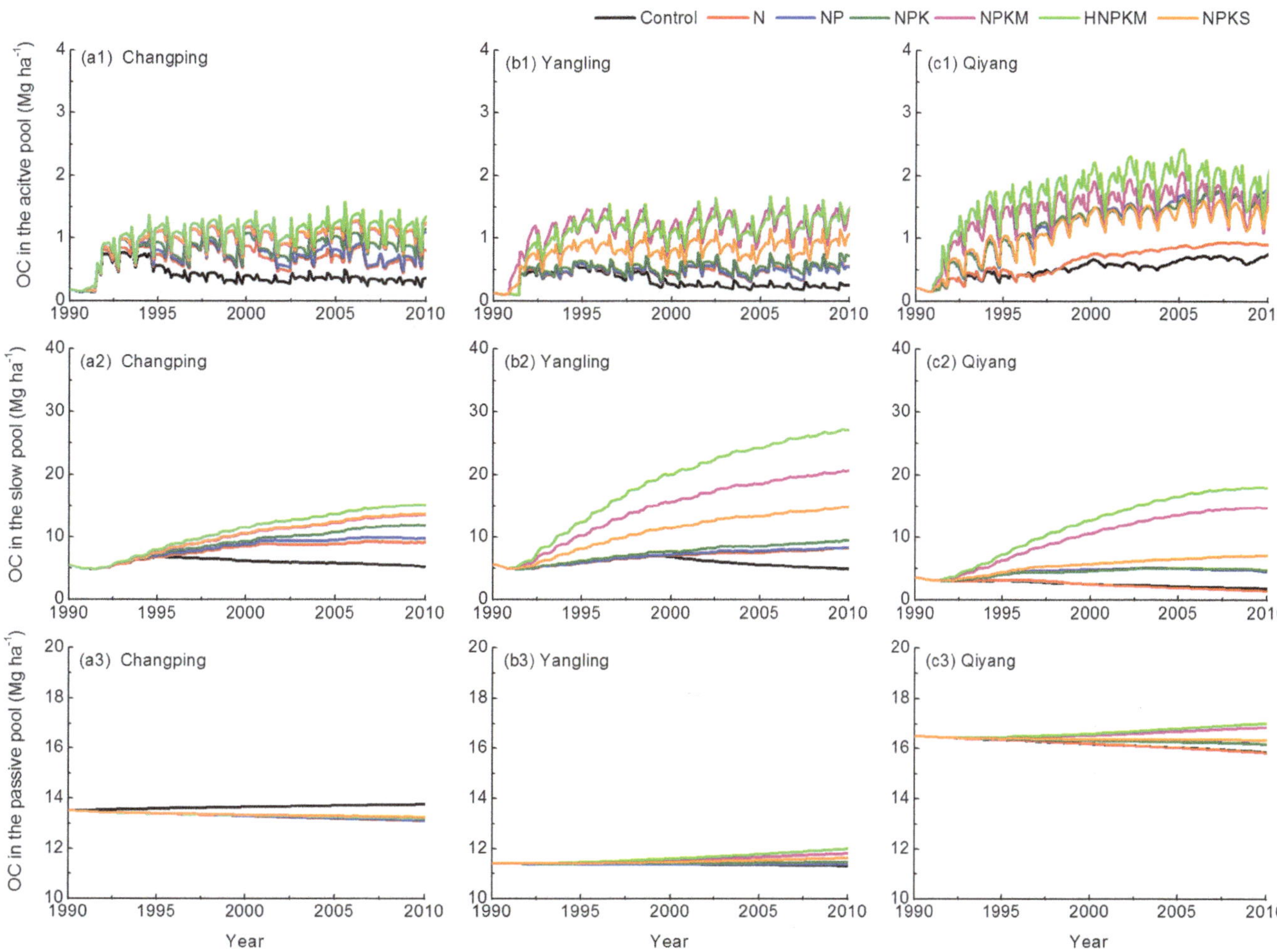

Figure 8. Soil organic carbon in the (1) active, (2) slow, and (3) passive pool under the control, N, NP, NPK, NPKM, hNPKM and NPKS treatments at (A) Changping, (B) Yangling, and (C) Qiyang sites.

R^2 would equal 100%, *RMSE* would equal zero and *d* would equal 1. In our study, *RMSE*<15%, *d*>0.5 and significant R^2 were used to bound the best case parameter sets for both calibration and validation simulations [38].

Results

Model Evaluation

We evaluated the model results for both grain yields and the SOC stock predictions. We were focusing on the SOC stock predictions of the model. In summary for the grain yield results (Fig. 3), we found that the model had correlation coefficients (R^2)

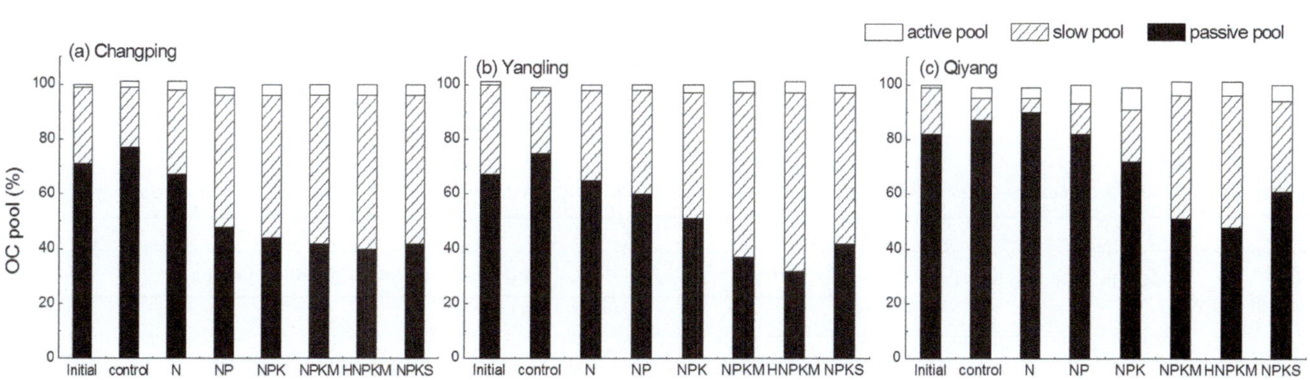

Figure 9. Proportion of changed SOC between the 1990 (initial) and 2010 in each soil organic matter pool under the control, N, NP, NPK, NPKM, hNPKM and NPKS treatments at (A) Changping, (B) Yangling, and (C) Qiyang sites.

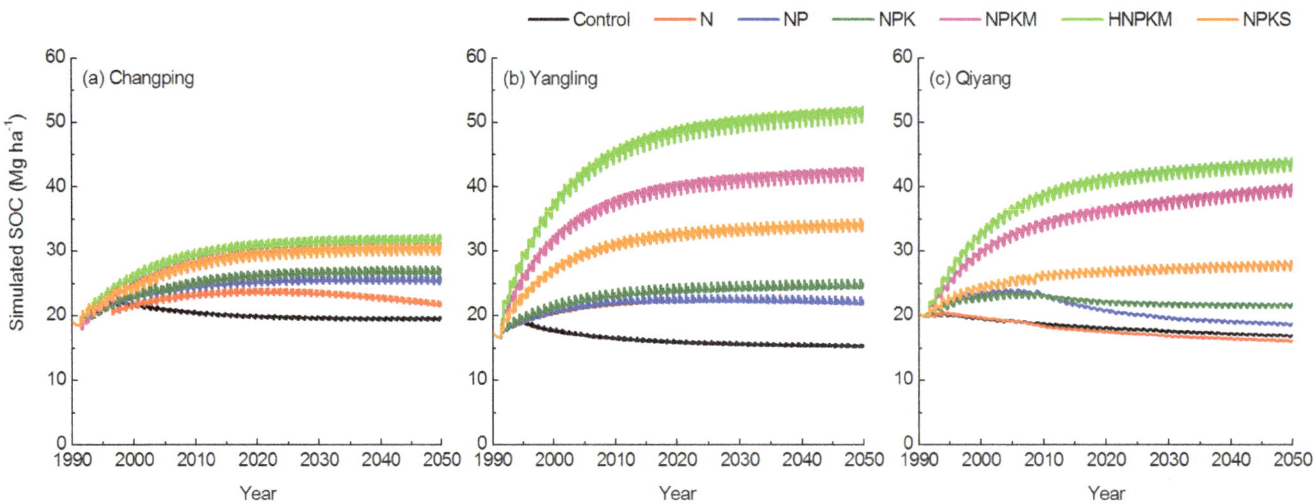

Figure 10. Soil organic carbon dynamics under the control, N, NP, NPK, NPKM, hNPKM and NPKS treatments at (A) Changping, (B) Yangling, and (C) Qiyang sites.

ranging from 0.57 to 0.92 ($p<0.05$), indicating that CENTURY model simulates the grain yields with reasonable accuracy.

The modeled SOC follows the trend of the measurements reasonably well in all treatments across the sites (Figs. 4, 5, and 6). Model results and measured SOC demonstrated that carbon levels decrease in the control treatment at all sites but increase under the fertilization treatments, except N only. For the N treatment, SOC is maintained at a stable level at most sites but did decline at the Qiyang site. The reason was related to a low soil pH that caused a decline in crop production and thus decline in residue input to the soil [25].

Linear regressions of observed vs. simulated SOC stocks (Fig. 7) are all highly significant ($P<0.001$) under various fertilization treatments. Approximately 77%–84% of the variability in the observed SOC stocks can be explained by the simulation. Therefore, in our experiment, the CENTURY model performs well and is suitable for predicting SOC dynamics for different types of N fertilizer (i.e., manure *versus* mineral) and N rates. The coefficients of determination (R^2) are higher at the Yangling and Qiyang sites than the Changping site.

RMSE is useful for evaluation of precision in model results [33]. Several studies have emphasized the need to estimate the precision in model predictions to evaluate model uncertainty [14,39,40]. In this study, we used several parameters for quantitative statistical analysis of measured and simulated SOC (Table 4). The *n-RMSE* ranges from 4% to 7% at the Changping site, 9%–14% at the Yangling site, and 6%–11% at the Qiyang site, respectively. Similarly, Fallon and Smith [41] calculated *n-RMSE* ranging from 1.8% to 16.4% when they modeled SOC in several long-term experiments in Europe. Also Álvaro et al. [20] had a low *n-RMSE* (below 6%) when they used the CENTURY model to evaluate tillage effects on SOC. Overall, the *n-RMSE* value is less than 15% for all the treatments in our study, demonstrating that the model is relatively precise in predicting the trends for different fertilization practices that are common in Chinese agricultural systems.

The *d* values and coefficients of determination (R^2) are highest in the NPKM, hNPKM and NPKS treatments at most sites. However, the R^2 value was much lower in the control (~0.07) or N (~0.19) treatments. The reason was that inter-annual fluctuations in crop yield were larger in the control and N treatments due to the long-term nutrient deficiency [25]. Therefore, carbon input from biomass was fluctuated and thus the simulation was not satisfied as other treatments.

Fertilization Effects on Organic Carbon Pools

Long-term fertilization has a variety of effects on SOC pools (Fig. 8 and 9). Generally, the active pool is the smallest carbon pool [16], and our results are consistent with past studies with an active pool containing only 1–3 Mg ha^{-1} across the sites (first row in Fig. 8). By 2010, the largest active pool is at the Qiyang site (Fig. 8_c1), followed by Changping (Fig. 7_a1) and Yangling (Fig. 8_b1) sites.

Long-term manure fertilization significantly increased the proportion of slow soil organic matter at all sites (Fig. 8). By the

Table 5. Estimated carbon sequestration potential (0–20 cm) under long-term hNPKM and NPKS fertilization treatments.

Site	SOC in 1990 (Mg ha^{-1})	SOC in 2050 (Mg ha^{-1})		Carbon sequestration potential (Mg ha^{-1})	
		NPKS	hNPKM	NPKS	hNPKM
Changping	22.4	30.5	31.8	8.1	9.4
Yangling	16.4	34.3	52.1	17.9	35.7
Qiyang	20.5	28.2	44.2	7.7	23.7

end of 2010, the slow pool accounted for ~42% of the total SOC at the Changping site, ~65% at the Yangling site, and ~48% at the Qiyang site (Fig. 9). In addition, trends in the slow pool are consistent with the overall change in total SOC stock at each of the sites (Figs. 4, 5, and 6). These results suggest that the carbon dynamics associated with the slow pool are key drivers of the trend in total SOC.

The percentage of SOC in the passive soil organic matter is 75%–87% under the control treatment but only 32%–48% under the hNPKM treatment (Fig. 9). Our modeling results suggest that a higher proportion of the organic carbon is contained in the passive pool with the slowest turnover rate if no fertilizer is added. However, long-term fertilization generally has a smaller effect on the passive pool compared to the slow pool according to the experimental data (third row in Fig. 8). The reason would be corresponded to turnover times of 25 years for the slow pool but 1000 years for the passive pool in the CENTURY [28].

Soil Carbon Dynamics and Carbon Sequestration Potential

Long-term fertilization has different effects on soil organic carbon across the three sites (Fig. 10). For example, the SOC stock (0–20 cm) increases and then decreases to a relative stable level in the control and N treatments at Changping site (Fig. 10a). Plots with balanced fertilization or manure application and straw return (i.e., NPK, NPKM, hNPKM and NPKS) have an increase in SOC and then remain relatively stable. By 2050, the SOC stocks reach 30.5–31.8 Mg ha^{-1} with organic fertilization and 26.8 Mg ha^{-1} with chemical fertilization (NPK).

For the Yangling site (Fig. 10b), SOC contents increase over time in the NPK, NPKM, hNPKM, and NPKS treatments. By the end of simulated period, SOC reaches 42.6 Mg ha^{-1} and 52.1 Mg ha^{-1} under the NPKM and hNPKM treatments, respectively. With straw return, the SOC level reaches 34.3 Mg ha^{-1}. Balanced chemical fertilization (NPK) also yields relatively high SOC contents at 24.6 Mg ha^{-1} by 2050. There is little difference in SOC stocks between N and NP treatments. SOC increases to 20.1 Mg ha^{-1} in 1994, then declines to 15.3 Mg ha^{-1} in 2050 for the control treatment.

Similar to the Yangling site, SOC levels increase through the entire period of simulation under the NPKM, hNPKM, and NPKS treatments for the Qiyang site (Fig. 10c). By 2050, SOC stocks are 40.2 Mg ha^{-1}, 44.2 Mg ha^{-1}, and 28.2 Mg ha^{-1} for the NPKM, hNPKM, and NPKS treatments, respectively. For all chemical fertilization treatments, we assume no decrease in pH after year 2010. Therefore, soil pH has not negative impact on carbon input due to a reduction in crop production associated with pH, and applying NPK fertilizers maintain SOC levels at 21.7 Mg ha^{-1}.

The SOC trends are relatively stable by the end of 2050 in the different treatments (Fig. 10). In order to estimating the carbon sequestration potential (i.e., the SOC increment by the end of simulation) with manure addition and regular straw incorporation, we further evaluated the SOC change under the hNPKM and NPKS treatments in the period of 1990–2050 (Table 5). The initial SOC level is highest at the Changping site (22.4 Mg ha^{-1}) and lowest at the Yangling site (16.4 Mg ha^{-1}). In contrasts, the highest simulated SOC level in 2050 happens at the Yangling site and lowest at the Changping site. Compared with the NPKS treatment, carbon sequestration potential is greater with hNPKM management, ranging from 9.4 Mg ha^{-1} to 35.7 Mg ha^{-1}. For the NPKS treatment, SOC increases to 17.9 Mg ha^{-1} at the Yangling site but only 7.1 Mg ha^{-1}–8.1 Mg ha^{-1} at the other two sites.

Discussion

Studies has proved that optimizing fertilization managements can raise crop yield as well as biomass, which would enhance biomass input into the soil from crop straw and roots [28,42,43]. Halvorson et al [44] reported that increasing nitrogen fertilizer rate would increase SOC levels in the 0–7.5 cm soil depth after 11 years cropping. In our study, chemical fertilization (expect N treatment) would also increase crop yield and thus increase SOC in the first 20 years of crop rotation at two of three sites. However, most chemical fertilization treatment would only maintain the SOC level in the long-term run (Fig. 10).

Straw incorporation would also increase carbon input to the soil. Li et al [45] found that cropland soil lost 1.6% of SOC due the less (25%) aboveground residue's returning. Using the DNDC model, Tang et al. [46] predicted that the soil carbon loss in China's cropland would be partly reversed if straw return and no tillage practices were expanded widely. In our study, averaged carbon sequestration rate would reach to 118 kg ha^{-1} yr^{-1} to 298 kg ha^{-1} yr^{-1} with straw returning during the period of 1990–2050 under current returning rate of straw residue (i.e., 2000–4500 kg ha^{-1} yr^{-1}). The results are lower than the estimated rate (108–728 kg ha^{-1} yr^{-1}) in the similar region by Lu et al [47], in which the estimated straw returning rates were also higher (4530–8110 kg ha^{-1} yr^{-1}).

Moreover, there are differences in the influence of manure on SOC trends among the sites. At the Changping site, SOC stocks reach a relatively stable level (30.0 Mg ha^{-1}–31.0 Mg ha^{-1}) by 2013 (Fig. 10a). In contrast, plots from the other two sites continue to increase in SOC over the entire simulated period. The reason for the site difference in the simulated trends is related to the manure quality between the sites, and the associated decomposition rates of the organic matter [28,48]. For the Changping site, farmyard manure is livestock manure mixed with soil and/or crop residue. The lignin content in farmyard manure is lower and decompose at a faster rate [49]. When we enhanced the lignin content of farmyard manure in CENTURY, the simulation of SOC rose to the much higher level (data not shown).

Many modeling research also focus on the regional estimation of SOC sequestration. Follett [50] estimated soil carbon sequestration potential would be 11–56 Tg C yr^{-1} for residue management in the USA. In China, Yan et al [51] estimated 32.5 Tg C yr^{-1} to be sequestered in croplands if 50% of the crop residue was returned to soils, and no tillage was adopted on 50% of the arable lands. However, these results are criticized for the models that are calibrated against few or no field experimental data. Our study estimated carbon sequestration potential using long-term experiment data, providing credible data to evaluate carbon sequestration response to different fertilization managements.

Conclusion

We used the CENTURY model (version 4.5) to simulate soil organic carbon (SOC) dynamics for long-term fertilization experiments with wheat-corn cropping systems in China. Our study indicated that CENTURY can simulate fertilization effects on SOC trends for different climatic conditions and soil properties that are common for wheat-corn systems in China. For most sites, the model was more precise in predicting the SOC trends for treatments with balanced nutrient additions (NPK, NPKM, hNPKM, and NPKS) than the control and unbalanced fertilization treatments (N and NP). After simulating SOC dynamics from 1990–2050, the SOC levels increased from 31.8–52.1 Mg ha^{-1} across the four sites. We estimate that carbon sequestration potential ranges from 9.4–35.7 Mg ha^{-1} under the high manure application practice (i.e., hNPKM treatment). Analysis of the proportion of organic carbon in each of the soil organic matter pools indicates that long-term fertilization enhances the slow pool and leads to a decline in the proportion of carbon in the passive pool. Model results suggest that changes in slow organic matter drive the overall trends in SOC stocks associated with long-term fertilization.

Acknowledgments

The authors would like to thank Dr. Sigen Chen and Ms. Cynthia Keough for the technical support on the CENTURY model. We acknowledge all the colleagues for their unremitting efforts on the long-term experiments from these sites.

Author Contributions

Conceived and designed the experiments: RHC MGX XJW. Performed the experiments: RHC SMO WJP. Analyzed the data: RHC MGX XJW. Contributed reagents/materials/analysis tools: SMO WJP. Wrote the paper: RHC.

References

1. Schlesinger WH (1997) Biogeochemistry: an analysis of global change: New York Academic Press.
2. Schuman GE, Janzen HH, Herrick JE (2002) Soil carbon dynamics and potential carbon sequestration by rangelands. Environmental pollution 116: 391–396.
3. Manzoni S, Porporato A (2009) Soil carbon and nitrogen mineralization: Theory and models across scales. Soil Biology & Biochemistry 41: 1355–1379.
4. Lal R, Follett F, Stewart BA, Kimble JM (2007) Soil carbon sequestration to mitigate climate change and advance food security. Soil Science 172: 943–956.
5. Lal R (2004) Soil Carbon Sequestration Impacts on Global Climate Change and Food Security. Science 304: 1623–1627.
6. Körschens M (2006) The importance of long-term field experimentsfor soil science and environmental research - a review. Plant, soil and environment 52: 1–8.
7. Zhang W, Xu M, Wang X, Huang Q, Nie J, et al. (2012) Effects of organic amendments on soil carbon sequestration in paddy fields of subtropical China. Journal of Soils and Sediments: 1–14.
8. Powlson D, Zhao BQ, Li XY, Li XP, Shi XJ, et al. (2010) Long-Term Fertilizer Experiment Network in China: Crop Yields and Soil Nutrient Trends. Agronomy Journal 102: 216–230.
9. Edmeades DC (2003) The long-term effects of manures and fertilisers on soil productivity and quality: a review. Nutrient Cycling in Agroecosystems 66: 165–180.
10. Thomsen IK, Christensen BT (2004) Yields of wheat and soil carbon and nitrogen contents following long-term incorporation of barley straw and ryegrass catch crops. Soil Use and Management 20: 432–438.
11. Bruun S, Christensen BT, Hansen EM, Magid J, Jensen LS (2003) Calibration and validation of the soil organic matter dynamics of the Daisy model with data from the Askov long-term experiments. Soil Biology & Biochemistry 35: 67–76.
12. Shirato Y, Yokozawa M (2005) Applying the Rothamsted Carbon Model for long-term experiments on Japanese paddy soils and modifying it by simple mining of the decomposition rate. Soil Science and Plant Nutrition 51: 405–415.
13. Yang Y, Luo Y, Finzi AC (2011) Carbon and nitrogen dynamics during forest stand development: a global synthesis. New Phytologist 190: 977–989.
14. Smith P, Smith JU, Powlson DS, McGill WB, Arah JRM, et al. (1997) A comparison of the performance of nine soil organic matter models using datasets from seven long-term experiments. Geoderma 81: 153–225.
15. Chilcott CR, Dalal RC, Parton WJ, Carter JO, King AJ (2007) Long-term trends in fertility of soils under continuous cultivation and cereal cropping in southern Queensland. IX*. Simulation of soil carbon and nitrogen pools using CENTURY model. Australian Journal of Soil Research 45: 206–217.
16. Denef K, Six J, Merckx R, Paustian K (2004) Carbon sequestration in microaggregates of no-tillage soils with different clay mineralogy. Soil Science Society of America Journal 68: 1935–1944.
17. Kelly RH, Parton WJ, Crocker GJ, Grace PR, Klir J, et al. (1997) Simulating trends in soil organic carbon in long-term experiments using the century model. Geoderma 81: 75–90.
18. Falloon P, Smith P (2002) Simulating SOC changes in long-term experiments with RothC and CENTURY: model evaluation for a regional scale application. Soil Use and Management 18: 101–111.
19. Bhattacharyya T, Pal DK, Williams S, Telpande BA, Deshmukh AS, et al. (2010) Evaluating the Century C model using two long-term fertilizer trials representing humid and semi-arid sites from India. Agriculture, Ecosystems & Environment 139: 264–272.
20. Álvaro-Fuentes J, López MV, Arrúe JL, Moret D, Paustian K (2009) Tillage and cropping effects on soil organic carbon in Mediterranean semiarid agroecosystems: Testing the Century model. Agriculture, Ecosystems & Environment 134: 211–217.
21. Cerri CEP, Easter M, Paustian K, Killian K, Coleman K, et al. (2007) Simulating SOC changes in 11 land use change chronosequences from the Brazilian Amazon with RothC and Century models. Agriculture, Ecosystems & Environment 122: 46–57.
22. Feng XM, Zhao YS (2011) Grazing intensity monitoring in Northern China steppe: Integrating CENTURY model and MODIS data. Ecological Indicators 11: 175–182.
23. Fang D, Zhou G, Jiang Y, Jia B, Xu Z, et al. (2012) Impact of fire on carbon dynamics of Larix gmelinii forest in Daxingan Mountains of Northeast China:A simulation with CENTURY model. The Journal of Applied Ecology 23: 2411–2421.
24. Wang SH, Shi XZ, Zhao YC, Weindorf DC, Yu DS, et al. (2011) Regional Simulation of Soil Organic Carbon Dynamics for Dry Farmland in East China by Coupling a 1:500 000 Soil Database with the Century Model. Pedosphere 21: 277–287.
25. Cong RH, Wang XJ, Xu MG, Zhang WJ, Xie LJ, et al. (2012) Dynamics of soil carbon to nitrogen ratio changes under long-term fertilizer addition in wheat-corn double cropping systems of China. European Journal of Soil Science 63: 341–350.
26. Cong R, Xu M, Wang X, Zhang W, Yang X, et al. (2012) An analysis of soil carbon dynamics in long-term soil fertility trials in China. Nutrient Cycling in Agroecosystems 93: 201–213.
27. Walkley A, Black IA (1934) An examination of Degtjareff method for determining soil organic matter and a proposed modification of the chromic acid titration method. Soil Science 37: 29–38.
28. Parton WJ, Rasmussen PE (1994) Long-Term Effects of Crop Management in Wheat-Fallow: II. CENTURY Model Simulations. Soil Science Society of America Journal 58: 530–536.
29. Alister K Metherell, Laura A Harding, C. Vernon Cole, Parton WJ (1993) CENTURY Soil Organic Matter Model Environment.
30. van Heemst HDJ (1986) Physiological principles. In: Keulen; HV, Wolf J, editors. Modelling of agricultural production: weather, soils and crops. Pudoc Wageningen. pp. 13–26.
31. Metherell AK, Harding LA, Cole CV, Parton WJ (1993) Plant Production Submodels. CENTURY Soil Organic Matter Model Environment Technical Documentation Agroecosystem Version 40.
32. Zhang H, Wang B, Xu M (2008) Effects of Inorganic Fertilizer Inputs on Grain Yields and Soil Properties in a Long-Term Wheat-corn Cropping System in South China. Communications in Soil Science and Plant Analysis 39: 1583–1599.
33. Smith JU, Smith P, Addiscott TM (1996) Quantitative methods to evaluate and compare soil organic matter (SOM) models. In: Powlson DS, Smith P, Smith JU, editors. Evaluation of Soil Organic Matter Models Using Existing, Long-term Datasets. Heidelberg: Springer-Verlag. pp. 181–200.
34. Cheng WX, Zhang QL, Coleman DC, Carroll CR, Hoffman CA (1996) Is available carbon limiting microbial respiration in the rhizosphere? Soil Biology & Biochemistry 28: 1283–1288.
35. Kong AYY, Six J, Bryant DC, Denison RF, van Kessel C (2005) The Relationship between Carbon Input, Aggregation, and Soil Organic Carbon Stabilization in Sustainable Cropping Systems. Soil Sci Soc Am J 69: 1078–1085.
36. Domke GM, Woodall CW, Walters BF, Smith JE (2013) From Models to Measurements: Comparing Downed Dead Wood Carbon Stock Estimates in the U.S. Forest Inventory. PLoS One 8: e59949.
37. Willmott CJ (1982) Some Comments on the Evaluation of Model Performance. Bulletin of the American Meteorological Society 63: 1309–1313.
38. Tonitto C, David MB, Drinkwater LE, Li CS (2007) Application of the DNDC model to tile-drained Illinois agroecosystems: model calibration, validation, and uncertainty analysis. Nutrient Cycling in Agroecosystems 78: 51–63.
39. Blagodatskaya E, Kuzyakov Y (2008) Mechanisms of real and apparent priming effects and their dependence on soil microbial biomass and community structure: critical review. Biology and Fertility of Soils 45: 115–131.
40. Ogle SM, Breidt FJ, Easter M, Williams S, Paustian K (2007) An empirically based approach for estimating uncertainty associated with modelling carbon sequestration in soils. Ecological Modelling 205: 453–463.
41. Falloon P, Smith P (2003) Accounting for changes in soil carbon under the Kyoto Protocol: need for improved long-term data sets to reduce uncertainty in model projections. Soil Use and Management 19: 265–269.
42. Halvorson AD, Wienhold BJ, Black AL (2002) Tillage, nitrogen, and cropping system effects on soil carbon sequestration. Soil Science Society of America Journal 66: 906–912.

43. Dumanski J, Desjardins RL, Tarnocai C, Monreal C, Gregorich EG, et al. (1998) Possibilities for future carbon sequestration in Canadian agriculture in relation to land use changes. Climatic Change 40: 81–103.
44. Halvorson AD, Reule CA, Follett RF (1999) Nitrogen fertilization effects on soil carbon and nitrogen in a dryland cropping system. Soil Science Society of America Journal 63: 912–917.
45. Li C, Zhuang Y, Frolking S, Galloway J, Harriss R, et al. (2003) Modeling Soil Organic Carbon Change in Croplands of China. Ecological Applications 13: 327–336.
46. Tang HJ, Qiu JJ, Van Ranst E, Li CS (2006) Estimations of soil organic carbon storage in cropland of China based on DNDC model. Geoderma 134: 200–206.
47. Lu F, Wang XK, Han B, Ouyang ZY, Duan XN, et al. (2009) Soil carbon sequestrations by nitrogen fertilizer application, straw return and no-tillage in China's cropland. Global Change Biology 15: 281–305.
48. Paustian K, Parton W, Persson J (1992) Modeling Soil Organic Matter in Organic-Amended and Nitrogen-Fertilized Long-Term Plots. Soil Science Society of America Journal 56: 476–488.
49. Zhang WJ, Wang XJ, Xu MG, Huang SM, Liu H, et al. (2010) Soil organic carbon dynamics under long-term fertilizations in arable land of northern China. Biogeosciences 7: 409–425.
50. R.F F (2001) Soil management concepts and carbon sequestration in cropland soils. Soil and Tillage Research 61: 77–92.
51. Yan HM, Cao MK, Liu JY, Tao B (2007) Potential and sustainability for carbon sequestration with improved soil management in agricultural soils of China. Agriculture Ecosystems & Environment 121: 325–335.

20

Can Biochar and Phytoextractors Be Jointly Used for Cadmium Remediation?

Huanping Lu[1,2], Zhian Li[1]*, Shenglei Fu[1], Ana Méndez[3], Gabriel Gascó[4], Jorge Paz-Ferreiro[1,4]

1 Key Laboratory of Vegetation Restoration and Management of Degraded Ecosystems, South China Botanical Garden, Chinese Academy of Sciences, Guangzhou, China, **2** University of Chinese Academy of Sciences, Beijing, China, **3** Departamento de Ingeniería de Materiales, Universidad Politécnica de Madrid, Madrid, Spain, **4** Departamento de Edafología, Universidad Politécnica de Madrid, Madrid, Spain

Abstract

Phytoremediation of soils contaminated with cadmium was tested after liming (CaO) or biochar addition, using red amaranth (*Amaranthus tricolor* L.) as test plant species. Two biochars with contrasting characteristics were prepared from two feedstocks and added to the soil at a rate of 3% (w:w): Eucalyptus pyrolysed at 600°C (EB) and poultry litter at 400°C (PLB). Liming was carried out in two treatments (CaO1) and (CaO2) to the same pH as the treatments EB and PLB respectively. Total plant mass increased in soils amended with PLB and with a mixture of PLB and EB; however this was not sufficient to increase the efficiency of phytoextraction. Bioavailable and mobile fractions of Cd diminished after liming or biochar addition. Our study infers that, both the amount of Cd immobilized and the main mechanism responsible for this immobilization varies according to biochar properties.

Editor: Richard G. Haverkamp, Massey University, New Zealand

Funding: J. Paz-Ferreiro was sponsored by the Chinese Academy of Sciences (fellowship for young international scientists number 2012Y1SA0002). The authors acknowledge support from the National Natural Science Foundation of China (No. 40871221 and No. 41301571), and the Research Fund Program of Guangdong Provincial Key Laboratory of Environmental Pollution Control and Remediation Technology (2013K0008). The funders had no role in study design, data collection and analysis, decision to publish, or preparation of the manuscript.

Competing Interests: The authors have declared that no competing interests exist.

* E-mail: lizan@scbg.ac.cn

Introduction

Industrialisation and technical advances have led to an increase in the use of heavy metals. Anthropogenic activities, including smelting, mining, use of pesticides, fertilisers and sludge are responsible for the input of high levels of heavy metals into the soil. In the last years there is an increasing concern about soil heavy metal pollution. Soil heavy metal pollution poses a risk to the environment and to human health due to biomagnification (increases in metal concentration as the element passes from lower to higher trophic levels). An additional concern is that, contrary to organic substances, heavy metals are non-degradable and accumulate in the environment.

In particular, cadmium pollution is a concern in vast areas of China and some areas in Japan, Taiwan and other countries, rendering valuable farmland into non-cultivated areas. Cadmium is known to originate from lithogenic sources in areas with high abundance in zinc and lead ores. However, in the last years the agricultural use of fertilisers and the addition of sewage sludges to the soil have resulted in an increase in soil Cd concentrations. In general plants tolerate Cd better than human and animals and Cd toxicity in plants appears at higher concentrations than in animals or humans. In humans, excessive Cd intake can cause the "Itai-itai disease" and has also been related to some types of cancer or to kidney damage.

In the last years there has been an increasing interest in the use of biochar to tackle soil heavy metal pollution [1], [2], [3]. Biochar influences a number of biogeochemical processes and in general there has been reported a positive effect on plant productivity [4] while simultaneously mitigating greenhouse emissions (as CO_2 and N_2O) from soil. Several co-benefits may result from the use of biochar including positive effects on soil field capacity, nutrient availability, fertilizer use efficiency, pH ("liming effect"), cation exchange capacity (CEC) and soil biological properties [5]. Therefore, in the past few years, there has been a growing interest in the use of biochar as a soil amendment. The potential of biochar to immobilize toxic substances, in particular cadmium, from different types of soil is well documented [2], [3], however, more uncertainties prevail about the mechanisms involved in heavy metal immobilization and their relative importance.

Biochars using plant materials as feedstocks are considered a soil conditioner, due to their low leachable nutrient contents [6]. On the contrary, manure-derived biochar can act as soil fertilizer and conditioner due to NPK release [7].

On the other hand, phytoextraction consists in the use of hyperaccumulator plants to remove heavy metals from the environment. This is a relatively new technology developed from 1990 onwards. Because of its low costs, phytoremediation can be considered as a relatively attractive technique to restore or partially decontaminate a site compared to other options, e.g. [8], [9]. Other advantages of this technique would include its good perception among the general public. Plant species used for phytoextraction must not only accumulate high amounts of the target element but also have a high growth rate, tolerate toxic effects of heavy metals, be adapted to local soil and climate, be resistant to pathogen and pests, be easy to cultivate and repulse herbivores to avoid food chain contamination [9], [10]. Management of soil pH (liming) and soil nutrients have been utilised in the past in order to increase metal hyperaccumulation. In particular, liming is more effective when heavy metal concentrations in the soil are too elevated to reduce plant stress [11], [12] [13]. Organic matter amendments from several sources have been jointly used with phytoremediators and results showed that organic matter had either null or positive effect in enhancing phytoremediation [12], [14].

Of the more than 400 plants identified as phytoextractors most are Ni-hyperaccumulators and only a handful of plants accumulating Cd have been identified. Moreover, most Cd-accumulating plants present slow growth rates. Recently *Amaranthus tricolor* L. has been identified as a Cd-accumulator [15]. This species presents the main advantage of a rapid growth and a relative high yield compared to other Cd accumulating plants [15].

Simultaneous use of biochar and phytoremediation may seem contradictory, because until now most research results indicate that biochar reduce the bioavailability of heavy metals [2], [3] but plants need high concentrations of soluble metals to extract and to accumulate them. In fact, exploration of the efficiency in the joint use of biochar amendments and phytoremediation, has been attempted before [16], [17] for the case of Cd using *Brassica napus* L. as phytoextractor and Miscanthus biochar [16] or for the case of multicontaminated soils using different biochars and plant species [17].

Based on the above considerations we performed an experiment with the following aims: a) To explore the possibilities of biochars with contrasting characteristics on Cd remediation b) To explore the possibility of combining biochar with phytoremediation c) To deepen our mechanistic knowledge about how biochar contributes to heavy metal immobilization, using lime controls.

Material and Methods

Ethics statement

No specific permissions were required for collecting the soil, poultry litter or Eucalyptus used in this study.

Our study did not involve the use of endangered or protected species.

Soil sampling and characteristics

The soil was collected from the surface layer (0–20 cm) of a cropland area, near a waste landfill site in the suburb of Guangzhou, China (23° 07′ N and 113° 21′ E). Guangzhou is located in the subtropical humid area having an average annual temperature of 12.7°C and annual average precipitation of 1700 mm. According to FAO, the soil is classified as a fimic anthrosol.

Part of the soil was sieved to 2 mm to conduct general analyses, while the rest was sieved to 10 mm to conduct the experiment. Before the starting of the pot experiment, the soil had a total organic carbon content of 1.98%, total nitrogen content of 0.142%, pH of 6.00, total phosphorus of 690 mg kg^{-1}, available phosphorus of 126 mg kg^{-1} and total Cd 6.1 mg kg^{-1}.

Preparation and characterization of biochar

Two biochars were used in this experiment using poultry litter and eucalyptus as feedstocks. Poultry litter was collected at the Experimental Poultry Farm of South China Agricultural University (23° 09′ N and 113° 21′ E) located at Guangzhou, China. Chicken at this facility are organically bred and fed with a mixture of corn, wheat bran and soybean pulp. After collection, litter was dried in an oven (70°C) for 24 hours and sieved to 1 mm.

Eucalyptus (*Eucalyptus urophylla* S.T. Blake) wood was collected at Heshan Hilly Land Experiment Station of the Chinese Academy of Sciences in Heshan, China (22° 84′ N and 112° 54′ E). Wood was collected from branches of four different individuals, mixed together and brought to the South China Botanical Garden where it was ground to wood chips with a particle size smaller than 0.5 mm.

Biochars were prepared as described by [18] with samples being pyrolysed by increasing the temperature to 400°C (poultry litter biochar) or 600°C (Eucalyptus biochar) at a rate of 10°C min^{-1}. The final temperature was maintained for 2 h. Different temperatures were used in the preparation of the biochars as our main aim was to produce two biochars with contrasting characteristics.

Mesocosm experiments

On the 13th April of 2013, a mesocosm using a fully replicated randomised experiment was set up in a greenhouse in South China Botanical Garden. Each mesocosm consisted in a pot filled with 1.5 kg of soils in pots with plants and 500 g of soil in pots without crops. The reason for using different amounts of soil was to minimise the amount of biochar prepared, while allowing the crops to have enough soil to grow. The soils were watered to 60% of field capacity and watered daily to account for moisture losses.

The experiment studied two factors, namely the type of amendment and the presence/absence of phytoremediator, having 4 replications per treatment. The treatments in the amendment factor were: Control (C), poultry litter biochar (3% w/w) (PLB), eucalyptus biochar (3% w/w) (EB), poultry litter biochar (1.5% w/w) + eucalyptus biochar (1.5% w/w) (BB), liming with CaO as necessary to increase soil pH as much as EB (CaO1) and liming with CaO as necessary to increase soil pH as much as PLB (CaO2).

With respect to the factor involving presence/absence of phytoremediator, half of the mesocosm were planted with 3 seeds of red amaranth (Amaranth tricolor L.), while in the remainder half the vegetative cover was absent.

After 60 days soils and plant were collected for analyses.

Soil analysis

Soil pH was determined using a 1:2.5 soil:water ratio. Total carbon and total nitrogen were determined using using a vario ISOTOPE CUBE elemental analyzer (elementar, Germany). Plant-available metals were extracted from treated soil using diethylenetriaminepentaacetic acid–$CaCl_2$–triethanolamine (DTPA) as described elsewhere [19]. The mobile forms of heavy metals were extracted using 0.1 M $CaCl_2$ [20].

Plant analysis

The crops were harvested and separated into roots, stems and leaves. Then they were rinsed firstly with tap water followed by distilled water. Samples were oven-dried for 72 h at 65°C. Once dried, plant biomass was weighted and ground. For digestion, a microwave closed system (Multiwave3000, Anton Paar, Austria) was used. Samples (0.1 g) were digested with 5 mL of nitric acid and 1 mL of hydrogen peroxide in a microwave digestion system for 30 min and diluted with deionized water to 30 mL. Cadmium concentrations were analyzed using an ICP-MS spectrometer (Agilent 7700x, USA).

Biochar analysis

pH and TOC in the biochar were determined as described for the soil. Biochar nitrogen adsorption analysis to determine BET surface was carried out at 77 K in a Micromeritics Tristar 3000 (Instituto de Catálisis y Petroquímica, CSIC, Spain).

Total pore volume (V) was estimated from the amount of nitrogen adsorbed at a relative pressure. Average pore size (D) was estimated from V and BET surface using the following equation and assuming a cylindrical pore shape:

$D = 4V/S_{BET}$.

Proximate analysis was calculated by thermogravimetry using a Labsys Setaram equipment. Samples were heated up to 600°C under an N_2 atmosphere at a flux of 40 mL min^{-1} using a heat rate of 20°C min^{-1}. Moisture content (H) was determined as weight loss at 120°C and Volatile matter (VM) was determined as the weight loss from 120°C to 600°C. At 600°C, air flux was introduced until a constant weight was reached and ashes were determined as the final weight of the samples. Fixed carbon (FC) was calculated by difference.

Cadmium in the biochars was determined as described for the soil samples.

Phytoextraction indices

The amount of Cd transported from soil to shoots was calculated using the bioconcentration factor (BCF), [21].

Accumulation Factor = Cd concentration plant tissue/Cd concentration in soil.

The ability of each species to translocate Cd from the roots to the shoots was calculated by the translocation index (TI) [22].

TI (%) = Cd concentration in aerial parts/Cd concentration in roots x 100.

Statistical analysis

Statistical analyses (calculation of means and standard deviations, differences between treatments) were performed using SPSS 15.0 package. Differences of means were tested using a two-way ANOVA with the presence/absence of plant and the type of amendment (eucalyptus biochar, poultry litter biochar, CaO1, CaO2, mixture of biochar or none) as factors. Means were considered to be different when $P<0.05$ using the Tukey's test.

Results and Discussion

Biochar properties

The physical characteristics of both biochars were contrasted (Table 1). Eucalyptus biochar (EB) had a larger surface area and pore volume compared to the poultry litter biochar (PLB) due to its higher content in carbon. Biochar pore volume has been directed related before to surface area [23] and both, pore volume and surface area, are expected to have larger values in biochars prepared at higher temperatures [24]. Rouquerol et al [25] defined the internal diameter range of pores for porous solids, suggesting <2.0 nm as micropores, 2.0–50.0 nm as mesopores, and >50.0 nm as macropores. Our results showed clear evidence that the PLB was dominated by mesopores, while EB was dominated by micropores.

Biochar pHs are very basic according to other works that attributed this fact to the polymerisation/condensation reactions and acidic surface group releases during pyrolysis [26]. The total Cd content of biochars was negligible. Finally, the PLB presented a higher content in ash, VM and a lower content in FC than EB due to the different composition of the raw materials and pyrolysis conditions.

Soil properties

The addition of lime or biochar clearly modified soil pH (see Table 2). All treatments increased significantly the pH compared to the control ($P<0.001$). While the control soil had a pH of 6.03, this increased to 7.62 for the treatment PLB, 7.01 for EB and 7.45 for BB. The pH of PLB was significantly higher than that of EB.

Total carbon and total nitrogen increased in BB and EB compared to the control. However, in the case of PLB only an increase in total nitrogen with respect to the control was found.

DTPA extracted-Cd and $CaCl_2$ extracted-Cd can be considered respectively as the bioavailable and mobile forms of Cd present in soil. Our study showed that both fractions were affected by the type of amendment (in both cases $P<0.001$); however only $CaCl_2$ extracted-Cd was affected by the presence of red amaranth ($P<0.010$). Moreover, for $CaCl_2$ extracted-Cd there was an interaction phytoremediation x treatment ($P<0.010$). Thus, mobile forms of Cd were higher in the EB soil with amaranth with respect to the one without amaranth. This leads us to think that, at least to some extent, plants with a high affinity for heavy metals can mobilize part of the heavy metal stabilized by biochar.

Incorporation of biochar reduced $CaCl_2$ extracted Cd compared to the control by 97, 67 and 92% respectively for PLB, EB and BB (see Figure 1). The value for PLB was similar to the one found by [16] using Miscanthus biochar. These values were 40, 10 and 29% comparing the reduction with respect to the control in the case of DTPA extracted Cd (Figure 2).

In the last years a number of studies have shown that biochars have a potential for metal adsorption and immobilization, but there is no agreement about what drives biochar effects on soil heavy metals. While some authors postulate that this effect is

Table 2. General properties of the soils at the end of the experiment.

	pH	Carbon (%)	Nitrogen (%)
Control	6.03±0.03 a	1.76±0.03 a	0.123±0.005 a
PLB	7.62±0.03 d	2.35±0.12 a	0.171±0.006 b
EB	7.01±0.02 b	5.19±0.52 c	0.176±0.010 b
CaO1	6.81±0.06 b	1.79±0.23 a	0.128±0.012 a
CaO2	7.08±0.12 bc	1.73±0.12 a	0.122±0.009 a
BB	7.45±0.22 cd	3.80±0.40 b	0.172±0.020 b

Different letters in the same column indicate statistical significant differences ($P<0.05$).

Table 1. General characteristics of the biochars used in the experiment.

	pH	Carbon (%)	Nitrogen (%)	Cd (mg kg^{-1})	Surface area (m^2 g^{-1})	Average pore width (nm)	Pore volume (cm^3 g^{-1})	Ash (%)	VM (%)	FC (%)
PLB	10.02	16.77	1.37	n.a.	7.418	15.406	0.0286	74.95	7.66	17.39
EB	10.40	81.03	1.07	n.a.	334.560	1.928	0.1612	1.74	2.19	96.06

VM and FC stand for volatile matter and fixed carbon, respectively.

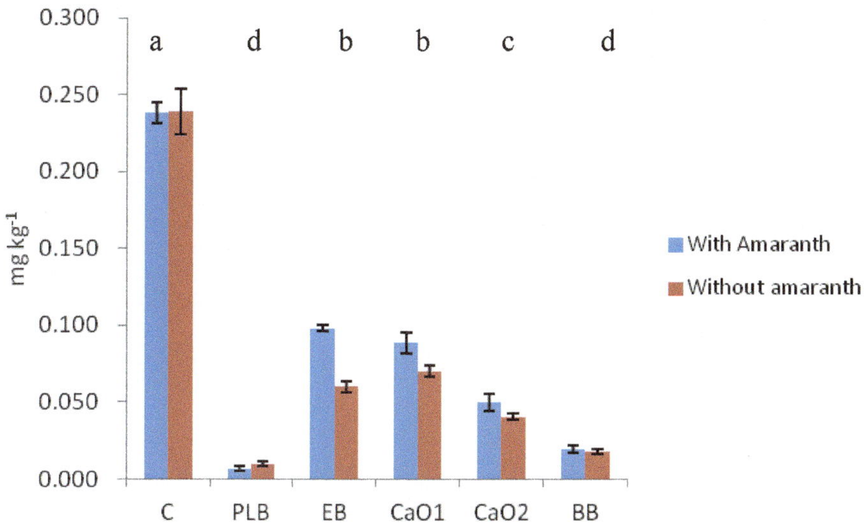

Figure 1. $CaCl_2$ extracted Cd for all the treatments at the end of the experiment. Different letters in the same indicate statistical significant differences for the factor "type of amendment" ($P<0.05$).

mainly pH mediated [27], other authors pointed out to high surface are and pore volume [23], [28], or the amount of oxygen functional groups [29]. Our study confirms the capacity of biochar to immobilize cadmium and points out to different mechanisms depending on biochar type. This experiment demonstrated that the EDTA and $CaCl_2$ extractable Cd in EB was similar to that of CaO1, pointing out to Cd immobilization as caused primarily by changes in soil pH for EB and pointing to negligible effects of biochar surface area, contrary to the suggestion of other authors [30]. However, this was not the case of PLB, which had a very different EDTA and $CaCl_2$ extractable Cd compared to CaO2. We would expect that Cd immobilisation in the PLB treatment was driven both by pH changes and by precipitation with mineral ash. Thus, PLB has a high content of ash, which must be enriched by mineral salts of K, Ca, Mg, C and P among other elements. Cadmium can precipitate as insoluble phosphate and carbonate salts, in particular and high pH values. Another mechanism that could explain Cd immobilization in the PLB treatment would be the presence of high oxygen-containing functional groups in the poultry litter biochar. These groups are particularly effective for heavy metal stabilization in soils with low organic carbon contents [29].

Plant growth and Cd uptake

Plant biomass increased as a consequence of liming and poultry litter biochar addition (Figure 3). This way, total plant biomass was 3.44 and 2.24 times higher under PLB and BB respectively than in the control. Treatments CaO1, CaO2 and EB showed no difference with the control. We would like to highlight that plant height was higher in all treatments with respect to the control for one month following the start of the experiment (data not shown), with the addition of two different biochars (BB) performing better than any other treatment. This suggests that at least in the initial state of growing there could be some synergistic effect between different types of biochar.

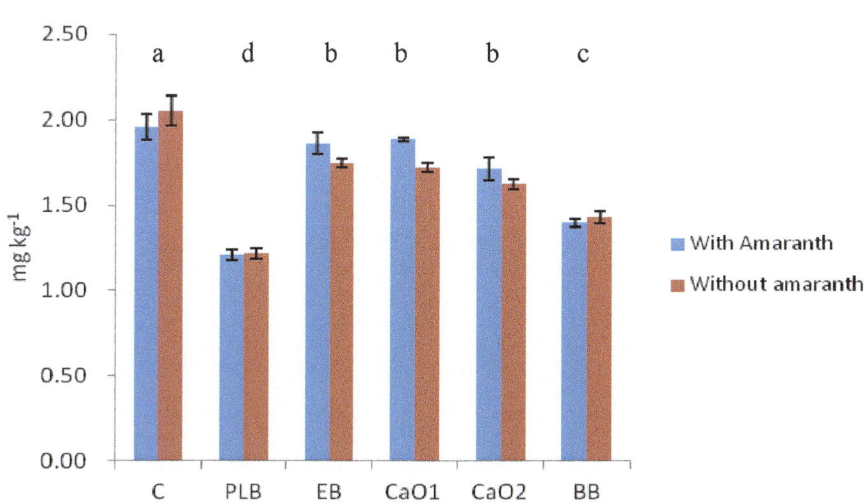

Figure 2. DTPA extracted Cd for all the treatments at the end of the experiment. Different letters in the same indicate statistical significant differences for the factor "type of amendment" ($P<0.05$).

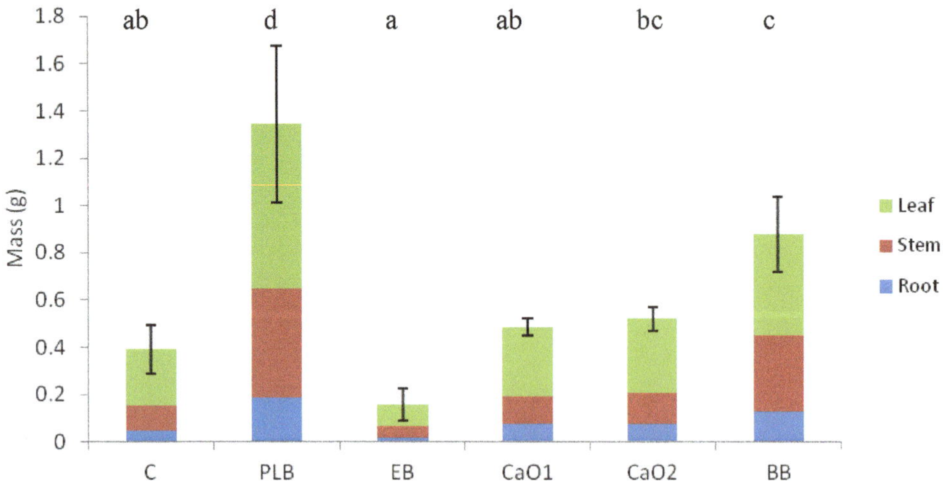

Figure 3. Total plant biomass at the end of the experiment. Different letters in the same indicate statistical significant differences for the factor "type of amendment" ($P<0.05$).

Cadmium concentration in the different plant tissues was reduced by every treatment. Concentrations diminished more for PLB and for BB (with both treatments having similar values) than in EB. The concentrations of Cd in the root, shoots and leaves of the plants were in all cases over the range of concentrations (see Table 3) that is usually considered as excessive or toxic (5–30 mg kg^{-1}, [31]) for humans, but we did not observe any signs of Cd toxicity in the plants.

Quantification of phytoextraction efficiency

All treatments extracted significantly less Cd from the soil than the control ($P<0.001$) as shown in Figure 4. EB treatment removed only 22% of the Cd accumulated by Amaranthus in the control, while this percentage was 49% in the case of PLB.

The translocation index (TI) was 120.5 in the control soil and dropped to 71.0 for PLB (see Table 3). The bioconcentration factor (BCF) varied from 14.76 in the control soil to 2.17 in PLB. The treatments EB, CaO1, CaO2 and BB also had a lower transfer factor than the control soil, being 8.12 5.67, 4.05 and 2.71 respectively.

The values of TI for EB and the control soil were similar probing that EB did not inhibit heavy metal transportation from roots to shoots, but did from soil to shoots (see BCF values). On the other hand, liming and PLB inhibited both, transportation from soil to roots and from roots to the aerial parts of the plant.

Both indices, TI and BCF are important in screening hyperaccumulators species. Yoon et al. [32] recommends the use of plants that have a TI of more than 100% and BCF values higher than 1. While all our treatments had BCF values higher than 1, the low values of TI would not make realistic to combine amaranth with biochar for phytoextraction purposes due to the excessive amount of time required. However, the increase of biomass observed with PLB and BB, could have some useful implications that will be next addressed.

Significance of our study

Based on our study, it seems that using biochar alone as a soil amendment is a promising technique to immobilize Cd in polluted areas, but a careful choice of feedstock and pyrolysis temperature should be made. A study on five sites concerning rice grown in polluted areas demonstrated that biochar can reduce rice Cd uptake through a reduction in Cd mobility [1]. However, PLB would have a greater potential than the wheat straw biochar used by [1] as its use resulted in greater heavy metal immobilization, as measured by $CaCl_2$ extraction.

Other experiments with Cd and biochar have shown that Cd mobility and plant availability can be reduced with increasing doses of biochar, at least up to an amendment of 10% biochar [16]. In addition to its use as a phytoremediator, amaranth has been used for as forage species or for human consumption. The values obtained in this study for Cd concentrations in the leaves of

Table 3. Phytoremediation parameters and Cd concentration in different plant tissues at the end of the experiment.

	BCF	TI	Root Cd concentration	Leaf Cd concentration	Stem Cd concentration
Control	12.76±1.90 a	120.27±5.42 a	76.40±4.65 a	108.57±7.87 a	52.15±7.87 a
PLB	2.17±0.42 d	70.74±20.57 b	17.79±0.58 d	15.65±1.78 d	7.31±1.78 d
EB	8.12±0.78 b	93.22±15.39 ab	53.62±4.90 b	57.77±2.47 b	30.33±2.47 b
CaO1	5.67±0.80 c	79.31±7.68 b	42.03±3.12 bc	39.27±3.12 c	17.57±3.12 c
CaO2	4.05±0.26 cd	80.21±9.89 b	29.91±1.45 cd	28.53±0.83 cd	12.07±0.83 cd
BB	2.71±0.20 d	71.80±10.27 b	22.05±1.67 d	21.64±0.73 d	7.77±0.73 d

Different letters in the same column indicate statistical significant differences ($P<0.05$).

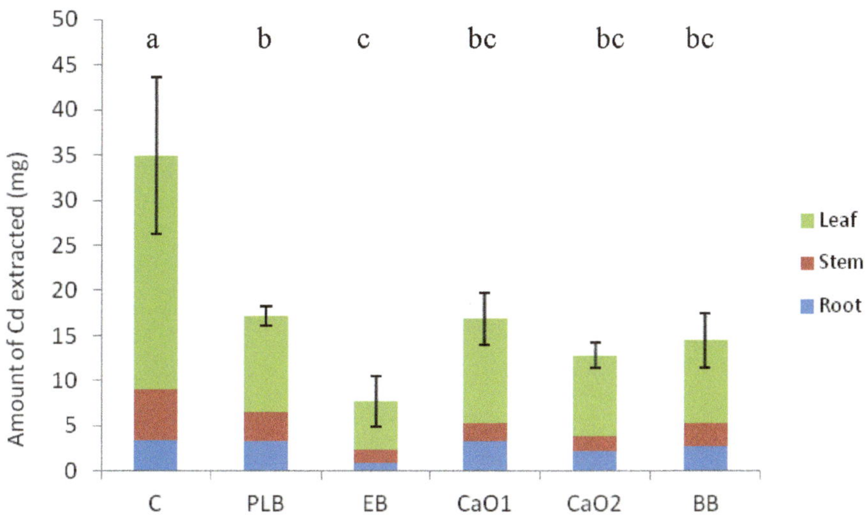

Figure 4. Total amount of Cd extracted by plants. Different letters in the same indicate statistical significant differences for the factor "type of amendment" ($P<0.05$).

plants (>10 mg kg^{-1}) would discourage the use of amaranth for this purpose, even in the treatments with lower leaf Cd concentration.

In the last years much research about biochar has been emerged and some unintended consequences have been identified including the potential presence of toxic compounds in biochars [33] and increasing difficulties for weed control due to pesticide adsoption on biochar surface [34]. Our study is, to our knowledge, pioneering (but see also [14], [17]) with respect to addressing a joint use of phytoremediation and biochar for the remediation of heavy metal polluted soils and raises awareness over a new unintended consequence (limitation on the use of phytoextractors), which can arise from biochar management. Our results point out that biochars could impede the transfer of heavy metals to phytoextractors and slow down remediation processes. However, these results should be carefully extrapolated to the field scale, as the yield metal extracted by phytoremediators can vary greatly among field and mesocosm studies [35].

Our study did not account for other potentially beneficial effects which are concomitant to the use of phytoremediation and are derived from the establishment of a plant cover. In this aspect, introduction of vegetation in a polluted area can help to prevent erosion or contaminant leaching. In addition, it could be expected that the vegetation will help the improvement of soil biological properties. Moreover, the harvested biomass of the phytoremediator could be used as feedstock for biochar production. This option would only be feasible if the feedstock biomass does not contain excessive levels of heavy metals. Indeed, Méndez et al. [2] have demonstrated that after pyrolysis of a sewage sludge containing heavy metals, the major part of the metal was immobilized in the biochar and as a consequence, addition of sewage sludge biochar did not result in increased mobile fractions of metals in soil. However, metal concentrations in a biochar prepared using a metallophyte as feedstock would presumable be higher than in the sewage sludge used by Méndez et al. [2]. Carbon sequestration is another benefit that could arise from the combined use of phytoextractors and biochar. Thus, we could devise some situations where it could be advantageous to combine phytoextraction with biochar amendment. This type of scenarios would include soils with low organic matter contents and with acidic pH values which would compromise the growth of phytoextractors.

There are two big interrogations regarding the future use of biochar and phytoremediation as joint strategies to address heavy metal pollution. Firstly, our experiment has only considered a single heavy metal (cadmium). According to other studies [3], [36] biochar has a moderate affinity for Cd which would make our results difficult to extrapolate to heavy metals that are not so likely to be immobilized by biochar or to multicontaminated soils. Secondly, the development of biochar-based fertilizers [37] could change the present paradigm of soil heavy metal remediation, as we could expect that with a careful choice of biochar lessened amounts would be needed to obtain a plant effect, resulting in fewer amounts of biochar added to the soil and consequently diminished pollutant immobilization. Another future line for this type of studies would be to use phytoremediators and biochars targeting at different heavy metals (or the biochar not to target any metal at all), as the biochar effect is strongly dependent on the chemical element immobilized.

Author Contributions

Conceived and designed the experiments: HPL ZAL SF AM GG JPF. Performed the experiments: HPL JPF. Analyzed the data: HPL ZAL SF AM GG JPF. Contributed reagents/materials/analysis tools: ZAL SF. Wrote the paper: HPL ZAL SF AM GG JPF.

References

1. Bian R, Chen D, Liu X, Cui L, Li L, et al. (2013) Biochar soil amendment as a solution to prevent Cd-tainted rice from China: Results from a cross-site field experiment. Ecol Eng 58: 378–383.
2. Méndez A, Gómez A, Paz-Ferreiro J, Gascó G (2012) Effects of biochar from sewage sludge pyrolysis on Mediterranean agricultural soils. Chemosphere 89: 1354–1359.
3. Uchimiya M, Lima IM, Klasson KT, Wartelle LH (2010) Contaminant immobilization and nutrient release by biochar soil amendment: Roles of natural organic matter. Chemosphere 80: 935–940.
4. Liu X, Zhang A, Ji C, Joseph S, Bian R, et al. (2013) Biochar's effect on crop productivity and the dependence on experimental conditions- a meta-analysis of literature data. Plant Soil 373: 583–594.

5. Paz-Ferreiro J, Fu S, Méndez A, Gascó G (2014) Interactive effect of biochar and the earthworm Pontoscolex corethrurus on plant productivity and soil enzyme activity. J Soil Sediment 14: 483–494.
6. Steiner C, Teixeira WG, Lehmann J, Nehls T, de Macedo JLV, et al. (2007) Long term effects of manure, charcoal and mineral fertilization on crop production and fertility on a highly weathered Central Amazonian upland soil. Plant Soil 291: 275–290.
7. Chan KY, Van Zwieten L, Meszaros I, Downie A, Joseph S (2008) Using poultry litter biochars as soil amendments Aust J Soil Res 46: 437–444.
8. Paz-Ferreiro J, Lu H, Fu S, Méndez A, Gascó G (2014) Use of phytoremediation and biochar to remediate heavy metal polluted soils: A review. Solid Earth 5: 65–75.
9. Ali H, Khan E, Sajad MA (2013) Phytoremediation of heavy metals-Concepts and applications. Chemosphere 91: 869–881.
10. McGrath SP, Zhao FJ (2003) Phytoextraction of metals and metalloids from contaminated soils. Curr Opin Biotechnol 14: 277–282.
11. Adriano DC, Wenzel WW, Vangrosveld J, Nolam NS (2004) Role of assisted natural remediation in environmental clean-up. Geoderma 122: 121–142.
12. de Abreu CA, Coscione AR, Pires AM, Paz-Ferreiro J (2012) Phytoremediation of a soil contaminated by heavy metals and boron using castor oil plants and organic matter amendments. J Geochem Explor 123: 3–7.
13. Komárek M, Vaněk A, Ettler V (2013) Chemical stabilization of metals and arsenic in contaminated soils using oxides-a review. Environ Pollut 172: 9–22.
14. Gabos MB, Casagrande G, Abreu CA, Paz-Ferreiro J (2011) Use of organic matter to mitigate multicontaminated soil and sunflower plants as phytoextractor. Rev Bras Eng Agr Amb 15: 1298–1306.
15. Watanabe T, Murata Y, Osaki M (2009) Amaranthus tricolor has the potential for phytoremediation of cadmium-contaminated soils. Commun Soil Sci Plant Anal 40: 3158–3169.
16. Houben D, Evrard L, Sonnet P (2013) Beneficial effects of biochar application to contaminated soils on the bioavailability of Cd, Pb and Zn and the biomass production of rapeseed (Brassica napus L.). Biomass Bioenerg 57: 196–204.
17. Fellet G, Marmiroli M, Marchiol L (2014) Elements uptake by metal accumulator species grown on mine tailings amended with three types of biochar. Sci Total Environ 468–469: 598–608.
18. Gascó G, Paz-Ferreiro J, Méndez A (2012) Thermal analysis of soil amended with sewage sludge and biochar from sewage sludge pyrolysis. J Thermal Anal Calorim 108: 769–775.
19. Lindsay WL, Norvell WA (1969) Development of a DTPA micronutrients soil test. Agron Abs 69: 84.
20. Ure AM, Davidson CM, Thomas RP (1995) Single and sequential extraction schemes for trace metal speciation in soil and sediment. Quality Assurance for Environmental Analysis, vol. 20. Elsevier Science B.V., pp. 505–523.
21. Zhuang P, Yang Q, Wang H, Shu W (2007) Phytoextraction of heavy metals by eight plant species in the field. Water Air Soil Pollut 184: 235–242.
22. Zacchini M, Pietrini F, Mugnozza GS, Iori V, Pietrosanti L, et al. (2009) Metal tolerance, accumulation and translocation in poplar and willow clones treated with cadmium in hydroponics. Water Air Soil Poll 197: 23–34.
23. Downie A, Crosky A, Munroe P (2009) Physical properties of biochar. J. Lehmann, S. Joseph (Eds.), Biochar for Environmental Management: Science and Technology, Earthscan, London, UK.
24. Kameyama K, Miyamoto T, Shiono T, Shinogi Y (2012) Influence of sugarcane bagasse-derived biochar application on nitrate leaching in calcaric dark red soil. J Environ Qual 41: 1131–1137.
25. Rouquerol F, Rouquerol I, Sing K (1999) Adsorption by Powders and Porous Solids, Academic Press, London, UK.
26. Méndez A, Gascó G, Freitas MMA, Siebielec G, Stuczynsky T, et al. (2005) Preparation of carbon-based adsorbents from pyrolysis and air activation of sewage sludges. Chem Eng J 108: 169–177.
27. Houben D, Evrard L, Sonnet P (2013) Mobility, bioavailability and pH-dependent leaching of cadmium, zinc and lead in a contaminated soil amended with biochar. Chemosphere 92: 1450–1457.
28. Mohan D, Pittman CU Jr, Brcika M, Smith F, Yancey B, et al. (2007) Sorption of arsenic, cadmium and lead by chars produced from fast pyrolysis of wood and bark during bio-oil production. J Colloid Interface Sci 310: 57–73.
29. Uchimiya M, Chang SC, Klasson KT (2011) Screening biochars for heavy metal retention in soil: Role of oxygen functional groups. J Hazard Mater 190: 432–444.
30. Park JH, Choppala GK, Bolan NS, Chung JW, Chuasavathi T (2011) Biochar reduces the bioavailability and phytotoxicity of heavy metals. Plant Soil 348: 439–451.
31. Kabata-Pendias A, Mukherjee AB (2007) Trace elements from soil to human. Berlin: Springer.
32. Yoon J, Cao X, Zhou Q, Ma LQ (2006) Accumulation of Pb, Cu, and Zn in native plants growing on a contaminated Florida site. Sci Total Environ 368: 456–464.
33. Kookana RS, Sarmah AK, Van Zwieten L, Krull E, Singh B (2011) Biochar application to Soil: Agronomic and environmental benefits and unintended consequences. Adv Agron 112: 103–144.
34. Yu XY, Ying GG, Kookana RS (2009) Reduced plant uptake of pesticides with biochar additions to soil. Chemosphere 76: 665–671.
35. Ji P, Sun T, Song Y, Ackland ML, Liu Y (2011) Strategies for enhancing the phytoremediation of cadmium-contaminated agricultural soils by *Solanum nigrum* L. Environ Pollut 159: 762–768.
36. Namgay T, Singh B, Singh BP (2010) Influence of biochar application to soil on the availability of As, Cd, Cu, Pb, and Zn to maize (*Zea mays* L.). Aust J Soil Res 48: 638–647.
37. Joseph SD, Graber ER, Chia C, Munroe P, Donne S, et al. (2013) Shifting paradigms Shifting paradigms: development of high-efficiency biochar fertilizers based on nano-structures and soluble components. Carbon Management 4: 323–343.

Phytoavailability of Cadmium (Cd) to Pak Choi (*Brassica chinensis* L.) Grown in Chinese Soils: A Model to Evaluate the Impact of Soil Cd Pollution on Potential Dietary Toxicity

Muhammad Tariq Rafiq[1,2], Rukhsanda Aziz[1], Xiaoe Yang[1], Wendan Xiao[1], Peter J. Stoffella[3], Aamir Saghir[4], Muhammad Azam[5], Tingqiang Li[1]*

1 Ministry of Education Key Laboratory of Environmental Remediation and Ecological Health, College of Environmental and Resource Sciences, Zhejiang University, Hangzhou, China, **2** Department of Environmental Science, International Islamic University, Islamabad, Pakistan, **3** Indian River Research and Education Center, Institute of Food and Agricultural Sciences, University of Florida, Fort Pierce, Florida, United States of America, **4** Institute of Statistics, Zhejiang University, Hangzhou, China, **5** College of Agriculture and Biotechnology, Zhejiang University, Hangzhou, China

Abstract

Food chain contamination by soil cadmium (Cd) through vegetable consumption poses a threat to human health. Therefore, an understanding is needed on the relationship between the phytoavailability of Cd in soils and its uptake in edible tissues of vegetables. The purpose of this study was to establish soil Cd thresholds of representative Chinese soils based on dietary toxicity to humans and develop a model to evaluate the phytoavailability of Cd to Pak choi (*Brassica chinensis* L.) based on soil properties. Mehlich-3 extractable Cd thresholds were more suitable for Stagnic Anthrosols, Calcareous, Ustic Cambosols, Typic Haplustalfs, Udic Ferrisols and Periudic Argosols with values of 0.30, 0.25, 0.18, 0.16, 0.15 and 0.03 mg kg^{-1}, respectively, while total Cd is adequate threshold for Mollisols with a value of 0.86 mg kg^{-1}. A stepwise regression model indicated that Cd phytoavailability to Pak choi was significantly influenced by soil pH, organic matter, total Zinc and Cd concentrations in soil. Therefore, since Cd accumulation in Pak choi varied with soil characteristics, they should be considered while assessing the environmental quality of soils to ensure the hygienically safe food production.

Editor: Wenju Liang, Chinese Academy of Sciences, China

Funding: This study was supported by the Ministry of Environmental Protection of China (grant no. 2011467057), the ministry of Science and Technology of China (grant no. 2012AA100605), and by the Fundamental Research Funds for the Central Universities of China. The funders had no role in study design, data collection and analysis, decision to publish, or preparation of the manuscript.

Competing Interests: The authors have declared that no competing interests exist.

* Email: litq@zju.edu.cn

These authors contributed equally to this work.

Introduction

Cadmium (Cd) is an important environmental pollutant toxic to animals and human beings. It is one of the most mobile elements, among all the toxic heavy metals [1]. Cadmium is not required for plants growth or reproduction, however its bioaccumulation and subsequent accrual in the food chain surpasses all other trace elements due to its high mobility in soil [2]. It is the most toxic element in the environment and even at low concentrations is very toxic to living cells and considered as carcinogenic [3]. In humans, Cd exposure can result in multiple adverse effects, such as testicular damage, renal and hepatic dysfunction, etc. [3]. Moreover, Cd is implicated in the development of cancer, phytotoxic at higher levels of concentrations [4] and classified as a type I carcinogen by the International Agency for Cancer Research [5]. Significant quantities of Cd can be transferred from contaminated soil to plants [6]. Therefore, crops produced from Cd contaminated soils may be unsuitable or even detrimental for animal and human consumption [7].

Vegetables are an important component of human diet since they contain proteins, carbohydrates as well as minerals and vitamins [8]. The proportion of vegetables consumed in the total diet has been increased with the improvement of living standards. However, vegetables are also one of the most important pathways through which heavy metals enter the food chain and affect human health. Leafy vegetables can accumulate higher concentrations of Cd than other crops [9,10]. Leafy vegetables are known to accumulate higher concentrations of Cd in the edible parts even when grown in soils containing low concentrations of Cd [11]. Pak choi (*Brassica chinensis* L.), also known as Chinese cabbage, is a popular leafy vegetable, grown and consumed worldwide. Therefore, it is imperative to control Cd concentrations in Pak choi, especially in its edible parts to ensure food safety. To limit the transfer of soil Cd into the edible parts of Pak choi, an understanding of its accumulation characteristics is required. Currently, there is an elevated concern over Cd accumulation in food and its potential risks to human health [12]. Cadmium accumulation and distribution varies among vegetable cultivars and tissues [13]. However, the accumulation and distribution of Cd in vegetables grown in a diversity of soil types were rarely studied [14].

About one fifth of agricultural land is contaminated by Cd, lead (Pb) and arsenic (As) in China [15]. Moreover, it was reported that about 20% of farm lands in China are contaminated with heavy metals and Cd contamination accounts for more than 1.3×10^5 ha of the total affected area [16,17]. Cadmium uptake by rice (*Oryza sativa* L.) and vegetables from soil is the initial source of exposure for human beings [18,19]. Therefore, there are environmental concerns of soils, food safety and human health for the present and future agricultural and environmental sustainability of world vegetable supplies. As, only a small fraction of total trace metals in soil is available for plant absorption, it is widely accepted that the total metal content in soils is neither a viable indicator of phytoavailability nor an adequate tool to assess the potential risk of dietary toxicity [12]. Tracy and Sheila [20] reported that extractable Cd content in soil may be an improved indicator of bioavailability and toxicity than the total contents and toxicity and availability of metals differed among soils types. Metal uptake and translocation studies were conducted for different crops under varying soil conditions, to further understanding uptake and the transport mechanisms [21,22].

To ensure the food safety and environmental quality of soils, guidelines for permissible concentrations of Cd in agricultural soils need to be established. Due to limited number of studies, the soil environmental quality guidelines for heavy metals in farmland soils developed and applied in the world are still based on total metal contents of soil. Minimal attention has been focused on metal accumulation differences among the edible parts of crops, and the relationship between total concentration and phytoavailability of heavy metals in different soil types [4]. Developing the linkage between the bioavailability of Cd in soil and its transfer into the edible plant parts is a key to improving existing soil environmental quality standards. Information is vital on the degree of translocation of heavy metals from soils to plants, which are used as food crops, and absorption of metals in food plants to concentration that does not cause phytotoxicity symptoms [23]. This study was conducted in seven Chinese soil types to establish direct relationship of Cd level in such contaminated soils and Cd uptake in Pak choi. The main objectives were to establish soil Cd thresholds for representative Chinese soils based on human dietary toxicity and to determine the relationships between several soil properties and Cd accumulation in Pak choi. This information will be useful in establishing soil protection guidelines to produce hygienically safe vegetables.

Materials and Methods

Ethics Statement

The soils used in this study were agricultural soils. No specific permissions were required for the described locations. We confirm that the field studies did not involve endangered or protected species.

Soil Collection and Analysis

Seven Chinese soils were selected for this study. Udic Ferrisols, Mollisols, Periudic Argosols, Typic Haplustalfs, Ustic Cambosols, Calcaric Regosols and Stagnic Anthrosols were collected from Chinese cities of Guilin (104°40′–119°45′E, 24°18′– 25°41′N), Harbin (125°42′– 130°10′E, 44°04′–46°40′N), Huzhou (119°14′–120°29′E, 30°22′–31°11′N), Zhanjiang (110°08′–110°77′E, 20°33′–21°62′N), Qufu (116°51′–117°13′E, 35°29′–35°49′N), Ya'an (102°37′–103°12′E, 29°23′–30°37′N) and Jiaxing (120°7′–121°02′E, 30°5′–30°77′N), respectively. Soils samples were taken at a depth of up to 20 cm from the upper horizon. Each sample was air-dried, ground, and screened through two mm sieve before laboratory analysis. Soil pH, cation exchange capacity, organic matter contents, and particle size density were measured by using previously described methods [24–27]. Physicochemical properties of these soils are reported (Table 1).

Cadmium Spiking and Aging

Soil samples of Mollisols, Periudic Argosols, Stagnic Anthrosols and Ustic Cambosols were spiked with Cd as $Cd(NO_3)_2$ in an aqueous solution at loading rates of 1.0, 2.0, 4.0, 6.0 and 8.0 mg Cd kg^{-1} soil along with an untreated control (Ck), the background values of Cd concentration was below 0.50 mg kg^{-1} in these soil. However, the Udic Ferrisols, Typic Haplustalfs and Calcaric Regosols soil samples, with the background values of Cd concentration above 0.50 mg kg^{-1}, were spiked with Cd to establish the contamination levels of 2.0, 4.0, 6.0 and 8.0 mg Cd kg^{-1} soil along with the untreated control (Ck). Soil moisture was maintained up to 70% of its water-holding capacity by using distilled water. All the spiked soils were aged for one year subsequent to greenhouse experimentation. After one year aging period, the concentrations of total Cd, and Mehlich-3 extractable Cd were determined in each of the spiked soils.

Containerized Experiment

A containerized experiment was performed in greenhouse by growing Pak choi (*Brassica chinensis* L.) during March – April, 2012 at Zhejiang University, Hangzhou, China. Seed of Pak choi was obtained from the Zhejiang Seed Co. Hangzhou, China. Seeds were washed with distilled water and air-dried prior to sowing. Seeds were germinated in dark at 25°C and transplanted into quartz sand bed to establish seedlings. Four healthy, uniform and 21-day-old seedlings were transplanted into plastic containers with a diameter of 18 cm and height of 17 cm. Each container had 3 kg of soil. Fertilizers were applied at the rates of 0.4 g of N as $CO(NH_2)_2$ and 0.2 g P as KH_2PO_4 per kg of soil. The experiment was carried out in a completely randomized design (CRD). Treatments were established in triplicate, and the containers were randomly arranged in a greenhouse bench under controlled conditions of 16 h of light at 30°C and 8 h of dark at 22°C. Plants were monitored daily and watered as necessary.

Plant Sample Collection

Pak choi was harvested after 30 days from transplanting. The plants of Pak choi were removed from each container and separated into root and shoots (including stems and leaves). Roots and shoots of Pak choi were first washed with tap water and then with ultrapure distilled water, to remove all visible soil particles. Clean plant samples were first blotted dry, and then dried at 70°C for 72 h in an oven. Dry shoot weight of samples was recorded. Dry plant samples were ground to pass through a 60 mm sieve using an agate mill prior to Cd concentration analysis.

Total Cd of Soil and Plant

For determination of total Cd concentration in soil, 0.20 g of soil samples was digested with HNO_3–HF –$HClO_4$ (5:1:1) [4]. For plant samples, 0.20 g of shoots for each treatment was digested with HNO_3–H_2O_2 (5:1). After cooling the digest was transferred to a volumetric flask, diluted with distilled water to 50 mL [28]. The concentrations of Cd in the filtrate were determined using inductively coupled plasma–mass spectrometry (ICP-MS, Agilent, 7500a, CA, USA). The ICP-MS was operated at the following

Table 1. Basic Chemical and Physical Characteristics of Seven Chinese soils.

Soil Types	Mollisols	Ustic Cambosols	Stagnic Anthrosols	Periudic Argosols	Typic Haplustalfs	Udic Ferrisols	Calcaric Regosols
pH	7.23±0.08	7.80±0.02	6.49±0.03	4.85±0.06	5.16±0.05	4.43±0.07	8.02±0.04
OM (g kg^{-1})	32.2±0.32	7.54±0.20	21.4±0.34	11.6±0.17	6.37±0.56	19.1±0.15	21.8±0.14
CEC (cmol kg^{-1})	34.0±2.51	15.8±1.62	20.2±1.41	12.6±1.52	8.33±2.14	17.3±1.96	25.5±1.46
Total Cd (mg kg^{-1})	0.51±0.02	0.59±0.06	0.79±0.04	0.47±0.02	0.92±0.01	1.06±0.09	0.96±0.06
Total Zn (mg kg^{-1})	31.18±1.47	26.93±0.43	41.36±1.71	15.17±0.88	25.3±1.44	24.6±0.23	28.59±1.38
Sand (%)	20.6±1.54	21.6±1.29	11.4±0.26	24.8±0.65	37.4±0.96	32.75±1.65	31.6±0.57
Silt (%)	60.2±2.21	65.4±2.62	73.0±2.41	58.2±1.04	40.8±1.66	39.8±1.26	44.0±1.26
Clay (%)	19.2±1.24	13.0±1.05	15.6±1.17	17.0±0.34	21.8±0.82	49.6±1.19	24.4±1.32

conditions: the radio frequency power at the torch 1.2 kW, the plasma gas flow 15 L min^{-1}, the auxiliary gas flow 0.89 L min^{-1}, and the carrier gas flow 0.95 L min^{-1} [28]. The same procedure without samples was used as control and three replications were conducted for each sample.

Mehlich-3 Extractable Cd in Soils

Mehlich-3 extractable Cd in soils was determined following the extraction method described by Mehlich [29]. Briefly, 5 g (0.2 mm sieved) of dry soil was shaken with 50 mL of Mehlich-3 solution (0.2 mol L^{-1} CH_3COOH, 0.25 mol L^{-1} NH_4NO_3, 0.015 mol L^{-1} NH_4F, 0.013 mol L^{-1} HNO_3, 0.001 mol L^{-1} EDTA) for 5 min (200 rpm) at 25°C. The suspension was centrifuged at 5000 rpm for 10 min and filtered through 0.45 μm filter paper. The same procedure without samples was used as control and three replications were conducted for each soil sample. The Cd concentration in the filtrate was analyzed by inductively coupled plasma–mass spectrometry (ICP-MS, Agilent 7500a, CA, USA).

Quality Control for Cd Analysis

Quality assurance and quality control (QA/QC) for Cd in soil and Pak choi were conducted by determining Cd contents in the certified reference materials (soil GSBZ 50013-88 and plant GBW-07402) respectively, approved by General Administration of Quality Supervision, Inspection and Quarantine of the People's Republic of China (AQSIQ) and National Center for Reference Materials. The analytical results showed a recovery rate of 97.3% and 102.1% respectively.

Derivation of Soil Cd Thresholds for Potential Dietary Toxicity in Pak choi

For ensuring the environmental and food safety for human beings, an effort has been made to develop guidelines for acceptable concentrations of potentially harmful Cd in seven agricultural soils types of China. In this context, the amounts of Cd in Pak choi above than threshold level of food safety are adversely affecting humans are critical. Since, Cd bioavailability differed among soil types, the focus was on the development of soil Cd thresholds for representative Chinese soils based on food safety, Provisional Tolerable Weekly Intake (PTWI) of Cd recommended by FAO/WHO Joint Expert Committee on Food Additives, is 7 μg kg^{-1} of body weight [30]. Estimated daily intake of metal (EDIM) was determined by the following equation.

$$\mathrm{EDIM} = \frac{C_{cadmium} \times C_{factor} \times D_{daily\ intake}}{B_{average\ weight}}$$

Where, Cc_{admium}, C_{factor}, $D_{food\ intake}$ and $B_{average\ weight}$ represent average Cd concentration in Pak choi (mg kg^{-1}), conversion factor, daily consumption of Pak choi (g) and average body weight (kg) of the adult consumers, respectively. Average daily consumption of Pak choi for adults was considered to be 0.345 kg person^{-1} d^{-1} [31] and a conversion factor 0.085 was used to convert fresh Pak choi weight to dry weight [32]. Average body weight of adult was considered to be 60 kg as motioned in previous reports [30]. According to the above equation of EDIM, the provisional tolerable daily tolerable intake of Cd for Pak choi was 2.04 mg kg^{-1} on a dry weight basis. Soil Cd threshold levels for potential dietary toxicity from Pak choi were calculated according to the tolerable daily dietary intake level of Cd (2.04 mg kg^{-1}) and the regression equations.

Statistical Analysis

Stepwise multiple regression analysis, single linear regression and one-way analysis of variance (ANOVA) were performed using the statistical software package SPSS (version 18.0). All values reported in this work are means of three independent replications. Treatment means were separated by least significant difference (LSD) test, at 5% level.

Results

Characteristics of Soils

Soils evaluated were representative of most of Chinese soil types, pH range of soils were strongly acidic to mild alkaline. Chemical and physical characteristics varied among the seven soils. Highest total Cd and Zn concentrations (background value) were observed in Udic Ferrisols and Stagnic Anthrosols respectively. Mollisols contained the highest amount of organic matter and exhibited an elevated cation exchange capacity as well (Table 1).

Mehlich-3 Extractable Cd in Soils after Aging of 1 Year

Mehlich-3 extractable Cd content increased significantly with increasing Cd spiking levels in all the seven soils. Mehlich-3 extractable Cd ranged from 0.16–3.95 mg kg^{-1} in these soils under different Cd levels (Table 2). The Cd contents varied significantly among these soils, decreasing in order: Periudic Argosols> Typic Haplustalfs> Udic Ferrisols> Stagnic Anthrosols> Mollisols> Ustic Cambosols> Calcaric Regosols. Mehlich-3 extractable Cd concentration was greater at higher rates of Cd spiking in each soil. These results indicated that minimum and maximum extractability of Cd was found in Calcaric Regosols and Periudic Argosols, respectively under the highest (8 mg kg^{-1}) level of Cd spiked. Mehlich-3 extractable concentrations were dependent on total Cd in each soil, however the extractability was significantly higher in low pH soils as compared to the medium and high pH soils (Table 2).

Biomass Yield of Pak choi

Generally, Pak choi had tolerance to Cd toxicity in Mollisols, Stagnic Anthrosols and Calcaric Regosols soils, indicating low phytoavailability of Cd in these soils. Shoot biomass of Pak choi under different Cd treatments of these soils did not decrease significantly as compared with their respective controls. However, the shoot biomass of Pak choi grown in Ustic Cambosols, Udic Ferrisols, Periudic Argosols and Typic Haplustalfs decreased significantly as compared with the control indicating higher phytoavailability of Cd in these soils (Table 3). The stimulating effect of Cd on shoot biomass of Pak choi occurred at 1 mg kg^{-1} and 2 mg kg^{-1} in Ustic Cambosols and Mollisols respectively, whereas in Stagnic Anthrosols, it occurred at 4 mg kg^{-1}. The dry weight of Pak choi shoots at 8 mg kg^{-1} Cd generally decreased in order of: Calcaric Regosols> Mollisols, Stagnic Anthrosols> Ustic Cambosols> Udic Ferrisols> Periudic Argosols> Typic Haplustalfs (Table 3).

Accumulation and Distribution of Cadmium in Pak choi

Cadmium concentration in the shoots and roots of Pak choi varied significantly among soils at different Cd levels and soil types. Roots exhibited the higher Cd contents as compared with Pak choi shoots. The content of Cd enhanced with increasing Cd loading rate in the soils. Cd concentration was high in the roots (2.42 to 169. 95 mg kg^{-1} DW), while low in the shoot (0.48 to 89.21 mg kg^{-1} DW) (Table 4). Cd uptake in Pak choi tissues was affected by soil type, primarily due to the variation in Cd

Table 2. Mehlich-3 Extractable Cd Contents (mg kg^{-1}) in Seven Chinese Soils at the onset of Containerized Experiment after Aging of 1 year.

Cd conc. (mg kg^{-1})	Mollisols	Ustic Cambosols	Stagnic Anthrosols	Periudic Argosols	Typic Haplustalfs	Udic Ferrisols	Calcaric Regosols
Ck	0.19±0.05c	0.17±0.07c	0.22±0.08d	0.16±0.06e	0.31±0.12d	0.42±0.09d	0.29±0.07c
1	0.21±0.04c	0.19±0.06c	0.30± 0.04d	0.41±0.13e	-	-	-
2	0.62±0.04bc	0.48±0.10c	0.56±0.27d	0.99±0.29d	0.69±0.17d	0.65±0.11d	0.35±0.06c
4	1.19±0.08b	1.05±0.90bc	1.17±0.24c	1.79±0.54c	1.52±0.26c	1.46±0.60c	0.71±0.10bc
6	2.21±0.91a	2.13±0.90b	2.01±0.47b	2.98±0.46b	2.44±0.64b	2.39±0.90b	1.38±0.29ab
8	2.89±1.04a	3.28±0.84a	3.29±0.73a	3.95±0.41a	3.58±0.53a	3.47 ±0.70a	2.01±0.94a

Mean values followed by different letters within the same column are significantly different at P <0.05.

Table 3. Dry Biomass (g plant^{-1}) of Pak choi Shoots Grown on Seven Chinese Soils with Different Loading Rates of Cd.

Cd conc. (mg kg^{-1})	Mollisols	Ustic Cambosols	Stagnic Anthrosols	Periudic Argosols	Typic Haplustalfs	Udic Ferrisols	Calcaric Regosols
Ck	1.95±0.18a	1.54±0.13a	1.13±0.31a	0.84±0.29a	0.73±0.20a	0.95±0.11a	2.07±0.46a
1	1.91±0.37a	1.66±0.59a	1.10±0.30a	0.63±0.35ab	-	-	-
2	2.22±0.58a	1.02±0.49a	1.14±0.53a	0.59±0.47ab	0.54±0.41ab	0.79±0.23ab	1.98±0.11a
4	1.89±0.49a	1.47±0.08a	1.25±0.28a	0.51±0.03ab	0.43±0.38ab	0.58±0.06abc	1.95±0.06a
6	1.86±0.48a	1.35±0.13ab	1.09±0.09a	0.36±0.02ab	0.38±0.11ab	0.50±0.24bc	1.90±0.15a
8	1.54±0.28a	0.95±0.15b	1.11±0.51a	0.17±0.06b	0.10±0.01b	0.39±0.29c	1.75±0.17a

Mean values followed by different letters within the same column are significantly different at $P<0.05$.

bioavailability. The lowest and highest Cd concentrations in the Pak choi tissues were at the highest (8 mg kg^{-1}) level of Cd in Calcaric Regosols and Periudic Argosols, respectively. Cadmium concentrations in Pak choi followed an order of: Periudic Argosols> Typic Haplustalfs> Udic Ferrisols> Ustic Cambosols> Mollisols> Stagnic Anthrosols> Calcaric Regosols at 8 mg kg^{-1} soil Cd level (Table 4).

Relationship between Mehlich-3 Extractable Cd in Soils and Pak choi Cd Content

Cadmium concentrations in shoots of Pak choi were significantly correlated to total Cd and Mehlich-3 extractable Cd contents in soils ($R^2 = 0.95$ to 0. 99, and 0.97 to 0.99 respectively). Cadmium concentrations in Pak choi shoots were best related to total Cd content in Mollisols ($R^2 = 0.99$). Whereas, the Cd concentrations of Pak choi shoots were best correlated to Mehlich-3 extractable Cd in Ustic Cambosols, Stagnic Anthrosols Periudic Argosols, Udic Ferrisols, Typic Haplustalfs and Calcaric Regosols with $R^2 = 0.97$, 0.99, 0.99,0.98,0.99 and 0.98, respectively (Table 5).

Soil Cd Thresholds for Potential Dietary Toxicity in Pak choi

Total soil Cd thresholds for potential dietary toxicity from the consumption of Pak choi conformed to an order of: Calcaric Regosols> Stagnic Anthrosols> Ustic Cambosols> Mollisols> Udic Ferrisols> Typic Haplustalfs> Periudic Argosols, and were 1.25, 1.16, 1.02, 0.86, 0.72, 0.70 and 0.12 mg kg^{-1}, respectively. Mehlich-3 extractable Cd thresholds were 0.30, 0.25, 0.23, 0.18, 0.16, 0.15 and 0.03 mg kg^{-1} and decreased in the following order of: Stagnic Anthrosols> Calcareous> Mollisols> Ustic Cambosols> Typic Haplustalfs> Udic Ferrisols>Periudic Argosols, respectively (Table 6).

Discussion

Biomass Yield of Pak choi

Dry weight of Pak choi did not decrease significantly under different Cd levels (Ck to 8.0 mg kg^{-1}) in Mollisols, Stagnic Anthrosols and Calcaric Regosols and even increased at 1, 2 and 4.0 mg kg^{-1}of treatment levels. Similar stimulatory responses of biomass to Cd exposure have also been reported in several plant species [33,34]. The stimulatory effect of Cd on plant biomass may be explained by various mechanisms, for examples, metal ions can serve as enzyme activators in cytokinins metabolism, which stimulates the growth of plants, [35] and a low dose of metal exposure may cause changes in cytokinins and plant hormones that regulate growth and development of plants [36]. Kaminek [36] reported that cytokinins may delay senescence by maintaining chlorophyll production and photosynthetic activity in plant leaves.

Cd exposure may cause changes to various physiological and biochemical processes in plant tissues, such as, reduction in dry biomass may be due to the negative effects of Cd on the roots, and plants could not take up nutrients to continue their normal activities. It has been well reported that Cd can reduce plant growth and development by interfering in various metabolic processes, such as, inhibition of the proton pump, reduction in root elongation, and damage to photosynthetic activity [37,38]. The excess amount of Cd in soil may be responsible for causing disturbances in mineral nutrition and carbohydrate metabolism [39].

Shoot biomass of Pak choi grown in Ustic Cambosols, Udic Ferrisols, Periudic Argosols and Typic Haplustalfs decreased significantly as compared with the control. The inhibitory effect of

Table 4. Cd Concentration (mg kg^{-1} DW) in Pak choi Grown under Different Cd Levels in Seven Chinese Soils.

Cd (mg kg^{-1})	Mollisols		Ustic Cambosols		Stagnic Anthrosols		Periudic Argosols		Typic Haplustalfs		Udic Ferrisols		Calcaric Regosols	
	Root	Shoot	Root	Shoot	Root	Shoot	Root	Shoot	Root	Shoot	Root	Shoot	Root	Shoot
Ck	3.41±0.64e	1.00±0.28e	2.42±0.45f	0.48±0.22e	2.73±0.28e	0.85±0.12e	7.26±1.10f	1.84±0.56f	12.35±1.68e	4.37±1.02e	11.81±0.78e	6.77±0.92e	3.11±0.66e	1.92±0.58b
1	5.09±0.91e	2.53±0.76e	6.69±0.65e	2.06±0.98e	4.12±0.75e	1.92±0.94e	21.3±2.01e	11.00±1.57e	-	-	-	-	-	-
2	8.92±1.09d	4.31±0.54d	9.30±1.29d	4.77±0.22d	8.34±1.11d	3.94±0.83d	39.33±3.86d	28.51±2.61d	25.41±1.79d	12.09±1.34d	19.41±1.27d	10.06±1.10d	4.24±0.99d	2.70±0.99b
4	14.21±0.90c	8.34±1.11c	14.00±1.43c	9.01±1.06c	13.21±1.55c	7.06±1.06c	68.43±3.71c	39.12±3.06c	39.11±2.28c	32.51±2.51c	33.55±1.71c	20.89±1.46c	10.37±1.21c	3.46±1.10b
6	21.24±1.20b	11.75±1.17b	24.36±1.11b	17.23±1.50b	20.21±1.80b	11.63±1.54b	117.76±6.53b	68.21±4.08b	68.00±3.96b	41.11±3.01b	55.53±2.67b	31.45±2.48b	14.41±1.03b	6.57±0.93a
8	38.96±1.39a	18.11±1.12a	43.21±2.21a	22.22±1.51a	28.56±1.42a	17.63±1.37a	169.95±8.11a	89.21±5.70a	118.21±6.24a	69.21±4.11a	85.32±4.28a	53.73±2.88a	18.73±1.54a	8.13±0.87a

Mean values followed by different letters within the same column are significantly different at $P<0.05$.

Cd on shoot growth is consistent with earlier reports of three Chinese cabbage cultivars exposed to different soil Cd levels of 1, 2.5 and 5 mg kg^{-1}. A significant decrease in the shoot biomass was observed at 2.5 and 5 mg kg^{-1} levels of Cd as compared to their respective controls [40]. Shentu *et al.* [4] found a 46% reduction in root dry weight of radish at 6.31 mg kg^{-1} Cd exposure in red yellow soil, which is in accordance with our results as we also noticed a shoot dry weight reduction of 58.9%, 79.7%, and 86.3% in Udic Ferrisols, Periudic Argosols and Typic Haplustalfs respectively at 8 mg kg^{-1} level of soil Cd as compared to their respective controls.

Accumulation and Distribution of Cadmium in Pak choi

Variations of Cd accumulation in Pak choi grown in different soils with different pH may be due to the difference in bioavailability of Cd in each soil. Liang *et al.* [13] stated that Cd content of spinach plants was highly dependent upon the soil pH being highest at pH 5.3. Lai and Chen [41] reported that Cd concentration in Pak choi shoots was up to 85 mg kg^{-1} DW with an application of soil Cd up to 20 mg kg^{-1}. Moreover, it was observed that accumulation of the Cd in rice shoot ranged from 67.9 to 241.7 μg/pot in different rice genotypes at 5 mg kg^{-1} soil Cd level [42].

Relationship between Mehlich-3 Extractable Cd in Soils and Pak choi Cd Content

Mehlich-3 extraction technique appeared efficient to assess Cd phytoavailability to Pak choi, grown in seven textured soils, as evidenced by high correlation coefficients (R^2>0.97). This is in agreement with our previous studies, [43,28] which reflected a high linear correlation (R^2>0.98) between Mehlich-3 Cr and Cr contents in Pak choi and rice grown under six different textured soils. These results are similar to those reported in which extractable soil metal was an improved indicator for Cd phytoavailability in several vegetable crops [4]. Mehlich-3 extraction method is applicable to a large range of soil types, from acidic to alkaline, and makes it ideal for application at a wide scale [44]. Generally, the extraction techniques are assumed to have a relationship between the extractable fraction of metals and the phytoavailability of the metals to plants, and these metals such as exchangeable, soluble, and loosely adsorbed metals are labile and thus readily available to plants [12,45]. The efficiency of Mehlich-3 extraction method was compared with the EPA 3050 B method (a strong acid digestion method) to assess the predictive capabilities through a lettuce (green specie) bioassay. Mehlich-3 extraction was positively correlated with the more costly EPA test, and could be developed as a less expensive and easily conduct able technique [46].

Soil Cd Thresholds for Potential Dietary Toxicity in Pak choi

Cadmium concentrations in the shoots of Pak choi were significantly correlated to total Cd and Mehlich-3 extractable Cd contents in soils, with R^2 values of 0.95 to 0.99, and 0.97 to 0.99, respectively. From this investigation, Cd contents in Pak choi shoots were correlated to total Cd content in Mollisols (R^2 values of 0.99). Cadmium concentrations of Pak choi shoots were highly correlated to Mehlich-3 extractable Cd in Ustic Cambosols, Stagnic Anthrosols, Periudic Argosols, Udic Ferrisols, Typic Haplustalfs and Calcaric Regosols with R^2 values of 0.97, 0.99, 0.99, 0.98, 0.99 and 0.98, respectively. Total Cd threshold levels for potential dietary toxicity conformed to an order of: Calcaric Regosols> Stagnic Anthrosols> Mollisols> Ustic Cambosols>

Table 5. Regression Correlation between Cd Contents in the Edible Shoots of Pak choi and Different Forms of Cd in Various Soils.

Soil type	Form of Soil Cd	Regression equation	R^2
Mollisols	Total Cd	y = 2.1706x +0.1669	0.99
	Mehlich-3 extractable Cd	y = 5.7207x +0.7037	0.98
Ustic Cambosols	Total Cd	y = 2.9358x −0.9586	0.96
	Mehlich-3 extractable Cd	y = 7.0409x +0.7648	0.97
Stagnic Anthrosols	Total Cd	y = 2.2344x −0.5651	0.98
	Mehlich-3 extractable Cd	y = 5.3848x +0.4458	0.99
Periudic Argosols	Total Cd	y = 11.061x +0.6961	0.98
	Mehlich-3 extractable Cd	y = 22.326x +1.3968	0.99
Typic Haplustalfs	Total Cd	y = 8.7318x −4.0825	0.97
	Mehlich-3 extractable Cd	y = 19.123x −0.8046	0.98
Udic Ferrisols	Total Cd	y = 6.6697x −2.7793	0.97
	Mehlich-3 extractable Cd	y = 14.873x −0.3773	0.99
Calcaric Regosols	Total Cd	y = 0.8936x +0.924	0.95
	Mehlich-3 extractable Cd	y = 3.5937x +1.1512	0.98

Typic Haplustalfs> Udic Ferrisols> Periudic Argosols and were 1.25, 1.16, 1.02, 0.86, 0.72, 0.70 and 0.12 mg kg^{-1}, respectively. Mehlich-3 extractable Cd thresholds decreased in the following order of: Stagnic Anthrosols> Calcareous> Mollisols> Ustic Cambosols> Udic Ferrisols>Typic Haplustalfs> Periudic Argosols and were 0.30, 0.25, 0.23, 0.18, 0.16, 0.15 and 0.03 mg kg^{-1}, respectively (Table 6).

Cadmium concentrations in Pak choi shoots, were highly correlated to total Cd content in Mollisols with the threshold levels of 0.86 mg kg^{-1} with a $R^2 = 0.99$. However, the Cd concentrations of Pak choi shoots were best related to Mehlich-3 extractable Cd in Stagnic Anthrosols, Calcaric Regosols, Mollisols, Ustic Cambosols, Typic Haplustalfs and Periudic Argosols with thresholds values of 0.30, 0.25, 0.23, 0.18, 0.16, 0.15 and 0.03 mg kg^{-1}, (R^2 values of 0.97, 0.99, 0.99, 0.98, 0.98 and 0.98), respectively. Based on the wide range of applicability and the simplicity of extraction method, it is proposed that Mehlich-3 extractable Cd is more suitable to be used as soil Cd thresholds for potential dietary toxicity in Pak choi. Our previous study evaluated the phytoavailability of Cd to rice, and demonstrated the suitability of Mehlich-3 extraction method in different textured soils [47]. Similar to our results, Murakami *et al.* [48] reported that Mehlich-3 extractable Cd was an improved indicator than total soil Cd and HCl-extractable Cd to predict the grain Cd content of japonica rice varieties. Our results are also in agreement with Shentu *et al.* [4] who also concluded that extractable Cd was a better soil test index for Cd phytoavailability of several vegetables and could be used as soil Cd thresholds for food safety. Among the predicted thresholds (total soil Cd) the lowest value (0.12 mg kg^{-1}) was observed for the Periudic Argosols, an acidic soil. Bioavailability and uptake of Cd are very high in this soil. The leafy vegetables like Pak choi can accumulate large quantities of Cd as compared to other crops [9,10]. Therefore the predicted threshold is even lower than background value of Cd in soil; it means that there is a risk for dietary toxicity from Pak choi grown on it, even with the background value of total Cd in soil. This kind of information has been reported in our previous study. The threshold of total soil Cd for rice was 0.21 mg kg^{-1} which was also lower than background value of total Cd in soil [47].

Cd levels (Ck, 1, 2, 4, 6, 8 mg kg^{-1}) used in this investigation represented uncontaminated, lightly contaminated, and moderately Cd polluted soils. Therefore, these levels of Cd contamination are realistic, comparable to those applied in other soil safety risk assessment studies, and thus, the results are applicable in field conditions as well.

Table 6. Soil Cd Threshold Levels for Potential Dietary Toxicity in Edible Part of Pak choi Calculated from the Permissible Limit of Cd (2.04 mg kg^{-1} DW) in Leafy Vegetables and Regression Equations.

Soil Type	Total Cd (mg kg^{-1})	Mehlich-3 extractable Cd (mg kg^{-1})
Mollisols	0.86	0.23
Ustic Cambosols	1.02	0.18
Stagnic Anthrosols	1.16	0.30
Periudic Argosols	0.12	0.03
Udic Ferrisols	0.72	0.16
Typic Haplustalfs	0.70	0.15
Calcaric Regosols	1.25	0.25

doi:10.1371/journal.pone.0111461.t006

Table 7. Stepwise Regression Model for Predicting Cd Concentration (Y) in Edible Part of Pak choi based on Soil Properties.

Stepwise regression model	R^2	F value	T value and R^2 of partial regression coefficient		
				T value	R^2
$Y = -39.256 - 15.516\ pH - 0.944\ Zn_T - 0.379\ OM + 26.752\ Cd_T$	0.977	138.808*	pH	−18.682**	0.964
			Zn_T	−3.788**	0.524
			OM	−3.652*	0.376
			Cd_T	2.67*	0.354

[a]Cd_T and Zn_T refer to the total Cadmium and Zinc concentrations.
[b]Superscripts * and ** indicate significant levels of probability at 0.05 and 0.01, respectively.

Stepwise Regression Model for Predicting Cd Phytoavailability to Pak choi

Many physicochemical properties of soils can influence the heavy metal accumulation in vegetables. For example, the amount of heavy metal uptake from soils was influenced by soil pH, organic matter (OM) content, cation exchange capacity (CEC) and soil texture [49]. The combinations of basic soil properties may explain Cd uptake by plants [50]. By considering this aspect, soil pH, OM, CEC, total soil Cd, total Zn and clay contents were integrated to simulate the combined effects of soil environment on Cd phytoavailability to Pak choi. Stepwise linear regression was conducted and four independent variables pH, total Zn, OM and total Cd significantly influenced the accumulation of Cd in Pak choi plants (Table 7). Both the multiple correlation and partial regression coefficients reached the statistically significant levels at least the 0.05. For multiple linear regression analyses, R^2 values could be used to explain variation of the dependents [12]. It was found that R^2 value was above 0.97, which means that more than 97% of variation in Cd concentration in Pak choi shoots could be attributed to soil pH, total Zinc, OM and total Cd contents in soils (Table 7).

The influence of each factor on Cd concentration of Pak choi (Y) shoots could be further explained by the values of each coefficient [12]. Stepwise regression model revealed that Cd concentration in the Pak choi was enhanced by lower soil pH (negative coefficients showed negative effect and vice versa), total Zinc, OM contents and higher total soil Cd. Lower soil pH, zinc, OM and higher soil total Cd are among the factors which enhance the bioavailability Cd contents in soils. Therefore, these three variables had the contradictory effect on Cd phytoavailability to Pak choi. Wang *et al.* [12] reported that soil characteristics (e.g. pH, CEC and OM) affected the phytoavailability of different heavy metals in soils, and such influences could be considered in the assessment of phytoavailability of heavy metals. There are four parameters involved in this model and then interactions between them were obvious (e.g. Cd concentration in the extractable fraction was correlated with lower soil pH, soil zinc and OM content). Furthermore, the coefficients obtained in the present model can regulate these cross effects and result in an improved model fitting. For example, there was a negative correlation between the soil pH, Zinc and OM, these factors had an inverse effect on Cd phytoavailability and soil Cd was the leading factor influencing Cd phytoavailability to Pak choi (coefficient of soil Cd was greater than those of pH, Zinc and OM). Our results about soil Cd and pH are in accordance with our recent study which developed an empirical model to correlate the Cd phytoavailability to rice with several soil properties. Soil pH and bioavailable soil Cd were major influencing factors which (pH negatively and soil Cd positively) correlate with the Cd phytoavailability, however total Zn and OM were not included in our previously developed model [47]. Eriksson and Soderstrom [51] reported that the Cd concentration of wheat grain grown on non-calcareous soils of Sweden was positively correlated to soil total Cd and negatively to extractable Zn. A study was conducted on Cd contaminated soils in Taiwan, whereas regression equation was developed to predict Cd concentrations in rice roots by available fractions of Cd and Zn in soil [52]. The negative coefficient of Zn indicated that soil Zn suppressed the uptake of Cd by rice roots in all varieties as Zn has an antagonistic effect on Cd uptake by root [53]. Oliver *et al.* [54] also observed a significant decrease of Cd up to 50% in wheat grain when 2.5–5.0 kg Zn ha^{-1} was applied to Cd contaminated Australian soils.

Organic matter content was negatively correlated with the accumulation of Cd in Pak choi shoots (Table 7). Organic matter plays an important role in determining the bioavailability and mobility of heavy metals in soils. Organic matter is involved in supplying organic chemicals to the soil solution, which may act as chelates and increase metal bioavailability to plants [55]. However, OM could reduce the bioavailability of heavy metals in soils by adsorption or forming stable complexes with humic substances [56]. Halim *et al.* [57] reported that addition of humic acid demonstrated a decrease in extractable heavy metal fraction in metal contaminated soils. This could partially explain the negative correlation of organic matter contents and Cd uptake observed in our present study.

Conclusions

The present study concludes that Cd concentration in Pak choi tissues was dependent on soil type. To establish the soil Cd thresholds of potential dietary toxicity from Pak choi, both Cd bioavailability in garden soils and Pak choi tissues should be taken into consideration. The selection of proper soil types for vegetable production can help us to avoid the toxicity of Cd in our daily diet. Stepwise regression model demonstrated that soil pH, organic matter, total Cd and Zinc contents may be the major factors having influence on the phytoavailability of Cd in different textured soils.

Acknowledgments

M.T. Rafiq acknowledges International Islamic University Islamabad Pakistan for providing a PhD scholarship under HEC-IIUI faculty development program.

Author Contributions

Conceived and designed the experiments: MTR XY TL. Performed the experiments: MTR RA WX. Analyzed the data: RA AS MA. Contributed reagents/materials/analysis tools: XY. Wrote the paper: MTR RA PJS.

References

1. Liu JG, Qian M, Cai GL, Yang JC, Zhu QS (2007) Uptake and translocation of Cd in different rice cultivars and the relation with Cd accumulation in rice grain. J Hazard Mater 143: 443–447.
2. Mahler RJ, Bingham FT, Page AL (1978) Cadmium-enriched sewage sludge application to acid and calcareous soils: Effect on yield and cadmium uptake by lettuce and Swiss chard. J Environ Qual 7: 274–281.
3. Stohs SJ, Bagchi D, Hassoun E, Bagchi M (2000) Oxidative mechanisms in the toxicity of chromium and cadmium ions. J Environ Pathol Toxicol Oncol 19: 201–213.
4. Shentu J, He Z, Yang XE, Li TQ (2008) Accumulation properties of cadmium in a selected vegetable –rotation system of south eastern China. J Agric Food Chem 56: 6382–6388.
5. IARC (1993) (International agency for research on cancer), Monographs on the evaluation of the carcinogenic risks to humans beryllium, cadmium, mercury and exposures in the glass manufacturing industry. IARC, Scientific Publications. Lyon, France. 119–238.
6. Li ST, Liu RL, Wang M, Wang XB, Shan H, et al. (2006) Phytoavailability of cadmium to cherry-red radish in soil applied composted chicken or pig manure. Geoderma 136: 260–271.
7. Lebeau T, Bagot D, Jezequel K, Fabr B (2002) Cadmium biosorption by free and immobilized microorganisms cultivated in a liquid soil extract medium: Effects of Cd, pH, and techniques of culture. Sci Total Environ 291: 73–83.
8. Abdola M, Chmtelnicka J (1990) New aspects on the distribution and metabolism of essential trace elements after dietary exposure to toxic metals. Biol Trace Element Res 23: 25–53.
9. Yang JX, Guo HT, Ma YB, Wang LQ, Wei DP, et al. (2010) Genotypic variations in the accumulation of exhibited by different vegetables. J Environ Sci 22: 1246–1252.
10. Yang Y, Zhang FS, Li HF, Jiang RF (2009) Accumulation of cadmium in the edible parts of six vegetable species grown in Cd-contaminated soils. J Environ Manag 90: 1117–1122.
11. Chen HS, Huang QY, Liu LN, Cai P, Liang W, et al. (2010) Poultry manure compost alleviates the phytotoxicity of soil cadmium: Influence on growth of Pak choi (*Brassica chinensis* L.). Pedosphere 20: 63–70.
12. Wang XP, Shan XQ, Zhang SZ, Wen B (2004) A model for evaluation of the phytoavailability of trace elements to vegetables under field conditions. Chemosphere 55: 811–822.
13. Liang Z, Ding Q, Wei D, Li J, Chen S, et al. (2013) Major controlling factors and predictable equations for Cd transfer factor involved in soil-spinach system. Ecotox Environ Saf 93: 180–185.
14. Ge Y, Murray P, Hendershot WH (2000) Trace metal speciation and bioavailability in urban soils. Environ. Pollut. 107, 137–144.
15. Gu JG, Zhou QX, Wang X (2003) Reused path of heavy metal pollution in soils and its research advance. J Basic Sci Eng 11: 143–151 (in Chinese).
16. Gu JG, Zhou QX (2002) Cleaning up through phytoremediation: A review of Cd contaminated soils. Ecol Sci 21: 352–356. (in Chinese with English abstract)
17. Du TP (2005) Food safety and strategy in China. Productivity Res: 6, 139–141. (in Chinese with English abstract)
18. Franz E, Römkens P, Van Raamsdonk L, Van Der Fels-Klerx I (2008) A chain modeling approach to estimate the impact of soil cadmium pollution on human dietary exposure. J Food Protect: 71, 2504–13.
19. Kobayashi E, Suwazono Y, Dochi M, Honda R, Nishijo M, et al. (2008) Estimation of benchmark doses as threshold levels of urinary cadmium, based on excretion of β_2-microglobulin in cadmium polluted and non-polluted regions in Japan. Toxicol Lett 179: 108–12.
20. Tracy S, Sheila M (2006) Cadmium and zinc accumulation in soybean: A threat to food safety. Sci Total Environ 371: 63–73.
21. Ide G, Becker B (1995) Relationship between the arsenic concentration in soil and in the cropped vegetables. International Conference on Heavy Metal in the Environment, Hamburg, vol. 2, CEP Consultants Ltd, Norwich, UK 302–304
22. Szteke B, Jedrzejczak R (1995) The variability of heavy metal contents in plant and soil samples from fields of one farm. International Conference on Heavy Metal in the Environment, Hamburg, vol. 2, CEP Consultants Ltd., Norwich, UK 228–231.
23. Salvatore M, Carratù G, Carafa A (2009) Assessment of heavy metals transfer from a moderately polluted soil into the edible parts of vegetables. J Food Agric Environ 7: 683–688.
24. Chaturvedi R, Sankar K, (2006) Laboratory manual for the physicochemical analysis of soil, water and plant. Wildlife Institute of India, Dehradun, India.
25. Hendershot WH, Duquette M (1986) A simple barium chloride method for determining cation exchange capacity and exchangeable cations. Soil Sci Soc Am J 50: 605–608.
26. Rashid A, Ryan J, Estefan G (2001) Soil and plant analysis laboratory manual. International center for agricultural research in the dry areas (ICARDA), Aleppo, Syria.
27. Day PR (1965) Particle fractionation and particle-size analysis. In: Klute, A. (eds.), Methods of soil analysis. ASA and SSSA, Madison, WI, pp. 545–567.
28. Xiao W, Yang XE, He Z, Rafiq MT, Hou D, et al. (2013) Model for evaluation of the phytoavailability of chromium (Cr) to rice (*Oryza sativa* L.) in representative Chinese soils. J Agric Food Chem 61: 2925–2932.
29. Mehlich A (1984) Mehlich-3 soil test extractant a modification of Mehlich-2 extractant. Commun Soil Sci Plan 15: 1409–1416.
30. FAO/WHO (2003) Report of the sixty first meeting of Joint FAO/WHO expert committee on food additives. Rome.
31. Wang XL, Sato T, Xing BS, Tao S (2005) Health risks of heavy metals to the general public in Tianjin, China via consumption of vegetables and fish. Sci Total Environ 350: 28–37.
32. Rattan R, Datta S, Chhonkar P, Suribabu K, Singh A (2005) A Long-term impact of irrigation with sewage effluents on heavy metal content in soils, crops and groundwater: A case study. Agric Ecosyst Environ 109: 310–322.
33. Peter MC (2002) Ecological risk assessment (ERA) and hormesis. Sci Total Environ 288: 131–140.
34. Liu X, Peng K, Wang A, Lian CL, Shen ZG (2010) Cadmium accumulation and distribution in populations of *Phytolacca Americana* L. and the role of transpiration. Chemosphere 78: 1136–1141.
35. Peter N, Karoly B, Laszlo G (2003) Characterization of the stimulating effect of low-dose stressors in maize and bean seedlings. J Plant Physiol 160: 1175–1183.
36. Kaminek M (1992) Progress in cytokinin research. TIBTECH 10: 159–162.
37. Ali B, Tao QJ, Zhou YF, Gill RA, Ali S, et al. (2013) 5-aminolevolinic acid mitigates the cadmium-induced changes in *Brassica napus* as revealed by the biochemical and ultra-structural evaluation of roots. Ecotox Environ Safe 92: 271–280.
38. Ali B, Wang B, Ali S, Ghani MA, Hayat MT, et al. (2013) 5-Aminolevulinic acid ameliorates the growth, photosynthetic gas exchange capacity and ultrastructural changes under cadmium stress in *Brassica napus* L. J Plant Growth Regul 32: 604–614.
39. Moya JL, Ros R, Picazo I (1993) Influence of cadmium and nickel on growth, net photosynthesis and carbohydrate distribution in rice plants. Photosyn Res 36: 75–80.
40. Weitao L, Zhou Q, Ana J Suna Y, Liu R (2010) Variations in cadmium accumulation among Chinese cabbage cultivars and screening for Cd-safe cultivars. J Hazard Mater 173: 737–743.
41. Lai HY, Chen BC (2013) The dynamic growth exhibition and accumulation of cadmium of Pak choi (*Brassica campestris* L. ssp. chinensis) grown in contaminated soils. Int J Environ Res Public Health 10: 5284–5298.
42. Cui YJ, Zhu YG, Smith SA, Smith SE (2004) Cadmium uptake by different rice genotypes that produce white or dark grains. J Environ Sci 16: 962–967.
43. Xiao W, Yang XE, Zhang Y, Rafiq MT, He Z, et al. (2013) Accumulation of chromium in Pak choi (*Brassica chinensis* L.) grown on representative Chinese soils. J Environ Qual 42: 758–765.
44. De Villiers S, Thiart C, Basson NC (2010) Identification of sources of environmental lead in South Africa from surface soil geochemical maps. Environ Geochem Health 32: 451–459.
45. Kabata-Pendias A, Pendias H (2001) Trace Elements in Soils and Plants. Boca Raton, FloridaCRC Press.
46. Laura W, Wander M, Phillips E (2011) Testing and educating on urban soil lead: A case of Chicago community gardens. J Agric Food Sys Comm Develop, ISSN: 2152–0801 online.
47. Rafiq MT, Aziz R, Yang XE, Wendan X, Rafiq MK, et al. (2014) Cadmium phytoavailability to rice (*Oryza sativa* L.) grown in representative Chinese soils. A model to improve soil environmental quality guidelines for food safety. Ecotox Environ Safe 103:101–107.
48. Murakami M, Nakagawa F, Ae N, Ito M, Arao T (2009) Phytoextraction by rice capable of accumulating Cd at high levels: Reduction of Cd content of rice grain. Environ Sci Technol 43: 5878–5883.
49. Jung MC, Thornton I (1996) Heavy metal contamination of soils and plants in the vicinity of a lead-zinc mine, Korea. Appl Geochem 11: 53–59.
50. McBride M (2002) Cadmium uptake by crops estimated from soil total Cd and pH. Soil Sci 15: 84–92.
51. Eriksson JE, Sderstrom M (1996) Cadmium in soil and winter wheat grain in southern Sweden. Factors influencing Cd levels in soils and grain. Acta Agric Scand Sect B 46: 240–248.
52. Romkens PFAM, Guo HY, Chu CL, Liu TS, Chiang CF, et al. (2009) Prediction of cadmium uptake by brown rice and derivation of soil-plant transfer models to improve soil protection guidelines. Environ Pollut 157: 2435–2444.
53. Giordano PM, Mays DA, Behel AD (1979) Soil temperature effects on the uptake of cadmium and zinc by vegetables grown on sludge amended soil. J Environ Qual 8: 232–236.
54. Oliver DP, Hannam R, Tiller KG, Wilhelm NS, Merry RH, et al. (1994) The effects of zinc fertilization on cadmium concentration in wheat grain. J Environ Qual 23: 705–711.
55. McCauley A, Jones C, Jacobsen J (2009) Soil pH and Organic Matter. Nutrient management modules 8, #4449-8. Montana State University Extension Service, Bozeman, Montana, pp. 1–12.
56. Liu LN, Chen HS, Cai P, Liang W, Huang QY (2009) Immobilization and phytotoxicity of Cd in contaminated soil amended with chicken manure compost. J Hazard Mater 163: 563–567.
57. Halim M, Conte P, Piccolo A (2003) Potential availability of heavy metals to phytoextraction from contaminated soils induced by exogenous humic substances. Chemosphere 52: 265–275.

22

Root Interactions in a Maize/Soybean Intercropping System Control Soybean Soil-Borne Disease, Red Crown Rot

Xiang Gao[1], Man Wu[1], Ruineng Xu[1], Xiurong Wang[1], Ruqian Pan[1], Hye-Ji Kim[2], Hong Liao[1]*

1 State Key Laboratory for Conservation and Utilization of Subtropical Agro-bioresources, Root Biology Center, South China Agricultural University, Guangzhou, China, 2 Department of Tropical Plants and Soil Sciences, College of Tropical Agriculture and Human Resources, University of Hawaii at Manoa, Honolulu, Hawaii, United States of America

Abstract

Background: Within-field multiple crop species intercropping is well documented and used for disease control, but the underlying mechanisms are still unclear. As roots are the primary organ for perceiving signals in the soil from neighboring plants, root behavior may play an important role in soil-borne disease control.

Principal Findings: In two years of field experiments, maize/soybean intercropping suppressed the occurrence of soybean red crown rot, a severe soil-borne disease caused by *Cylindrocladium parasiticum* (*C. parasiticum*). The suppressive effects decreased with increasing distance between intercropped plants under both low P and high P supply, suggesting that root interactions play a significant role independent of nutrient status. Further detailed quantitative studies revealed that the diversity and intensity of root interactions altered the expression of important soybean *PR* genes, as well as, the activity of corresponding enzymes in both P treatments. Furthermore, 5 phenolic acids were detected in root exudates of maize/soybean intercropped plants. Among these phenolic acids, cinnamic acid was released in significantly greater concentrations when intercropped maize with soybean compared to either crop grown in monoculture, and this spike in cinnamic acid was found dramatically constrain *C. parasiticum* growth *in vitro*.

Conclusions: To the best of our knowledge, this study is the first report to demonstrate that intercropping with maize can promote resistance in soybean to red crown rot in a root-dependent manner. This supports the point that intercropping may be an efficient ecological strategy to control soil-borne plant disease and should be incorporated in sustainable agricultural management practices.

Editor: Ching-Hong Yang, University of Wisconsin-Milwaukee, United States of America

Funding: This work was supported by the National Natural Science Foundation of China (grant no. 31025022) and National Key Basic Research Special Funds of China (grant no. 2011CB100301). The funders had no role in study design, data collection and analysis, decision to publish, or preparation of the manuscript.

Competing Interests: The authors have declared that no competing interests exist.

* E-mail: hliao@scau.edu.cn

Introduction

Reducing disease severity through increasing the within-field multiple crop species, intercropping has been widely reported in the literature and commercially implemented [1]. For example, disease-susceptible rice varieties have been grown in combination with resistant varieties to achieve higher yield and greatly suppress blast severity compared to growth in monoculture [2]. Also, growing mixtures of resistant oats and susceptible barley results in decreased stem rust severity, and thereby increases yield [3]. However, the mechanisms underlying these heterogeneous genotype and crop species effects on disease control remain unclear.

Soil-borne diseases are caused by phytopathogenic microbes in the soil, which may damage plants upon penetration of the root or basal stem [3]. With a potential to seriously harm the agroecosystem, and with broad geographic ranges, soil-borne diseases can significantly impact global food production [4]. Furthermore, eradication of the pathogen is difficult as long as soil-borne disease pathogens accumulate in cultivated soils. To date, effective methods to control soil-borne diseases remain lacking. Therefore, the simple and economic alternative of agricultural management is often a preferred method for preventing crop losses caused by soil-borne diseases. It has long been recognized that crop heterogeneity suppresses soil-borne disease infection and appears to be a more promising solution than petrochemical-based pesticide application. Among the successful examples of this strategy are inhibition of tomato bacterial wilt caused by *Pseudomonas solanacearum* in a Chinese chive-tomato intercropping system [5], suppression of Fusarium wilt in rice with watermelon intercropping [6,7], and alleviation of tomato early blight disease by intercropping with marigold [8].

One potential target for control through crop heterogeneity is red crown rot, which is one of the most severe soil-borne diseases for plants. It is caused by the fungus *Cylindrocladium parasiticum* (*C. parasiticum*, teleomorph *Calonectria ilicicola*) and is named after the reddish brown basal stalk tissue of infected plants [9–11]. This pathogen has a wide host range, and red crown rot has been found in many countries with severe crop losses possible throughout its distribution. Soybean (*Glycine max* L. Merr.), an economically important host of red crown rot, is a major cash crop and has been planted world wide. In South China, estimates for soybean yield loss due to red crown rot range to as high as 50% [11]. At present, there are no effective pesticides to control red crown rot.

Intercropping has been widely used in Asia, Latin America and Africa to enhance crop productivity through improved nutrient uptake, enhanced land equivalent ratio (LER), increased yield, and better disease control [12,13]. Legume/cereal intercropping is the most common intercropping system, which not only allows crops to take up more nutrients than in monoculture, but also reduces disease occurrence [14,15]. However, little attention has been paid to the belowground environment of the plant root systems. As the main organ to contact the soil and perceive signals from neighboring plants [16], root function and behavior should be particularly important in soil-borne disease control. In one case, Fang et al. [17] reported that root interactions in the maize/soybean intercropping can integrate responses to P status and root behaviors among neighboring plants, and thereby impact maize P nutrition. A previous study by the current authors found that co-inoculation with rhizobia and mycorrhizal fungi can inhibit soybean red crown rot [18]. However, uncertainty remains about what happens when only mycorrhizal fungi, which favor maize over soybean, are introduced. In this study, effects of maize intercropping with soybean on red crown rot in soybean were studied in two years of field experiments. Further in-door experiments were then carried out to evaluate the underlying mechanisms for inhibition of soybean red crown rot by intercropping with maize.

Methods and Materials

1. The Boluo (E114.28°, N23.18°) field site is the experimental base of the South China Agricultural University Root Biology Center. Therefore, the authority that issued the permit for this location is South China Agricultural University.
2. No specific permissions were required for this location. I am here to confirm that the field studies did not involve endangered or protected species.

Field trials

Field experiments were carried out on acidic lateritic red soils at the Boluo field site (E114.28°, N23.18°) in the Guangdong Province of China from July to October in 2009 and 2010 with the average day/night temperature around 25–33°C and monthly rainfall around 150 mm. Basic soil chemical characteristics were as follows: pH, 5.37; organic matter, 17.63 g kg^{-1}; available P (Bray I method), 15.68 mg P kg^{-1}; available N, 86.64 mg N kg^{-1}; available K, 75.28 mg K kg^{-1}. The field site had a history of up to 10 years of continuous soybean cultivation, in which severe soybean red crown rot (RCR) damage occurred [11]. Two years of field trials were conducted in a split-block design with two P supplies (high P and low P), and four cultivation modes (soybean monoculture, maize monoculture, two maize/soybean intercropping systems) for a total of 8 treatments (Figure A in File S1). High P (HP) was implemented by adding calcium superphosphate (SSP) at the rate of 80 kg P_2O_5 ha^{-1}, and low P (LP) plots had no P fertilizer added. The two maize/soybean intercropping systems differed in planting distances between soybean and maize plants, which were 20 cm for ISC1 and 5 cm for ISC2. Urea and KCl were added to all plots as 80 kg N ha^{-1} and 60 kg K ha^{-1} for supplementing N and K supplies. Each treatment had four replicates with a total of 32 plots (i.e. 2 P × 4 mode × 4 replicates). The planting area of each plot was 54 m^2. Soybean (*Glycine max*) variety HN89 and maize (*Zea mays*) variety ZD958 were employed in the field experiment.

Seventy days after planting, 100 plants were selected from each plot to investigate the severity of RCR. Disease incidence and severity caused by *C. parasiticum* were determined according to Gao et al. [18]. Disease incidence was defined as the percentage of infected subterranean stems. Disease severity was recorded on a 0–5 scale: i.e., 0, no visible symptoms; 1, small necrotic lesions on the subterranean stem; 2, necrotic lesions extending around the subterranean stem, 3, necrotic lesions extending to the soil surface; 4, severe necrosis on the subterranean stem and roots, as well as severe chlorosis appearing in leaves; 5, plant death. The disease index was summarized within each plot as $\{[(n_1\times1)+(n_2\times2)+(n_3\times3)+\ldots+(n_N\times N)]/[N\times(n_1+n_2+n_3\ldots+n_N)]\}\times100$, where $n_1\ldots n_N$ was the number of subterranean stems in each of the respective disease category, and N was the highest disease severity score [2,18].

Seventy days after planting, two representative healthy soybean plants from each plot were harvested for measuring root length and symbiotic traits. Total root length in the 20-cm upper soil layer of the 5-cm strip between intercropped plants was measured to represent the intensity of interacting roots between the two crop species in the field as described by Fang et al. [16]. In short, trenches were dug by shoveling between rows of intercropped plants. The walls were carefully scraped with a screwdriver to reveal the tips of the roots. Plastic transparent sheets (25×30 cm) were positioned adjacent to the exposed soil wall. The roots in the 20 cm upper soil layer were marked on the sheets, and the root length in the 5 cm section between the two plants was quantitatively measured with root image analysis software (WinRhizo Pro, Régent Instruments). Arbuscular mycorrhizal fungi (AMF) colonization rates and nodule numbers on soybean roots were measured according to Wang et al. [19].

At maturity, soybean biomass and yield were average for 20 healthy and 20 infected plants within each soybean monoculture plot. Growth reduction was calculated as: reduction (%) = (yield/biomass of healthy plants - yield/biomass of infected plants) yield/biomass of healthy plants × 100. This value was determined for each of the four replicates separately.

Root barrier sand culture

Root barrier sand culture experiments were conducted in the greenhouse of Root Biology Center of South China Agricultural University in 2011 using soybean variety HN89 and maize variety ZD958. *Cylindrocladium parasiticum* (*C. parasiticum*, GenBank Accession No. GU073284) was isolated from infected soybean roots at the Boluo field site and used as the pathogen inoculant [11]. The sand culture experiment consisted of 3 factors (P level, root barrier and pathogen inoculation), including two P levels, three root barrier arrangements and two pathogen inoculations, for a total of 12 treatments. There were 6 replicates for each treatment. The two P levels were HP (500 μM P) and LP (15 μM P). Plastic pots with two equal compartments were used to provide three types of root barriers as illustrated in Figure A in File S1 and described as follows: solid barrier to eliminate root interaction and exudates movement, a nylon mesh barrier (30 μm pores) to prevent interspecies root intermingling while permitting root exudates exchange, and no barrier to allow roots and exudates to completely interact. Soybean plants were inoculated with *C. parasiticum* infected wheat seeds or sterilized media as a control before sowing. Wheat seeds of infected *C. parasiticum* were obtained from 14-day-old media inoculated with *C. parasiticum*. The spore concentration was determined using a hemocytometer and adjusted to 1×10^5 spores per mL [18]. Before sowing, infected wheat seeds were placed at depth of 5 cm in sand culture pots. Two independent biological experiments were conducted, with very similar results between experiments, so only the results from one experiment were presented.

Plants were irrigated daily using modified 1/2 strength Hoagland nutrient solution with either of the two P additions listed above [20]. Sixty days after planting, incidence and severity of red crown rot was measured as described above. In order to determine the root infection by *C. parasiticum*, root colony forming units (CFUs) were measured from sand culture plants. Randomly selected soybean roots were cut into 1 cm segments and washed thoroughly and surface-sterilized with 0.5% (v/v) NaClO for 3 min, rinsed three times in sterilized water, and then blotted on sterilized filter paper. About 1 g root segments were ground in 50 mL of sterile deionized water using a blender at high speed for 1 min. The homogenized root suspension was diluted 10 fold with sterile deionized water. 100 μL aliquots of diluted solutions were evenly spread on 4 plates containing Rose Bengal Medium using a sterile glass rod. There were 4 replicates and thus 16 plates for each treatment. All plates were incubated at 28°C in the dark for 4 days. Colonies of *C. parasiticum* were identified and counted on each plate to determine the CFUs per gram of roots for each treatment [18].

Another sand culture experiment was carried out for pathogen-related (*PR*) gene expression analysis. Except for the inoculation process, all the treatments were the same as above. Plants were grown in sand for thirty days without inoculation, and then inoculated with *C. parasiticum*. Spore suspensions of *C. parasiticum* were obtained from 14-day-old V8-juice media which were collected by adding 10 mL of sterile water to each Petri dish and rubbing the surface with a sterile L-shaped spreader. The suspension was subsequently filtered through 3-layers of cheesecloth. The spore concentration was determined using a hemacytometer and adjusted to 1×10^5 spores per mL [18]. Thirty days after planting, plant stem base was infected with 20 mL *C. parasiticum* spore suspensions each pot. Three soybean roots from each pot were randomly harvested at 1 and 5 days after inoculation (DAI) to detect the expression pattern of eight related *PR* genes in response to inoculation using quantitative real time PCR (RT-PCR) described by Gao et al. [18]. Specific primer sequences and putative functions of tested plant *PR* genes were listed in the Table S1 in File S2. Simultaneously, the activities of polyphenol oxidase (PPO) and phenylalanine ammonia-lyase (PAL) were determined based on methods reported in Song et al. [21].

Root exudates composition and *in vitro C. parasiticum* assays

A hydroponic experiment was carried out to collect root exudates in 2012. Soybean and/or maize seedlings were grown in half strength Hoagland nutrient solution with either of the two P additions (LP and HP) as described above. The cultivation modes included monoculture maize (MC) or soybean (MS), and intercropping of soybean and maize (ISC). There were six treatments (three cultivation modes by 2 P levels) with four replicates. Thirty days after planting, root exudates from each treatment were collected and processed, and a bioassay for colony diameter growth and sporulation was conducted as described in Gao et al. [18]. Roots were gently removed from pots and washed with deionized water. Cleaned roots were submerged in a plastic cup containing 500 mL of 0.5 μM $CaCl_2$ to collect exudates for 6 hours. Root exudates were filtered through a 0.45 μm Millipore membrane and stored at −20 °C. During collection, each cup containing 3 plants was covered by a black plastic lid to avoid contamination and light [7,22]. Pathogen growth diameter and sporulation was quantified by adding 2 mL root exudates to V8-juice medium before it solidified to yield a total volume of 20 mL per Petri dish. Plates were incubated at 28 °C in the dark. Colony diameter and sporulation of *C. parasiticum* were determined 5 and 14 days after incubation using ruler and hemocytometer [18]. Phenolic compounds in the root exudates were identified using an HPLC system (SPD-20A, Shimadzu, Tokyo) with gallic acid, p-coumaric acid, phthalic acid, vanillic acid, syringic acid, ferulic acid, salicylic acid and cinnamic acid included as standard phenolic compounds [22]. According to the above HPLC analyses results, the dominant phenolic acids, including ferulic, gallic, *p*-coumaric, cinnamic and salicylic acids (Sigma, USA), were subsequently applied exogenously at five concentrations, 0, 10, 20, 30 and 40 mg/L, in V8-juice media to test for allelopathic effects on growth and sporulation of *C. parasiticum* as described above.

Data analysis

Data were statistically analyzed by Two-way ANOVA using Excel 2003 (Microsoft Corporation, 1985–2003) and SAS 9.1 (SAS Inc., Cary, NC, USA) for multiple comparisons.

Results

Reduction of soybean growth and yield caused by red crown rot in the field

Red crown rot significantly inhibited soybean growth, and subsequently reduced soybean yield in comparison to uninfected plants in field experiments (Figure B in File S1). Compared to healthy plants, infected soybean biomass in the monoculture treatment was reduced in 2009 by 23% and 27%, and in 2010 by 28% and 43% at LP and HP, respectively. Soybean yield was reduced in 2009 by 21% and 33%, and in 2010 by 17% and 44% at LP and HP, respectively. These observations illustrate the severe damage potential of red crown rot on soybean, particularly with high P supply.

Disease severity of soybean red crown rot in the field

Intercropping significantly reduced disease incidence and severity index of soybean red crown rot, and high P slightly enhanced red crown rot in both monoculture and intercropping (Table 1). The planting distance of soybean and maize plants also significantly affected the disease severity as indicated by the lowest disease incidence and index occurring when soybean was grown in close proximity with maize (ISC2, 5 cm apart). Compared to monoculture soybean (MS), the disease incidence in ISC2 was reduced by 53% and 59% in 2009, and by 43% and 47% in 2010 at LP and HP, respectively. The disease index in ISC2 was decreased by 49% and 46% in 2009, and by 57% and 50% in 2010, at LP and HP, respectively (Table 1). This suggests that the occurrence and development of red crown rot on soybean growth can be alleviated by intercropping with maize, with particular attention paid to planting distance.

Mycorrhization, nodulation and root interactions in the field

Co-inoculation with AMF and rhizobia has been shown to alleviate soybean red crown rot [18]. However, in this study, the mycorrhization rate and nodule numbers were not affected by intercropping or planting distance (Table S2 in File S2), which

Table 1. Disease severity of soybean caused by *C. parasiticum* as affected by cultivation mode, P level and planting distance in the two years of field experiments on acid soils.

		Incidence (%)		Disease index	
		LP	HP	LP	HP
2009	MS	43±3.4Ab	62±5Aa	33±3Aa	40±3Aa
	ISC1	32±2Ba	39±4Ba	26±1Ab	32±2Ba
	ISC2	20±2Ca	26±2Ca	17±1Ba	22±3Ca
2010	MS	45±3Ab	58±4Aa	36±3Ab	48±2Aa
	ISC1	33±1Bb	41±2Ba	26±2Bb	36±2Ba
	ISC2	26±2Ca	31±1Ca	16±1Cb	24±2Ca

Note: Disease incidence and index were measured as described in Materials and Methods. HP: 80 kg P_2O_5 ha^{-1} added as calcium superphosphate, LP: no P fertilizer added. MS: soybean monoculture, ISC1: maize/soybean intercropping with 20 cm spacing; ISC2: maize/soybean intercropping with 5 cm spacing. All the data are the mean of four replicates ± SE. The same upper-case letter after numbers in the same column for the same trait in the same year indicates no significant difference among cultivation modes at 0.05 ($P<0.05$); The same lower-case letter after numbers in the same row for the same trait in the same year indicates no significant difference between two P levels at 0.05 ($P<0.05$).

indicates that neither mycorrhization nor nodulation are major contributors to the variation of soybean red crown rot observed in the field. In contrast, intermingled roots between the two different crop species were significantly affected by planting distance (Fig. 1). The total root length in the upper 20 cm soil layer of the 5 cm section centrally located between plants in ISC2 was greater in 2009 by 179% and 234% compared to ISC1 (20 cm apart), and in 2010 by 183% and 181% at LP and HP, respectively. Moreover, low P supply also affected root interactions. The root length in the upper, middle area at low P compared to high P increased by 33% and 24% in ISC1, and by 11% and 25% in ISC2 in 2009 and 2010, respectively (Fig. 1). This suggests that the intensity of intermingled roots between soybean and maize plants might be the major player in reducing the disease severity of soybean red crown rot by maize/soybean intercropping.

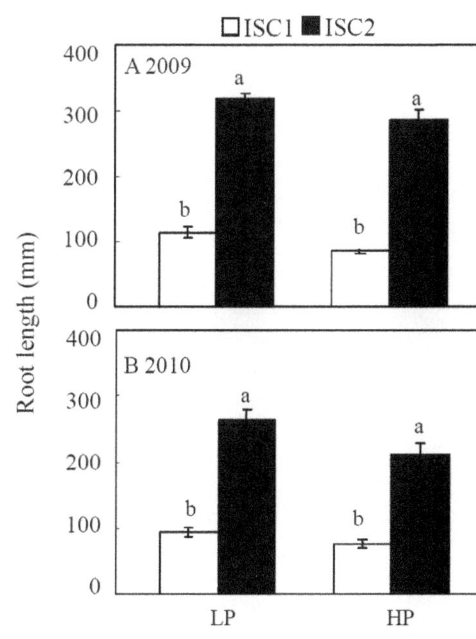

Figure 1. Root interactions between soybean and maize plants in the field. Root interactions were measured as the total root length in the upper 20 cm soil layer of the 5 cm section centrally located between the maize and soybean plants. HP: 80 kg P_2O_5 ha^{-1} added as calcium superphosphate, LP: no P fertilizer added. ISC1: maize/soybean intercropping with 20 cm spacing; ISC2: maize/soybean intercropping with 5 cm spacing. All the data are the mean of four replicates ± SE. F value from Two-way ANOVA: 8.35 for P treatment ($P<0.05$), 159.08 for cultivation mode ($P<0.001$), 1.99 for interaction (not significant). Bars with different letter(s) vary significantly among treatments as determined by Duncan's multiple range test ($P<0.05$).

Disease severity of soybean red crown rot in sand culture

The disease severity of soybean red crown rot was dramatically decreased with increasing root interactions in sand culture. Compared to the solid barrier treatment, both mesh and no barrier treatments had significantly lower disease incidences and indices, and fewer pathogen CFUs at both P levels (Fig. 2, Figure C in File S1). When soybean and maize roots completely interacted with each other (no barrier), the lowest disease incidences and indices and fewest pathogen CFUs were observed. An intermediate level of disease incidence was observed in plants grown with the mesh barrier treatment, and the highest level of disease was seen in plants separated by solid barriers. This is consistent with field results that an increase in root interactions reduced the occurrence of soybean red crown rot.

Expressions of important *PR* genes and activities of corresponding enzymes in sand culture

The results displayed in Figure 3 showed that transcript abundances of all tested *PR* genes in soybean roots were induced to a high level by inoculation with *C. parasiticum*, and this induction was enhanced by increasing root interactions. Except *PR4* on day 5 and *PR3*, *PR10*, *PR12* and *PAL* on day 1, the relative expression values of most tested *PR* genes were highest in the no barrier treatment, followed by the mesh barrier treatment. For example, *PPO* transcription abundance increased by 7.3- and 7.4- fold, and *PAL* by 12.9- and 4.1- fold in LP and HP, respectively, at 5 DAI in a comparison of the no barrier treatment with the solid barrier treatment (Fig. 3G and H). Furthermore, the enzyme activities of PPO and PAL in soybean roots were significantly enhanced by root interactions with maize (Figure D in File S1). PPO activity increased in the no barrier treatment compared to the solid barrier treatment at 1 DAI by 114% and 99%, and in the mesh treatment by 66% and 43% in LP and HP, respectively. Also, PAL activity increased in the no barrier treatment compared to the solid barrier

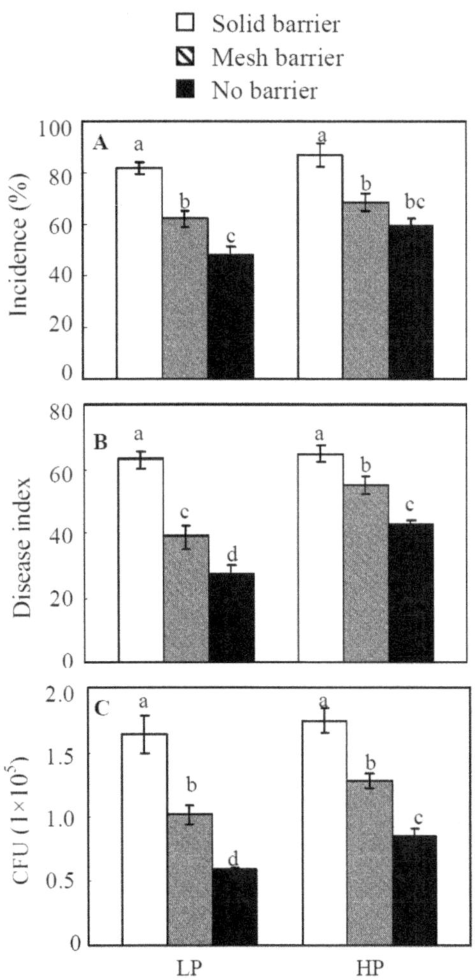

Figure 2. Disease severity of soybean red crown rot in sand culture. A, disease incidence, B, disease index, C, CFU. LP, 15 μM P; HP, 500 μM P. All of the data are the means of four replicates ±SE. Bars with different letter(s) vary significantly among the different inoculation treatments as determined by Duncan's multiple range test ($P<0.05$).

treatment at 1 DAI by 65% and 227%, and in the mesh treatment by 39% and 105% in LP and HP, respectively.

In vitro C. parasiticum assays using root exudates

In comparison with soybean monoculture (MS), the addition of root exudates from the maize/soybean intercropping (ISC) or maize monoculture (MC) dramatically inhibited pathogen growth (Fig. 4). At LP, colony diameter of *C. parasiticum* decreased by 54% and 38%, and sporulation declined by 37% and 27% when growth in MS exudates was compared to that in ISC or MC exudates, respectively. Furthermore, P supply also affected pathogen growth as indicated by the observed 34%, 31% and 49% increases in colony diameter in HP compared to LP when adding the root exudates from MS, MC and ISC, respectively. This suggests that maize roots can be triggered to secrete antagonistic substances against soybean pathogen growth in the intercropping system, which might be enhanced under low P supply.

HPLC analysis of root exudates

We investigated the composition of root exudates using HPLC analysis, and found that the composition of root exudates varied significantly among the two crop species and the different cultivation modes (i.e., monoculture of soybean or maize, intercropping with soybean and maize). Roots of maize plants released greater amounts of phenolic acids than those of soybean (Figure E in File S1). We further analyzed the presence and levels of 8 common phenolic acids in the root exudates using HPLC with chemical standards. This analysis identified 5 phenolic acids, including ferulic, gallic, *p*-coumaric, cinnamic and salicylic acids, in the root exudates from MC or ISC, but only gallic acid was detected in soybean monoculture root exudates (Table 2). The concentration of the phenolic acids other than gallic acid in root exudates increased when maize was intercropped with soybean. For example, cinnamic acid concentration in exudates from the ISC treatment compared to the MC treatment increased by 20% and 43% at LP and HP, respectively. Phosphorus supply also influenced the concentrations of most phenolic acids. Ferulic, *p*-coumaric and cinnamic acids in root exudates of the ISC grown plants exhibited 68%, 58% and 22% increases at LP compared to HP.

Effects of phenolic acids on *C. parasiticum* growth

To further investigate the effects of the above five phenolic acids on *C. parasiticum* growth and sporulation, we conducted an *in vitro C. parasiticum* assay using individual phenolic acids. The results showed that when the concentration reached 20 mg/L, all of the tested phenolic acids inhibited the growth and sporulation of *C. parasiticum* (Fig. 5). At lower concentrations, only cinnamic and ferulic acids inhibited *C. parasiticum* growth, while, lower concentration gallic and salicylic acids slightly stimulated *C. parasiticum* growth. Furthermore, the inhibitory effects of cinnamic acid on *C. parasiticum* growth increased with increasing concentration, and even completely stopped *C. parasiticum* growth when the concentration reached 40 mg/L. These results suggest that cinnamic acid is the most effective phenolic acid in suppressing *C. parasiticum* growth and development.

Discussion

It has been well documented that increasing crop species diversity can yield benefits in plant disease control [2,3]. However, most studies have focused on above-ground plant functions and performances. Few plant disease studies pay attention to the root, the main organ of plants to sense environmental attacks and related signals in the soil. To the best of our knowledge, this is the first report on interspecies signaling between root systems that results in improved control of soil-borne disease, with evidence coming both from field trials and greenhouse-laboratory experiments at both physiological and molecular levels.

Intercropping is the intermingled growth of at least two crops grown together in the same field, which is the most popular cultivation model for increasing crop genetic diversity [23]. It has been widely noted that intercropping systems can reduce disease damage [4,15,24], probably due to fewer attacks by disease organisms compared to when crops are grown in monoculture [2,24]. In the current study, based on two years of field experiments where soybean red crown rot was severe in monoculture, we found that intercropping soybean with maize significantly reduced disease severity of soybean red crown rot (Table 1). This was further confirmed in sand culture with different root barrier treatments. Most significantly, when root interactions were completely blocked, severe red crown rot still occurred in soybean plants, even when intercropped with maize (Fig. 2). This indicates that above-ground interactions in the maize/soybean intercropping system do not contribute to control of soybean red

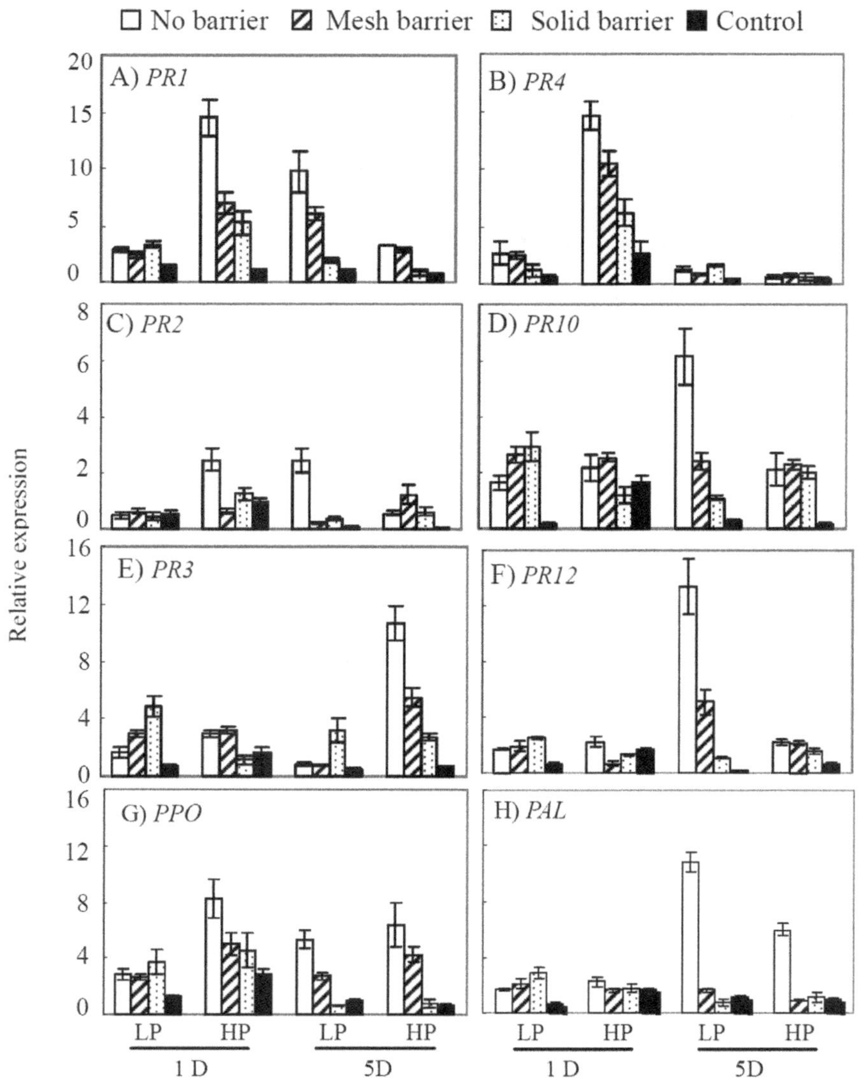

Figure 3. Expression of eight defense-related (*PR*) genes in soybean roots. LP: 15 µM P; HP: 500 µM P. Soybean roots were inoculated with *C. parasiticum* (see Materials and Methods for details). Solid barrier: eliminate root interactions and exudates movement between the roots of the two plant species; mesh barrier: prevent root intermingling of two species while permitting root exudates exchange; no barrier: allow roots and exudates to completely interact. Each bar represents the mean of three replicates ± SE.

crown rot. In contrast, the least severe red crown rot was recorded when the roots freely interacted with each other in the no barrier treatment, indicating that interactions among roots in intercropping is important in control of this soil-borne disease. Moreover, we also found that in the field, the reduction of soybean red crown rot was obviously enhanced with decreasing planting distance between soybean and maize plants, which increased interaction intensity between the two different crop species (Table 1 and Fig. 1). This finding suggests that besides genetic diversity, the intensity of root interaction is also important for control of soybean red crown rot in the maize/soybean intercropping system. This point is also supported by reports that in order to effectively control disease spread, roots of mixed crops need to be sufficiently intermingled [24,25].

Root exudates have been recognized to play vital roles in the rhizosphere [26–28] in large part due to continuous production and secretion of allelochemical compounds into the rhizosphere [29,30]. There are many reports attributing defense against phytopathogens to root exudates released by neighboring non-host plants [6–7,31]. In the present sand culture experiment, the mesh barrier treatment had less effects on disease control than the no barrier treatment, yet still significantly reducing the severity of soybean red crown rot (Fig. 2), implying that root exudates alone can inhibit soybean red crown rot. This was supported by *in vitro* assays where the significant growth inhibition of *C. parasiticum* was observed when root exudates were applied. Root exudates collected from intercropped soybean and maize inhibited *C. parasiticum* growth more effectively than from maize alone (Fig. 4), suggesting that intercropped maize releases more allelopathic compounds than in monoculture.

Phenolic acids have been well established as major rhizosphere allelochemicals that suppress phytopathogens [22,32], and they are detectable in root exudates by HPLC [7,29]. In order to identify the specific compounds inhibitory to *C. parasiticum* growth from the maize/soybean intercropping system, the 8 most common phenolic acids in root exudates were initially tested, with 4 phenolic acids specifically released by maize roots (Figure E in File S1). Among these 4 compounds, release of cinnamic acid

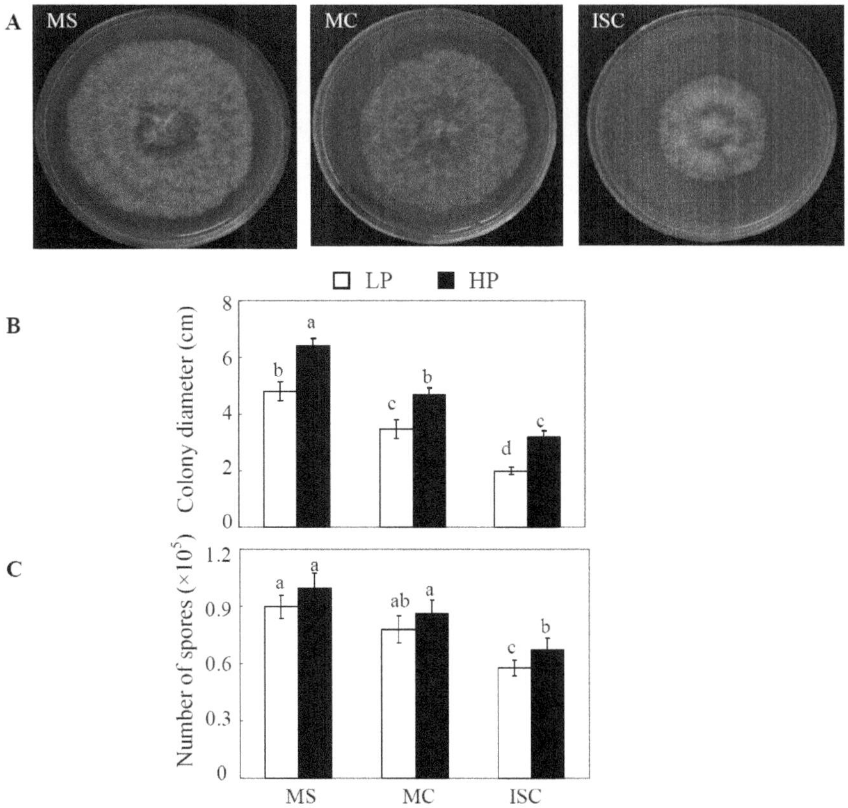

Figure 4. *C. parasiticum* growth as affected by root exudates. A, growth performance, B, colony diameter, C, sporulation. LP, 15 μM P; HP, 500 μM P. MS: soybean monoculture; MC: maize monoculture; ISC: maize/soybean intercropping. F value from Two-way ANOVA: colony diameter, 685.62 for P treatment ($P<0.001$), 1119.75 for cultivation mode ($P<0.001$), 12.73 for interaction ($P<0.001$); sporulation, 38.33 for P treatment ($P<0.001$), 116.49 for cultivation mode ($P<0.001$), 1.02 for interaction (not significant). Each bar represents the mean of four replicates ± SE. Bars with different letter(s) vary significantly among treatments as determined by Duncan's multiple range test ($P<0.05$).

was significantly enhanced by maize/soybean intercropping (Table 2), and this compound dramatically constrained *C. parasiticum* growth *in vitro* (Fig. 5). Cinnamic acid has been demonstrated to inhibit phytopathogen *Fusarium oxysporum* f.sp. *niveum in vitro* [7,22]. Therefore, we speculate that cinnamic acid is the primary player of interest in the tested root exudates, and it is directly responsible for the protection of soybean roots from the attack of *C. parasiticum* in the maize/soybean intercropping system. Since plant roots can exude a broad range of compounds, a more thorough search for more efficient phytotoxins from soybean and/or maize needs to be carried out in the future.

The *PR* related genes included pathogenesis-related, antimicrobial protein genes *PR1* (PR1a precursor), *PR2* (β 1–3 endoglucanase), *PR3* (chitinase class I), *PR4* (wound-induced protein), *PR10* (ribonuclease-like protein), *PR12* (defending-like protein), *PPO* (polyphenol oxidase) and *PAL* (phenylalanine ammonia-lyase) that are involved in phytoalexin biosynthesis [33,34]. Here we shown that the defense responses elicited by inoculation of pathogen and intercropping with maize. In legumes, many genes encoding PR

Table 2. Concentrations of phenolic acids ($\mu g \cdot pot^{-1}$) in root exudates from different treatments at the flowering stage as detected by HPLC.

Treatment		Ferulic acid	Gallic acid	*P*-coumaric acid	Cinnamic acid	Salicylic acid
	MS	-	6.53±1.59c	-	-	-
LP	MC	12.78±2.34b	29.08±0.80a	13.06±1.91b	20.98±2.51b	7.51±1.41a
	ISC	22.13±3.24a	14.72±2.84b	20.33±2.78a	25.11±2.52a	7.88±2.72a
	MS	-	11.12±2.88b	-	-	-
HP	MC	13.72±1.05a	13.09±1.58b	8.51±0.26a	14.41±2.86b	11.00±2.21a
	ISC	13.18±1.73a	21.65±2.99a	12.86±2.28a	20.59±2.78a	9.32±0.47a

Note: LP, 15 μM P; HP, 500 μM P. MS, soybean monoculture; MC, maize monoculture; ISC, maize/soybean intercropping. -, not detectable. All the data are the mean of four replicates ± SE. Data with the same letter represent no significant differences among the different cultivation modes at the same P level as determined by Duncan's multiple range test or a *t*-test ($P<0.05$).

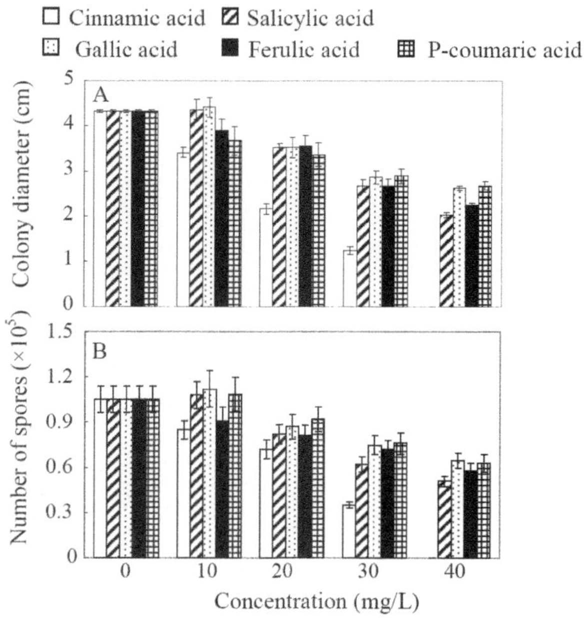

Figure 5. Effect of 5 different phenolic acids on the growth and sporulation of *C. parasiticum*. A, colony diameter, B, number of spores. Each bar represents the mean of four replicates ± SE.

proteins have been shown to be upregulated in plants following inoculation with several pathogens [33]. The upregulation of defense-related genes, such as *PR1* and *PR2* was associated with lesion limitation in soybean seedling roots inoculated with *P. sojae* and to the high level of partial resistance of cultivar Conrad [33]. Defense-related genes are upregulated in both compatible and incompatible interactions, but more rapidly and/or to higher levels in plant varieties with *PR* gene resistance [34]. In addition to producing allelopathic compounds directly, intercropped plants can promote specific plant defense reactions in neighboring plants to protect against phytopathogens [2,31]. Consistent with this, the results herein show that direct interactions with maize roots significantly induce the expression of important soybean *PR* genes and increase the activities of corresponding PR enzymes in soybean roots (Fig. 3, Figure C in File S1). This elicitation, via root exudates or signal compounds for specific plant defense reactions can predispose soybean plants to an early response to attack by root pathogens [24,31]. Infecting soybean roots after pre-treatment with maize enhances PR gene expression and enzyme activity, and therefore, might improve soybean's capability to defend against disease invasion. Furthermore, salicylic acid is a reported enhancer of plant defense pathways [35]. In this study, more salicylic acid was measured in root exudates from the maize/soybean intercropping system than in monoculture, which suggests that salicylic acid or a related component might also contribute to disease control through promotion of defense pathways.

Although adding additional P fertilizer in field and sand culture experiments increased plant growth, it also promoted disease incidence and index of red crown rot (Table 1 and Fig. 2). Likewise, the application of P has been shown to increase the severity of diseases in many plant species [18,36]. Here, we demonstrate that higher P addition is associated with more severe soybean red crown rot (Table 1 and Fig. 2). Furthermore, high P induces colony *C. parasiticum* growth and proliferation *in vitro* (Fig. 4). Plant roots and phytopathogens most likely compete directly for resources (e.g. P) in the rhizosphere, and therefore low P supply could inhibit phytopathogen growth, especially if the roots more effectively acquire the soil P. Consequently, optimal fertilizer application should be carefully considered, especially in acid soils where P is limiting.

To the best of our knowledge, this study is the first report to demonstrate that intercropping with maize can promote disease resistance in soybean to red crown rot in a root interaction dependent manner. Direct root interactions of soybean with maize may inhibit pathogen growth and reproduction, while enhancing expression of *PR* genes and the activities of corresponding enzymes in host plants. This study suggests that soybean and maize intercropping is an effective tool to sustainably control soil-borne disease and should be incorporated in agricultural management practices.

Supporting Information

File S1 Supporting figures. Figure A. Planting arrangement of diagram and actual pictures in monoculture or intercropping of soybean and maize. **A**, Planting arrangement in the field. **B**, Three root barriers were set-up in sand culture, either with a solid barrier (left) to eliminate root interactions and exudates movement, a nylon mesh barrier (middle) to prevent root intermingling of two species while permitting root exudates exchange, and no barrier (right) to allow roots and exudates to completely interact. **Figure B.** Soybean growth and grain yield as affected by red crown rot. HP: 80 kg P_2O_5 ha^{-1} added as calcium superphosphate, LP: no P fertilizer added. The reduction of growth parameters were calculated as follows: reduction (%) = (yield/biomass of healthy plants- yield/biomass of infected plants) yield/biomass of healthy plants ×100. Each bar represents the mean of four replicates ± SE. **Figure C.** Disease severity of soybean red crown rot in sand culture (Second round). A, disease incidence, B, disease index, C, CFU. LP, 15 μM P; HP, 500 μM P. All of the data are the means of four replicates ±SE. Bars with different letter(s) vary significantly among treatments as determined by Duncan's multiple range test ($P<0.05$). **Figure D.** Activities of the enzyme PPO (A) and PAL (B) in soybean roots. LP, 15 μM P; HP, 500 μM P. Except for control, all the roots were inoculated with *C. parasiticum* (see Materials and Methods for details). The solid barrier eliminated root contact and exudates movement, the nylon mesh (30 μm) barrier prevented root intermingling for the two species while permitting root exudates exchange, and no root barrier permitted roots and exudates to completely interact. Each bar represents the mean of three replicates ± SE. **Figure E.** HPLC scan of soybean and/or maize root exudates. **(a and b)** HPLC scan of soybean root exudates, (**c and d**) HPLC scan of maize root exudates, (**e and f**) HPLC scan of the root exudates from maize/soybean intercropping, (**g**) HPLC analysis of eight standard phenolic acids. LP, 15 μM P (a, c and e); HP, 500 μM P (b, d and f). 1, Gallic acid; 2, *P*-coumaric; 3, Phthalic acid; 4, Vanillic acid; 5, Syringic acid; 6, Ferulic acid; 7, Salicylic acid; 8, Cinnamic acid.

File S2 Supporting tables. Table S1. Real time PCR primers designed for this study. **Table S2.** AMF colonization rate and nodule number of soybean in the field.

Acknowledgments

We thank Mingfang Guan, Haiyan Zhang, Kai Li, Hui Jiang, Zuotong He, Hanxiang Wu, Yiyong Zhu and Renshen Zeng for their helps in field and lab experiments; Zide Jiang and Shuxian Li for guidance on pathogen analysis, and Thomas C Walk for critical comments.

Author Contributions

Conceived and designed the experiments: XG HL. Performed the experiments: XG MW RX. Analyzed the data: XG XW HK HL. Contributed reagents/materials/analysis tools: RP. Wrote the paper: XG HL.

References

1. Ratnadass A, Fernandes P, Avelino J, Habib R (2012) Plant species diversity for sustainable management of crop pests and diseases in agroecosystems: a review. Agronomy for Sustainable Development 32, 273–303.
2. Zhu Y, Chen H, Fan J, Wang Y, Li Y, et al. (2000) Genetic diversity and disease control in rice. Nature 406, 718–722.
3. Browning JA, Frey KJ (1969) Multiline cultivars as a means of disease control. Annual Review of Phytopathology 7, 355–382.
4. Smedegaard-Petersen V (1985) The limiting effect of disease resistance on yield. Annual Review of Phytopathology 23, 475–490.
5. Yu JQ (1999) Allelopathic suppression of Pseudomonas solanacearum infection of tomato (*Lycopersicon esculentum*) in a tomato-chinese chive (*Allium tuberosum*) intercropping system. Journal of Chemical Ecology 25, 2409–2417.
6. Ren LX, Su SM, Yang XM, Xu YC, Huang QW, et al. (2008) Intercropping with aerobic rice suppressed Fusaium wilt in watermelon. Soil Biology Biochemistry 40, 834–844.
7. Hao WY, Ren LX, Ran W, Shen QR (2010) Allelopathic effects of root exudates from watermelon and rice plants on *Fusarium oxysporum* f.sp. *niveum*. Plant and Soil 336, 485–497.
8. Gómez RO, Zavaleta M, González H, Livera M, Cárdenas S (2003) Allelopathy and microclimatic modification of intercropping with marigold on tomato early blight disease development. Field Crops Research 83, 27–34.
9. Bell DK, Sobers EK (1966) A peg, pod and root necrosis of peanuts caused by a species of *Calonectria*. Phytopathology 56, 1361–1364.
10. Kuruppu P, Schneider R, Russin J (2004) Factors affecting soybean root colonization by *Calonectria ilicicola* and development of red crown rot following delayed planting. Plant Disease 88, 613–619.
11. Guan M, Pan R, Gao X, Xu D, Deng Q, et al. (2010) First report of red crown rot caused by *Cylindrocladium parasiticum* on soybean in Guangdong, Southern China. Plant Disease 94, 485.
12. Li L, Zhang F, Li X, Peter C, Sun J, et al (2003) Interspecific facilitation of nutrient uptake by intercropped maize and faba bean. Nutrient Cyclling in Agroecosystems 65, 61–71.
13. He Y, Ding N, Shi JC, Wu M, Liao H, et al. (2013) Profiling of microbial PLFAs: Implications for interspecific interactions due to intercropping which increase phosphorus uptake in phosphorus limited acidic soils. Soil Biology Biochemistry 57, 625–634.
14. Ae N, Arihara J, Okada K, Yoshihara T, Johansen C (1990) Phosphorus uptake by pigeon pea and its role in cropping systems of Indian subcontinent. Science 248, 477–480.
15. Li C, He X, Zhu S, Zhou H, Wang Y, et al. (2009) Crop diversity for yield increase. PLoS ONE 4, e8049.
16. Fang SQ, Gao X, Deng Y, Chen XP, Liao H (2011) Crop root behavior coordinates phosphorus status and neighbors: from field studies to three-dimensional in situ reconstruction of root system architecture. Plant Physiology 155, 1277–1285.
17. Fang S, Clark RT, Zheng Y, Iyer-Pascuzzia AS, Weitzf JS, et al. (2013) Genotypic recognition and spatial responses by rice roots. Proceedings of the National Academy of Sciences, USA 110, 2670–2675.
18. Gao X, Lu X, Wu M, Zhang HY, Pan RQ, et al. (2012) Co-inoculation with rhizobia and AMF inhibited soybean red crown rot: from field study to plant defense-related gene expression analysis. PLoS ONE 7, e33977.
19. Wang XR, Pan Q, Chen FX, Yan XL, Liao H (2011) Effects of co-inoculation with arbuscular mycorrhizal fungi and rhizobia on soybean growth as related to root architecture and availability of N and P. Mycorrhiza 21, 173–181.
20. Liao H, Wan HY, Shaff J, Wang XR, Yan XL, et al. (2006) Phosphorus and aluminum interactions in soybean in relation to aluminum tolerance. Exudation of specific organic acids form different regions of the intact root system. Plant Physiology 141, 674–684.
21. Song YY, Zeng RS, Xu JF, Li J, Shen X, et al. (2010) Interplant communication of tomato plants through underground common mycorrhizal networks. PLoS ONE 5, e13324.
22. Ling N, Huang QW, Guo SW, Shen QR (2011) Paenibacillus polymyxa SQR-21 systemically affects root exudates of watermelon to decrease the conidial germination of *Fusarium oxysporum* f.sp. *niveum*. Plant and Soil 341, 485–493.
23. Li L, Li SM, Sun JH, Zhou LL, Bao XG, et al. (2007) Diversity enhances agricultural productivity via rhizosphere phosphorus facilitation on phosphorus-deficient soils. Proceedings of the National Academy of Sciences, USA 104, 11192–11196.
24. Trenbath BR (1993) Intercropping for the management of pests and diseases. Field Crops Research 34, 381–405.
25. Meynard JM, Dore T, Lucas P (2003) Agronomic approach: cropping systems and plant disease. Comptes Rendus Biologies 326, 37–46.
26. Hoffland E, Findenegg G, Nelemans J, van den Boogaard R (1992) Biosynthesis and root exudation of citric and malic acids in phosphate-starved rape plants. New Phytologist 122, 675–680.
27. Bardgett RD, Denton CS, Cook R (1999) Below-ground herbivory promotes soil nutrient transfer and root growth in grassland. Ecology Letters 2, 357–360.
28. Bais HP, Prithiviraj B, Jha AK, Ausubel FM, Vivanco JM (2005) Mediation of pathogen resistance by exudation of antimicrobials from roots. Nature 434, 217–221.
29. Bais HP, Vepachedu R, Gilroy S, Callaway RM, Vivanco JM (2003) Allelopathy and exotic plant invasion: from molecules and genes to species interactions. Science 301, 1377–1380.
30. Weir TL, Park SW, Vivanco JM (2004) Biochemical and physiological mechanisms mediated by allelochemicals. Current Opinion in Plant Biology 7, 472–479.
31. Bais HP, Weir TL, Perry LG, Gilroy S, Vivanco JM (2006) The role of root exudates in rhizosphere interactions with plants and other organisms. Annual Review of Plant Biology 57, 233–266.
32. Wu HS, Liu DY, Ling N, Bao W, Ying RR, et al. (2009) Influence of root exudates of wantermelon on *Fusarium oxysporum* f. sp. *niveum*. Soil Biology Biochemistry 73, 1150–1156.
33. Robert GU, Martha ER (2010) Defense-related gene expression in soybean leaves and seeds inoculated with *Cercospora kikuchii* and *Diaporthe phaseolorum* var. *Meridionalis*. Physiological and Molecular Plant Pathology 75, 64–70.
34. Van Loon LC, Rep M, Pirterse CMJ (2006) Significance of inducible defense-related proteins in infected plants. Annual Review of Phytopathology 44, 135–162.
35. Wees S, Swart E, Pelt J, Loon L, Pieterse C (2000) Enhancement of induced disease resistance by simultaneous activation of salicylate- and jasmonate-dependent defense pathways in Arabidopsis thaliana. Proceedings of the National Academy of Sciences, USA 97, 8711–8716.
36. Dordas C (2008) Role of nutrients in controlling plant disease in sustainable agriculture: a review. Agronomy for Sustainable Development 28, 33–46.

23

Illumina Amplicon Sequencing of 16S rRNA Tag Reveals Bacterial Community Development in the Rhizosphere of Apple Nurseries at a Replant Disease Site and a New Planting Site

Jian Sun, Qiang Zhang, Jia Zhou, Qinping Wei*

Institute of Forestry and Pomology, Beijing Academy of Agriculture and Forestry Sciences, Beijing, China

Abstract

We used a next-generation, Illumina-based sequencing approach to characterize the bacterial community development of apple rhizosphere soil in a replant site (*RePlant*) and a new planting site (*NewPlant*) in Beijing. Dwarfing apple nurseries of 'Fuji'/SH6/Pingyitiancha trees were planted in the spring of 2013. Before planting, soil from the apple rhizosphere of the replant site (*ReSoil*) and from the new planting site (*NewSoil*) was sampled for analysis on the Illumina MiSeq platform. In late September, the rhizosphere soil from both sites was resampled (*RePlant* and *NewPlant*). More than 16,000 valid reads were obtained for each replicate, and the community was composed of five dominant groups (Proteobacteria, Acidobacteria, Bacteroidetes, Gemmatimonadetes and Actinobacteria). The bacterial diversity decreased after apple planting. Principal component analyses revealed that the rhizosphere samples were significantly different among treatments. Apple nursery planting showed a large impact on the soil bacterial community, and the community development was significantly different between the replanted and newly planted soils. Verrucomicrobia were less abundant in *RePlant* soil, while *Pseudomonas* and *Lysobacter* were increased in *RePlant* compared with *ReSoil* and *NewPlant*. Both *RePlant* and *ReSoil* showed relatively higher invertase and cellulase activities than *NewPlant* and *NewSoil*, but only *NewPlant* soil showed higher urease activity, and this soil also had the higher plant growth. Our experimental results suggest that planting apple nurseries has a significant impact on soil bacterial community development at both replant and new planting sites, and planting on new site resulted in significantly higher soil urease activity and a different bacterial community composition.

Editor: Gabriele Berg, Graz University of Technology (TU Graz), Austria

Funding: This work was supported by China Agriculture Research System, CARS-28, Minister of Agriculture (PRC), (http://english.agri.gov.cn/) (QPW), and Beijing Technology Foundation for Selected Overseas Chinese Scholar, Human Resource and Social Security Bureau of Beijing (http://www.bjld.gov.cn/) (JS). The funders had no role in study design, data collection and analysis, decision to publish, or preparation of the manuscript.

Competing Interests: The authors have declared that no competing interests exist.

* Email: qinpingwei@gmail.com

Introduction

Apple trees are among the most important fruit trees in the world. China has the world's highest apple tree acreage (2.060 million hectares, 42.54% of the world's supply; FAOSTAT, 2012) and production (37.00 million tons, 48.44% of the world's supply; FAOSTAT, 2012). However, apple trees in most of China's dominant production areas experience a full fruit period and senescence phase; currently, approximately 70% apple orchards in China are over 20 years old. The need for apple orchard renewal is more than 140 thousand hectares per year. Complicating this renewal is apple replant disease (ARD), which, because of a lack of land resources, is becoming a serious problem in fruit tree nurseries and old orchards.

The term ARD refers to the poor growth of young apple trees, which occurs after replanting on a site that was previously planted with apple. The phenomenon is common to all major apple growing regions of the world, including Asia, Europe, North America, Africa etc. Compared with new planting sites, directly replanting nurseries on old sites can result in decreased tree growth and a significantly lower seedling survival rate [1]. Trees may take several years to recover from the initial growth depression and eventually reach the size and annual yields of unaffected trees, and the cumulative yields and profitability in ARD-affected orchards are usually much lower than in unaffected orchards [2]. In China, the traditional method to avoid ARD is allowing a fallow period of over three years, while using plants from the grass family such as wheat to 'clean' the soil [3]. The efficacy of growing wheat for reducing apple root infection by species of *Rhizoctonia* and *Pythium* was confirmed by greenhouse and field trials [4,5]. However, the lost productivity is unaffordable for famers because land is limited and expensive.

Typically, in ARD-affected orchards, the root systems of apple seedlings are small, with discolored feeder roots and few functional root hairs [6]. In a previous field trial, leaf analysis for macro- and micronutrients showed most elements in ARD-affected and ARD-unaffected orchards exhibited no significant difference; thus, the observed growth responses were not associated with any nutritional effect [7]. Replant disease of fruit trees has been studied for

Table 1. The invertase, urease and cellulase activity of soil of different treatments and plant growth mass of *RePlant* and *NewPlant*.

	Invertase	Urease	Cellulase	New shoots (DW, g)	New roots (DW, g)
RePlant	8.93±0.93[c]	0.73±0.06[a]	0.164±0.032[b]	150.5±22.9	22.3±6.2
NewPlant	6.59±0.14[b]	1.02±0.22[b]	0.053±0.015[a]	261.4±34.6	44.1±9.7
ReSoil	8.87±0.22[c]	0.55±0.18[a]	0.182±0.064[b]	NA	NA
NewSoil	5.16±0.65[a]	0.60±0.13[a]	0.073±0.040[a]	NA	NA

"DW" means dry weight. Averages of replicates ± standard error; means followed by different letters are significantly different at $P<0.05$.
NA: not applicable.

many decades and although reported to be attributed to certain abiotic elements including phytotoxins [8], nutrient imbalance, low or high pH, and lack or excess of moisture [9], the preponderance of evidence indicates that the disease in large part is due to biotic factors [10–12]. Soil sterilization and fumigation with methyl bromide were effective in treating ARD [13,14], thus demonstrating the role of biotic components of soil as prominent determinants of ARD in orchards.

Soil microorganisms are recognized as the key factor inducing ARD, but it is a challenging task to fully characterize soil microbial communities. The information from culture-dependent methods is limited because only a small fraction of soil microorganisms is culturable [15]. Recently, culture-independent methods have been developed for investigating microbial communities, including molecular analyses of nucleic acids extracted from soil, including denaturing gradient gel electrophoresis (DGGE) and terminal restriction fragment length polymorphism (T-RFLP) after PCR [16,17], as well as community profiling based on fatty acid methyl esters (FAME) and phospholipid fatty acid (PLFA) [18]. Next-Generation Sequencing (NGS) is a new DNA sequencing method, which relies on the detection of pyrophosphate release upon nucleotide incorporation, rather than chain termination with dideoxynucleotides. Compared with DGGE and T-RFLP, NGS may provide more detailed information about the community because each DNA molecule is sequenced as an individual read and because the identification of individual species group is more accurate [19]. Of NGS, both Pyrosequencing and Illumina Miseq are used for characterize soil microbial community structure including rhizosphere microbiome [16,20], however in recent studies, Illumina Miseq platform has been more frequently used since the around 10-fold increase in read depth, similar sequencing quality together with much lower cost [21].

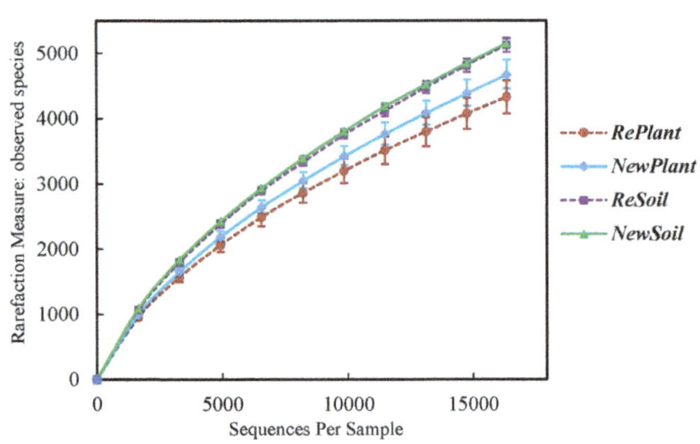

Figure 1. Rarefaction on species-abundance data. Average value of 3 replicates and error bar were showed.

Growth of fruit plants is affected by soil enzyme activity [22,23], and soil enzyme activity is highly related to soil microbes [24], which are involved in nutrient cycling and plant nutrient availability, and are in turn influenced by plant species. Sun et al. [25] used pyrosequencing to characterize the bacterial community structure of apple rhizosphere soil with different manure ratios, finding that certain levels of manure treatment resulted in significantly higher soil enzyme activity and a more diverse bacterial community composition. Past surveys of microbial communities of apple replant sites in specialized growing areas in Europe [13], Australia [26] and the USA [2] have confirmed a complex of biotic pathogens as causal agents of this etiology. It is also suggested that genotype-specific interactions with soil microbial consortia are linked with apple rootstock tolerance or susceptibility to ARD [27], while some CG series dwarfing rootstock showed high tolerance to replant disease compared to M26 [28]. However, apple orchards in China have significant differences compared with other areas in the world, including soil with low organic matter content and the specific rootstock, such as Pingyitiancha, which is commonly used in the Bohai Bay region, one of the two leading apple-producing areas in China. Little is known about the status of soil microbial communities in ARD orchards in China. For this reason, a specific survey of bacterial community development in the rhizosphere of apple trees at a replant disease site and a new planting site was conducted with the following objectives: (i) to compare tree growth rate and soil enzyme activity between a replant disease site and a new planting site, (ii) to reveal bacterial community development under ARD and non-ARD conditions and (iii) to define the relative importance of biotic components in replant disease etiology.

Methods

Ethics statement

The experiment was carried out in our scientific research field for pomology studies which is owned by our institute, therefore, no specific permissions were required for these locations/activities, and the field studies did not involve endangered or protected species.

Soil sampling

The test soil was collected from an orchard operated by the Institute of Pomology and Forestry, Beijing Academy of Agricultural Sciences. The soil type of local area was sandy loam soil, while the pH was 6.0–6.5. The orchard site was originally planted with apple trees approximately 1985 with trees grafted on *Malus Robusta* rootstock. In September 2012, the old trees were removed, and in April 4, 2013, the soil from a depth of 0–20

Table 2. Comparison of the estimated operational taxonomic unit (OTU) richness, diversity indexes and Pielous evenness of the 16S rRNA gene libraries for clustering at 97% identity as obtained from the pyrosequencing analysis.

Treatments	Observed OTUs	Shanon	Chao1	Pielous evenness (%)
RePlant	4328.1[a]	10.53[a]	9010.8[a]	87.17[a]
NewPlant	4509.9[a]	10.67[a]	9942.7[ab]	88.09[a]
ReSoil	5125.8[b]	11.02[b]	11371.8[c]	89.45[a]
NewSoil	5176.9[b]	11.12[b]	10614.9[bc]	88.53[a]

Averages of replicates ± standard error; means followed by different letters are significantly different at $P<0.05$.

cm was collected and mixed as replant soil sample (*ReSoil*). Another soil sample from alongside the orchard intensively cultivated with vegetables was collected as a new plant soil sample (*NewSoil*). Soil was put into 64 liter cubic Plexiglas boxes with rainproof shelter, and the nursery was planted right after the soil sampling. In late September, soil from a depth of 0–20 cm in three different locations at 20-cm distances from the center of trunk was collected from both sites and the youngest part of roots and the adjacent soil were resampled as *RePlant* and *NewPlant*.

Rootstock variety and tree growth

The two-year-old apple saplings were planted on April 4, 2013. Trees were of the scion variety 'Fuji' and were first grafted onto SH6 inter stock and then onto *Malus hupehensis* Var. Pingyiensis Jiang rootstock. The rootstock Pingyiensis Jiang has been widely used in Chinese orchards since the 1970 s. Trees were planted into either the replant soil or the new plant soil, with 10 replicates/treatment, and trunk heights were handed to 1.0–1.2 m. On September 20, 2013, new shoots and new roots of these trees were collected separately, dried at 105°C for 30 min and then dried at 70°C until a constant weight was reached in a forced-air oven.

Soil enzyme activity characterization

Soil samples were collected from depths of 0–20 cm in three different locations at 20-cm distances from the center of nurseries using a 5-cm diameter soil auger and transferred on ice to the laboratory both before nursery planting, and at the beginning of autumn 2013, just after the growth of autumn-shoot ceased. The soil samples were sieved through a 2-mm screen and homogenized prior to the analysis. One portion of the composite soil was stored

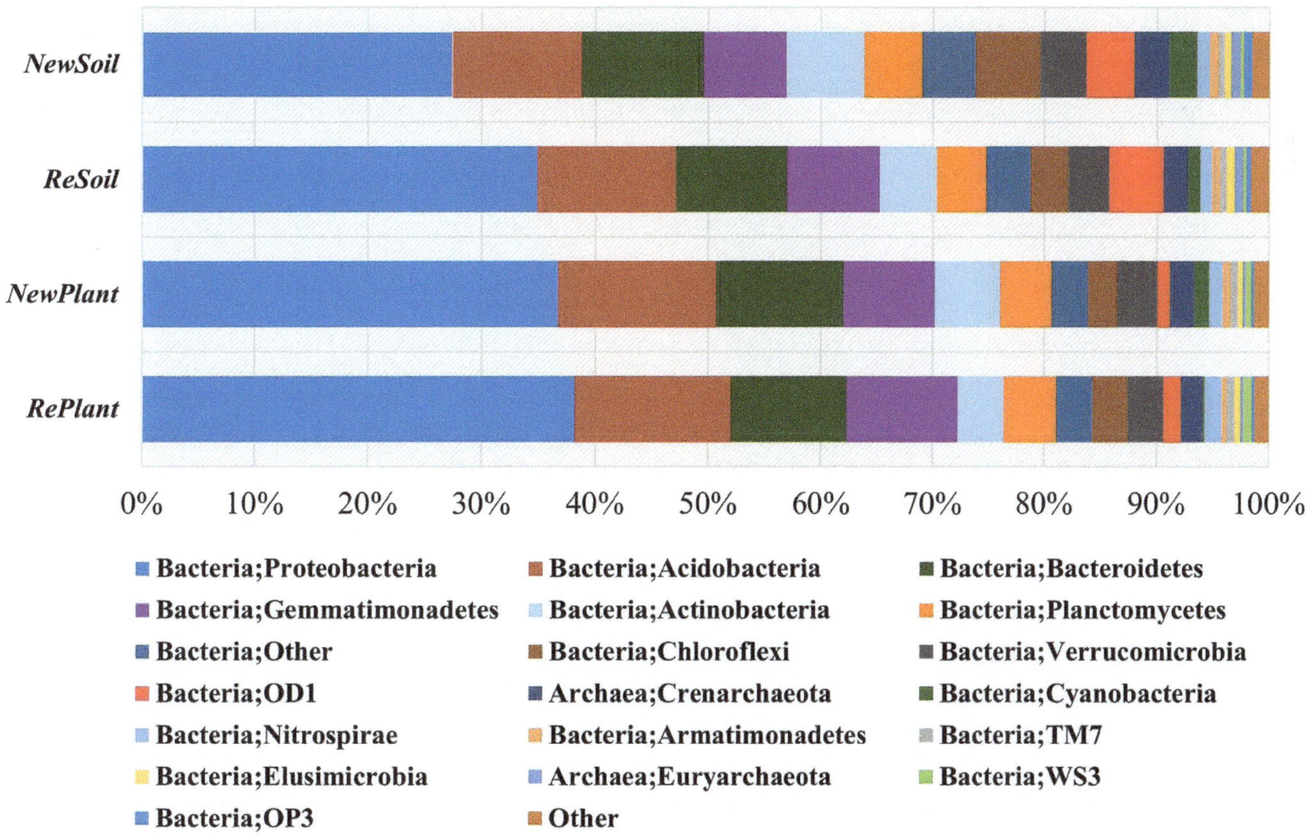

Figure 2. Comparison of the bacterial communities at the phylum level. Relative read abundance of different bacterial phyla within the different communities. Sequences that could not be classified into any known group were labeled "Other".

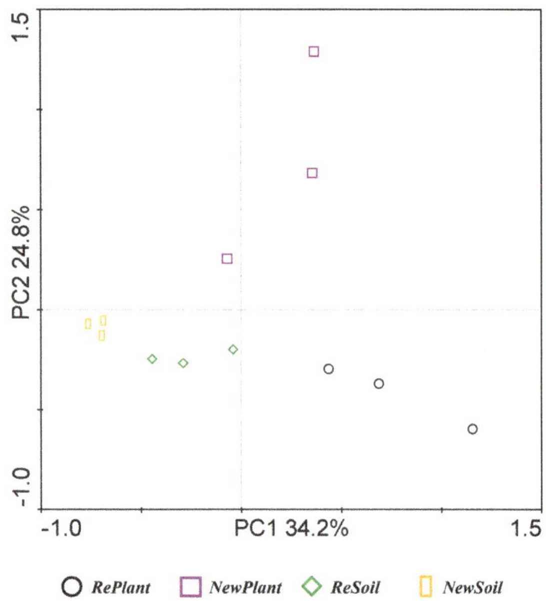

Figure 3. Principal component analysis (PCA) on the relative abundance of bacterial genera using Canoco 4.5. Principal components (PCs) 1 and 2 explained 34.2% and 24.8% of the variance, respectively.

in DNA-free polythene bags and kept on dry ice for the molecular analysis, while another portion was used for enzyme activity measurements.

Soil urease, invertase and cellulase activities were estimated according to a previous report [25]. Soil urease activity was detected using improved sodium phenate and sodium hypochlorite colorimetry. Invertase and cellulase activities were estimated colorimetrically by determining the reduction of 3,5-dinitrosalicylic acid from reducing sugars after the soil was incubated with a buffered sucrose and sodium carboxymethylcellulose solution and toluene at 37°C for 24 h and 72 h, respectively.

Soil DNA extraction

Three replicate samples were randomly picked for one treatment and used for DNA extraction. Soil DNA was extracted from the 1 g of soil after sieving using a Fast DNA SPIN Kit for soil (MP Biomedicals, Santa Ana, CA), according to the manufacturer's instructions. The extracted soil DNA was dissolved in 100 µl TE buffer, quantified by NanoDrop and stored at − 70°C before use.

Bacterial 16S rRNA gene amplification and Illumina Sequencing

To determine the diversity and composition of the bacterial communities in each of these samples, we used the protocol described in Caporaso et al. [29]. PCR amplifications were conducted in with the 515f/806r primer set that amplifies the V4 region of the 16S rDNA gene. The primer set was selected as it exhibits few biases should yield accurate phylogenetic and taxonomic information. The reverse primer contains a 6-bp error-correcting barcode unique to each sample. DNA was amplified following the protocol described previously [30]. Amplicon pyrosequencing was performed on the Illumina MiSeq platforms at Novogene Bioinformatics Technology Co., Ltd, Beijing, China. Complete data sets are submitted to the NCBI Short Read Archive under accession no. SRX337490.

Pairs of reads from the original DNA fragments were merged by using FLASH [29] -a very fast and accurate software tool which was designed to merge pairs of reads when the original DNA fragments were shorter than twice the length of reads. Sequencing reads was assigned to each sample according to the unique barcode of each sample. Sequences were analyzed with the QIIME [31] software package (Quantitative Insights Into Microbial Ecology) and UPARSE pipeline [32], in addition to custom Perl scripts to analyze alpha (within sample) and beta (between sample) diversity.

First, the reads were filtered by QIIME quality filters. Default settings for Illumina processing in QIIME was used ($r = 3$ $p = 0.75$ total read length; $q = 3$; $n = 0$).

(p) min_per_read_length: minimum number of consecutive high-qualitybase calls to retain read(as percentage of totalread length).

(r) max_bad_run_length: maximum number of consecutive low-quality base calls allowed before truncating a read.

(n) sequence_max_n: maximum number of ambiguous (N) characters allowed in a sequence.

(q) phred_quality_score: last quality score considered low quality.

Then we use UPARSE pipeline to picking operational taxonomic units (OTUs) through making OTU table. Sequences were assigned to OTUs at 97% similarity. We picked a representative sequences for each OTU and used the RDP classifier [33] to assign taxonomic data to each representative sequence. In order to compute Alpha Divesity, we rarified the OTU table and calculated three metrics: Chao1 metric estimates the richness, the Observed OTUs metric was simply the count of unique OTUs found in the sample, and shannon index. Rarefaction curves were generated based on these three metrics.

Statistical and bioinformatics analysis

Heatmap figures were generated using custom R scripts. Canoco 4.5 was used to run principal component analysis (PCA). Analysis of variance and Spearman's rank correlations were performed using SPSS Statistics 18 (IBM, Armonk, New York, USA). The community richness index, community diversity index, data preprocessing, operational taxonomic unit-based analysis and hypothesis tests were performed using mothur (http://www.mothur.org/). The histogram was created using Microsoft Excel 2010 (Microsoft, Redmond, Washington, USA). Significance was accepted at $p<0.05$, unless otherwise noted.

Results

Seedling biomass accumulation

In the spring of 2013, dwarfing apple nurseries of 'Fuji'/SH6/Pingyitiancha trees were planted a replant site (*RePlant*) and a new planting site (*NewPlant*). Before planting, soil from the replant site (*ReSoil*) and from the new planting site (*NewSoil*) was sampled. After 1 year of growth, significant differences ($P<0.05$) were found between *RePlant* and *NewPlant* sites in dry mass accumulation (**Table 1**). Seedling growth was significantly inhibited in replant soil comparing with non-replant soil; the inhibition levels on root dry weight and shoot dry weight were 49.4% and 42.3%, respectively.

Soil enzyme activity

Soil urease activity was highest in *NewPlant* soil, whereas there was no significant difference in urease activity in soil from

Table 3. The genera showing significant differences among the samples.

Taxon	*RePlant* (%)	*NewPlant* (%)	*ReSoil* (%)	*NewSoil* (%)
Pseudomonas	0.207 ± 0.048^{b}	0.120 ± 0.041^{a}	0.063 ± 0.031^{a}	0.057 ± 0.009^{a}
Lysobacter	0.941 ± 0.260^{b}	0.471 ± 0.065^{a}	0.276 ± 0.061^{a}	0.277 ± 0.037^{a}
Phenylobacterium	0.337 ± 0.007^{c}	0.258 ± 0.045^{b}	0.317 ± 0.022^{c}	0.135 ± 0.013^{a}
Ramlibacter	0.276 ± 0.128^{a}	0.728 ± 0.109^{b}	0.345 ± 0.061^{a}	0.336 ± 0.046^{a}
Chthonomonas	0.051 ± 0.013^{a}	0.171 ± 0.048^{b}	0.12 ± 0.055^{ab}	0.202 ± 0.038^{b}
Flavisolibacter	0.217 ± 0.060^{a}	0.847 ± 0.350^{b}	0.364 ± 0.095^{a}	0.354 ± 0.048^{a}
Opitutus	0.378 ± 0.022^{a}	0.400 ± 0.057^{a}	0.644 ± 0.120^{b}	0.524 ± 0.060^{ab}
Bdellovibrio	0.096 ± 0.014^{a}	0.161 ± 0.058^{ab}	0.256 ± 0.025^{b}	0.239 ± 0.031^{b}
Novosphingobium	0.185 ± 0.136^{ab}	0.266 ± 0.049^{b}	0.063 ± 0.018^{a}	0.058 ± 0.009^{a}
Planctomyces	0.167 ± 0.025^{a}	0.152 ± 0.032^{a}	0.195 ± 0.032^{ab}	0.239 ± 0.002^{b}
Kaistobacter	2.463 ± 0.555^{b}	2.302 ± 0.461^{b}	2.17 ± 0.311^{b}	0.978 ± 0.083^{a}
Cellvibrio	0.124 ± 0.173^{a}	0.100 ± 0.064^{a}	0.144 ± 0.051^{a}	0.378 ± 0.06^{b}
Kribbella	0.012 ± 0.012^{a}	0.031 ± 0.011^{a}	0.026 ± 0.004^{a}	0.081 ± 0.025^{b}
Pseudonocardia	0.035 ± 0.014^{a}	0.045 ± 0.004^{ab}	0.045 ± 0.028^{ab}	0.095 ± 0.034^{b}
Rubrobacter	0.118 ± 0.066^{a}	0.226 ± 0.082^{ab}	0.144 ± 0.023^{a}	0.311 ± 0.092^{b}
Luteimonas	0.388 ± 0.071^{b}	0.268 ± 0.158^{ab}	0.325 ± 0.052^{ab}	0.120 ± 0.029^{a}
Steroidobacter	0.902 ± 0.136^{ab}	0.701 ± 0.085^{a}	0.996 ± 0.104^{b}	0.761 ± 0.065^{ab}
Candidatus Nitrososphaera	1.282 ± 0.400^{a}	2.054 ± 0.372^{ab}	2.04 ± 0.357^{ab}	3.044 ± 0.683^{b}
Cenarchaeaceae;g__	0.476 ± 0.180^{b}	0.043 ± 0.043^{a}	0.061 ± 0.064^{a}	0.017 ± 0.010^{a}
Hyphomonadaceae;g__	0.721 ± 0.078^{b}	0.360 ± 0.069^{a}	0.646 ± 0.055^{b}	0.382 ± 0.117^{a}
Rhodospirillaceae;g__	0.721 ± 0.104^{c}	0.348 ± 0.012^{a}	0.577 ± 0.099^{bc}	0.455 ± 0.129^{ab}
Haliangiaceae;g__	1.520 ± 0.475^{b}	0.793 ± 0.028^{a}	0.811 ± 0.143^{a}	0.879 ± 0.108^{a}
Syntrophobacteraceae;g__	0.914 ± 0.046^{b}	0.614 ± 0.212^{a}	0.563 ± 0.095^{a}	0.587 ± 0.051^{a}
Rhodothermaceae;g__	0.033 ± 0.018^{a}	0.102 ± 0.037^{b}	0.033 ± 0.013^{a}	0.031 ± 0.016^{a}
Saprospiraceae;g__	0.366 ± 0.114^{a}	0.671 ± 0.142^{b}	0.378 ± 0.054^{a}	0.807 ± 0.093^{b}
Erythrobacteraceae;g__	0.553 ± 0.150^{ab}	0.469 ± 0.080^{a}	0.815 ± 0.229^{b}	0.309 ± 0.042^{a}
Polyangiaceae;g__	0.033 ± 0.015^{a}	0.061 ± 0.044^{a}	0.073 ± 0.016^{a}	0.145 ± 0.012^{b}
NB1-j;g__	0.341 ± 0.056^{b}	0.157 ± 0.113^{a}	0.083 ± 0.044^{a}	0.108 ± 0.013^{a}
Chromatiales;g__	0.106 ± 0.025^{b}	0.020 ± 0.020^{a}	0.047 ± 0.029^{a}	0.027 ± 0.003^{a}
Sphingomonadales;g__	0.159 ± 0.048^{ab}	0.240 ± 0.015^{c}	0.220 ± 0.043^{a}	0.099 ± 0.009^{a}
Sphingomonadaceae;g__	0.679 ± 0.076^{a}	1.026 ± 0.234^{b}	0.689 ± 0.064^{a}	0.423 ± 0.051^{a}
Micrococcaceae;g__	0.183 ± 0.062^{a}	0.563 ± 0.167^{b}	0.228 ± 0.013^{a}	0.673 ± 0.105^{b}

"g__" represents genus not grouped into any known genera within these families/groups. Averages of replicates ± standard error; means followed by different letters are significantly different at $P<0.05$.

RePlant, *ReSoil* and *NewSoil* samples, which were 28.4%~46.1% lower than *NewPlant* soil (**Table 1**). Cellulase activities in *NewPlant* and *NewSoil* were lower than in *RePlant* and *ReSoil*, but planting apple nurseries did not significantly change soil cellulase activity. Regardless of whether a new apple nursery had yet been planted, the soil from the old orchard had about three-fold higher cellulase activity than soil from a site that had never been planted. Invertase activity was lower in *NewSoil* and *NewPlant* samples than in *RePlant* and *ReSoil* samples.

Richness

More than 16,000 valid reads were obtained for each replicate through a sequence optimization process, and the bacterial community richness index was calculated as shown in **Table 2**. After quality filtering, median sequence length of each read was 252 bp. In *ReSoil* and *NewSoil* samples, more than 600 additional OTUs were observed compared with *RePlant* and *NewPlant* soil (**Fig. 1**). Higher Shannon and Chao 1 indices before planting indicated that planting of apple nurseries reduced diversity within the bacterial community. However, Pielou's evenness values were indicating approximately equally distributed OTU abundances among community members (**Table 2**).

Taxonomic coverage

All of the sequences were classified into 26 phyla or groups by the mothur program. The overall bacterial composition of the different samples was similar, while the distribution of each phylum or group varied (**Fig. 2**). In all samples, Proteobacteria, Acidobacteria, Bacteroidetes, Gemmatimonadetes and Actinobacteria were the five most dominant phyla, accounting for >60% of the reads. Significantly more unclassified species were detected in *ReSoil* and *NewSoil* samples, which was in accordance with their higher diversity indices. Compared with other samples, *NewSoil* had a significantly higher percentage of GN02 (2.7–5.3-fold), OP3

(1.8–3.1-fold), Chloroflexi (1.8–2.4-fold), and Verrucomicrobia (1.1–1.3-fold), and a lower percentage of Proteobacteria (*NewSoil*: 27.2%, *ReSoil*: 34.9%, *RePlant*: 38.2% and *NewPlant*: 36.8%). The percentages of CyanoBac and Verrucomicrobia were lowest in *RePlant*. More Firmicutes were detected in *NewPlant* than *RePlant* or *ReSoil* samples, and the lowest level was found in *NewSoil*. The OD1 group was approximately two- to three-fold higher in *RePlant* and *NewPlant* compared with *ReSoil* and *NewSoil*.

On a genus level, all 260 detected genera were shared by the four samples, except for *Chryseobacterium*, which was not detected in *ReSoil*, and *Marinobacter*, which was not detected in *RePlant*. Several cold-tolerant species belong to *Chryseobacterium* [34], while *Marinobacter* has been reported to be halophilic and is found in seawater [35].

To further compare the microbiota among the different samples, we performed PCA on the relative abundance of bacterial genera using Canoco 4.5 (**Fig. 3**). Data are presented as a 2D plot to better illustrate the relationship. *ReSoil* and *NewSoil* were relatively similar, but planting of an apple nursery had a significant impact on the soil microbial community. *RePlant* had a significantly higher PC1 value, and *NewPlant* had a higher PC2 value. 32 genera showing significant differences among the samples are listed in **Table 3**.

Pseudomonas, Lysobacter and *Phenylobacterium* were significantly higher in *RePlant* compared with *NewPlant*, while *Ramlibacter, Chthonomonas* and *Flavisolibacter* were higher in *NewPlant*. *Phenylobacterium* and *Kaistobacter* were higher in *ReSoil* compared with *NewSoil*, while *Cellvibrio, Kribbella* and *Rubrobacter* were higher in *NewSoil*. On both the replant site and the new plant site, apple nurseries showed significant impacts on the bacterial community structure. On the new plant site, *Phenylobacterium, Ramlibacter, Flavisolibacter, Novosphingobium* and *Kaistobacter* had increased abundance, and *Planctomyces, Cellvibrio* and *Kribbella* decreased in abundance. On the replant site, only *Pseudomonas* and *Lysobacter* increased, while *Opitutus* and *Bdellovibrio* decreased. Compared with the *NewPlant* sample, planting in the replant site also resulted in a greater abundance of certain genera of Cenarchaeaceae, Hyphomonadaceae, Rhodospirillaceae, Haliangiaceae and Syntrophobacteraceae families, however these genera weren't grouped into known genera within these families.

Discussion

Replanted soil significantly inhibited root and shoot development and exhibited different soil enzyme activity and a different bacterial community pattern. The levels of inhibition on root dry weight and shoot dry weight were 49.4% and 42.3%. Rosette disease and decreasing photosynthetic efficiency were also observed in the replant site, as were fewer main branches (data not shown). This observation is in accordance with the negative impact of replanting on apple growth that is widely reported [9,13,36].

Soil enzymes are involved with biological cycling and the development of fertility, so they are crucial indicators of the soil biochemistry. Urease catalyzes the hydrolysis of urea to produce ammonia and carbamate, and it is thus recognized as an important indicator of soil health. In this study, rhizosphere soil of *NewPlant* showed significantly increased urease activity, but rhizosphere soil of RePlant did not, indicating that in new planting sites, the root exudates might support a new and different functional microbial community which was responsible for this apparent increase in mineralization and result in a better supply of available nutrients. This result was supported by data on plant dry mass (**Table 1**), where *NewPlant* exhibited more than 40% greater plant growth compared with *RePlant*. Xun et al. [22] reported that soil urease activity increased during apple orchard maturation, and in our previous study on manure refinement of apple orchards, urease was the key indicator of soil health and highly correlative to tree growth, no matter whether the soil type was sandy [24] or loam [25]. Therefore, soil urease is an ideal indicator of apple orchard maturation and the lack of a significant increase in urease activity at the replant site may explain the decreased growth and late fruiting. However, soil invertase, which is an important factor affecting hydrolysis of sucrose into glucose and fructose, and cellulase, which is involved in breaking down cellulase, were higher in replant soil rather than new soil, and planting of apple nurseries had no further impact on these two enzyme activities. This could be related to the residual small root tissues of previous trees. Invertase also increased with an overdose of manure refinement in apple orchards, but was not closely associated with tree growth [37].

The soil microbial community composition of the replant site and the new plant site were distinct, and planting of an apple nursery significantly increased the difference. The diversity within the bacterial community was reduced after planting an apple nursery in both replanted and newly planted soil, which was in accordance with the fewer unclassified species observed after nurseries had been growing for one year. It has been reported that many plant species reduce the microbial diversity of rhizosphere soil compared with surrounding sites, including maize [38] and switchgrass [39]. Such a reduced bacterial richness in the plant rhizosphere is known as the 'rhizosphere effect'. This is typically characterized by a selective enrichment of root specialized guilds and reduction of rhizosphere bacterial richness in comparison to unplanted soil. However, although the diversity at both sites decreased after planting, there was no significant difference in bacterial diversity between the replant and new plant sites, either before or after the nursery had been planted, which means the rhizosphere effect of apple trees is the critical factor determining bacterial community diversity.

Although Proteobacteria, Acidobacteria, Bacteroidetes, Gemmatimonadetes and Actinobacteria predominated in all of the samples, *NewSoil* had a unique phyla composition compared with the other samples. Higher percentages of GN02, OP3, Chloroflexi and Verrucomicrobia and lower percentages of Proteobacteria were observed. This could also be explained by the rhizosphere effect' because there must be some root tissue left in the *ReSoil* site, even without a new nursery being planted. It is worth noticing that the OD1 group was approximately two- to three-fold higher in the *RePlant* and *NewPlant* samples as compared with *ReSoil* and *NewSoil*, and the *RePlant* soil had a higher percentage of WS3 compared with the other treatments. Verrucomicrobia were less common in the *RePlant* sample compared with *NewPlant*. Cultivation-independent approaches detect representatives of the Verrucomicrobia phylum in a wide range of environments, including soils, water and human feces [40], suggesting the phylum is widespread, but it is still poorly characterized. A few species that are extremely acidophilic [41] or ectosymbionts of protists [42] belong to this phylum. In our previous study [25], on loam soil of apple orchards with manure refinement, an optimal manure ratio resulted in an increase of Verrucomicrobia compared with soil with no manure applied, with the abundance increasing from 1.10% to 1.51%. However, on sandy soil [24], Verrucomicrobia decreased monotonically from 2.62% to 1.71%, 1.36% and 0.97% following application of 5%, 10% and 15% (which was optimal for tree growth) manure, respectively, and increased back to 1.28% when 20% manure applied. More research is required to determine whether this phylum is related to the rhizosphere effect of apple trees. Because most species of GN02, OP3, OD1 and WS3 groups are known only from metagenomics study and remain uncultivated, little information is available for functional discussion.

Principal component analysis on the genus level showed that apple nurseries had significant impacts on the soil microbial community, and the changes from *ReSoil* to *RePlant* and from *NewSoil* to *NewPlant* differed significantly. The bacterial community difference between *RePlant* and *NewPlant* was much greater than between *ReSoil* and *NewSoil*.

Phenylobacterium and *Kaistobacter* were higher in *ReSoil* compared with *NewSoil*, while *Cellvibrio, Kribbella* and *Rubrobacter* were higher in *NewSoil*. The genus *Phenylobacterium* comprises a single species called *P. immobile*, which has previously been described as growing optimally only on artificial compounds such as chloridazon [43]. More *P. immobile* in replant soil could be related to herbicide applied in the orchard. Unfortunately, little information is available in the literature concerning the genus *Kaistobacter*. *Cellvibrio* is a genus of gamma proteobacteria, which can oxidize cellulose to oxycellulose, but *NewSoil* showed lower cellulase than *ReSoil*, probably because in an aerobic environment, soil cellulase mainly derives from fungi, rather than bacteria. Some species of *Kribbella* have been isolated from the rhizosphere of *Typhonium giganteum* [44] and from tissues such as apricot leaves [45] and roots of *Lupinus angustifolius* [46]. *Kribbella antibiotic* [47] of the genus was reported to have a strong inhibitory activity toward *Botrytis* sp., *Rhizoctonia solani* and *Pyricularia oryzae*. *Rubrobacter* is a genus of Actinobacteria, which are radiotolerant [48], and a novel DNA repair enzyme was isolated from *Rubrobacter radiotolerans* [49], but little information is available regarding its presence in orchard soil.

On the replant site, *Lysobacter* and *Pseudomonas* increased in abundance, while *Opitutus* and *Bdellovibrio* decreased. *Pseudomonas* and *Lysobacter* were significantly higher in *RePlant* compared with *NewPlant*, while *Ramlibacter, Chthonomonas* and *Flavisolibacter* were more abundant in the *NewPlant* sample. *Pseudomonas* has been proposed to play a role in replant disease etiology of peach [50] and apple [51] trees through the production of hydrogen cyanide (HCN). However, *Pseudomonas* also includes several soil bacterial species with plant growth promoting activity, including *P. fluorescens* [10,52], and *Pseudomonas* and *Bacillus* are two of the most common biocontrol agent sources [53]. *Lysobacter* [54] has been a rich source for novel antibiotics, and some species have potential as biological control agents for plant diseases. Considering the previous study of manure refinement of apple orchard, in which an optimal manure ratio for nursery growth resulted in a decrease of *Pseudomonas* and *Lysobacter* [24,37], we speculate that in the *RePlant* sample, the higher percentage of *Pseudomonas* was related to replant disease and the increase of *Lysobacter* and was probably induced by the high percentages of Pathogenic fungi. However, further culture-based experiments would be needed to confirm this. There was only one species of *Chthonomonas*, *G. calidirosea*, which is an aerobic, pigmented, thermophilic micro-organism [55], while two species of *Flavisolibacter* were isolated from ginseng cultivating soil [56]. However, little was known about their function in soil or their relationship with planting until now.

Compared with *NewPlant*, planting in the replant site also resulted in more of certain genera of Cenarchaeaceae, Hyphomonadaceae, Rhodospirillaceae, Haliangiaceae and Syntrophobacteraceae families; however, these genera weren't grouped into any known genera within these families. This is in accordance with the fact that pyrosequencing can detect many uncultured microbes.

Conclusions

Our study has documented that replanting has a large negative impact on growth of apple nurseries in Beijing, China. Planting apple nurseries raised soil urease activity at the new planting site but not the replant site, while no significant impact on invertase and cellulase was observed. Apple nurseries had a significant impact on the soil bacterial community. *Lysobacter* and *Pseudomonas* were increased at the replant site, and the bacterial communities of the new and replant sites responded differently, resulting in more distinct community patterns.

Author Contributions

Conceived and designed the experiments: JS QPW. Performed the experiments: JS QZ JZ. Analyzed the data: JS QPW. Contributed reagents/materials/analysis tools: QZ JZ QPW. Contributed to the writing of the manuscript: JS QPW.

References

1. Mazzola M, Granatstein DM, Elfving DC, Mullinix K, Gu Y (2002) Cultural management of microbial community structure to enhance growth of apple in replant soils. Phytopathology 92: 1363–1366.
2. Mazzola M (1998) Elucidation of the microbial complex having a causal role in the development of apple replant disease in Washington. Phytopathology 88: 930–938.
3. Fan H, Zhao Z, Liu H, Zhao G, Zhang X, et al. (2008) Changes of Soil Nutrition in Root Zone and Their Effects on Growth of the Replanted Apple. Acta Horticulturae Sinica 12: 3.
4. Mazzola M, Gu Y (2000) Impact of wheat cultivation on microbial communities from replant soils and apple growth in greenhouse trials. Phytopathology 90: 114–119.
5. Mazzola M, Mullinix K (2005) Comparative field efficacy of management strategies containing *Brassica napus* seed meal or green manure for the control of apple replant disease. Plant Dis 89: 1207–1213.
6. Wilson S, Andrews P, Nair TS (2004) Non-fumigant management of apple replant disease. Sci Hortic-Amsterdam 102: 221–231.
7. Hofmann A, Wittenmayer L, Arnold G, Schieber A, Merbach W (2012) Root exudation of phloridzin by apple seedlings (*Malus* x *domestica* Borkh.) with symptoms of apple replant disease. Journal of Applied Botany and Food Quality 82: 193–198.
8. Utkhede RS, Smith EM (1993) Biotic and abiotic causes of replant problems of fruit trees. III International Symposium on Replant Problems 363. 25–32.
9. Rumberger A, Yao S, Merwin IA, Nelson EB, Thies JE (2004) Rootstock genotype and orchard replant position rather than soil fumigation or compost amendment determine tree growth and rhizosphere bacterial community composition in an apple replant soil. Plant Soil 264: 247–260.
10. Mazzola M, Manici LM (2012) Apple replant disease: role of microbial ecology in cause and control. Annu Rev Phytopathol 50: 45–65.
11. Yim B, Smalla K, Winkelmann T (2013) Evaluation of apple replant problems based on different soil disinfection treatments-links to soil microbial community structure? Plant Soil 1–15.
12. Covey Jr RP, Benson NR, Haglund WA (1979) Effect of soil fumigation on the apple replant disease in Washington [USA]. Phytopathology 69.
13. Manici LM, Kelderer M, Franke-Whittle IH, Rühmer T, Baab G, et al. (2013) Relationship between root-endophytic microbial communities and replant disease in specialized apple growing areas in Europe. Appl Soil Ecol 72: 207–214.
14. Cardenas E, Tiedje JM (2008) New tools for discovering and characterizing microbial diversity. Curr Opin Biotech 19: 544–549.
15. Smalla K, Oros-Sichler M, Milling A, Heuer H, Baumgarte S, et al. (2007) Bacterial diversity of soils assessed by DGGE, T-RFLP and SSCP fingerprints of PCR-amplified 16S rRNA gene fragments: do the different methods provide similar results? J Microbiol Meth 69: 470–479.
16. Hirsch PR, Mauchline TH, Clark IM (2010) Culture-independent molecular techniques for soil microbial ecology. Soil Biology and Biochemistry 42: 878–887.
17. Drenovsky RE, Elliott GN, Graham KJ, Scow KM (2004) Comparison of phospholipid fatty acid (PLFA) and total soil fatty acid methyl esters (TSFAME)

for characterizing soil microbial communities. Soil Biology and Biochemistry 36: 1793–1800.
18. Ronaghi M, Uhlén M, Nyrén P (1998) A sequencing method based on real-time pyrophosphate. Science 281: 363–365.
19. Lee OO, Wang Y, Yang J, Lafi FF, Al-Suwailem A, et al. (2010) Pyrosequencing reveals highly diverse and species-specific microbial communities in sponges from the Red Sea. The ISME journal 5: 650–664.
20. Mazzola M, Strauss SL (2013) Resilience of orchard replant soils to pathogen re-infestation in response to Brassicaceae seed meal amendment. Aspects of Applied Biology 119: 69–77.
21. Gołębiewski M, Deja-Sikora E, Cichosz M, Tretyn A, Wróbel B (2014) 16S rDNA Pyrosequencing Analysis of Bacterial Community in Heavy Metals Polluted Soils. Microbial Ecol 1–13.
22. Qian X, Gu J, Sun W, Li Y, Fu Q, et al. (2014) Changes in the soil nutrient levels, enzyme activities, microbial community function, and structure during apple orchard maturation. Appl Soil Ecol 77: 18–25.
23. Jia T, Chao S, Mingyan Y, Xiaoqi Z (2012) Studies on the activities of three kinds of soil enzyme, organic matters, microbes and the yields and quality of apple in different tree-aged apple orchards in Loess plateau. Journal of Agricultural Science and Technology (Beijing) 14: 115–122.
24. Zhang Q, Sun J, Liu S, Wei Q (2013) Manure Refinement Affects Apple Rhizosphere Bacterial Community Structure: A Study in Sandy Soil. PloS one 8: e76937.
25. Sun J, Zhang Q, Zhou J, Wei Q (2014) Pyrosequencing technology reveals the impact of different manure doses on the bacterial community in apple rhizosphere soil. Appl Soil Ecol 78: 28–36.
26. Dullahide SR, Stirling GR, Nikulin A, Stirling AM (1994) The role of nematodes, fungi, bacteria, and abiotic factors in the etiology of apple replant problems in the Granite Belt of Queensland. Animal Production Science 34: 1177–1182.
27. St. Laurent A, Merwin I, Fazio G, Thies J, Brown M (2010) Rootstock genotype succession influences apple replant disease and root-zone microbial community composition in an orchard soil. Plant Soil 337: 259–272.
28. Atucha A, Emmett B, Bauerle T (2014) Growth rate of fine root systems influences rootstock tolerance to replant disease. Plant Soil 376: 337–346.
29. Caporaso JG, Kuczynski J, Stombaugh J, Bittinger K, Bushman FD, et al. (2010) QIIME allows analysis of high-throughput community sequencing data. Nature methods 7: 335–336.
30. Magoč T, Salzberg SL (2011) FLASH: fast length adjustment of short reads to improve genome assemblies. Bioinformatics 27: 2957–2963.
31. Wang Q, Garrity MG, Tiedje MJ, James RC (2007) Naïve Bayesian Classifier for Rapid Assignment of rRNA Sequences into the New Bacterial Taxonomy. Applied and Environmental Microbiology 73: 5261–5267.
32. Edgar RC (2013) UPARSE: highly accurate OTU sequences from microbial amplicon reads. 10: 996–998.
33. Caporaso JG, Lauber CL, Walters WA, Berg-Lyons D, Lozupone CA, et al. (2011) Global patterns of 16S rRNA diversity at a depth of millions of sequences per sample. Proc Natl Acad Sci U S A 108 Suppl 1: 4516–4522.
34. Kämpfer P, Dreyer U, Neef A, Dott W, Busse H (2003) Chryseobacterium defluvii sp. nov., isolated from wastewater. Int J Syst Evol Micr 53: 93–97.
35. Gauthier MJ, Lafay B, Christen R, Fernandez L, Acquaviva M, et al. (1992) *Marinobacter hydrocarbonoclasticus* gen. nov., sp. nov., a new, extremely halotolerant, hydrocarbon-degrading marine bacterium. International Journal of Systematic Bacteriology 42: 568–576.
36. Sewell G (1981) Effects of *Pythium* species on the growth of apple and their possible causal role in apple replant disease. Ann Appl Biol 97: 31–42.
37. Sun J, Zhang Q, Liu S, Wang X, Liu J, et al. (2013) Comparison of soil microbe community structure between organic and conventional apple orchards. Journal of Fruit Science 30: 230–234.
38. García Salamanca A, Molina Henares MA, Dillewijn P, Solano J, Pizarro Tobías P, et al. (2013) Bacterial diversity in the rhizosphere of maize and the surrounding carbonate-rich bulk soil. Microbial biotechnology 6: 36–44.
39. Mao Y, Li X, Smyth EM, Yannarell AC, Mackie RI (2014) Enrichment of specific bacterial and eukaryotic microbes in the rhizosphere of switchgrass (*Panicum virgatum* L.) through root exudates. Environmental Microbiology Reports.
40. Hou S, Makarova KS, Saw JH, Senin P, Ly BV, et al. (2008) Complete genome sequence of the extremely acidophilic methanotroph isolate V4, *Methylacidiphilum infernorum*, a representative of the bacterial phylum Verrucomicrobia. Biol Direct 3: 11.
41. Dunfield PF, Yuryev A, Senin P, Smirnova AV, Stott MB, et al. (2007) Methane oxidation by an extremely acidophilic bacterium of the phylum Verrucomicrobia. Nature 450: 879–882.
42. Sato T, Kuwahara H, Fujita K, Noda S, Kihara K, et al. (2013) Intranuclear verrucomicrobial symbionts and evidence of lateral gene transfer to the host protist in the termite gut. The ISME journal.
43. Kanso S, Patel BK (2004) *Phenylobacterium lituiforme* sp. nov., a moderately thermophilic bacterium from a subsurface aquifer, and emended description of the genus *Phenylobacterium*. Int J Syst Evol Micr 54: 2141–2146.
44. Xu Z, Xu Q, Zheng Z, Huang Y (2012) *Kribbella amoyensis* sp. nov., isolated from rhizosphere soil of a pharmaceutical plant, *Typhonium giganteum* Engl. Int J Syst Evol Micr 62: 1081–1085.
45. Kaewkla O, Franco CM (2013) *Kribbella endophytica* sp. nov., an endophytic actinobacterium isolated from the surface-sterilized leaf of a native apricot tree. Int J Syst Evol Micr 63: 1249–1253.
46. Trujillo ME, Kroppenstedt RM, Schumann P, Martínez-Molina E (2006) *Kribbella lupini* sp. nov., isolated from the roots of *Lupinus angustifolius*. Int J Syst Evol Micr 56: 407–411.
47. Wang D, Zhang Y, Jiang Y, Wu W, Jiang C (2003) A study on polyphasic taxonomy of one antifungal actinomycete strain YIM31530~(T). Journal of Yunnan University (Natural Sciences) 26: 265–269.
48. Kausar J, Ohyama Y, Terato H, Ide H, Yamamoto O (1997) 16S rRNA gene sequence of *Rubrobacter radiotolerans* and its phylogenetic alignment with members of the genus *Arthrobacter*, gram-positive bacteria, and members of the family Deinococcaceae. International journal of systematic bacteriology 47: 684–686.
49. Asgarani E, Terato H, Asagoshi K, Shahmohammadi HR, Ohyama Y, et al. (2000) Purification and characterization of a novel DNA repair enzyme from the extremely radioresistant bacterium *Rubrobacter radiotolerans*. J Radiat Res 41: 19–34.
50. Yang J, Ruegger PM, McKenry MV, Becker JO, Borneman J (2012) Correlations between root-associated microorganisms and peach replant disease symptoms in a California soil. PloS one 7: e46420.
51. Rumberger A, Merwin IA, Thies JE (2007) Microbial community development in the rhizosphere of apple trees at a replant disease site. Soil Biology and Biochemistry 39: 1645–1654.
52. Himani S, Mohinder K, Prakash VP, Sheetal R (2011) Phenotypic and genotypic characterization of plant beneficial fluorescent *Pseudomonas* species associated with apple to overcome replant problem. Plant Disease Research 26.
53. Santoyo G, Orozco-Mosqueda MDC, Govindappa M (2012) Mechanisms of biocontrol and plant growth-promoting activity in soil bacterial species of *Bacillus* and *Pseudomonas*: a review. Biocontrol Sci Techn 22: 855–872.
54. Xie Y, Wright S, Shen Y, Du L (2012) Bioactive natural products from *Lysobacter*. Nat Prod Rep 29: 1277–1287.
55. Lee KC, Dunfield PF, Morgan XC, Crowe MA, Houghton KM, et al. (2011) *Chthonomonas calidirosea* gen. nov., sp. nov., an aerobic, pigmented, thermophilic micro-organism of a novel bacterial class, *Chthonomonadetes classis* nov., of the newly described phylum Armatimonadetes originally designated candidate division OP10. Int J Syst Evol Micr 61: 2482–2490.
56. Yoon M, Im W (2007) *Flavisolibacter ginsengiterrae* gen. nov., sp. nov. and *Flavisolibacter ginsengisoli* sp. nov., isolated from ginseng cultivating soil. Int J Syst Evol Micr 57: 1834–1839.

SoilGrids1km — Global Soil Information Based on Automated Mapping

Tomislav Hengl[1]*, Jorge Mendes de Jesus[1], Robert A. MacMillan[2], Niels H. Batjes[1], Gerard B. M. Heuvelink[1,3], Eloi Ribeiro[1], Alessandro Samuel-Rosa[4], Bas Kempen[1], Johan G. B. Leenaars[1], Markus G. Walsh[5], Maria Ruiperez Gonzalez[1]

1 ISRIC — World Soil Information, Wageningen, the Netherlands, **2** LandMapper Environmental Solutions Inc., Edmonton, Canada, **3** Wageningen University, Wageningen, the Netherlands, **4** Federal Rural University of Rio de Janeiro, Rio de Janeiro, Brazil, **5** The Earth Institute, Columbia University, New York, New York, United States of America, and Selian Agricultural Research Inst., Arusha, Tanzania

Abstract

Background: Soils are widely recognized as a non-renewable natural resource and as biophysical carbon sinks. As such, there is a growing requirement for global soil information. Although several global soil information systems already exist, these tend to suffer from inconsistencies and limited spatial detail.

Methodology/Principal Findings: We present SoilGrids1km — a global 3D soil information system at 1 km resolution — containing spatial predictions for a selection of soil properties (at six standard depths): soil organic carbon (g kg−1), soil pH, sand, silt and clay fractions (%), bulk density (kg m−3), cation-exchange capacity (cmol+/kg), coarse fragments (%), soil organic carbon stock (t ha−1), depth to bedrock (cm), World Reference Base soil groups, and USDA Soil Taxonomy suborders. Our predictions are based on global spatial prediction models which we fitted, per soil variable, using a compilation of major international soil profile databases (ca. 110,000 soil profiles), and a selection of ca. 75 global environmental covariates representing soil forming factors. Results of regression modeling indicate that the most useful covariates for modeling soils at the global scale are climatic and biomass indices (based on MODIS images), lithology, and taxonomic mapping units derived from conventional soil survey (Harmonized World Soil Database). Prediction accuracies assessed using 5-fold cross-validation were between 23–51%.

Conclusions/Significance: SoilGrids1km provide an initial set of examples of soil spatial data for input into global models at a resolution and consistency not previously available. Some of the main limitations of the current version of SoilGrids1km are: (1) weak relationships between soil properties/classes and explanatory variables due to scale mismatches, (2) difficulty to obtain covariates that capture soil forming factors, (3) low sampling density and spatial clustering of soil profile locations. However, as the SoilGrids system is highly automated and flexible, increasingly accurate predictions can be generated as new input data become available. SoilGrids1km are available for download via http://soilgrids.org under a Creative Commons Non Commercial license.

Editor: Ben Bond-Lamberty, DOE Pacific Northwest National Laboratory, United States of America

Funding: ISRIC is a non-profit organization primarily funded by the Dutch government. The authors are especially thankful for support from the Africa Soil Information Service (AfSIS) project, funded by the Bill and Melinda Gates foundation and the Alliance for a Green Revolution in Africa (AGRA). The funders had no role in study design, data collection and analysis, decision to publish, or preparation of the manuscript.

Competing Interests: RAM is owner and retired principle of LandMapper Environmental Solutions Inc. There are no patents, products in development or marketed products to declare. This does not alter our adherence to all the PLOS ONE policies on sharing data and materials.

* Email: tom.hengl@wur.nl

Introduction

There is increasing recognition of the urgent need to improve the quality, quantity and spatial detail of information about soils to respond to challenges presented by growing pressures on soils to support a large variety of critical functions [1–4]. Arrouays et al. [3] argue that existing soils information is not well suited to addressing vital questions related to mapping, monitoring or modelling soil processes that are driven by fluxes or changes in soils of water, nutrients, carbon, solutes or energy. Conventional models of soil variation describe variation in the horizontal dimension using polygons comprising classes of named soils [5]. In the vertical dimension, variation is described in terms of classes of horizons or layers that vary in their properties, thickness and depth. These conceptual models of discrete variation of classes of soil in horizontal and vertical directions are not well suited for use in many of the (global) simulation models and decision making systems currently used to describe and interpret soil functions and processes, such as supporting crop growth modelling, modelling hydrological and climatological processes, soil carbon dynamics or erosion [2,5]. Most modern spatial models that require information about soils as an input need accurate numerical information about continuous variation in soil properties. Models also require input data layers that are complete, consistent and as correct and current as possible. These requirements are not well met by current sources of soils information, especially sources of global extent.

Soil is probably one of the least well described thematic layers at the global scale, and existing global soil maps are often of undocumented or unknown accuracy [5]. At the moment, only coarse scale soil maps of the world are available at an effective resolution of about ~20 km [10]. The most commonly used global soil maps include [2,5]: Harmonized World Soil Database (HWSD) [11], USGS-produced soil property maps (http://soils.usda.gov/use/worldsoils/mapindex/) and ISRIC-WISE based soil property maps [12].

While widely used and cited, these various coarse resolution soil maps tend to suffer from artefacts due to use of different soil mapping concepts between countries and regions, from variation in the underlying soil mapping scale (usually between 1:0.5 M to 1:5 M) and from differences in reliability of source data within and between continents [2,5]. They can also not easily be updated with new information and often lack any measure of uncertainty, which is assumed to be significant. In summary, currently available global soil maps are not comparable in level of detail, spatial accuracy and usability with other global environmental layers such as global land cover and climatic products (Figure 1).

In this paper, we present and describe SoilGrids1km — a global 3D soil information system at 1 resolution — as a first response to the need for a new, consistent and coherent, global soil information. SoilGrids1km was produced using the Global Soil Information Facilities (GSIF), which was recently developed at ISRIC as a framework and platform to support widespread, open collaboration in the assembly, collation and production of global soil information.

Materials and Methods

Global Soil Information Facilities

ISRIC — World Soil Information has a mandate to serve the international community with information about the world's soil resources to help addressing major global issues. Over the last four years, in collaboration with a growing number of international partners and with a direct support from the Bill and Melinda Gates Foundation (AfSIS project; http://africasoils.net), ISRIC has been developing a cyberinfrastructure called Global Soil Information Facilities (GSIF).

GSIF has a particular emphasis on supporting the assembly and collation of geo-registered soil profile descriptions with associated analytical data, and on supporting the production of new maps of 3D continuous soil properties and soil classes at global to regional scales. GSIF consists of several components: data portals for assembling and hosting soil profile data and covariate data, software for global soil data analysis and mapping, and facilities for documenting data and methods and for automating workflows.

One of these components is "SoilGrids" — an automated system for global soil mapping. SoilGrids is an implementation of model-based geostatistics [13,14] for the purpose of predicting soil properties (in 2D or 3D) and soil classes for a global soil mask (see further Figure 3c) using automated mapping. Automated mapping is the computer- aided generation of maps from point observations and covariate layers, with minimal human intervention, so that map updating is easy. In the context of geostatistical mapping, automated mapping implies that model fitting, prediction and visualization are run using fully automated and reproducible workflows [14,15]. The current implementation of SoilGrids focuses on producing predictions at 1 km spatial resolution and for a selection of soil properties and classes of interest to modelers and to international organizations such as FAO, Intergovernmental Panel on Climate Change (IPCC), the Consultative Group on International Agricultural Research (CGIAR) and similar.

We have imagined GSIF as a crowd-sourcing system, largely inspired by systems such as OpenStreetMap, Geo-wiki [16] and the R Open Source environment for statistical computing [17]. In this context, GSIF follows the "Agile" approach to software/IT development [18] meaning that we support rapid development, integration of soil field data, output validation, and rapid publishing of results. A new development cycle with new outputs (in principle of improved accuracy) is implemented in succession within an automated processing framework until the desired target specifications have been reached.

Input data for SoilGrids1km

The main input data sources for SoilGrids1km are global compilations of publicly available (shared) soil profile data and environmental layers at 1 km resolution; both are freely accessible via portals (http://worldsoilprofiles and http://www.worldgrids.org). The main sources of soil profile data used to produce the first version of SoilGrids1km are: the USA National Cooperative Soil Survey Soil Characterization database (http://ncsslabdatamart.sc.egov.usda.gov/) and profiles from the USA National Soil Information System (http://soils.usda.gov/technical/nasis/), LUCAS Topsoil Survey database [19], Africa Soil Profiles database [20], Mexican National soil profile database [21], Brazilian national soil profile database [22], Chinese soil profile database [23], and the soil profile archive from the Canadian Soil Information System [24]. Other significant sources of profile data used are: ISRIC-WISE [25], SOTER [26], SPADE [27], and Russian soil reference profiles [28].

The compilation of points shown in Figure 2 is possibly the largest compilation of soil ground-truth data in the world. It can be compared, for example, to a compilation of meteorological station data used to generate the WorldClim dataset [29]. A large part of the soil profile data used to generate SoilGrids1km can be accessed via the WorldSoilProfiles.org data portal, however some data sets such as LUCAS [19] have strict data use policies and can only be obtained from the original data provider.

As covariates for SoilGrids1km we used a selection of GIS layers (75): mainly MODIS images, but also climate surfaces [29], Global Lithological Map (GLiM) [30], HWSD mapping units [11], and SRTM DEM-derived surfaces. These layers (apart for the GLiM) are all available via the WorldGrids.org data portal. The actual number of covariates used during the analyses is different for each soil variable as these are iteratively selected for each soil attribute, based on their statistical significance to help predict the specific attribute.

Before model fitting, the original covariates were converted to principal components ($n = 95$) to reduce data overlap and help remove noise and artefacts [7]. Number of components is larger than the number of original covariates because covariates such as lithology and land form classes are converted to indicators before the principal component analysis.

Soil mask map

We make no spatial predictions for global land cover categories that represent non-active soil areas, such as: artificial surfaces and associated areas (>50% of pixel covered with urban areas), bare rock areas, water bodies [31], shifting sands, permanent snow and ice. The global mask map of soils with vegetation cover and world deserts is shown in Figure 3c.

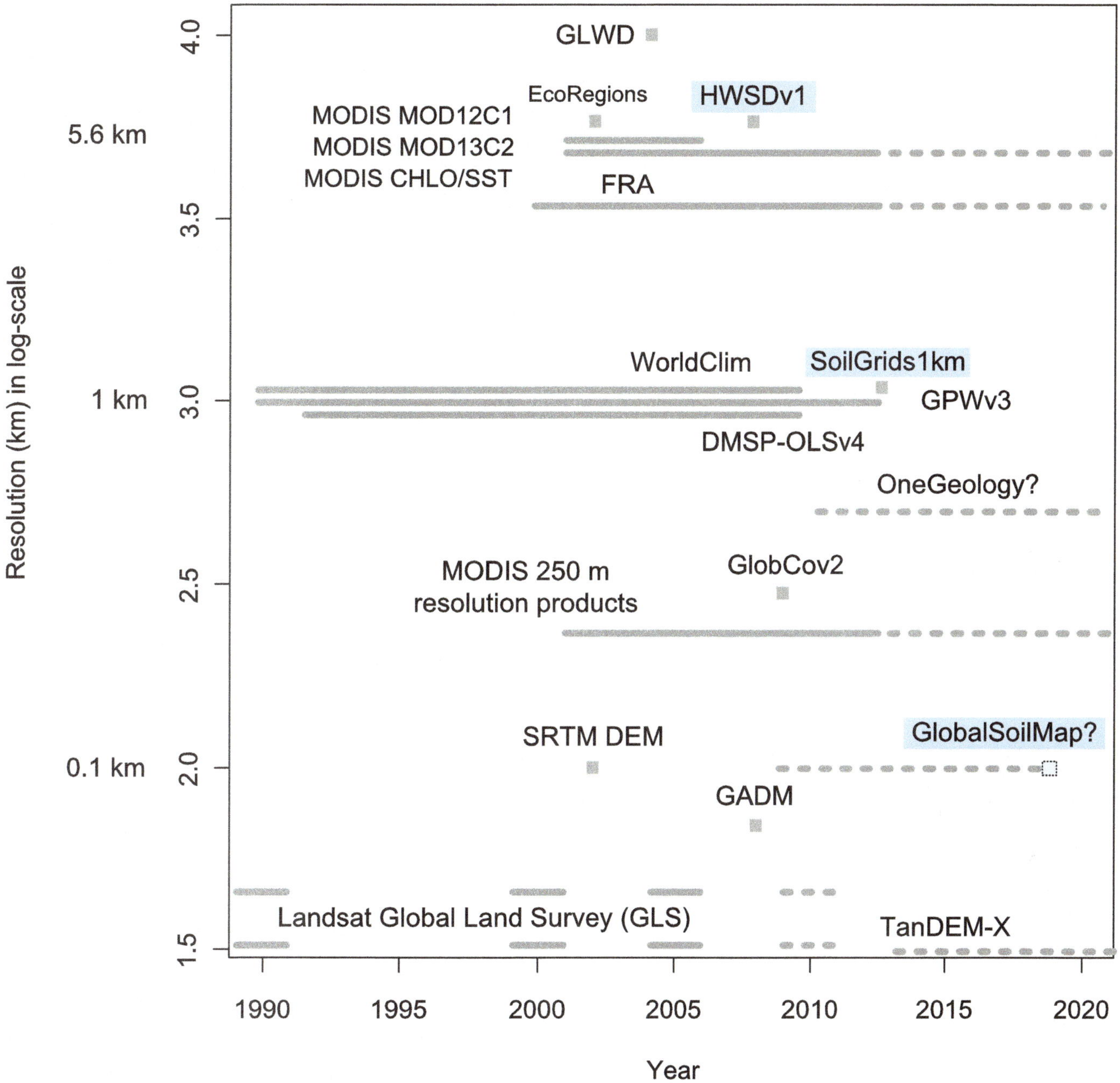

Figure 1. Spatial resolution and temporal coverage/publication time of some widely used global environmental data layers (global soil layers have been highlighted): GLWD — Global Lakes and Wetlands Database, HWSD — Harmonized World Soil Database, MOD12C1 — MODIS Land Cover Type Yearly L3, MOD13C2 — Vegetation Indices Monthly L3, CHLO/SST — MODIS Aqua Level-3 annual Chlorophyll/mid-IR Sea Surface Temperature, FRA — Forest Resources Assessment, GPW — Gridded Population of the World, DMSP-OLS — Nighttime Lights Time Series, GlobCov — Land Cover classes based on the MERIS FR images, GADM — Global Administrative Areas, TanDEM-X — Germany's topographic radar mission. Key agenda setters in the terms of production and dissemination of remote sensing and thematic environmental layers at the beginning of the 21st century include: NASA's MODIS (Moderate-resolution Imaging Spectroradiometer) and Landsat products — in terms of thematic content and usability [6–8], and Germany's TanDEM-X new global 12 m resolution DEM with ±2 m vertical accuracy [9]. Based on information retrieved on February 15th 2014. was produced using the Global Soil Information Facilities (GSIF), which was recently developed at ISRIC as a framework and platform to support widespread, open collaboration in the assembly, collation and production of global soil information.

The soil mask map was derived using the long term MODIS LAI images (MOD15A2), MODIS land cover product (MOD12Q1) [6], and global water mask [31] products. We distinguish three classes in the soil mask:

1. soils with vegetation cover — pixels with MODIS LAI>0 for at least one month in the last 12+ years (2000–2011),
2. urban areas — equal to the MODIS land cover product "*Urban and built-up*" class,

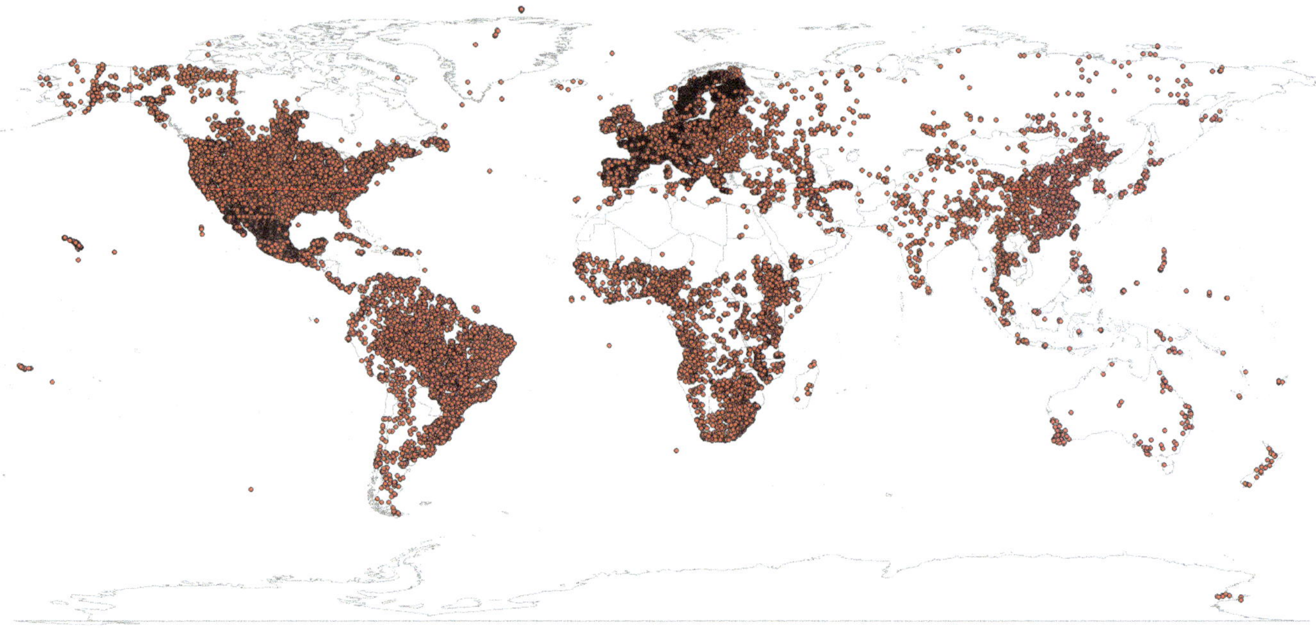

Figure 2. World distribution of soil profiles used to generate the SoilGrids1km product (about 110,000 points). Courtesy of various national and international agencies (see: Acknowledgments).

3. bare soil areas — areas without any biological activity but classified as "*Barren or sparsely vegetated*" in the MODIS land cover product.

Spatial prediction models

Two groups of spatial prediction models were implemented:

1. 2D or 3D regression and/or regression-kriging [32,33] combined with splines for numerical properties as implemented in the GSIF package for R. Here, the regression part is fitted using either:

   - Multiple linear regression [34] (for predicting pH, sand, silt and clay percentages and bulk density),
   - General Linear Models (GLM's) with log-link function [35,36] (for predicting organic carbon content and CEC),
   - Zero-inflated models [37] (for predicting coarse fragments and depth to bedrock; Figure 4),

2. Multinomial logistic regression (as implemented in the nnet package for R) for predicting distribution of soil classes [36].

As a general framework for mapping soil properties and classes we use the regression-kriging method commonly used in geostatistical mapping of soil properties [32,33,38]. We extend the existing 2D regression-kriging method to 3D space i.e. to predict values at voxels (Figure 4 right). In addition, we combine regression with splines, so that relationships between the soil property and covariates as well as soil-depth are modelled simultaneously:

$$\hat{z}(\mathbf{s}_0,d_0)=\sum_{j=0}^{p}\hat{\beta}_j\cdot X_j(\mathbf{s}_0,d_0)+\hat{\mathbf{g}}(d_0)+\sum_{i=1}^{n}\lambda_i(\mathbf{s}_0,d_0)\cdot e(\mathbf{s}_i,d_i) \quad (1)$$

where $\hat{z}$ is the predicted soil property, $\mathbf{s}_i$ are geographical coordinates, d_i is depth expressed in meters below land surface. Note that $\hat{\beta}_j\cdot X_j$ and $\hat{\mathbf{g}}(d_0)$ are the trend part of the model, where $X_j(\mathbf{s}_0,d_0)$ are covariates at the target location $\mathbf{s}_0$ and depth d_0, $\hat{\mathbf{g}}(d_0)$ is the predicted vertical trend, modelled by a spline function, and $e(\mathbf{s}_i,d_i)$ are residuals interpolated using 3D kriging using kriging weights $\lambda_i(\mathbf{s}_0,d_0)$. Because all covariates in the current version of SoilGrids1km are in fact 2D (i.e. values available at surface or for top-soil only), we copy the values of covariates for all depths in the regression matrix, which is a simplification. With the increasing availability of gamma radiometrics and similar, we anticipate that also 3D covariates will be used more in the near future with values differing per depth, although many covariates (e.g. elevation) will always remain 2D by definition.

3D regression and/or regression-kriging can be considered novel approaches to modeling soil variation. For comparison, the GlobalSoilMap project (http://globalsoilmap.net) proposes that soil-depth spline functions and spatial prediction functions should be fitted separately [3,40]. This spatial prediction system can be considered 2.5D because 2D models need to be fitted for each standard depth, i.e. each depth is modelled using a separate model that includes different combinations of covariates and in which data from predictions at one depth do not influence predictions at another. In the case of 3D modelling, a single model (Eq.1) is used for predicting in both X,Y and d for any property or class of interest, and fitting of the regression equation and residuals occurs at the same time as part of a single step. Another advantage of using a full 3D spatial prediction system, in comparison to the 2.5D, is also that it allows for producing spatial predictions and confidence intervals at any 3D location and not only at standard depths.

For each soil property, we have evaluated which version of the model in Eq.(1) would be most applicable. For example, initial tests showed that, for some soil properties e.g. soil organic carbon content and bulk density, the soil-depth relationship ($\hat{\mathbf{g}}(d_0)$) can often be better modelled using a log-log relationship. Consider for example:

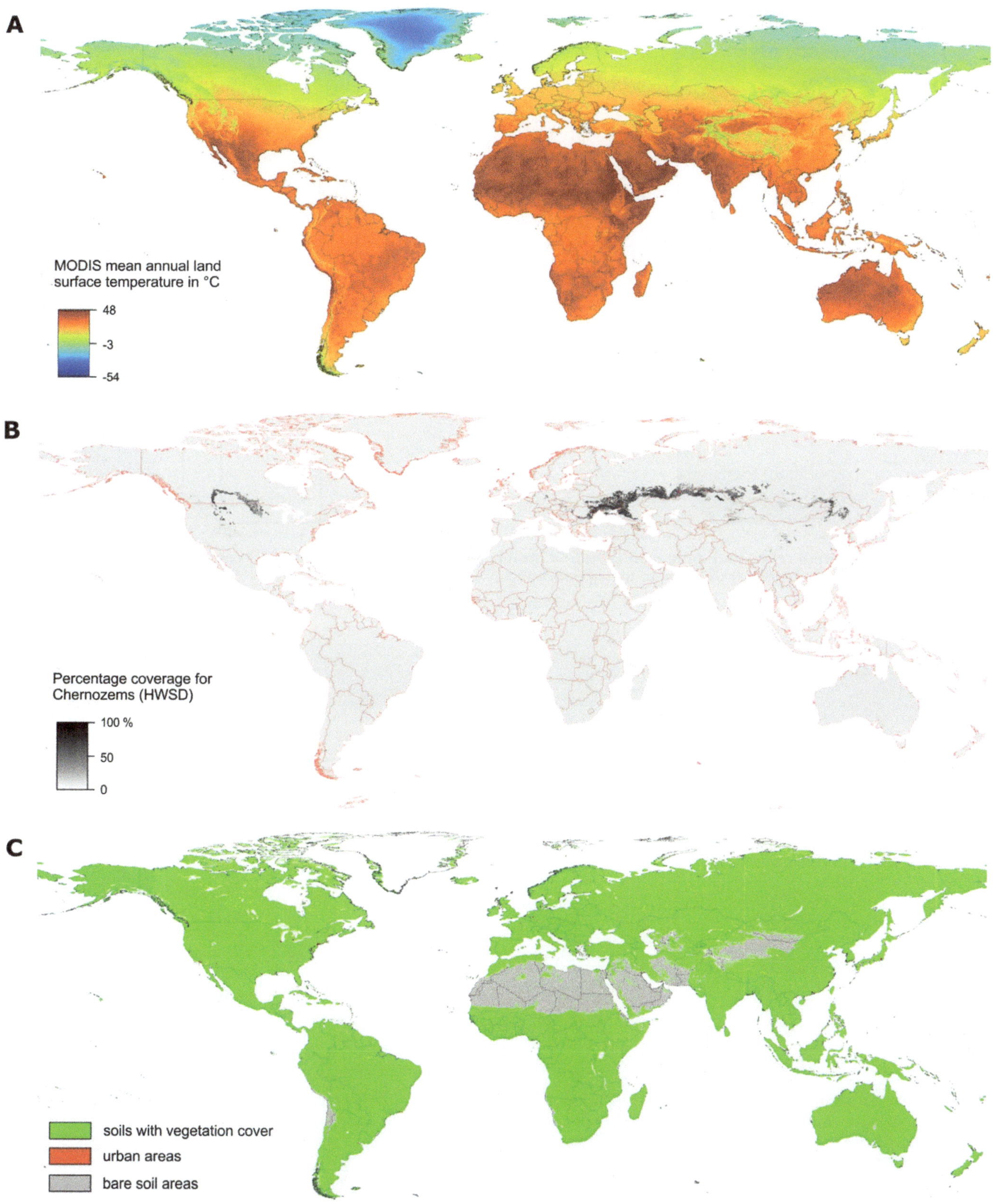

Figure 3. Examples of input layers used to generate SoilGrids1km: (a) long-term day-time MODIS land surface temperature, (b) percent cover Chernozems (based on the HWSD data set), and (c) global soil mask map. The spatial prediction domain of SoilGrids1km are the areas with vegetation cover and urban areas, while bare soil areas have been masked out. See text for more explanation.

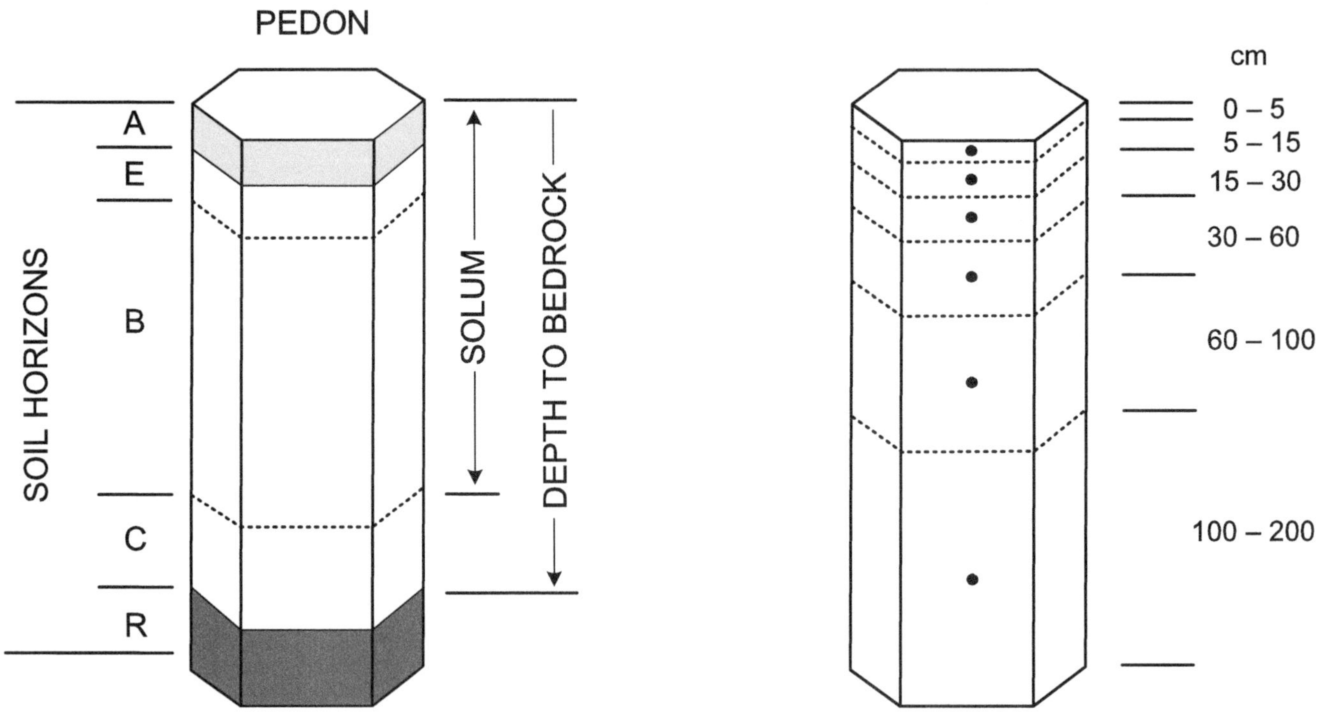

Figure 4. Standard stratification and designation of a soil profile: (left) soil horizons, solum thickness and depth to bedrock ('R' layer), and (right) six standard depths used in the *GlobalSoilMap* project [3].

$$\widehat{\mathrm{ORC}}(d) = \exp(\tau_0 + \tau_1 \cdot \log(d)) \quad (2)$$

where $\widehat{\mathrm{ORC}}(d)$ is the predicted soil organic carbon content at depth d and τ_1 is the rate of decrease with depth. The model fitted using the global compilation of soil profiles (Figure 5b) has $\tau_0 = 4.1517$ (standard error 0.005326) and $\tau_1 = -0.60934$ (standard error 0.00145). This model explains 36% of the variation in the log-transformed ORC, which is a significant portion. This illustrates that any global soil property model can significantly profit from including depth into the statistical modelling. For other soil properties that do not show a monotonic vertical trend, higher order splines implemented via the ns function in the package splines [35] have been used to account for complex, non-linear relationships.

Further, soil covariate layers (X_j) used to produce SoilGrids1km were selected to represent the CLORPT model originally presented by Jenny [38,41]:

$$S = f(\mathrm{cl,o,r,p,t}) \quad (3)$$

where S stands for soil (properties and classes), cl for climate, o for organisms (including humans), r is relief, p is parent material or geology and t is time. Most of the cl,o,r,p,t covariates are now publicly available and can be obtained at low cost thanks to NASA's/USGS Earth Observation projects such as MODIS and SRTM. We have also included soil class information (WRB reference groups) extracted from the HWSD (Figure 3b). These are basically traditional soil polygon delineations, comparable to other categorical covariates e.g. land cover classes or geological units.

The 3D regression function used for modelling changes of the of soil organic carbon content in 3D was thus (in R syntax):

```
formulaString=(ORCDRC+1)~PC1+PC2+...+PC95
+ns(altitude, df=2) glm(formula=formulaString,
family=gaussian(link=log),
data=rmatrix)
```

where `ORCDRC` is the organic carbon content, `PC1` to `PC95` are the principal components derived from some 75 covariate layers representing Jenny's soil forming factors, `altitude` is depth in meters from the soil surface, `rmatrix` is the regression matrix with values of target variable and predictors, ns is the natural spline function and `df = 2` sets the number of allowed breakpoints (in this case two breakpoints to allow for curvilinear relationship). Soil classes are useful *'carriers of soil information'* [42], hence for SoilGrids1km we also provide global predictions for standard soil classes classified according to the two most widely used international soil classification systems:

- FAO's World Reference Base (WRB) — with focus on mapping soil groups e.g. Chernozem, Luvisols, Gleysols and similar. The current system [43] defines 32 reference soil groups.
- United States Department of Agriculture (USDA) Soil Taxonomy — with focus on mapping the soil suborders. The current system [44] defines 67 soil suborders (subdivision of 12 orders: Alfisols, Andisols, Aridisols, Entisols, Gelisols, Histosols, Inceptisols, Mollisols, Oxisols, Spodosols, Ultisols and Vertisols).

Models for predicting WRB soil groups and USDA soil orders were fitted using the nnet package (fits multinomial log-linear models via neural networks) using the default settings of 100

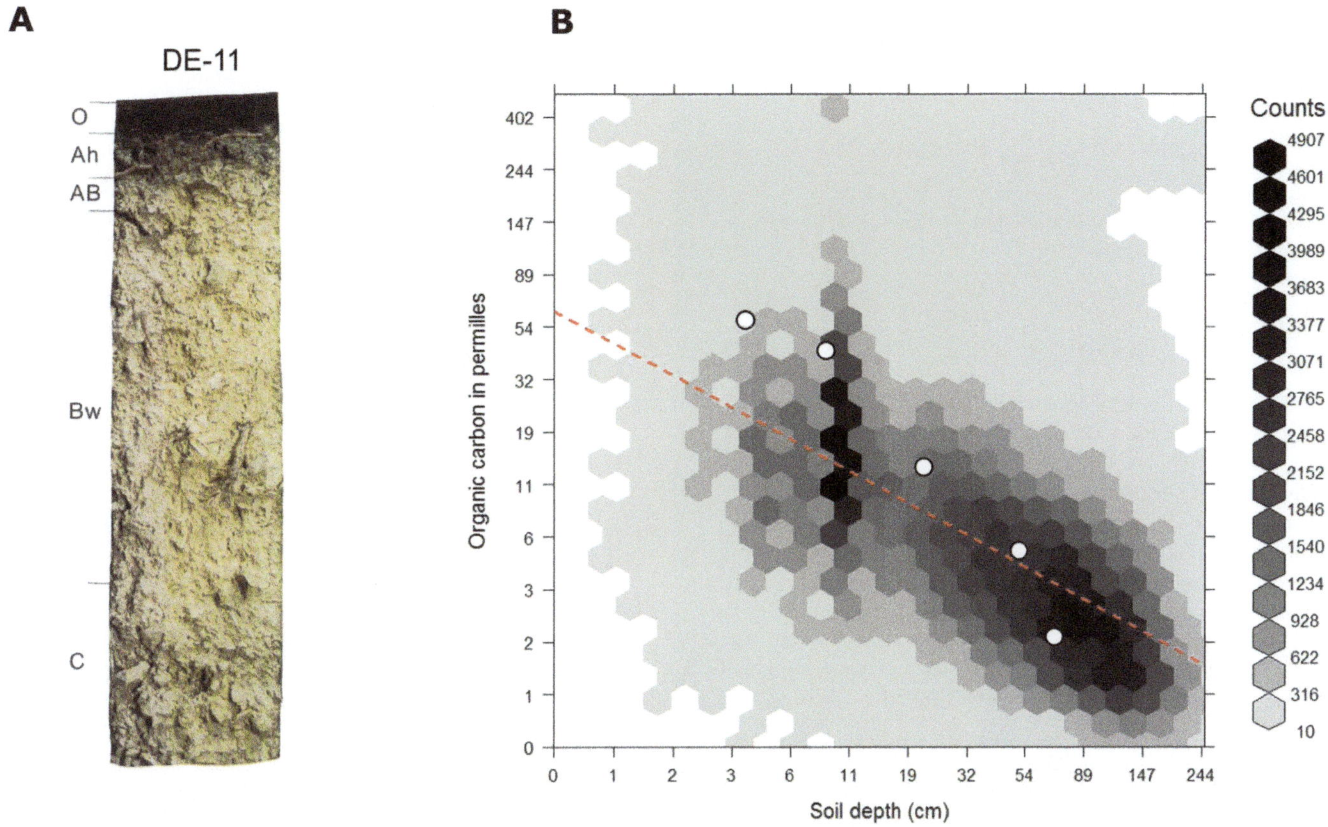

Figure 5. Individual soil profile from the ISRIC soil monolith collection (a) and globally fitted regression model for predicting soil organic carbon using depth only (b). The individual profile horizons are described by Mokma and Buurman [39]. Adjusted R-square for the model on the right is 0.363. Open circles show measured values for the profile on the left.

maximum iterations [36]. Soil classes are modeled as 2D variables i.e. the model does not include depth component, e.g.:

```
formulaString = TAXGWRB ~ PC1 + PC2 + ... + PC95
nnet :: multinom(formula = formulaString,
data = rmatrix, MaxNWts = 7000)
```

where `TAXGWRB` is the field observed WRB soil group, `nnet::multinom` is the function to fit a multinomial logistic regression and `MaxNWts` sets the maximum allowable number of weights high enough for such a large regression data (regression model with ca. 100 covariates).

Note that all predictions in the initial version of SoilGrids1km were made using regression modelling alone. 3D kriging on a sphere at almost one billion locations (130 million pixels times 6 depths) was beyond our technical capacities in 2013/2014. Efforts to use full 3D regression-kriging to produce the first version of SoilGrids1km were abandoned in response to two main issues. Firstly, the computational load to undertake global kriging was too demanding for the processing resources and time we initially had at our disposal. We are working to both increase our processing power and to make the global kriging algorithms more efficient so we can run them globally for subsequent versions of SoilGrids1km. Secondly, there are very large areas of the world (e.g. Russia, northern Canada) that presently have almost no point profile data. These areas lack a sufficient number and density of point observations to successfully compute residuals, which can then be kriged (otherwise kriging leads to serious artifacts). Since we were unable to produce residuals for large parts of the world, we decided not to try to krige residuals globally at first, at least until we obtain enough new point data to support computing and kriging residuals for all major portions of the globe. A full implementation of the 3D regression-kriging model built for SoilGrids has been run successfully at the continental level in Africa but, for the present (February 2014), we have not been able to apply full 3D regression-kriging globally. As soon as these technical limitations are solved, future versions of SoilGrids1km will likely also include a 3D kriging component.

Quality control

Resulting spatial predictions in SoilGrids1km are evaluated using two groups of methods:

- Cross-validation: We used 5-fold cross-validation to estimate the average mapping accuracy for each target variable. For continuous soil properties, we evaluate the amount of variation explained by the models [45]; and for soil classes we evaluate the map purity (i.e. proportion of observations correctly classified) and kappa statistic.
- Visual checking and overlay analysis: Because there is a large amount of spatial data, we have requested users to visually explore maps and look for artefacts and inconsistencies. Inconsistencies and artefacts in maps can be continuously reported through a Global Soil Information mailing list.

To derive amount of variation explained by the models for numeric variables we first derive Root Mean Square Error [46]:

$$RMSE = \sqrt{\frac{1}{l} \cdot \sum_{i=1}^{l} [\hat{z}(\mathbf{s}_i) - z(\mathbf{s}_i)]^2} \quad (4)$$

where l is the number of validation points. Amount of the variation explained by the model is then:

$$\Sigma_{\%} = \left[1 - \frac{SSE}{SSTO}\right] = \left[1 - \frac{RMSE^2}{\sigma_z^2}\right] [0 - 100\%] \quad (5)$$

where SSE is the sum of squares for residuals at cross-validation points (i.e. $RMSE^2 \cdot n$), and $SSTO$ is the total sum of squares.

Derivation of secondary soil properties: soil organic carbon stock

The SoilGrids1km output maps can be further used for estimation of secondary soil properties which are typically not measured directly in the field and need to be derived from primary soil properties. For instance, consider estimation of the global carbon stock (in t ha^{-1}). This secondary soil property can be derived from a number of primary soil properties [47]:

$$\begin{aligned} \text{OCS [kg m}^{-2}] &= \frac{\text{ORC}}{1000} \text{ [kg kg}^{-1}] \cdot \frac{\text{HOT}}{100} \text{ [m]} \\ &\cdot \text{BLD [kg m}^{-3}] \cdot \frac{100 - \text{CRF [\%]}}{100} \end{aligned} \quad (6)$$

where OCS is soil organic carbon stock, ORC is soil organic carbon mass fraction in permilles, HOT is horizon thickness in cm, BLD is soil bulk density in kg m^{-3} and CRF is volumetric fraction of coarse fragments (>2 mm) in percent (see also Figure 6).

The propagated error of the soil organic carbon stock (Eq.6) can be estimated using the Taylor series method [48]:

$$\begin{aligned} \sigma_{\text{OCS}} &= \frac{1}{10{,}000{,}000} \cdot \text{HOT} \cdot \\ &\sqrt{\text{BLD}^2 \cdot (100 - \text{CRF})^2 \cdot \sigma_{\text{ORC}}^2 + \sigma_{\text{BLD}}^2 \cdot (100 - \text{CRF})^2 \cdot \text{ORC}^2 + \text{BLD}^2 \cdot \sigma_{\text{CRF}}^2 \cdot \text{ORC}^2} \end{aligned} \quad (7)$$

where σ_{ORC}, σ_{BLD} and σ_{CRF} are standard deviations of the predicted soil organic carbon content, bulk density and coarse fragments, respectively. Note that we first predict OCS values for all depths/horizons, then aggregate values for the whole profile (0–2 m). We further use a map of predicted depth to bedrock to remove all predictions outside the effective soil depth (areas where soil is shallower than 2 m). A more robust way to estimate the propagated uncertainty of deriving OCS would be to use geostatistical simulations (e.g. derive standard error from a large number of realizations 100) that incorporate spatial and vertical correlations. Because we are dealing with massive data sets, running geostatistical simulations for millions of pixels was not yet considered as an option.

Software implementation

SoilGrids1km predictions are generated via the GSIF package for R, which makes use of a large number of other basic and contributed packages — gstat, raster, rgdal and other R packages for spatial analysis [49]. GSIF package for R contains most of the functions required to produce SoilGrids, and will remain the main platform in the future to obtain global model parameters and access SoilGrids through an API.

As previously mentioned, the target resolution of SoilGrids1km is relatively coarse, nevertheless, the compu- tational intensity and memory required to produce SoilGrids1km is high: one run of SoilGrids1km takes about 12–16 hours on a 12-core HP Z420 workstation with 64 GiB RAM running on a Windows 7 64-bit system. Note also that since we produce predictions at six depths and uncertainty for each depth, the quantity of GeoTIFF maps produced is in the order of 250×912MiB≈250 GiB. To deal with processing such large data sets we used a combination of tiling and parallel processing, as implemented via the snowfall package for R [50], to maximize the CPU usage and minimize the time required to produce predictions.

The spatial prediction process consists of four main steps:

1. preparation of gridded covariates (principal component analysis),
2. preparation of point data,
3. model fitting and
4. spatial prediction and construction of GeoTiffs.

From the steps listed above, spatial prediction take the longest computing time, which is often in the order of 20 or more hours using the computer specification listed above. As a rule of thumb, we look for mapping frameworks that can generate outputs within 48 hrs. If the whole process from model fitting to prediction and export of maps to GeoTiffs consumes >48 hrs of computing, we consider the system to be impractical for routine operational use.

Results

Model fitting

The results of model fitting (Table 1) indicate that the distribution of soil organic carbon content is mainly controlled by climatic conditions, i.e. monthly temperatures and rainfall [51], while the distribution of texture fractions (sand, silt and clay) is mainly controlled by topography and lithology. These key predictors agree with expectations based on existing knowledge. The regression models account for between ca. 20–50% of observed variability in the target variables (Table 1). Detailed model parameters can be obtained from the SoilGrids1km homepage at http://soilgrids.org.

Figure 7 illustrates two examples of spatial predictions for soil organic carbon content and pH. As mentioned previously, soil organic carbon clearly decreases with depth (see also the soil-depth curves shown in Figure 8). Areas mapped as having elevated values of organic carbon are typically associated with cooler and wetter climate regimes and boreal-tundra type vegetation [51–54]. Note that several soil variables have skewed distributions hence also the output predictions are skewed, so that we use log-transformed legends to maximize contrast in the map (Figure 7).

Figure 8 shows predicted values for organic carbon and pH (mean value and confidence intervals) for the same location shown in Figure 5. The prediction intervals are rather wide (see also Figure 11), which is connected to the fact that the models explain only 23–51% of the variation. However, it is important to note that these are global maps of predictions made using relatively coarse resolution covariates. We assume that is unlikely that any effort to map the distribution of soils at a resolution of 1 km could explain a much larger proportion of the total variation in soil properties, as much of this variation occurs over distances less than 1 km [55].

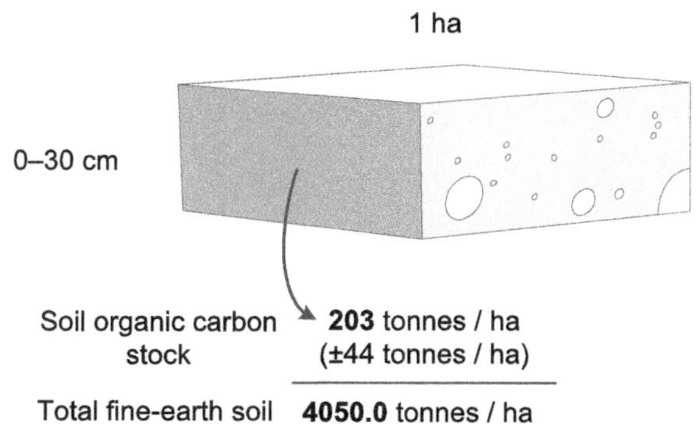

Bulk density (BLD):	1500 kg / m^3 (s.d. = ±100)
Organic carbon (ORC):	50‰ (s.d. = ±10)
Coarse fragments (CRF):	10% (s.d. = ±5)
Total volume of the block (HOT):	30 cm (· 1 ha)
Soil organic carbon stock (OCS):	203 tonnes / ha (±44)

$$\mathbf{OCS} = ORC/1000 \cdot BLD \cdot (100\text{-}CRF)/100 \cdot HOT/100$$
$$= 1/10{,}000{,}000 \cdot ORC \cdot BLD \cdot (100\text{-}CRF) \cdot HOT$$
$$= 1/10{,}000{,}000 \cdot 50 \cdot 1500 \text{ kg / m}^3 \cdot (100\text{-}10) \cdot 30 \text{ cm}$$
$$= 20.25 \text{ kg / m}^2 = 203 \text{ tonnes / ha}$$

$$\mathbf{OCS.sd} = 1/10{,}000{,}000 \cdot HOT \cdot \mathrm{sqrt}(\, BLD^2 \cdot (100 - CRF)^2 \cdot ORC.sd^2 +$$
$$+ BLD.sd^2 \cdot (100 - CRF)^2 \cdot ORC^2 + BLD^2 \cdot CRF.sd^2 \cdot ORC^2 \,)$$
$$= 4.4 \text{ kg / m}^2 = 44.1 \text{ tonnes / ha}$$

Figure 6. Soil organic carbon stock calculus scheme. Example of how total soil organic carbon stock (OCS) and its propagated error can be estimated for a given volume of soil using organic carbon content (ORC), bulk density (BLD), thickness of horizon (HOT), and percentage of coarse fragments (CRF). See text for more detail.

Also note that SoilGrids1km predictions are not capable of representing abrupt changes in values through depth e.g. due to buried horizons, textural heterogeneity or similar. Because we have used linear or close to linear models (plus smoothing splines) to predict values of targeted soil properties and not e.g. regression-trees, these models have smoothed out a significant amount of the variability in the point data, so that it is not realistic to expect abrupt changes in soil properties; at least not vertically (as illustrated previously in Figure 8).

Figure 9 (with a zoom in on Italy) shows that the SoilGrids1km predictions exhibit an order of magnitude greater spatial detail than previous global soil information products e.g. HWSD. This is mainly because a large stack of fine resolution remote sensing based covariate layers have been used to generate SoilGrids1km, and many of these have shown to be significantly correlated with soil properties and classes. Spatial classification accuracy for mapped soil classes, when evaluated using kappa statistics (Table 1), shows a somewhat better match between what was observed on the ground for the USDA classification system (ground-truth classification available for 16,212 profiles) than for the WRB system (classification available for 37,015 profiles).

For many WRB classes our models predicted occurrences in areas that are inconsistent with a strict definition of geographic areas where these classes can occur. The most difficult to map seem to be WRB classes such as Andosols, Solonchaks, Calcisols and Cryosols. These classes are strictly defined (e.g. Andosols are connected with volcanic activities and specific geology) and we need to explore ways to prepare covariates that will prevent

Table 1. Mapping performance of SoilGrids1km — amount of variation explained (from 100%) or purity/kappa for categorical variables — for eight targeted soil properties and two soil classes distributed via SoilGrids1km.

Variable name	Type	GSIF code	Units	Range (observed)	Amount of var. explained
Soil organic carbon (dry combustion)	3D	ORCDRC	g kg−1	0–450	22.9%
pH index (H2O solution)	3D	PHIH5X	10−1	2.1–11.0	50.5%
Sand content (gravimetric)	3D	SNDPPT	kg kg−1	1–94	23.5%
Silt content (gravimetric)	3D	SLTPPT	kg kg−1	2–74	34.9%
Clay content (gravimetric)	3D	CLYPPT	kg kg−1	2–68	24.4%
Coarse fragments (volumetric)	3D	GRAVOL	cm3 cm−3	0–89	-
Bulk density (fine earth fraction)	3D	BLDVOL	kg m−3	250–2870	31.8%
Cation-exchange capacity (fine earth fraction)	3D	CEC	cmol+/kg	0–234	29.4%
Depth to bedrock	2D	DBR	cm	0–240	-
Soil group (WRB taxonomy)	2D	TAXGWRB	-	-	28.1% (kappa)
Soil suborder (USDA taxonomy)	2D	TAXOKST	-	-	40.3% (kappa)

WRB = "World Reference Base"; USDA = "United States Department of Agriculture".
Amount of variation explained by the models (Eq.5) i.e. kappa statistics for soil types was determined using 5-fold cross-validation.

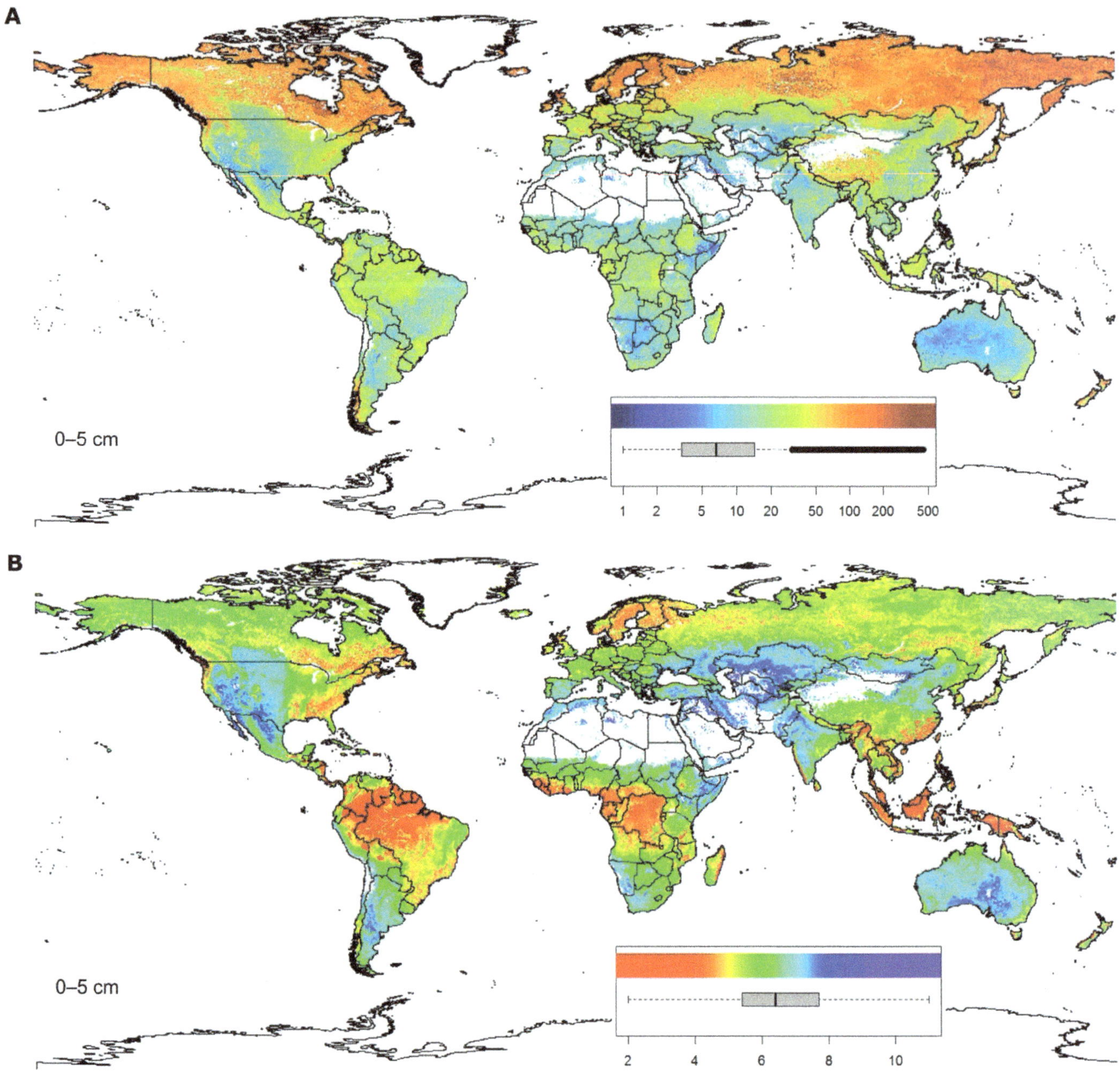

Figure 7. Example of SoilGrids1km layers: (A) soil organic carbon content in permille, and (B) soil pH for the topsoil (0–5 centimetres). Boxplots show the sampled distribution of the soil property based on the present compilation of global soil profile data.

prediction of those classes in areas where, by definition, they should not occur. Likewise, USDA suborders are based on soil moisture and climate regimes, for which we did not currently have global covariate maps, and consequently strictly defined classes such as Xerolls (Mollisols in Mediterranean climate; xeric moisture regime) were predicted in Brazil, which probably does not match the definition of the class.

Multinomial logistic regression is a purely data-driven method, so that the overall mapping performance highly depends on representation of environmental conditions by soil samples. All classes that are poorly represented in the environmental space, due to under-sampling, are understandably difficult to map accurately using a purely data- driven model [56]. Nevertheless, the final results of automated extraction of soil classes using multinomial logistic regression are promising, especially for mapping the USDA classes. The mapping accuracy could probably be improved by adding more classification-related covariates and more field observations of soil taxonomy, hopefully through crowd-sourcing, in areas where the accuracy is critically low.

Figure 10 shows derived total soil organic carbon stock based on Eq.(6). According to this map, the total (baseline) amount of soil organic carbon (up to 2 m depth; excluding deserts, bare rock areas and ice caps) is about 330 t ha^{-1} on average. The highest concentrations of soil organic carbon are in areas of cooler climate and high rainfall, i.e. northern parts of Canada and Russia seem to be pools for most of the world's soil organic carbon. This largely agrees with results by Hugelius et al. [53] and Scharlemann et al. [57].

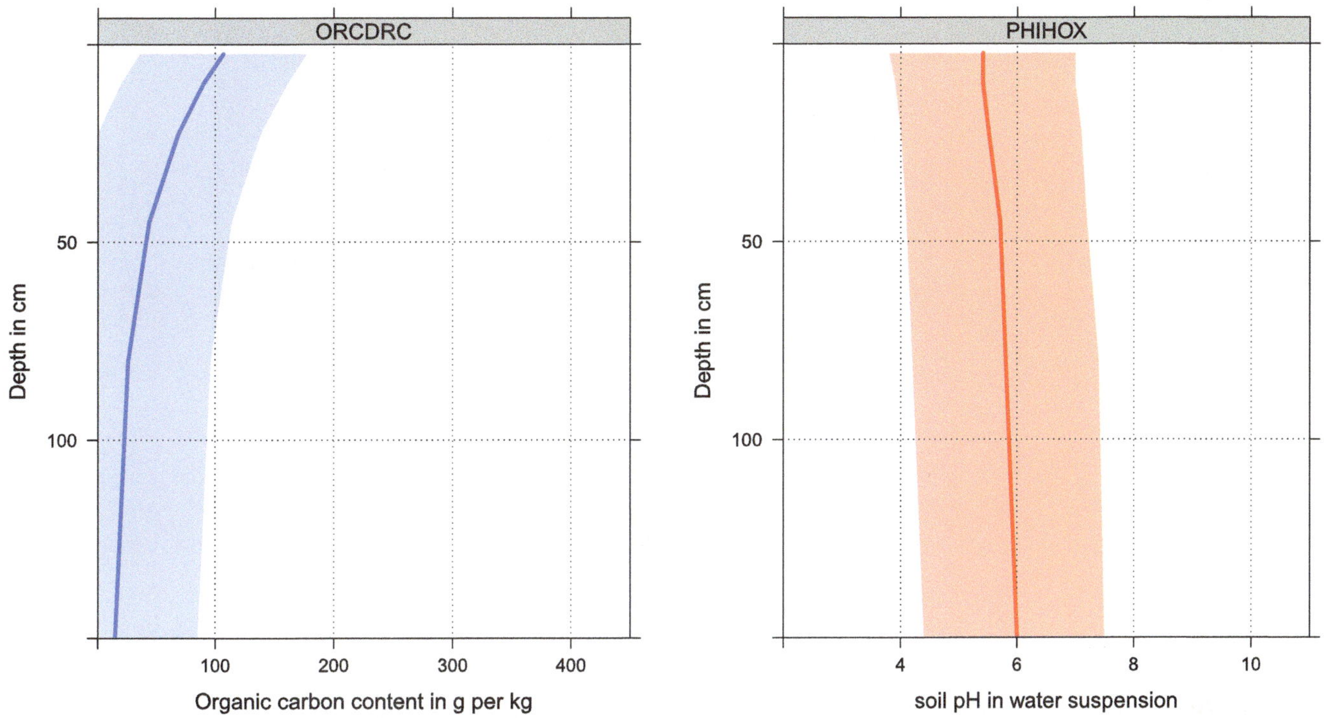

Figure 8. SoilGrids1km-derived soil-depth curves for the profile shown in Figure 5. Location of the profile: 6.3831°E, 50.479167°N. The shaded background indicates the 90% prediction interval for each depth. ORCDRC = soil organic carbon content in permilles; PHIHOX = soil pH in water suspension. See also Table 1.

The map shown in Figure 10 can be used to supplement maps of total aboveground biomass (see e.g. Ruesch and Gibbs [58] and Scharlemann et al. [57]). Our results also confirm that, overall, the amount of organic carbon below ground is greater than held in biomass above ground [51].

Quality issues

The results of cross-validation are shown in Table 1. The cross-validation results, as expected, largely reflect the model fitting success — properties that can be modeled successfully can also be mapped with higher accuracy. The soil properties that were most difficult to map are soil texture fractions, CEC and WRB soil groups. Although the accuracies of the predictions rarely exceed 50% of the total variation, all statistical models are significant showing clear spatial patterns (see e.g. Figure 7). Low cross-validation percentages are common in soil mapping [38,55], these numbers were not unexpected. Nevertheless, these can be considered promising initial results considering the complexity of harmonization of input point data (see further discussion).

Based on the feedback we received to date from users visiting the project homepage at http://soilgrids. org, the main limitations of SoilGrids1km are:

1. problems arising from poor relationships between covariates and dependent variables e.g. covariates can only explain part of the variability, which could possibly be improved by using more sophisticated statistical models;
2. problems arising from high spatial clustering of sampling locations (see Figure 2; observations are too sparse to improve on the regression using a kriging step);
3. problems associated with using partially-harmonized soil profile data;
4. problems arising from use of HWSD soil mapping units that are of too coarse scale and often not completely harmonized so that the country borders are still visible (obvious artefact);
5. limitations in the usability of SoilGrids1km for spatial planning at county or farm scale due to coarse resolution of the maps;
6. inability to consider and model significant sources of variability e.g. temporal variability due to changes in land use and/or land cover [59];
7. limitations arising from insufficient use of higher quality and finer resolution conventional soil maps prepared at national to regional scales.

Discussion

SoilGrids1km were released on December 5th 2013 (World Soil Day) at the FAO Rome, as a proposed contribution of the Netherlands to the Global Soil Partnership [60]. The system, at the moment, includes predicted values for (Table 1): soil organic carbon (g kg^{-1}), soil pH, sand, silt and clay fractions (%), bulk density (kg m^{-3}), cation-exchange capacity (cmol+/kg) of the fine earth fraction, coarse fragments (%), soil organic carbon stock (t ha^{-1}), depth to bedrock (in cm; see Figure 4), World Reference Base soil groups [43], and USDA Soil Taxonomy suborders [44]. We focused on generating spatial predictions at six standard depths (0–5 cm, 5–15 cm, 15–30 cm, 30–60 cm, 60–100 cm and 100–200 cm), for which spatially distributed estimates of upper and lower level 90% prediction intervals are presented. As such, we follow the corresponding specifications of the GlobalSoilMap project [3].

Initial predictions of soil classes were made at higher (more general) taxonomic levels for both WRB (soil groups) and Soil

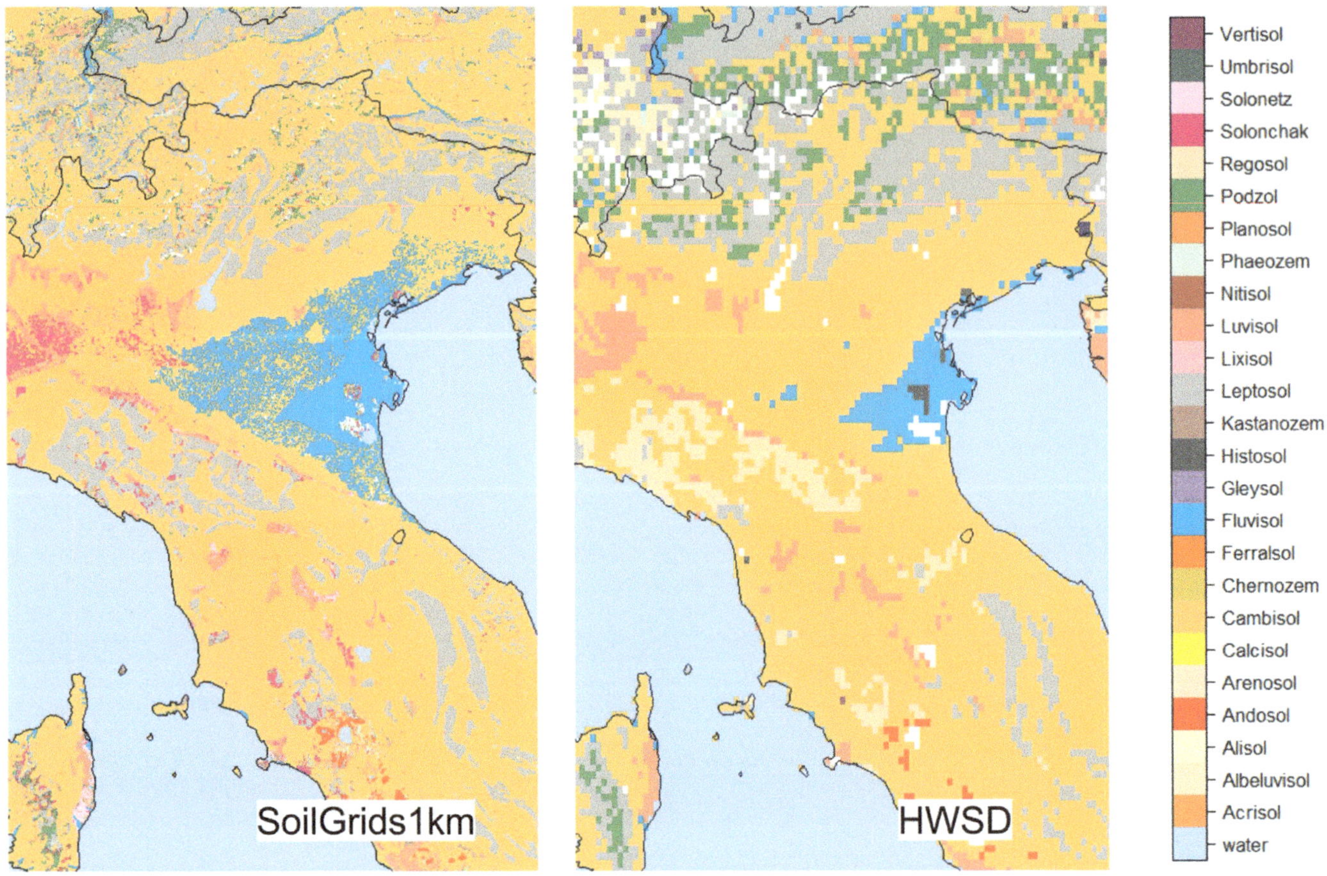

Figure 9. Spatial predictions of WRB soil groups for SoilGrids1km (left) and HWSD data set representing conventional soil maps (right). A zoom in on North of Italy. White pixels indicate missing values.

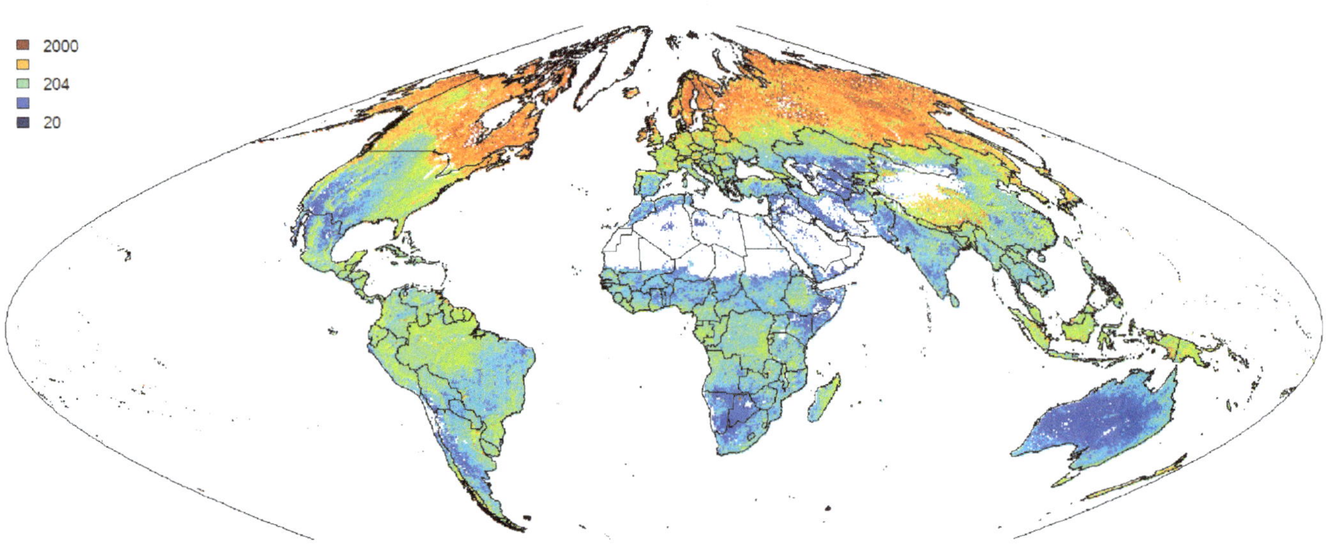

Figure 10. Predicted global distribution of the soil organic carbon stock in tonnes per ha for 0–200 centimetres. Total soil organic carbon stock (here displayed on a log-scale) was estimated as a sum of soil organic carbon stocks for six standard depths and adjusted for the depth to bedrock. Projected in the Sinusoidal equal area projection to give a realistic presentation of areas. Vast deserts (e.g. Sahara or Gobi) can be assumed to contain close to zero organic carbon stock. See also Figure 11.

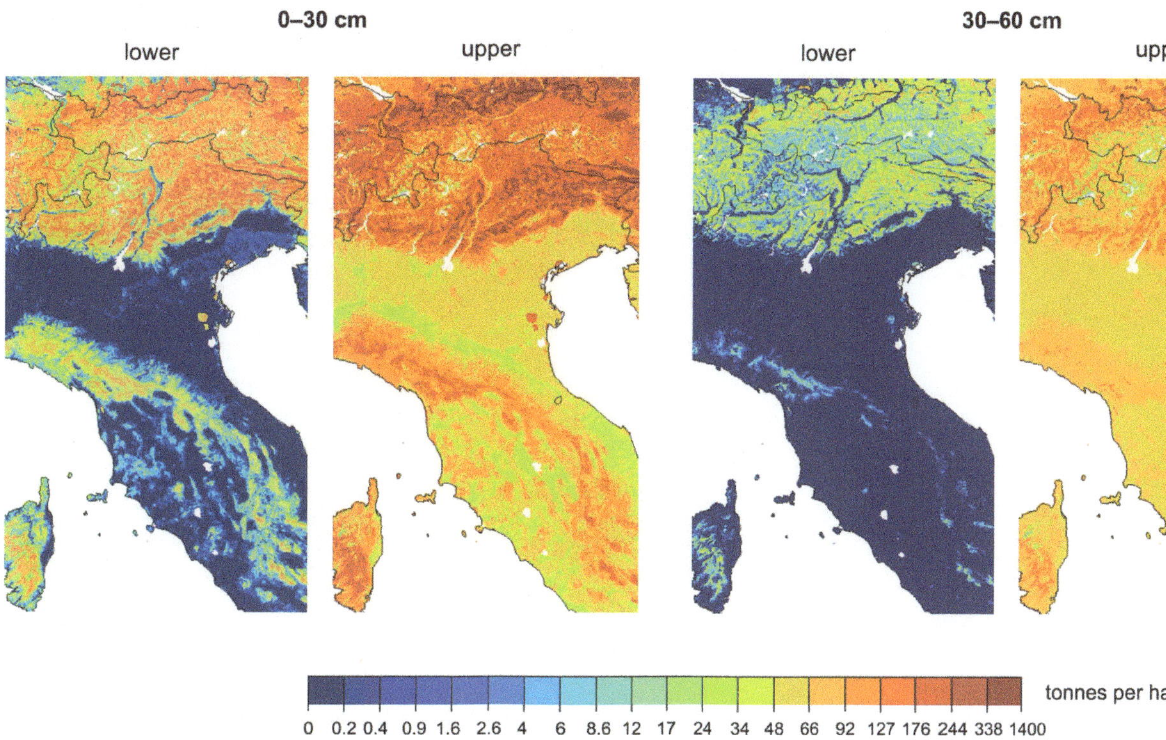

Figure 11. Lower and upper confidence limits (90% probability) of estimated soil organic carbon stock (tonnes per ha) for standard depths 0–30 and 30–60 centimeters for the same area as shown in Figure 9. Derived using the procedure explained in Figure 6.

Taxonomy (suborders). This was done because the available point profile data sets do not provide a sufficient number of locations representative of all of the lower levels of classification in each system. Without a sufficient number of examples for all lower classes, distributed fully across all of the feature space within which each class can occur, it is not possible to successfully predict many of the lower classes defined for either system. Once we have more point observations that encompass the full range of lower level classes across the entire environmental and geographic spectrum of their distribution, we will be able to predict at a more detailed taxonomic level for both classification systems.

The main purpose of SoilGrids1km is to provide initial, fully worked, examples of how complete and consistent global maps of soil properties, and soil classes, can be produced using currently available legacy soil profile data, freely available gridded maps of global covariates and an on-line automated soil mapping system (GSIF). Additionally, we want to use these initial example maps to implement and demonstrate procedures and systems for supporting free and unrestricted access to what we consider to be the best possible current, globally-complete, estimates of soil properties and soil classes. It is hoped that the production, distribution and use of these new, initial, global soil maps will stimulate additional efforts to both improve these maps and to launch new efforts to collect and use new soils information in new soil mapping and monitoring projects. We especially aim at supporting countries in Africa, and large parts of Asia and Latin America, that often have limited infrastructures to produce soil information at fine resolution [2,5]. We think that there is a great potential in using the existing field observations and Open Source software to map spatial and spatio-temporal patterns, i.e. without doing any major financial investments.

A number of legitimate concerns exist relative to the initial SoilGrids1km outputs. Probably the most immediate and significant concern has to do with the accuracy and usability of the initial predictions of soil property and class values. We acknowledge that the accuracy of these initial predictions rarely exceeds 50% of the total variation and, for many properties, is often closer to 20–30% (Table 1). The results of cross-validation are informative but need to be taken with caution because most of the soil profiles (Figure 2) were not collected using probability sampling, so that the cross-validation results possibly carry the same sampling bias as the original data [61]. Also note that the accuracy of mapping WRB groups is likely lower than the accuracy of mapping USDA soil suborders because over 40% of the soil profiles that were used for the WRB classification were actually classes translated from national systems. Translation i.e. harmonization of international soil records probably introduces additional noise that cannot be solved by regression modelling.

We argue that it is unreasonable to expect any global map of variation in soil properties to explain much more than 50% of the total observed variation. It is well known that a significant proportion of spatial variation in soil properties occurs over relatively short distances of meters to tens of meters [55,56]. It is therefore unreasonable to expect that a map of global variation in soil properties, portrayed at a spatial resolution of 1 km, will be able to capture and portray the 50% or more of total variation that occurs at resolutions shorter than 1000 m. Our hope and plan is to gradually improve the accuracy of the predictions by addressing these issues and concerns one by one, in a systematic way (Figure 13). This should be done primarily by working with national and regional soil data agencies, i.e. by adding additional covariates at increasingly finer spatial resolutions and by adding more field/point data from areas that are under-represented.

Although millions of soil profile records have undoubtedly been collected throughout the world, they are often unequally distributed (Figure 2). Likewise, many soil profiles funded by

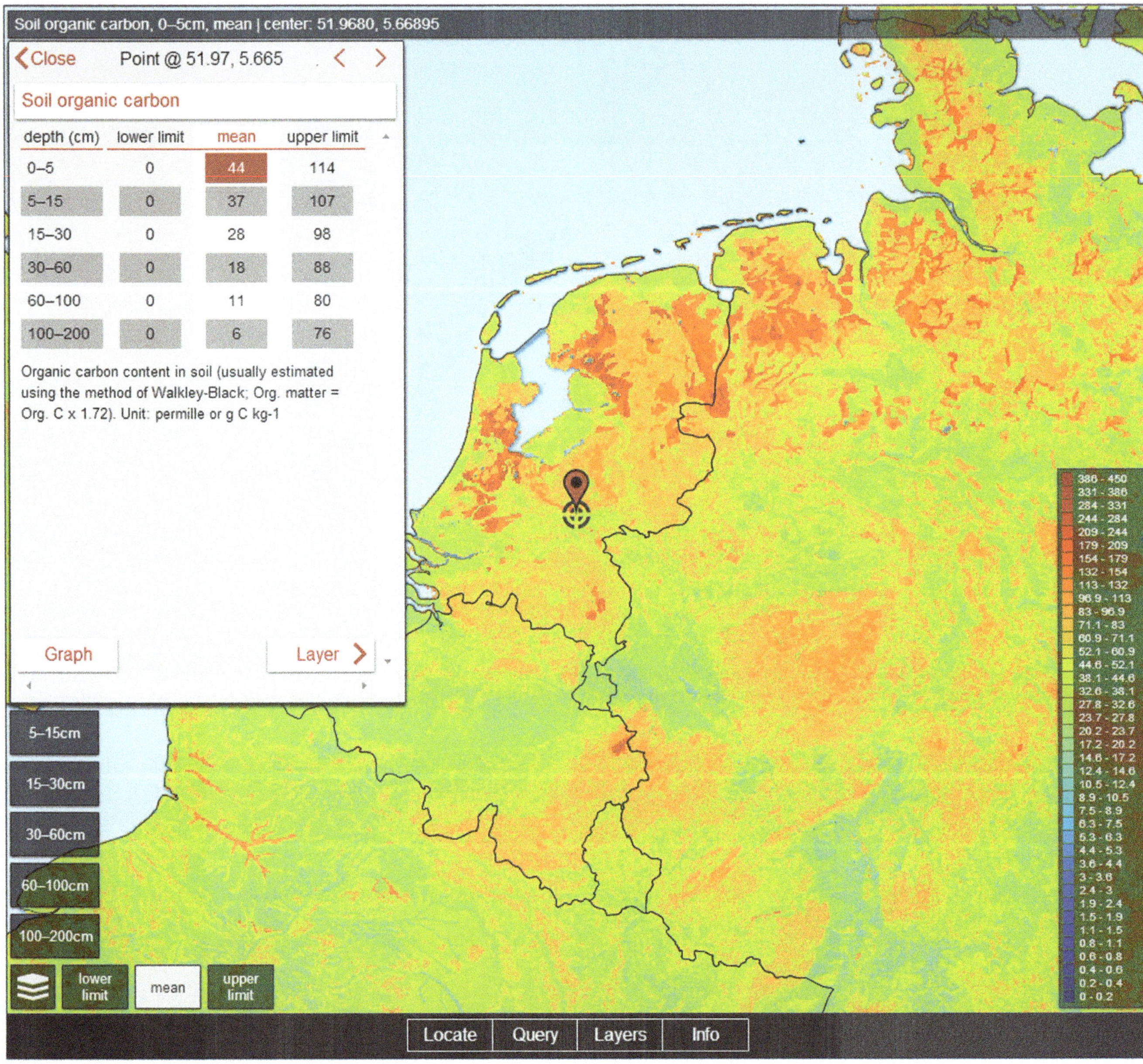

Figure 12. Accessing SoilGrids1km from the SoilInfo app for mobile devices. SoilInfo app is available for download via http://soilinfo.isric.org.

public money are not publicly available or are available in paper format only. Due to unbalanced representation and spatial clustering, predictions in the current version of SoilGrids1km are largely controlled by point data sets available for the USA and Europe. Most of these are from agricultural soils, which inflicts additional bias. Our predictions are therefore likely to exhibit lower accuracy for poorly represented areas such as most of the former Russian Federation, the northern Circumpolar Region, semi-arid and arid areas.

We have also purposely excluded all areas that show no evidence of historical vegetative cover. Our predictions are hence not globally complete. This is a definite drawback for use in global modelling and we acknowledge a need to use either expert judgment or data from other mapping sources to provide alternative predictions for areas with missing values. Again, for deserts and bare rock areas it is perfectly valid to assume a 0 value for soil organic carbon, but it is not as straightforward to estimate soil pH for shifting sand areas for example. For the present, we argue that it is inappropriate to try to make predictions for areas that completely lack vegetative cover e.g. shifting sands of Sahara. These areas have very few to zero point profile observations which can be used to calibrate statistical prediction models. In addition, even if they did have a sufficient number of point profile measurements, the environments of extreme climatic conditions are so different from vegetated ones so that any prediction model is likely to be very different from ones we develop for vegetated areas. We recommend that SoilGrids1km users who require values for the complete land mask fill in the gaps by using expert knowledge or best regional estimates as available from conventional soil mapping (e.g. HWSD, ISRIC-WISE).

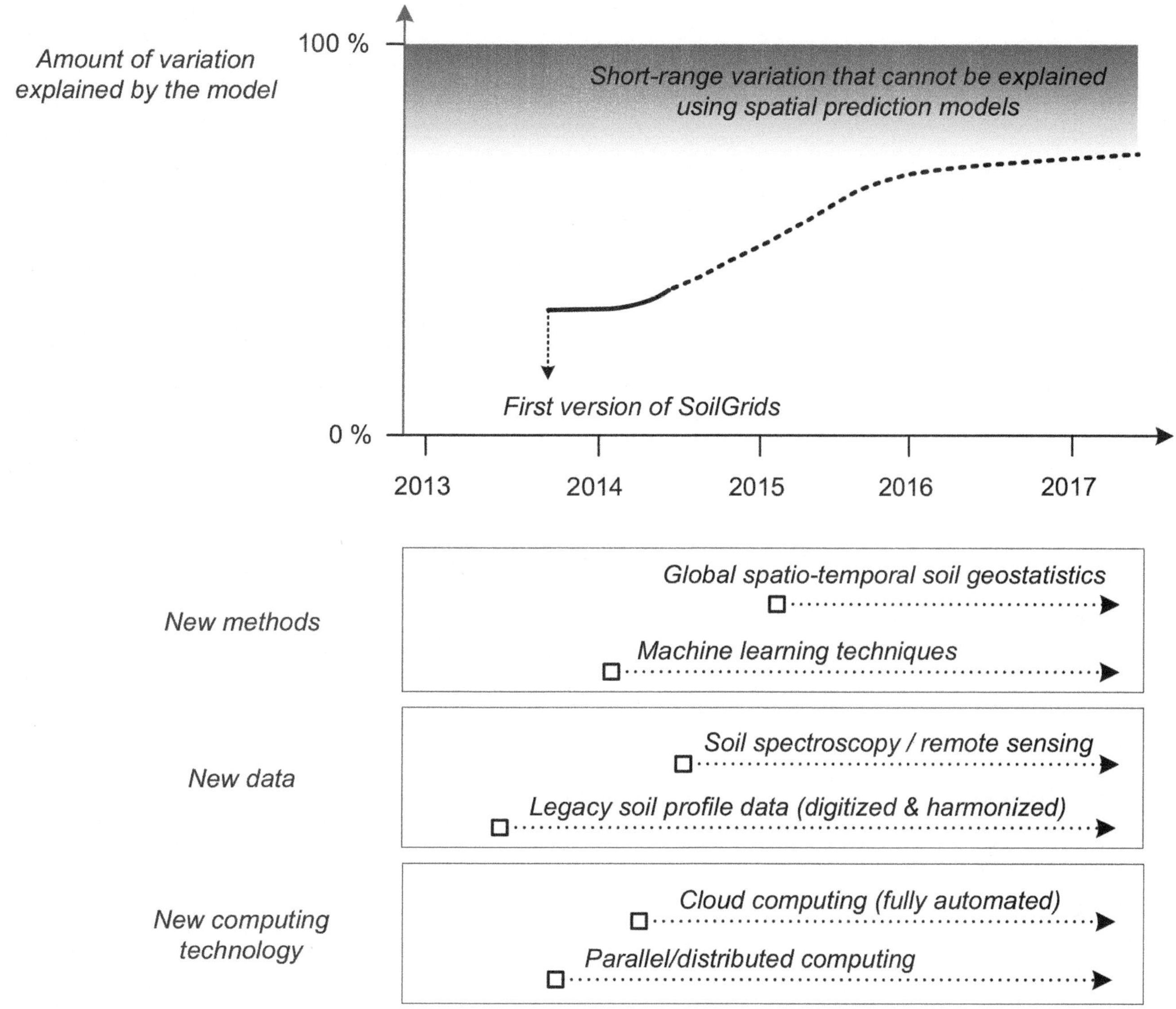

Figure 13. Projected evolution of SoilGrids in the years to come. We anticipate that the main drivers of success of SoilGrids will be use of machine learning methods for model fitting, development of spatio-temporal geostatistical models, use of new sources of field and remote sensing data and use of faster and more powerful computing capacities. Amount of variation explained by these models will eventually reach a *'natural limit'* (short-range variation that cannot be explained using spatial prediction models), until there is a technological jump in soil remote sensing technology e.g. ground penetrating scanners.

It is worth emphasizing that we designed GSIF as a flexible framework with respect to the choice of depths, dimensions (2D or 3D spatial predictions), spatial support size, soil properties and classes and prediction models. Outputs from GSIF are reproducible as a result of use of scripting. Consequently, all maps can be easily updated as new inputs (point and covariate data) become available. We used the GSIF system to generate SoilGrids1km maps for the standard depths defined by the GlobalSoilMap project, but basically one could use the same system for any depth and also for any new property. GSIF is therefore scalable and can be used to produce spatial predictions for virtually any soil property, at any depth and at any spatial or temporal resolution. This, of course, assumes the existence of a sufficient number of point soil observations of appropriate quality and of sufficient covariate layers at sufficiently fine spatial resolution to support modelling at a given spatial resolution.

All methods and models fitted for the purpose of producing SoilGrids1km are available via an Open Source platform (GSIF package for R) and could be adapted for both regional and local mapping. As with input data, the models used to make predictions in GSIF can be improved or replaced in subsequent iterations once better performing models are identified. Prediction models that could be considered in the future include those based on hierarchical Bayes models, regression trees, Random Forests and other machine learning techniques. Regression- trees and similar models could help model better abrupt changes in values vertically, and Random Forests could help emphasize relative importance of specific covariates. The actual modelling approach used to produce any set of predictions will be reviewed continuously to identify and apply the approach that produces the most correct, consistent and usable outputs.

Because the SoilGrids1km maps can be easily updated (or changed) the process used to produce the map (i.e. SoilGrids system) becomes more important than the map itself. Previously, the map product was seen as more important than the process used to produce it, because any map had to be considered as valid and useful for an extended period, as it took so long, and cost so much, to revise or update the map. Under the GSIF model, the final (or most current) map is no longer the most important output and any system that only provides a final map is considered deficient. We hence argue that it is more important to provide access to all data and models needed to produce (and reproduce) the map than to simply provide the final map itself.

In the future, we hope that GSIF will be used by an increasing number and variety of interested parties, including national and regional soil mapping agencies, commercial consulting agencies, advocacy groups and non-governmental organizations. We envisage GSIF as a platform for cooperation, collaboration, innovation and sharing. It will become so if interested parties decide to participate and contribute as committed partners. The number of soil profiles freely shared by the soil science community is constantly growing and national agencies and other data providers are encouraged to contribute their point data to help improve the prediction accuracy locally for specific countries/ regions, for the benefit of the global user community and in support of the global UN conventions.

SoilGrids1km are available for download under a Creative Commons non-Commercial license via http://soilgrids.org. SoilGrids1km are also accessible via a Representational State Transfer (http://rest. soilgrids.org) service and via a mobile phone app "SoilInfo App" (http://soilinfo-app.org; Figure 12).

Acknowledgments

SoilGrids were developed as a part of the Global Soil Information Facilities (GSIF) initiative — tools for collating and serving global soil data developed jointly by ISRIC — World Soil Information and collaborators. ISRIC is a non-profit organization primarily funded by the Dutch government. SoilGrids1km were possible due to generous contributions of point and covariate data by various national and international agencies and organizations to which we are sincerely grateful: USDA National Resources Conservation Service, European Commission's Joint Research Centre, Africa Soil Information Service project, Mexican Instituto Nacional de Estadística y Geografía/CONABIO, University of São Paulo, Agriculture and Agri-Food Canada, Institute of Soil Science, Chinese Academy of Sciences, Tehran University (National Database of Iran), and others. We are especially thankful for support from the Africa Soil Information Service (AfSIS) project, funded by the Bill and Melinda Gates foundation and the Alliance for a Green Revolution in Africa (AGRA), and to the Food and Agriculture Organization (FAO) of the United Nations for supporting the international soil standards and initiatives developed at ISRIC (SOTER, GSIF). The funders had no role in study design, data collection and analysis, decision to publish, or preparation of the manuscript. We gratefully acknowledge the past and present leadership of ISRIC (Prem Bindraban, Peter de Ruiter and Hein van Holsteijn) for recognizing the potential value of GSIF and SoilGrids1km and for consistently and actively supporting development both within ISRIC and in interactions with external agencies and partner organizations. We also gratefully acknowledge the significant contributions of Dr. H.I. Reuter, formerly with ISRIC, presently statistical officer at EUROSTAT, in developing the initial concepts and design for GSIF, in helping to conceptualize and assemble the global maps of covariates and in implementing some of the early algorithms.

Author Contributions

Conceived and designed the experiments: TH. Contributed to the writing of the manuscript: TH BM GH BK MW. Prepared input profile and covariate data: NB JL MG AR. Web mapping services and user interfaces: JJ ER. Cross-validation and quality control: AR GH BK NB RAM. Design of the system, programming and preparation of input data: TH JJ.

References

1. Sanchez PA, Ahamed S, Carré F, Hartemink AE, Hempel J, et al (2009) Digital Soil Map of the World. Science 325: 680–681.
2. Omuto C, Nachtergaele F, Vargas Rojas R (2012) State of the Art Report on Global and Regional Soil Information: Where are we? Where to go? Global Soil Partnership technical report. Rome: FAO, 66 pp.
3. Arrouays D, Grundy MG, Hartemink AE, Hempel JW, Heuvelink GB, et al. (2014) Chapter Three — GlobalSoilMap: Toward a Fine-Resolution Global Grid of Soil Properties. In: Sparks DL, editor, Soil carbon, Academic Press, volume 125 of Advances in Agronomy. pp. 93–134.
4. Viscarra Rossel RA, Webster R, Bui EN, Baldock JA (2014) Baseline map of organic carbon in Australian soil to support national carbon accounting and monitoring under climate change. Global Change Biology 20(9): 2953–2970.
5. Grunwald S, Thompson JA, Boettinger JL (2011) Digital Soil Mapping and Modeling at Continental Scales: Finding Solutions for Global Issues. Soil Science Society of America Journal 75: 1201–1213.
6. Savtchenko A, Ouzounov D, Ahmad S, Acker J, Leptoukh G, et al. (2004) Terra and Aqua MODIS products available from NASA GES DAAC. Advances in Space Research 34: 710–714.
7. Scharlemann JPW, Benz D, Hay SI, Purse BV, Tatem AJ, et al. (2008) Global data for ecology and epidemiology: a novel algorithm for temporal Fourier processing MODIS data. PLoS One 3: e1408.
8. Hansen MC, Potapov PV, Moore R, Hancher M, Turubanova SA, et al. (2013) High-resolution global maps of 21st-century forest cover change. Science 342: 850–853.
9. González JH, Bachmann M, Krieger G, Fiedler H (2010) Development of the TanDEM-X calibration concept: analysis of systematic errors. Geoscience and Remote Sensing, IEEE Transactions on 48: 716–726.
10. Hartemink AE, Krasilnikov P, Bockheim J (2013) Soil maps of the world. Geoderma 207/208: 256–267.
11. FAO/IIASA/ISRIC/ISS-CAS/JRC (2012) Harmonized World Soil Database (version 1.2). Rome: FAO.
12. Batjes NH (2012) ISRIC-WISE derived soil properties on a 5 by 5 arc-minutes global grid (ver. 1.2). Report 2012/01. Wageningen: ISRIC — World Soil Information, 57 pp.
13. Diggle PJ, Ribeiro Jr PJ (2007) Model-based Geostatistics. Springer Series in Statistics. Springer, 288 pp.
14. Kilibarda M, Hengl T, Heuvelink GB, Graeler B, Pebesma E, et al. (2014) Spatio-temporal interpolation of daily temperatures for global land areas at 1 km resolution. Journal of Geophysical Research: Atmospheres, 119(5): 2294–2313.
15. Pebesma E, Cornford D, Dubois G, Heuvelink GB, Hristopulos D, et al. (2011) INTAMAP: The design and implementation of an interoperable automated interpolation web service. Computers & Geosciences 37: 343–352.
16. Fritz S, McCallum I, Schill C, Perger C, See L, et al. (2012) Geo-Wiki: An online platform for improving global land cover. Environmental Modelling & Software 31: 110–123.
17. R Development Core Team (2009) R: A language and environment for statistical computing. Vienna, Austria: R Foundation for Statistical Computing, 409 pp. ISBN 3-900051-07-0.
18. Shore J (2007) The Art of Agile Development. Theory in practice. O'Reilly Media, 440 pp.
19. Tóth G, Jones A, Montanarella L, editors (2013) LUCAS Topsoil Survey. Methodology, data and results. JRC Technical Reports EUR 26102. Luxembourg: Publications Office of the European Union.
20. Leenaars J (2012) Africa Soil Profiles Database, Version 1.0. A compilation of geo-referenced and standardized legacy soil profile data for Sub Saharan Africa (with dataset). Wageningen, the Netherlands: Africa Soil Information Service (AfSIS) project and ISRIC — World Soil Information, 45 pp. Available: http://www.isric.org. ISRIC report 2012/03.
21. Instituto Nacional de Estadística y Geografía (INEGI) (2000) Conjunto de Datos de Perfiles de Suelos, Escala 1: 250 000 Serie II. (Continuo Nacional). Aguascalientes, Ags. México: INEGI. INEGI2000.
22. Cooper M, Mendes LMS, Silva WLC, Sparovek G (2005) A national soil profile database for brazil available to international scientists. Soil Science Society of America Journal 69: 649–652.
23. Shangguan W, Dai Y, Liu B, Zhu A, Duan Q, et al. (2013) A China data set of soil properties for land surface modeling. Journal of Advances in Modeling Earth Systems 5: 212–224.

24. MacDonald KB, Valentine KWG (1992) CanSIS/NSDB. A general description. Ottawa: Centre for Land and Biological Resources Research, Research Branch, Agriculture Canada.
25. Batjes NH (2009) Harmonized soil profile data for applications at global and continental scales: Updates to the WISE database. Soil Use and Management 25: 124–127.
26. Van Engelen V, Dijkshoorn J, editors (2012) Global and National Soils and Terrain Digital Databases (SOTER), Procedures Manual, version 2.0. ISRIC Report 2012/04. Wageningen, the Netherlands: ISRIC — World Soil Information, 192 pp.
27. Hollis JM, Jones RJA, Marshall CJ, Holden A, Van de Veen JR, et al. (2006) SPADE-2: The soil profile analytical database for Europe, version 1.0. Luxembourg: Office for official publications of the European Communities. EUR22127EN.
28. Stolbovoi V, McCallum I (2002) Land Resources of Russia (CD-ROM). Vienna: IIASA and RAS.
29. Hijmans RJ, Cameron SE, Parra JL, Jones PG, Jarvis A (2005) Very high resolution interpolated climate surfaces for global land areas. International Journal of Climatology 25: 1965–1978.
30. Hartmann J, Moosdorf N (2012) The new global lithological map database GLiM: A representation of rock properties at the Earth surface. Geochemistry, Geophysics, Geosystems 13: n/a–n/a.
31. Carroll M, Townshend J, DiMiceli C, Noojipady P, Sohlberg RA (2009) A new global raster water mask at 250 m resolution. International Journal of Digital Earth 2: 291–308.
32. Odeh I, McBratney AB, Chittleborough D (1995) Further results on prediction of soil properties from terrain attributes: heterotopic cokriging and regression-kriging. Geoderma 67: 215–226.
33. Hengl T, Heuvelink G, Rossiter DG (2007) About regression-kriging: from equations to case studies. Computers & Geosciences 33: 1301–1315.
34. Kutner M, Neter J, Nachtsheim C, Li W (2005) Applied Linear Statistical Models. Operations and decision sciences series. McGraw-Hill Irwin.
35. Hastie TJ (1992) Statistical Models in S, Wadsworth & Brooks/Cole, Chapter: Generalized additive models. pp. 249–307.
36. Venables WN, Ripley BD (2002) Modern applied statistics with S. New York: Springer-Verlag, 4th edition, 481 pp.
37. Agarwal DK, Gelfand AE, Citron-Pousty S (2002) Zero-inflated models with application to spatial count data. Environmental and Ecological statistics 9: 341–355.
38. McBratney AB, Minasny B, MacMillan RA, Carré F (2011) Digital soil mapping. In: Li H, Sumner M, editors, Handbook of Soil Science, CRC Press, volume 37. pp. 1–45.
39. Mokma DL, Buurman P (1982) Podzols and podzolization in temperate regions. ISM monograph. Wageningen: International Soil Museum, 126 pp.
40. Minasny B, McBratney AB (2010) Methodologies for Global Soil Mapping. In: Boettinger JL, Howell DW, Moore AC, Hartemink AE, Kienast-Brown S, editors, Digital Soil Mapping: Bridging Research, Environmental Application, and Operation, Springer, volume 2 of Progress in Soil Science, chapter 34. pp. 429–436.
41. Jenny H (1994) Factors of soil formation: a system of quantitative pedology. Dover books on Earth sciences. Dover Publications.
42. Bouma J, Batjes NH, Groot JJR (1998) Exploring land quality effects on world food supply. Geoderma 86: 43–59.
43. IUSS Working Group WRB (2006) World reference base for soil resources 2006: a framework for international classification, correlation and communication. World soil resources reports No. 103. Rome: Food and Agriculture Organization of the United Nations.
44. US Department of Agriculture (2010) Keys to Soil Taxonomy. U.S. Government Printing Office, 11th edition.
45. Hengl T, Nikolić M, MacMillan RA (2013) Mapping efficiency and information content. International Journal of Applied Earth Observation and Geoinformation 22: 127–138.
46. Goovaerts P (2001) Geostatistical modelling of uncertainty in soil science. Geoderma 103: 3–26.
47. Nelson DW, Sommers L (1982) Total carbon, organic carbon, and organic matter. In: Page A, Miller R, Keeney D, editors, Methods of soil analysis, Part 2, Madison, WI: ASA and SSSA, Agron. Monogr. 9. 2nd edition, pp. 539–579.
48. Heuvelink G (1998) Error propagation in environmental modelling with GIS. London, UK: Taylor & Francis.
49. Bivand R, Pebesma E, Rubio V (2013) Applied Spatial Data Analysis with R. Use R Series. Heidelberg: Springer, 2nd edition, 401 pp.
50. Knaus J, Porzelius C, Binder H, Schwarzer G (2009) Easier parallel computing in R with snowfall and sfCluster. The R Journal 1: 54–59.
51. Jobbágy EG, Jackson RB (2000) The vertical distribution of soil organic carbon and its relation to climate and vegetation. Ecological applications 10: 423–436.
52. Todd-Brown KEO, Randerson JT, Post WM, Hoffman FM, Tarnocai C, et al. (2013) Causes of variation in soil carbon simulations from CMIP5 Earth system models and comparison with observations. Biogeosciences 10: 1717–1736.
53. Hugelius G, Strauss J, Zubrzycki S, Harden JW, Schuur E, et al. (2014) Improved estimates show large circumpolar stocks of permafrost carbon while quantifying substantial uncertainty ranges and identifying remaining data gaps. Biogeosciences Discussions 11: 4771–4822.
54. Minasny B, McBratney AB, Malone BP, Lacoste M, Walter C (2014) Quantitatively Predicting Soil Carbon Across Landscapes. In: Hartemink AE, McSweeney K, editors, Soil Carbon, Springer International Publishing, Progress in Soil Science. pp. 45–57.
55. Heuvelink GBM, Webster R (2001) Modelling soil variation: past, present, and future. Geoderma 100: 269–301.
56. Antonić O, Pernar N, Jelaska SD (2003) Spatial distribution of main forest soil groups in Croatia as a function of basic pedogenetic factors. Ecological modelling 170: 363–371.
57. Scharlemann JPW, Tanner EVJ, Hiederer R, Kapos V (2014) Global soil carbon: understanding and managing the largest terrestrial carbon pool. Carbon Management 5: 81–91.
58. Ruesch A, Gibbs HK (2008) New IPCC Tier-1 global biomass carbon map for the year 2000. Oak Ridge National Laboratory, Tennessee: Carbon Dioxide Information Analysis Center.
59. Verburg PH, Neumann K, Nol L (2011) Challenges in using land use and land cover data for global change studies. Global Change Biology 17: 974–989.
60. Montanarella L, Vargas R (2012) Global governance of soil resources as a necessary condition for sustainable development. Current Opinion in Environmental Sustainability 4: 559–564.
61. Brus D, Kempen B, Heuvelink G (2011) Sampling for validation of digital soil maps. European Journal of Soil Science 62: 394–407.

PERMISSIONS

All chapters in this book were first published in PLOS ONE, by The Public Library of Science; hereby published with permission under the Creative Commons Attribution License or equivalent. Every chapter published in this book has been scrutinized by our experts. Their significance has been extensively debated. The topics covered herein carry significant findings which will fuel the growth of the discipline. They may even be implemented as practical applications or may be referred to as a beginning point for another development.

The contributors of this book come from diverse backgrounds, making this book a truly international effort. This book will bring forth new frontiers with its revolutionizing research information and detailed analysis of the nascent developments around the world.

We would like to thank all the contributing authors for lending their expertise to make the book truly unique. They have played a crucial role in the development of this book. Without their invaluable contributions this book wouldn't have been possible. They have made vital efforts to compile up to date information on the varied aspects of this subject to make this book a valuable addition to the collection of many professionals and students.

This book was conceptualized with the vision of imparting up-to-date information and advanced data in this field. To ensure the same, a matchless editorial board was set up. Every individual on the board went through rigorous rounds of assessment to prove their worth. After which they invested a large part of their time researching and compiling the most relevant data for our readers.

The editorial board has been involved in producing this book since its inception. They have spent rigorous hours researching and exploring the diverse topics which have resulted in the successful publishing of this book. They have passed on their knowledge of decades through this book. To expedite this challenging task, the publisher supported the team at every step. A small team of assistant editors was also appointed to further simplify the editing procedure and attain best results for the readers.

Apart from the editorial board, the designing team has also invested a significant amount of their time in understanding the subject and creating the most relevant covers. They scrutinized every image to scout for the most suitable representation of the subject and create an appropriate cover for the book.

The publishing team has been an ardent support to the editorial, designing and production team. Their endless efforts to recruit the best for this project, has resulted in the accomplishment of this book. They are a veteran in the field of academics and their pool of knowledge is as vast as their experience in printing. Their expertise and guidance has proved useful at every step. Their uncompromising quality standards have made this book an exceptional effort. Their encouragement from time to time has been an inspiration for everyone.

The publisher and the editorial board hope that this book will prove to be a valuable piece of knowledge for researchers, students, practitioners and scholars across the globe.

LIST OF CONTRIBUTORS

Wyatt H. Hartman and Curtis J. Richardson
Duke University Wetland Center, Nicholas School of the Environment, Duke University, Durham, North Carolina, United States of America

C. Ryan Penton and Amanda Pham
Center for Microbial Ecology, Michigan State University, East Lansing, Michigan, United States of America

V. V. S. R. Gupta, Paul Harvey, David K. Roget
CSIRO Ecosystem Sciences, Glen Osmond, South Australia, Australia

Stephen M. Neate
Department of Agriculture, Fisheries and Forestry, Queensland, Leslie Research Centre, Towoomba, Queensland, Australia

Kathy Ophel-Keller
SARDI, Glen Osmond, South Australia, Australia

Michael Gillings
Department of Biological Sciences, Macquarie University, North Ryde, New South Wales, Australia

Mohammad Haneef Khan, Mukesh K. Meghvansi, Rajeev Gupta, Vijay Veer
Defence Research Laboratory, Defence Research & Development Organisation, Post Bag 2, Tezpur, Assam, India

Lokendra Singh
Directorate General of Life Sciences, Defence Research & Development Organisation, New Delhi, India

Mohan C. Kalita
Department of Biotechnology and Bioengineering, Gauhati University, Guwahati, Assam, India

Guo-Chun Ding, Sven Jechalke, Holger Heuer and Kornelia Smalla
Julius Kühn-Institut - Federal Research Centre for Cultivated Plants (JKI), Institute for Epidemiology and Pathogen Diagnostics, Braunschweig, Germany

Viviane Radl, Brigitte Schloter-Hai and Michael Schloter
Helmholtz Zentrum Mu¨nchen, German Research Center for Environmental Health, Research Unit for Environmental Genomics, Neuherberg, Germany

Pingping Zhang
State Key Laboratory of Soil Erosion and Dryland Farming on the Loess Plateau, Northwest A & F University, Yangling, China

Minǵan Shao
State Key Laboratory of Soil Erosion and Dryland Farming on the Loess Plateau, Northwest A & F University, Yangling, China

Key Laboratory of Ecosystem Network Observation and Modeling, Institute of Geographical Science and Natural Resources, Chinese Academy of Sciences, Beijing, China

Shengzuo Fang , Dong Liu and Ye Tian
College of Forest Resources and Environment, Nanjing Forestry University, Nanjing, People's Republic of China

Shiping Deng and Xulan Shang
Department of Plant and Soil Sciences, Oklahoma State University, Stillwater, Oklahoma, United States of America

Reinildes Silva-Filho and JoséCola Zanuncio
Departamento de Biologia Animal, Universidade Federal de Vic¸osa, Vic¸osa, Brazil

Ricardo Henrique Silva Santos and Wagner de Souza Tavares
Departamento de Fitotecnia, Universidade Federal de Vic¸osa, Vic¸osa, Brazil

Germano Leão Demolin Leite
Instituto de Ciências Agrárias, Universidade Federal de Minas Gerais, Montes Claros, Brazil

Carlos Frederico Wilcken
Departamento de Protec¸ão Vegetal, Universidade Estadual Paulista, UNESP campus of Botucatu, Botucatu, Brazil

JoséEduardo Serrão
Departamento de Biologia Geral, Universidade Federal de Vic¸osa, Vic¸osa, Brazil

Samuel Dubelman, Joshua Fischer, Fatima Zapata, Kristin Huizinga, Changjian Jiang, Joshua Uffman and Steven Levine, David Carson
Regulatory Division, Monsanto Company, St. Louis, Missouri, United States of America

Ping Zhuan, Huanping Lu1 Zhian Li and Bi Zou
Key Laboratory of Vegetation Restoration and Management of Degraded Ecosystems, South China Botanical Garden, Chinese Academy of Sciences, Guangzhou, China

Murray B. McBride
Department of Crop and Soil Sciences, Cornell University, Ithaca, New York, United States of America

Huanjun Zhang, Weixin Ding, Hongyan Yu, Jianling Fan and Deyan Liu
State Key Laboratory of Soil and Sustainable Agriculture, Institute of Soil Science, Chinese Academy of Sciences, Nanjing, China

Xinhua He
State Key Laboratory of Soil and Sustainable Agriculture, Institute of Soil Science, Chinese Academy of Sciences, Nanjing, China
Forests NSW, NSW Department of Primary Industries, West Pennant Hills, New South Wales, Australia, 3 School of Plant Biology, University of Western Australia, Crawley, Western Australia, Australia

Sally Hilton, Amand J. Bennetta, Gary Keane, Gary D. Bending, David Chandler and Peter Mills
School of Life Sciences, University of Warwick, Wellesbourne, Warwickshire, United Kingdom

Ron Stobart
The National Institute of Agricultural Botany, Wymondham, Norfolk, United Kingdom

Lan Mu, Yinli Liang and Ruilian Han
Institute of Soil and Water Conservation, Northwest A&F University, Yangling Shaanxi Province, China
Institute of Soil and Water Conservation, Chinese Academy of Sciences and Ministry of Water Resources, Yangling, Shaanxi Province, China

Peng Zhang, Ting Wei, Zhikuan Jia, Qingfang Han and Xiaolong Ren
The Chinese Institute of Water-Saving Agriculture, Northwest A&F University, Yangling, Shaanxi, China
Key Laboratory of Crop Physi-Ecology and Tillage Science in Northwestern Loess Plateau, Ministry of Agriculture, Northwest A&F University, Yangling, Shaanxi, China

Yongping Li
Guyuan Institute of Agricultural Sciences, Guyuan, Ningxia, China

Liangang Mao, Taotao Ma, Pengfei Liu and Jin Shen
Department of Pesticides, Institute of Plant Protection, Chinese Academy of Agricultural Sciences, Ministry of Agriculture, Beijing, People's Republic of China

Qiuxia Wang, Dongdong Yan, Yuan Li, Canbin Ouyang, Meixia Guo and Aocheng Cao
Department of Pesticides, Institute of Plant Protection, Chinese Academy of Agricultural Sciences, Ministry of Agriculture, Beijing, People's Republic of China
State Key Laboratory for Biology of Plant Disease and Insect Pests, Beijing, People's Republic of China

Yidong Wang, Changcheng Guo and Qing Chen
Tianjin Key Laboratory of Water Resources and Environment, Tianjin Normal University, Tianjin, China

Zhong-Liang Wang
Tianjin Key Laboratory of Water Resources and Environment, Tianjin Normal University, Tianjin, China

State Key Laboratory of Environmental Geochemistry, Institute of Geochemistry, Chinese Academy of Sciences, Guiyang, China

Xiaoping Feng
Tianjin Key Laboratory of Water Resources and Environment, Tianjin Normal University, Tianjin, China
College of Urban and Environmental Sciences, Tianjin Normal University, Tianjin, China

Chunye Lin, Ruiping Li, Hongguang Cheng and Xiao Shao
State Key Joint Laboratory of Environmental Simulation and Pollution Control, School of Environment, Beijing Normal University, Beijing, China

Jing Wang
Key Laboratory of Land Use, China Land Surveying and Planning Institute, Ministry of land and Resources, Beijing, China

Guocheng Wang, Tingting L, Wen Zhang and Yongqiang Yu
State Key Laboratory of Atmospheric Boundary Layer Physics and Atmospheric Chemistry, Institute of Atmospheric Physics, Chinese Academy of Sciences, Beijing, China

Binu M. Tripathi and Jonathan M. Adams
Department of Biological Sciences, Seoul National University, Seoul, Republic of Korea

Dorsaf Kerfahi
Department of Biological Sciences, Seoul National University, Seoul, Republic of Korea
School of Chemical and Biological Engineering, Interdisciplinary Program of Bioengineering, Seoul National University, Seoul, Republic of Korea

Junghoon Lee
School of Chemical and Biological Engineering, Interdisciplinary Program of Bioengineering, Seoul National University, Seoul, Republic of Korea

David P. Edwards
Department of Animal and Plant Sciences, University of Sheffield, Sheffield, United Kingdom

Minggang Xu
Ministry of Agriculture Key Laboratory of Crop Nutrition and Fertilization, Institute of Agricultural Resources and Regional Planning, Chinese Academy of Agricultural Sciences, Beijing, China

Rihuan Cong
Ministry of Agriculture Key Laboratory of Crop Nutrition and Fertilization, Institute of Agricultural Resources and Regional Planning, Chinese Academy of Agricultural Sciences, Beijing, China
College of Resources and Environment, Huazhong Agricultural University, Wuhan, China

Xiujun Wang
State Key Laboratory of Desert and Oasis Ecology, Xinjiang Institute of Ecology and Geography, Chinese Academy of Sciences, Urumqi, China
Earth System Science Interdisciplinary Center, University of Maryland, College Park, Maryland, United States of America

Stephen M. Ogle and William J. Parton
Natural Resource Ecology Laboratory, Colorado State University, Fort Collins, Colorado, United States of America

Zhian Li and Shenglei Fu
Key Laboratory of Vegetation Restoration and Management of Degraded Ecosystems, South China Botanical Garden, Chinese Academy of Sciences, Guangzhou, China

Jorge Paz-Ferreiro
Key Laboratory of Vegetation Restoration and Management of Degraded Ecosystems, South China Botanical Garden, Chinese Academy of Sciences, Guangzhou, China
Departamento de Edafologı´a, Universidad Polite´cnica de Madrid, Madrid, Spain

Huanping Lu
Key Laboratory of Vegetation Restoration and Management of Degraded Ecosystems, South China Botanical Garden, Chinese Academy of Sciences, Guangzhou, China
University of Chinese Academy of Sciences, Beijing, China

Ana Méndez
Departamento de Ingenierı´a de Materiales, Universidad Polite´cnica de Madrid, Madrid, Spain

Gabriel Gascó
Departamento de Edafologı´a, Universidad Polite´cnica de Madrid, Madrid, Spain

Rukhsanda Aziz, Xiaoe Yang, Wendan Xiao and Tingqiang Li
Ministry of Education Key Laboratory of Environmental Remediation and Ecological Health, College of Environmental and Resource Sciences, Zhejiang University, Hangzhou, China

Muhammad Tariq Rafiq
Ministry of Education Key Laboratory of Environmental Remediation and Ecological Health, College of Environmental and Resource Sciences, Zhejiang University, Hangzhou, China
Department of Environmental Science, International Islamic University, Islamabad, Pakistan

Peter J. Stoffella
Indian River Research and Education Center, Institute of Food and Agricultural Sciences, University of Florida, Fort Pierce, Florida, United States of America

Aamir Saghir
Institute of Statistics, Zhejiang University, Hangzhou, China

Muhammad Azam
College of Agriculture and Biotechnology, Zhejiang University, Hangzhou, China

Xiang Gao, Man Wu, Ruineng Xu, Xiurong Wang, Ruqian Pan and Hong Liao
State Key Laboratory for Conservation and Utilization of Subtropical Agro-bioresources, Root Biology Center, South China Agricultural University, Guangzhou, China

Hye-Ji Kim
Department of Tropical Plants and Soil Sciences, College of Tropical Agriculture and Human Resources, University of Hawaii at Manoa, Honolulu, Hawaii, United States of America

Jian Sun, Qiang Zhang, Jia Zhou and Qinping Wei
Institute of Forestry and Pomology, Beijing Academy of Agriculture and Forestry Sciences, Beijing, China

Tomislav Hengl, Jorge Mendes de Jesus, Niels H. Batjes, Eloi Ribeiro, Maria Ruiperez Gonzalez, Bas Kempen and Johan G. B. Leenaars
ISRIC – World Soil Information, Wageningen, the Netherlands

Gerard B. M. Heuvelink
ISRIC – World Soil Information, Wageningen, the Netherlands Wageningen University, Wageningen, the Netherlands

Robert A. MacMillan
LandMapper Environmental Solutions Inc., Edmonton, Canada

Alessandro Samuel-Rosa
Federal Rural University of Rio de Janeiro, Rio de Janeiro, Brazil

Markus G. Walsh
The Earth Institute, Columbia University, New York, New York, United States of America, and Selian Agricultural Research Inst., Arusha, Tanzania

Index

A

Acidic Mine Drainage, 81
Actinobacteria, 26, 40-41, 92, 94-96, 98, 100, 213, 218-219
Amf Inoculum, 28
Animal Manure, 35, 157
Anova, 10, 17, 19, 29, 48, 60, 94, 107, 118-119, 127, 137, 143, 190, 198, 206-207, 210
Antibiotic Sulfadiazine, 35, 43
Aphid Arrival, 68
Arbuscular Mycorrhizal Fungi, 27, 34, 94, 212

B

Bioavailable, 35, 42, 86, 188, 190, 202
Biochar, 188-194
Biological Stoichiometry, 1-2, 7, 12, 34
Bohai Rim, 142-143
Bulk Density, 45-46, 48, 50, 55, 75-77, 93-94, 116, 126, 133, 159, 177, 179-180, 221, 224, 228-229

C

C Global Cycle, 45
Cd Accumulation, 195-196, 200, 203
Chinese Soils, 195-201, 203
Coastal Wetlands, 142
Compost, 27-28, 34-35, 92-102, 203, 219
Corn Rootworm, 74-77, 79
Cylindrocladium Parasiticum, 204-205, 212

D

Disease-susceptible Rice, 204

E

Ecm Fungi, 167, 171-173
Ecological Risk Assessments, 74, 79
Ectomycorrhizal (ecm) Fungal Abundance, 167
Experimental Semivariograms, 45

F

Farmland, 56, 115-119, 121, 123, 125, 146, 148, 175, 186, 188, 196
Fgoals, 157, 159, 165
Food Chain Contamination, 188, 195
Forestland, 51, 115-119, 121, 123
Fumigant Treatments, 135-136, 138
Fungal Community Structure, 15, 17, 19, 21, 23-25, 167, 173

G

Ginger Germination Levels, 138, 140
Ginger Production, 135-136, 139, 141
Glasshouse, 16, 103, 109, 111-112
Global Carbon Cycling, 1-2
Gobi Desert, 45-47, 49, 51, 53-56
Grain-for-green, 115-116, 121
Green Peach Aphid, 68-71, 73
Greenhouse Gas, 102, 157, 164, 166
Growth Rate Hypothesis, 6-7, 12

H

Heavy Metal Concentrations, 83-85, 89-90, 189
Human Hair Analysis, 81
Hydrologic Alterations, 142, 147
Hygienically Safe Food Production, 195

I

Initial Soil, 35, 37, 121, 159-160
Inorganic Fertilizer, 92-94, 186
Insect-protected Maize Biomass, 74

K

Kale Plants, 68-72

L

Low Organic Matter, 15, 23, 193, 214

M

Macroaggregates, 92, 94-96, 98-101, 124, 127, 129-130, 132-133
Metabolic Relationships, 1
Methyl Bromide, 135, 137, 141, 214
Microaggregates, 92, 95-96, 98-99, 102, 186
Microbial Biomass, 1-14, 16, 57, 59-67, 80, 94, 96, 99, 101-102, 124, 133, 148, 166, 186
Microcosm Experiment, 35-36
Monoculture, 57, 60, 62-66, 103, 113, 133, 204-211
Monsanto Genetically Modified (gm) Maize, 74

N

N Fertilization, 101, 157-161, 163-165
Natural Biological Suppression, 15
North China Plain, 148, 157-159, 165-166

O

Oilseed Rape, 103-105, 107-109, 111-113

Operational Taxonomic Units, 36, 168
Organic Carbon, 16, 23, 28, 34, 45, 47-53, 55-56, 60, 92-93, 95, 97, 99, 101-102, 115-119, 121, 123-124, 126, 128, 130, 132-134, 136, 141-144, 148, 157, 159, 161, 163, 165-166, 175, 177-181, 183-187, 189, 191, 221, 224, 226-234, 236-237
Organic Carbon Management, 115, 119
Organic Carbon Sink, 115, 117, 119, 121, 123
Organic Mulches, 68, 72-73

P
Pak Choi, 195-203
Phytoremediation, 188-190, 192-194, 203
Pic Concentrations, 138
Pipette, 136, 152
Polyethylene Film, 135-137
Poultry Manure, 69, 203
Protected Species, 36, 69, 82, 125, 136, 143, 151, 189, 205, 214

Q
Quantigene, 74-79

R
Ralstonia Solanacarum, 135, 137, 139, 141
Rdp Fungal Classifier, 15, 18, 24
Red Amaranth, 188-190
Red Crown Rot, 204-209, 211-212
Rhizosphere, 15-16, 21, 23, 25-27, 34, 42-44, 57-67, 103-113, 133, 174, 186, 209, 211-215, 217-220
Root Biomass, 103, 109, 112, 119, 121, 163

S
Soil Microbial Biomass C, 3, 6, 13, 57, 65, 67
Soil Organic Carbon, 34, 45, 47-53, 55-56, 101, 115-119, 121, 123-124, 126, 128, 132-134, 142, 144, 148, 157, 159, 161, 163, 165-166, 175, 177, 179-181, 183-184, 186-187, 221, 226-234, 237
Soil-borne Disease, 15, 23, 25, 136, 204-205, 207-209, 211
Stratification Ratios, 142, 144-146
Straw Incorporation, 124-134, 185
Stubble Retention, 15-16, 23, 157-164

T
Terminal Restriction Fragment Length Polymorphism, 103, 113
Topsoil Samples, 36
Total Nitrogen, 56, 60, 142-144, 159-160, 177, 189-190
Total Phosphorus, 60, 142-143, 189
Total Plant Mass, 188
Tree Species Coexistence, 57, 65-66
Triphenyltetrazolium Chloride (ttc), 136
Tropical Forests, 26, 55, 167-168, 173-174
Tungsten (w) Contamination, 149

V
Ventilated Greenhouse Conditions, 27
Vermiwash, 27-33
Vermiwash Foliar Spray, 27

W
W Bioaccumulation, 149
W Chemical Forms, 149, 154
W Mines, 149, 151-152, 155
Water Deficiency, 124

Y
Yayasan Sabah, 168, 173